VOLUME ONE HUNDRED AND THIRTY FIVE

PROGRESS IN MOLECULAR BIOLOGY AND TRANSLATIONAL SCIENCE

Molecular and Cellular Regulation of Adaptation to Exercise

VOLUME ONE HUNDRED AND THIRTY FIVE

Progress in Molecular Biology and Translational Science

Molecular and Cellular Regulation of Adaptation to Exercise

Edited by

CLAUDE BOUCHARD
Human Genomics Laboratory,
Pennington Biomedical Research Center,
Baton Rouge, Louisiana, USA

ELSEVIER

AMSTERDAM • BOSTON • HEIDELBERG • LONDON
NEW YORK • OXFORD • PARIS • SAN DIEGO
SAN FRANCISCO • SINGAPORE • SYDNEY • TOKYO
Academic Press is an imprint of Elsevier

Academic Press is an imprint of Elsevier
225 Wyman Street, Waltham, MA 02451, USA
525 B Street, Suite 1800, San Diego, CA 92101-4495, USA
The Boulevard, Langford Lane, Kidlington, Oxford OX5 1GB, UK
125 London Wall, London, EC2Y 5AS, UK

First edition 2015

ISBN: 978-0-12-803991-5
ISSN: 1877-1173

For information on all Academic Press publications
visit our website at store.elsevier.com

CONTENTS

CONTRIBUTORS

Nadia Agha
Department of Health and Human Performance, Laboratory of Integrated Physiology, University of Houston, Houston, Texas, USA

Jacob Allen
Department of Kinesiology and Community Health, University of Illinois at Urbana-Champaign, Urbana, Illinois, USA

Philip J. Atherton
MRC-ARUK Centre of Excellence for Musculoskeletal Ageing Research, Clinical, Metabolic and Molecular Physiology, University of Nottingham, Royal Derby Hospital Centre, Derby, United Kingdom

Frank W. Booth
Department of Biomedical Sciences; Department of Nutrition and Exercise Physiology; Department of Medical Pharmacology and Physiology, and Dalton Cardiovascular Research Center, University of Missouri, Columbia, Missouri, USA

Marni D. Boppart
Department of Kinesiology and Community Health, and Beckman Institute for Advanced Science and Technology, University of Illinois, Urbana, Illinois, USA

Claude Bouchard
Human Genomics Laboratory, Pennington Biomedical Research Center, Baton Rouge, Louisiana, USA

Heather Carter
Muscle Health Research Centre, School of Kinesiology and Health Science, York University, Toronto, Ontario, Canada

Chris Chen
Muscle Health Research Centre, School of Kinesiology and Health Science, York University, Toronto, Ontario, Canada

Matthew J. Crilly
Muscle Health Research Centre, School of Kinesiology and Health Science, York University, Toronto, Ontario, Canada

Michael De Lisio
Department of Kinesiology and Community Health, University of Illinois, Urbana, Illinois, USA

Christian A. Drevon
Department of Nutrition, Institute of Basic Medical Sciences, Faculty of Medicine, University of Oslo, Oslo, Norway

Kristin Eckardt
Department of Nutrition, Institute of Basic Medical Sciences, Faculty of Medicine, University of Oslo, Oslo, Norway

Jürgen Eckel
Paul-Langerhans-Group for Integrative Physiology, German Diabetes Center (DDZ), Auf'm Hennekamp, and German Center for Diabetes Research (DZD e.V.), Düsseldorf, Germany

Brian S. Ferguson
Department of Biomedical Sciences, University of Missouri, Columbia, Missouri, USA

Nuria Garatachea
Faculty of Health and Sport Science, University of Zaragoza, Huesca, Spain

Laurie J. Goodyear
Section on Integrative Physiology and Metabolism, Joslin Diabetes Center, and Department of Medicine, Brigham, and Women's Hospital, Harvard Medical School, Boston, Massachusetts, USA

Rachel Graff
Department of Health and Human Performance, Laboratory of Integrated Physiology, University of Houston, Houston, Texas, USA

Sven W. Görgens
Paul-Langerhans-Group for Integrative Physiology, German Diabetes Center (DDZ), Auf'm Hennekamp, Düsseldorf, Germany

Anthony C. Hackney
Department of Exercise and Sport Science; Department of Nutrition, Gillings School of Public Health, and Curriculum in Human Movement Science, Department of Allied Health Sciences, University of North Carolina, Chapel Hill, North Carolina, USA

Gilian F. Hamilton
Department of Psychology, The Beckman Institute, University of Illinois at Urbana-Champaign, Urbana, Illinois, USA

Mark Hargreaves
Department of Physiology, The University of Melbourne, Melbourne, Australia

Katja Heinemeier
Institute of Biomedical Sciences, Faculty of Health and Medical Sciences and Institute of Sports Medicine, Bispebjerg Hospital, Copenhagen, Denmark

Michael F. Hirshman
Section on Integrative Physiology and Metabolism, Joslin Diabetes Center, Harvard Medical School, Boston, Massachusetts, USA

David A. Hood
Muscle Health Research Centre, School of Kinesiology and Health Science, York University, Toronto, Ontario, Canada

Jørgen Jensen
Department of Physical Performance, Norwegian School of Sport Sciences, Oslo, Norway

Niklas Rye Jørgensen
Department of Clinical Chemistry, Glostrup Hospital and University of Southern Denmark, Glostrup, Denmark

Michael Kjaer
Institute of Sports Medicine, Department of Orthopedic Surgery, Bispebjerg Hospital and Centre for Healthy Aging, Faculty of Health and Medical Sciences, University of Copenhagen, Copenhagen, Denmark

Karsten Krüger
Department of Sports Medicine, University of Giessen, Giessen, Germany

Hawley Kunz
Department of Health and Human Performance, Laboratory of Integrated Physiology, University of Houston, Houston, Texas, USA

Séverine Lamon
Centre for Physical Activity and Nutrition (C-PAN) Research, School of Exercise and Nutrition Sciences, Deakin University, Burwood, Victoria, Australia

Amy R. Lane
Curriculum in Human Movement Science, Department of Allied Health Sciences, University of North Carolina, Chapel Hill, North Carolina, USA

M. Harold Laughlin
Department of Biomedical Sciences; Department of Medical Pharmacology and Physiology, and Dalton Cardiovascular Research Center, University of Missouri, Columbia, Missouri, USA

Alejandro Lucia
European University and Research Institute of Hospital 12 de Octubre ("i+12"), Madrid, Spain

S. Peter Magnusson
Musculoskeletal Rehabilitation Research Unit, Department of Physiotherapy and Institute of Sports Medicine, Bispebjerg Hospital, Faculty of Health and Medical Sciences, University of Copenhagen, Copenhagen, Denmark

Chris McGlory
Department of Kinesiology, McMaster University, Hamilton, Ontario, Canada

Jonathan Memme
Muscle Health Research Centre, School of Kinesiology and Health Science, York University, Toronto, Ontario, Canada

Frank C. Mooren
Department of Sports Medicine, University of Giessen, Giessen, Germany

Joram D. Mul
Section on Integrative Physiology and Metabolism, Joslin Diabetes Center, Harvard Medical School, Boston, Massachusetts, USA

Robert C. Noland
Pennington Biomedical Research Center, Louisiana State University, Baton Rouge, Louisiana, USA

T. Dylan Olver
Department of Biomedical Sciences, University of Missouri, Columbia, Missouri, USA

Helios Pareja-Galeano
European University and Research Institute of Hospital 12 de Octubre ("i+12"), Madrid, Spain

Marion Pauly
Muscle Health Research Centre, School of Kinesiology and Health Science, York University, Toronto, Ontario, Canada

Bethan E. Phillips
MRC-ARUK Centre of Excellence for Musculoskeletal Ageing Research, Clinical, Metabolic and Molecular Physiology, University of Nottingham, Royal Derby Hospital Centre, Derby, United Kingdom

Stuart M. Phillips
Department of Kinesiology, McMaster University, Hamilton, Ontario, Canada

Justin S. Rhodes
Department of Psychology, The Beckman Institute, University of Illinois at Urbana-Champaign, Urbana, Illinois, USA

Gregory N. Ruegsegger
Department of Biomedical Sciences, University of Missouri, Columbia, Missouri, USA

Aaron P. Russell
Centre for Physical Activity and Nutrition (C-PAN) Research, School of Exercise and Nutrition Sciences, Deakin University, Burwood, Victoria, Australia

Richard J. Simpson
Department of Health and Human Performance, Laboratory of Integrated Physiology, University of Houston, Houston, Texas, USA

Kristin I. Stanford
Section on Integrative Physiology and Metabolism, Joslin Diabetes Center, Harvard Medical School, Boston, Massachusetts, USA

Yi Sun
Department of Kinesiology and Community Health, University of Illinois at Urbana-Champaign, Urbana, Illinois, USA

Ryan G. Toedebusch
Department of Biomedical Sciences, University of Missouri, Columbia, Missouri, USA

Elijah Trefts
Department of Molecular Physiology and Biophysics, Vanderbilt University School of Medicine, Nashville, Tennessee, USA

Liam D. Tryon
Muscle Health Research Centre, School of Kinesiology and Health Science, York University, Toronto, Ontario, Canada

Thomas Tsiloulis
Biology of Lipid Metabolism Laboratory, Department of Physiology, Monash University, Clayton, Victoria, Australia

Anna Vainshtein
Muscle Health Research Centre, School of Kinesiology and Health Science, York University, Toronto, Ontario, Canada

David H. Wasserman
Department of Molecular Physiology and Biophysics, Vanderbilt University School of Medicine, Nashville, Tennessee, USA

Matthew J. Watt
Biology of Lipid Metabolism Laboratory, Department of Physiology, Monash University, Clayton, Victoria, Australia

Daniel J. Wilkinson
MRC-ARUK Centre of Excellence for Musculoskeletal Ageing Research, Clinical, Metabolic and Molecular Physiology, University of Nottingham, Royal Derby Hospital Centre, Derby, United Kingdom

Ashley S. Williams
Department of Molecular Physiology and Biophysics, Vanderbilt University School of Medicine, Nashville, Tennessee, USA

Sarah Witkowski
Department of Kinesiology, School of Public Health and Health Sciences, University of Massachusetts, Amherst, Massachusetts, USA

Jeffrey A. Woods
Department of Kinesiology and Community Health, University of Illinois at Urbana-Champaign, Urbana, Illinois, USA

Zhen Yan
Department of Medicine; Department of Pharmacology; Department of Molecular Physiology and Biological Physics, and Center for Skeletal Muscle Research, University of Virginia, Charlottesville, Virginia, USA

PREFACE

This volume in the series *Progress in Molecular Biology and Translational Science* is devoted to the mechanisms regulating molecular and cellular adaptation to acute and chronic exercise in a variety of settings. *Progress in Molecular Biology and Translational Science* provides a forum for discussion of new discoveries, approaches, and ideas in molecular biology which is what we aimed for in the development of the volume. We believe that it is a timely contribution to our understanding of exercise biology. We have been fortunate in being able to secure contributions from leading scientists and most major laboratories that are actively engaged in the study of the molecular mechanisms at play when people and other living organisms are physically active. The publication is particularly timely as it occurs just a few months after the leadership of the National Institutes of Health announced that the Common Fund of NIH will support a 6-year plan to uncover the molecular transducers of adaptation to physical activity in various tissues and organs.

As the editor of the volume, I am extremely pleased by the distinguished panel of authors that was assembled for the publication. Sixty-one authors and coauthors from seven countries have contributed to the volume. I am very grateful for their willingness to participate in this effort. I would like to express my gratitude to them not only for their outstanding science but also for the timely delivery of their contributions. They have been a delight to work with. Unfortunately, some topics had to be left out due to the page number limitation but the vast majority of the relevant topic found a home in the volume.

The leadership of the *PMBTS* publication series and the staff at Elsevier have been a delight to work with. I would like to express my thanks to Dr. Michael Conn, Editor of the *PMBTS* serial, from Texas Tech University Health Sciences Center who supported the concept of having a full volume dedicated to the molecular biology of adaptation to exercise. I also benefited greatly from the support of Mary Ann Zimmerman, Acquisition Editor, and Helene Kabes, Senior Editorial Project Manager, all at the Elsevier publishing house. I also want to recognize the diligent work of Roshmi Joy, Project Manager in the Book Publishing Division at Elsevier. They were all very supportive at various stages of the development of the publication, and I would like to express my most sincere thanks to them.

Finally, I would not have been able to undertake the task of serving as editors for this volume without the outstanding and competent support initially of Allison Templet and then later of Robin Post of the Pennington Biomedical Research Center. They worked diligently with each author in order to ensure that the instructions were well understood by the contributing authors and that their manuscripts met all the requirements of the publisher. During the last phase of the production of the volume, Robin worked diligently on complex scientific material with a dedication to excellence that made a difference in our ability to deliver a high-quality volume. I feel greatly indebted to both of them. However, if errors are later discovered in the volume, they are entirely my responsibility.

CLAUDE BOUCHARD
July 2015

CHAPTER ONE

Adaptation to Acute and Regular Exercise: From Reductionist Approaches to Integrative Biology

Claude Bouchard[1]
Human Genomics Laboratory, Pennington Biomedical Research Center, Baton Rouge, Louisiana, USA
[1]Corresponding author: e-mail address: bouchac@pbrc.edu

Contents

Abstract

This chapter serves as an introduction to the volume focused on the molecular and cellular regulation of adaptation to acute and chronic exercise exposure. It begins with a definition of the overall content of the "sedens–physical activity–exercise training–fitness" domain. One conclusion from this brief overview is that past and current studies have primarily dealt with very limited subsets of the traits and parameters of interest to exercise biologists. Molecular and cellular studies have focused more on adaptation to exercise and less on variable levels of cardiorespiratory fitness even though the latter is a powerful indicator of current and future health status and longevity. In this regard, molecular profiling of intrinsic versus acquired cardiorespiratory fitness would seem to be an area of research deserving more attention. Although molecular and cellular studies are clearly reductionist by nature, they constitute the primary material allowing systems biology to draw inferences about pathways, networks, and systems. Integrative physiology can be substantially enriched by taking advantage of the findings and lessons from molecular studies and systems biology approaches. DNA sequence variation within and between populations as well as recent advances in the definition of the functional elements in the human and other genomes offer unique opportunities to pursue new and more powerful molecular studies, and to reconcile reductionist and integrative approaches.

Progress in Molecular Biology and Translational Science, Volume 135
ISSN 1877-1173
http://dx.doi.org/10.1016/bs.pmbts.2015.07.009

1. INTRODUCTION

All tissues and organs of the human body are affected by exercise particularly when it is energetically demanding and sustained. There is an abundant literature on the metabolic and physiological changes taking place in response to acute endurance, high intensity, and resistance exercise even though much remains to be learned. Similarly, there is a growing body of data regarding adaptation of tissues, organs, and systems to regular exercise and exercise training, particularly with respect to endurance and resistance training. Although impressive advances have been made on the general topic of adaptation to exercise, there are still big gaps in knowledge that deserve our attention. One critical gap in the foundational body of knowledge of exercise biology is the limited understanding of the universe of molecular transducers involved in the regulation of adaptation to all forms of acute and chronic exercise and of the molecular pathways and networks associated with the health benefits of being physically active. There are many other gaps in knowledge and a few are of particular interest and are highlighted here.

One blatant weakness is that exercise biology studies by and large cover only a fraction of the sedens–physical activity–exercise–fitness domain. Figure 1 provides a schematic overview of the multiple dimensions of this domain. Included in the diagram are the sedens–physical activity–exercise training continuum, the fitness traits, the exercise exposure dimensions,

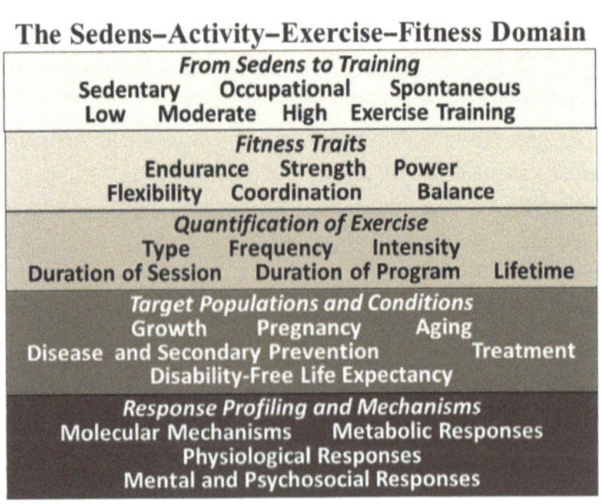

Figure 1 Schematic description of the sedentary behavior, physical activity level, exercise training, and fitness domain with its multiple dimensions and some of its implications. (See the color plate.)

the periods of life, health outcomes and aging, and the levels at which exercise biology scientists are investigating adaptation to acute and chronic exposures to exercise. When considering the global domain, it becomes apparent that exercise biology has thus far mainly focused on limited subsets of conditions and has fallen short of having comprehensively covered the multiple forms of exercise and fitness that deserve to be thoroughly investigated in multiple settings and a variety of clinical conditions. For instance, we know little regarding the impact of spontaneous physical activity or acute and chronic exposure to low-intensity exercise on multiple tissues and organs. One obvious conclusion from this quick overview is that the to-do-list of exercise biology research is extraordinarily long.

A specific area deserving more research is that of the general topic of the cellular and molecular adaptation to acute and chronic exposures to all types of exercise.[1] As this volume illustrates, we have some understanding of the cellular and molecular mechanisms associated with adaptation to some exercise exposures. But it is also clear from the numerous chapters, each contributed by world-class experts on a given topic, that we have gaping holes regarding our knowledge of these mechanisms and how they operate in multiple tissues and organs. Since the physiological responses to acute exercise exposure and to exercise training are often organ specific,[2] defining the mechanisms underlying tissue and organ specificity should shed light on the molecular pathways associated with adaptation, maladaptation, or health benefits. Importantly, even when some of the molecular mechanisms of adaptation to exercise have been evidenced, they have generally been investigated under a limited set of exercise conditions such as high-intensity exercise training or moderate exercise level meeting the current physical activity guidelines[3] and mainly in young adults of European descent. Thus, there is a need for a massive effort designed to uncover the molecular and cellular mechanisms underlying tissue and organ adaptation to all forms of exercise, particularly in light of the importance of regular exercise for the prevention of common diseases—including diabetes, cardiovascular diseases, cancer, and dementia—and premature death as well as healthy aging. The same conclusion seemed to have been reached recently by the leadership of the U.S. National Institutes of Health when they made public their new physical activity initiative to be funded by the Common Fund over a six-year period (2016–2022). The focus of this major effort will be to identify the populations of molecular transducers of adaptation to exercise in various tissues and organs using a combination of human and animal model studies.

2. SEDENTARY TIME, PHYSICAL ACTIVITY, AND FITNESS

The topic of sedentary behavior, low physical activity level, and low cardiorespiratory fitness is one that we have addressed in greater details recently.[3a] Professors Jeremy Morris (London bus drivers and conductors) and Ralph Paffenbarger (San Francisco Longshoremen and Harvard University Alumni studies) made the seminal observation that the level of physical activity on the job or during leisure time was inversely associated with mortality rates.[4–9] These observations have been repeated multiple times in large studies focusing on middle-aged adults as well as older people.[10,11] Prospective epidemiological studies have established over the last 60 years or so that the lower death rates resulting from a physically active lifestyle were seen for all-cause, cardiovascular, and cancer mortality. Regular exercise translates into multiple wide-ranging health benefits such that it has been defined by some as the equivalent of a "polypill" with favorable pleiotropic effects on all organs and systems.[12]

On the other hand, a number of studies reported in the last decade have highlighted the fact that sedentary behavior was also associated with mortality rates, with the most sedentary individuals exhibiting higher death rates. The first population study to focus on this question was a dose–response prospective study of participants of the 1981 Canada Fitness Survey, and it revealed a graded relationship between amount of sitting time and all-cause and cardiovascular mortality.[13] When the groups with the highest and lowest amount of daily sitting time were compared, the reduction in risk of death associated with less sitting time was about 15–20%, a risk reduction effect that persisted after adjustment for leisure time physical activity and body mass index. This observation has been confirmed in subsequent cohort studies from around the world.

Sedentary behavior and physical activity level have strong influences on mortality rates but so does cardiorespiratory fitness. This was well illustrated by reports from the laboratory of Professor Steven Blair based on the Aerobic Center Longitudinal Study starting in the 1980s.[14] The main findings from a series of papers published by Blair and colleagues are that low cardiorespiratory fitness, as estimated by time on a treadmill test to exhaustion, was associated with higher all-cause, cardiovascular, and cancer death rates and that this association was found to be present in overweight, diabetic, hypertensive, or hypercholesterolemic adults.[14] Interestingly, the same trend is observed in older adults in whom the powerful risk reduction impact of cardiorespiratory fitness on mortality was observed among male veterans from 65 to 90 years of age.[15]

In summary, a high altitude review of the evidence accumulated thus far strongly suggests that low cardiorespiratory fitness, low physical activity level, and increasing sedentary behavior are powerful predictors of all-cause, cardiovascular, and perhaps cancer mortality. These observations have considerable implications for the research agenda on exercise molecular mechanisms. Much energy is currently devoted to discovering the signaling pathways and molecular regulation of gene expression in relevant tissues (especially skeletal muscle) in response to acute and chronic exposure to exercise, particularly aerobic and resistance exercise. In contrast, little attention is being paid to tissue and organ molecular profiling of low versus moderate versus high cardiorespiratory fitness with the aim of discovering some of the molecular mechanisms at play in the relation between fitness, disease prevention, and longevity. Although the basic notion of targeting cardiorespiratory fitness for molecular studies appears to be simple on the surface, it would be in fact a complex undertaking for a number of reasons. For instance, it should be rather easy to identify adults with targeted cardiorespiratory fitness levels among subjects of existing long-term prospective cohorts, but accessibility of tissues, beyond skin, muscle, adipose tissue, blood, feces, and urine poses a major problem. A thorough molecular exploration should ideally include not only these tissues but also heart, lung, liver pancreas, kidney, bone, and brain to name but the most obvious ones. The only reasonable way to overcome this critical limitation would be to perform the same molecular and cellular studies on animal models. In this regard, there is solid evidence that the relationship between cardiorespiratory fitness and mortality rates described in humans is also observed in rodents. In a recent study, it was reported that, in rats kept sedentary all their life, those with a high intrinsic cardiorespiratory fitness, as measured by the distance they could run on a treadmill, lived 28–45% longer than the rats with a low cardiorespiratory fitness.[16]

One critical topic to address would be that of untangling the intrinsic and acquired component of the cardiorespiratory fitness phenotype at the individual level. An adult has an intrinsic level of cardiorespiratory fitness which can be observed in a direct manner by measuring maximal oxygen uptake adjusted for body mass and body composition in people who have a life history of being sedentary. For instance, among 174 sedentary young adult males, 17–35 years of age, measured twice (on separate days) for VO_2max at baseline in the HERITAGE Family Study, the mean value was 41 mL

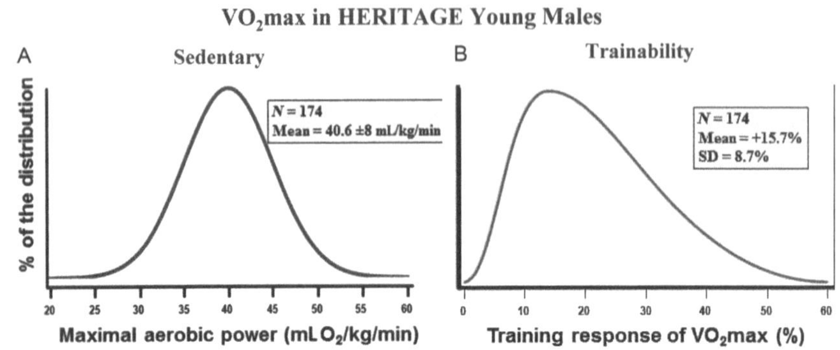

N = 174 males from 17 to 35 years of age

Figure 2 Distribution of VO_2max/kg body weight values in 174 sedentary men, 17–35 years of age, from the HERITAGE Family Study (A). Distribution of the VO_2max changes in % of baseline levels in response to a standardized endurance training program of 20 weeks in the same sedentary subjects (B). (See the color plate.)

O_2/kg/min with an SD of 8 mL (Fig. 2A). The distribution of VO_2max scores was almost perfectly Gaussian, which implies that about 7% had a VO_2max/kg of 29 mL or less (1.5 SD below the mean) and the same percentage exhibited a cardiorespiratory fitness level about 53 mL/kg and more, an extraordinary degree of heterogeneity in such a fundamental biological property among people who are confirmed sedentary with no significant amount of exercise training in their past. These data clearly show that there is a substantial fraction of sedentary adults who maintains a relatively high VO_2max despite the fact that they do not engage in any exercise program. Actually, some sedentary young adults maintain a VO_2max of 55 mL O_2/kg/min and more, a level of cardiorespiratory fitness that is even out of reach to many exercisers.

The importance of cardiorespiratory fitness from a biological point of view and the complexity of its interpretation with regard to mortality rates are augmented by the fact that the sedens level of VO_2max can be improved in most people by appropriate behavior, i.e., regular physical activity and especially exercise training. To illustrate this point, let us use again the same 174 young adult males of HERITAGE. They were trained for 20 weeks and achieved what we can call perfect adherence to the exercise training protocol. Maximal oxygen uptake was measured twice before the exercise program and twice again posttraining, i.e., 24 and 72 h after the last exercise session. The gains in VO_2max (expressed in % of baseline) are illustrated in Fig. 2B. We note from the figure that the mean gain calculated from the increase in mL O_2 was 16% with an SD of 9% with a distribution of scores clearly skewed to the right, i.e., skewed in the direction of the high gainers in response to the same exercise prescription. This extraordinary range of training responses occurred in spite of the fact that the program was fully standardized and that adherence to the exercise sessions, which were all performed in the laboratory under constant supervision, was deemed excellent. A substantial fraction of this group increased their indicator of cardiorespiratory fitness by 40% and more, whereas a large number gained 10% and less.

Personal characteristics, such as age and gender, are exerting major influences on intrinsic fitness level (sedens VO_2max) and on the absolute response (delta mL O_2) to an exercise program but not on the gains expressed in percentage of pretraining baseline level as the percentage VO_2max gain is the same on average in men and women and does not vary across age groups.[17–19] Ethnicity, defined here as blacks versus whites, is not contributing to either the intrinsic VO_2max level adjusted for body mass and body composition or its trainability when expressed as a percentage of baseline level.[17] The intrinsic cardiorespiratory fitness level adjusted for age, gender, body mass, and body composition is characterized by a heritability component of the order of 50%.[20] Similarly, the trainability of VO_2max, expressed in terms of gains in mL O_2, has a heritability level of about 45%.[19] Interestingly, there is no correlation between baseline, intrinsic fitness level and its trainability, with an r^2 ($\times 100$) of the order of 1%.[17,19]

The above observations raise many questions concerning the interpretation of the strong association between cardiorespiratory fitness level and mortality rate in prospective studies. They are too numerous to be all listed here but a number of examples will suffice to illustrate how critical the general topic of cardiorespiratory fitness, health, and longevity is to those with an interest in the study of the exercise biology and particularly the molecular basis of the causal relation between regular exercise and cardiorespiratory fitness. What are the biological differences between low and high fitness groups in molecular profiling at the level of the cardiovascular system, brain, lung, liver, kidneys, skeletal muscle, and adipose tissue? What are the molecular mechanisms accounting for the mortality rate difference between low and high cardiorespiratory fitness groups? Can the link between cardiorespiratory fitness and mortality rate in sedentary adults or in active adults be defined in terms of genomic, epigenomic, gene expression, and protein abundance differences in key tissues? What are the contributions to the fitness–mortality relationship of the sedentary levels of secreted myokines and adipokines, regulation of apoptosis, autophagy, stem cell populations, and subsets of miRNAs? An overarching question would be whether persons with a high intrinsic cardiorespiratory fitness level enjoy lower mortality rates comparable to those with more modest intrinsic fitness level but who are exercising regularly? If so, what are the molecular mechanisms driving these relationships to better health and longevity and are they identical in both groups?

3. REDUCTIONISM, SYSTEMS BIOLOGY, AND INTEGRATIVE PHYSIOLOGY

From time to time, we hear that integrative physiology is what we should focus on and that reductionist approaches are not contributing meaningful advances to our understanding of the adaptation of living organisms, especially humans, to acute and chronic exercise. Such views are not extremely frequent but they have been expressed by some of the most respected scientists in the field. For instance, Michael Joyner from the Mayo Clinic has provocatively affirmed that molecular biology and omics technologies have so far failed to deliver and asked whether physiology has the potential to fill the intellectual void left by reductionists.[21] According to him, reductionism is replete with "heroic narratives" and progress typically arising from reductionist research strategies is equivalent to "mirages," which are said to be stalling progress in physiology. Obviously, Joyner wants to provoke a debate and is pushing the limit in his expose of physiology as an antidote to molecular physiology.[21] But the basic question remains: Are the advances in our understanding of the molecular physiology of adaptation to exercise disconnected from progress in integrative biology? One could argue that the opposite is actually taking place. This volume provides an array of examples illustrating the fact that the science flows bidirectionally, i.e., from whole organism physiology to molecular studies and back to tissues, organs, and systems for further validation and potentially translational opportunities.

Molecular physiologists are particularly aware of the central observation that biological regulation of a given trait operates as a complex, multifactorial, and widely distributed system in all mammalian organisms. Adaptation to any behavior change or an environmental stimulus is in the end an integration of multiple mechanisms that are interactive, flexible, and redundant, the latter reflecting epistasis, pleiotropism, or independent mechanisms that come into play in response to upstream signaling events or feedback pathways. What reductionist scientists are guilty of is simply of trying to understand subsets of the molecular events taking place when whole-body changes occur with an acute or repeated exposure to exercise or other stimuli. I would venture to say that exercise molecular biologists as a group are of this school of thought and share the view that "every adaptation is an integration."[22]

It is difficult to understand how criticizing those who devote expertise, time, and energy to the study of the molecular mechanisms of adaptation to exercise can enhance our collective quest for the truth. One of the important advances of the last couple of decades has been the emergence of the field of "systems biology," which aims at integrating all the evidence generated at the molecular level into pathways, networks, and systems, which is simply and clearly a recognition by even hard core reductionists that adaptation can ultimately be understood only by attempting integration. One can perhaps conclude that system biology is likely to fail as it is still too close to the molecular and the omics.[21] An alternative view would simply recognize that systems biology aims at integrating the molecular evidence and that it constitutes a critical platform upon which integrative physiology and precision medicine will ultimately have a chance to thrive. One can only imagine how much stronger would the integrative physiology of exercise become if we had a comprehensive understanding of all molecular events taking place in response to acute and chronic exposure to exercise.

A productive path was laid out in a review by Greenhaff and Hargreaves[23] in which they recognize that molecular approaches, systems biology, and integrative physiology are conceptually different but they all strive for the same goal even though they rely on variable theoretical frameworks, technologies, and designs. Reductionist approaches are absolutely essential if we are to gain an in-depth understanding of the mechanisms by which the human organism as a whole adapts to the demands of acute and chronic exposure to exercise. One needs only to consult recent review papers on the molecular mechanisms driving the adaptation of skeletal muscle to acute and chronic exercise to develop a sense of excitement on the multitude of opportunities that the advances brought about largely by technologically intensive reductionist approaches represent for exercise biology.[24,25] This reality is clearly recognized by the American Physiological Society, the advocate-in-chief organization for integrative physiology, which advertises quite visibly on its website that APS stands for "Integrating the Life Sciences from Molecule to Organism," a position that should be sufficient to stop all dissenting voices about the merit of reductionist approaches. In this regard, the advances of the last 15 years on the coding and noncoding sequences and other features of the human genome have paved the way for a more profound understanding of the molecular regulation of adaptation in the broad sense.

4. GENOMIC AND ENCODE FACTS: A GOLD MINE FOR EXERCISE BIOLOGY

A powerful reason for the widespread use of reductionist approaches in the study of human variation for any traits, including those of interest to exercise biology, is that the human genome is extremely complex and cannot be apprehended by simple holistic methods and models. With the completion of the Human Genome Project, which gave us most of the sequence of the human genome and subsequently of the genomes of common animal models for the study of health and disease, the stage was set for exciting advances in our understanding of regulation at the molecular level.[26,27] Further progress in our knowledge of the complexity of the human genome was stimulated by the International HapMap and the 1000 Genomes Project which focused on sequence differences within and between populations and on patterns of human variation in the genome.[28,29] Since 2003, a large number of laboratories and scientists have been engaged in a massive effort to identify all the functional elements in the human genomic sequence. The effort is known as ENCODE, the Encyclopedia of DNA Elements. In 2012, in a series of papers published in *Nature* and other leading journals, ENCODE reported on functional products of the human genome.[30] More recently, the consortium presented evidence that combinations of biochemical, evolutionary, and genetic evidence provided complementary and more powerful evidence on the functionality of genomic regions.[31]

Among the most remarkable features that are of relevance to reductionist, systems biology, and integrative approaches in exercise biology, we will emphasize here just a few. Even though only about 1% of the human genome sequence encodes the estimated 20,687 protein-coding genes, 80% of the genome is transcribed and participates in the regulation of these genes and other cellular events. The human genome harbors almost 3 million protein-binding sites along its DNA. The 1800 or so transcription factors have been shown to bind at DNA sites representing about 8% of the genome. ENCODE along with other efforts has revealed that there are about 8800 small RNAs and more than 9600 long RNAs being transcribed in at least one type of cells. About 1000 of these small RNAs are known to be functionally relevant miRNAs. Many of these RNAs participate in the regulation of transcription and translation. One more set of numbers to show the complexity of the molecular regular regulation at the cellular level: human DNA encodes about 70,000 promoter regions and 400,000 enhancer regions, which can be at substantial genomic distance from the genes they are known to regulate. In brief, a whole web of regulatory molecules and DNA-binding sequences are involved in what can only be defined as a complex, widely distributed regulation of less than 21,000 protein-coding genes and other cellular functions.

In addition to the organizational complexity of the human genome, one needs to appreciate also the impact of variability in DNA sequence among people on biology in general and adaptation to exercise in particular. For instance, there are more than 40 million common single nucleotide polymorphisms (SNPs) in which the variant allele has a frequency of at least 1% in one human population. Whole-genome sequencing in thousands of individuals has shown that any given person carries from 3 to 4 million common SNPs. Among the latter, more than 10,000 translate into non-synonymous nucleotide changes, about 100 result in premature stop codons, more than 250 are loss-of-function variants, and up to 100 are DNA variants previously known to be disease causing even though the individuals carrying them do not exhibit such diseases at this point in time. Among other critical genomic features, any given individual carries more than 200 in-frame insertions or deletions, in excess of 1000 copy number variants at repeated DNA segments longer than 450 base pairs and even more polymorphisms in the number of copies for shorter repeats. One other source of variability: any given person carries as much as 500,000 rare variants that may be unique to the individual or the individual's family or pedigree.[32] In contrast to common polymorphisms, rare variants may exhibit larger effect sizes on the biology or the trait of interest. One striking example of the importance of rare variants for exercise biology is that of the Finnish skier legend, Aero Antero Matyranta, who won five gold medals, four silver medals, and three bronze medals at Olympic Games and World Championships in cross-country skiing events in the 1960s. It was shown that he had over the years hemoglobin levels in the range of 200–230 g/L with hematocrit around 68%.[33] Reports have documented that he had primary familial and congenital polycythemia due to a mutation in the erythropoietin (*EPOR*) gene. The EPOR mutation resulted in a truncation of 70 C-terminal amino acids of the gene. The G to A transition converted the TGG triplet encoding tryptophan to a TAG stop codon. In the Finnish pedigree composed of about 200 relatives, 29 were shown to harbor the same EPOR mutation.[34] It appears that he was the only one among all affected relatives who was able to compete at the international level in endurance events. He may have been the only one for which complex cellular regulatory systems allowed him to benefit from a very high oxygen-carrying capacity while not being unduly clinically affected by his polycythemia.

5. ABOUT THE CONTENT OF THE VOLUME

The volume is organized around 21 chapters. Chapters 2–4 focus on the molecular and cellular regulation of carbohydrates, lipids, and proteins, respectively, in relation to acute and chronic exposure to exercise. Chapter 5 reviews the evidence for mitochondrial biogenesis and degradation leading to expansion of the mitochondrial reticulum in response to repeated exposure to exercise. Chapter 6 covers the topic of the molecular regulation in skeletal muscle of the response to endurance exercise, while Chapter 7 focuses on the regulation of skeletal muscle hypertrophy. Chapter 8 deals with regulation of adipose tissue metabolism in response to exercise. Chapter 9 addresses the same issue but for the liver and hepatic metabolism. Chapter 10 covers the topics of exercise and the regulation of angiogenesis and vascular biology. Chapter 11 reviews the regulation of the response to exercise of bone, ligaments, cartilage, tendon, myotendinous junctions, and connective tissue. Chapter 12 covers the regulation of endocrine hormones and exercise. Chapter 13 is focused on the regulation of myokines, adipokines, and adipomyokines in adaptation to exercise. Chapter 14 reviews the topic of the regulation of inflammatory response and exercise. Chapter 15 deals with exercise and the regulation of immune functions. Chapter 16 examines the evidence for the role of exercise in the regulation of neurogenesis and brain functions. Chapter 17 addresses the rapidly evolving science of the changes taking place in leukocytes and skeletal muscle apoptosis and autophagy in response to acute and chronic endurance and resistance exercise. Chapter 18 provides an extensive summary of the rapidly growing evidence for a role of stem cell recruitment and biology in adaptation to exercise. Chapter 19 deals with the role of genomic and epigenomic markers in the complex regulation of gene expression when meeting the demands of acute and chronic exercise. Chapter 20 examines what is known about the emerging science of microRNAs in the adaptation to exercise. Finally, Chapter 21 was given the task of addressing the topic of exercise as the equivalent of a "polypill" against a number of common chronic ailments and it provides a broad coverage of this exciting concept.

6. SUMMARY AND CONCLUSIONS

In this chapter, a number of issues related to the content of the volume are raised. An attempt is made at defining the global field represented by the sedens–physical activity–exercise training–fitness domain. One major conclusion arising from the brief discussion of the topic is that many dimensions of this conceptual domain are not addressed in past and current portfolios of scientific research. Two behavioral traits (sedentary behavior and physical activity level) and one state (cardiorespiratory fitness) have been widely considered in studies pertaining to health indicators and longevity. A powerful predictor of health status and longevity is cardiorespiratory fitness but it is also one of the most challenging to investigate. In this regard, inherent cardiorespiratory fitness (in the sedentary state) and acquired fitness seem to be both important but no study has thus far attempted to identify their specific contributions in humans.

Molecular and cellular biologists are keenly aware that biological regulation is widely distributed and that adaptation is the result of an integration of multiple signals and mechanisms that are interactive, flexible, and redundant. Thus, opposing the science done at the ground level (reductionist approaches) against that performed on whole organisms (integrative physiology) is not likely to be a productive exercise as integrative physiology can only develop better and more powerful models when it incorporates all lines of evidence. We posit here that molecular studies, systems biology, and integrative physiology are intimately connected and ought to be seen as components of a comprehensive human biology research enterprise. With the growing completeness of the human genome sequence and understanding of the functional elements of the nonprotein coding sequences, as progressively revealed by the ENCODE project, it is an exciting time to be involved in the study of the molecular regulation of adaptation to acute and chronic exercise exposure. The last section of the chapter outlines the main topics covered by the other 20 chapters of the volume.

REFERENCES

1. Wackerhage H. *Molecular Exercise Physiology: An Introduction*. Oxen, UK: Routledge; 2014. xiv, 323 pages.
2. Heinonen I, Kalliokoski KK, Hannukainen JC, Duncker DJ, Nuutila P, Knuuti J. Organ-specific physiological responses to acute physical exercise and long-term training in humans. *Physiology (Bethesda)*. 2014;29(6):421–436.

3. U.S. Department of Health and Human Services. *2008 Physical Activity Guidelines for Americans: Be Active, Healthy, and Happy!.* Washington, DC: U.S. Department of Health and Human Services; 2008.
3a. Bouchard C, Blair SN, Katzmarzyk PT. Less sitting, more physical activity or higher fitness? *Mayo Clin Proc.* 2015. In press.
4. Morris JN, Heady JA, Raffle PA, Roberts CG, Parks JW. Coronary heart-disease and physical activity of work. *Lancet.* 1953;265(6796):1111–1120. conclusion.
5. Morris JN, Crawford MD. Coronary heart disease and physical activity of work; evidence of a national necropsy survey. *Br Med J.* 1958;2(5111):1485–1496.
6. Paffenbarger Jr RS, Laughlin ME, Gima AS, Black RA. Work activity of longshoremen as related to death from coronary heart disease and stroke. *N Engl J Med.* 1970;282(20):1109–1114.
7. Paffenbarger RS, Hale WE. Work activity and coronary heart mortality. *N Engl J Med.* 1975;292(11):545–550.
8. Paffenbarger Jr RS, Hyde RT, Wing AL, Hsieh CC. Physical activity, all-cause mortality, and longevity of college alumni. *N Engl J Med.* 1986;314(10):605–613.
9. Paffenbarger Jr RS, Wing AL, Hyde RT. Physical activity as an index of heart attack risk in college alumni. *Am J Epidemiol.* 1978;108(3):161–175.
10. Manini TM, Everhart JE, Patel KV, et al. Daily activity energy expenditure and mortality among older adults. *JAMA.* 2006;296(2):171–179.
11. Blair SN, Haskell WL. Objectively measured physical activity and mortality in older adults. *JAMA.* 2006;296(2):216–218.
12. Fiuza-Luces C, Garatachea N, Berger NA, Lucia A. Exercise is the real polypill. *Physiology (Bethesda).* 2013;28(5):330–358.
13. Katzmarzyk PT, Church TS, Craig CL, Bouchard C. Sitting time and mortality from all causes, cardiovascular disease, and cancer. *Med Sci Sports Exerc.* 2009;41(5):998–1005.
14. Blair SN, Kohl 3rd HW, Paffenbarger Jr RS, Clark DG, Cooper KH, Gibbons LW. Physical fitness and all-cause mortality. A prospective study of healthy men and women. *JAMA.* 1989;262(17):2395–2401.
15. Kokkinos P, Myers J, Faselis C, et al. Exercise capacity and mortality in older men: a 20-year follow-up study. *Circulation.* 2010;122(8):790–797.
16. Koch LG, Kemi OJ, Qi N, et al. Intrinsic aerobic capacity sets a divide for aging and longevity. *Circ Res.* 2011;109(10):1162–1172.
17. Skinner JS, Jaskolski A, Jaskolska A, et al. Age, sex, race, initial fitness, and response to training: the HERITAGE Family Study. *J Appl Physiol.* 2001;90(5):1770–1776.
18. Skinner JS, Wilmore KM, Krasnoff JB, et al. Adaptation to a standardized training program and changes in fitness in a large, heterogeneous population: the HERITAGE Family Study. *Med Sci Sports Exerc.* 2000;32(1):157–161.
19. Bouchard C, An P, Rice T, et al. Familial aggregation of VO(2max) response to exercise training: results from the HERITAGE Family Study. *J Appl Physiol.* 1999;87(3):1003–1008.
20. Bouchard C, Daw EW, Rice T, et al. Familial resemblance for VO2max in the sedentary state: the HERITAGE family study. *Med Sci Sports Exerc.* 1998;30(2):252–258.
21. Joyner MJ, Pedersen BK. Ten questions about systems biology. *J Physiol.* 2011;589(pt 5):1017–1030.
22. Joyner MJ, Limberg JK. Blood pressure regulation: every adaptation is an integration? *Eur J Appl Physiol.* 2014;114(3):445–450.
23. Greenhaff PL, Hargreaves M. 'Systems biology' in human exercise physiology: is it something different from integrative physiology? *J Physiol.* 2011;589(pt 5):1031–1036.
24. Hoppeler H, Baum O, Lurman G, Mueller M. Molecular mechanisms of muscle plasticity with exercise. *Compr Physiol.* 2011;1(3):1383–1412.

25. Egan B, Zierath JR. Exercise metabolism and the molecular regulation of skeletal muscle adaptation. *Cell Metab.* 2013;17(2):162–184.
26. Venter JC, Adams MD, Myers EW, et al. The sequence of the human genome. *Science (New York, NY).* 2001;291(5507):1304–1351.
27. Lander ES, Linton LM, Birren B, et al. Initial sequencing and analysis of the human genome. *Nature.* 2001;409(6822):860–921.
28. International HapMap Consortium. The International HapMap Project. *Nature.* 2003;426(6968):789–796.
29. Abecasis GR, Auton A, Brooks LD, et al. An integrated map of genetic variation from 1,092 human genomes. *Nature.* 2012;491(7422):56–65.
30. ENCODE Project Consortium. An integrated encyclopedia of DNA elements in the human genome. *Nature.* 2012;489(7414):57–74.
31. Kellis M, Wold B, Snyder MP, et al. Defining functional DNA elements in the human genome. *Proc Natl Acad Sci USA.* 2014;111(17):6131–6138.
32. Lupski JR, Belmont JW, Boerwinkle E, Gibbs RA. Clan genomics and the complex architecture of human disease. *Cell.* 2011;147(1):32–43.
33. Juvonen E, Ikkala E, Fyhrquist F, Ruutu T. Autosomal dominant erythrocytosis caused by increased sensitivity to erythropoietin. *Blood.* 1991;78(11):3066–3069.
34. de la Chapelle A, Traskelin AL, Juvonen E. Truncated erythropoietin receptor causes dominantly inherited benign human erythrocytosis. *Proc Natl Acad Sci USA.* 1993;90(10):4495–4499.

CHAPTER TWO

Exercise and Regulation of Carbohydrate Metabolism

Joram D. Mul*, Kristin I. Stanford*, Michael F. Hirshman*, Laurie J. Goodyear*,†,1

*Section on Integrative Physiology and Metabolism, Joslin Diabetes Center, Harvard Medical School, Boston, Massachusetts, USA

†Department of Medicine, Brigham, and Women's Hospital, Harvard Medical School, Boston, Massachusetts, USA

[1]Corresponding author: e-mail address: laurie.goodyear@joslin.harvard.edu

Contents

Abstract

Carbohydrates are the preferred substrate for contracting skeletal muscles during high-intensity exercise and are also readily utilized during moderate intensity exercise. This use of carbohydrates during physical activity likely played an important role during the survival of early *Homo sapiens*, and genes and traits regulating physical activity, carbohydrate metabolism, and energy storage have undoubtedly been selected throughout evolution. In contrast to the life of early *H. sapiens*, modern lifestyles are predominantly sedentary. As a result, intake of excessive amounts of carbohydrates due to the easy and continuous accessibility to modern high-energy food and drinks has not only become unnecessary but also led to metabolic diseases in the face of physical inactivity. A resulting metabolic disease is type 2 diabetes, a complex endocrine disorder characterized by abnormally high concentrations of circulating glucose. This disease now affects millions of people worldwide. Exercise has beneficial effects to help control impaired glucose homeostasis with metabolic disease, and is a well-established tool

Progress in Molecular Biology and Translational Science, Volume 135
ISSN 1877-1173
http://dx.doi.org/10.1016/bs.pmbts.2015.07.020

to prevent and combat type 2 diabetes. This chapter focuses on the effects of exercise on carbohydrate metabolism in skeletal muscle and systemic glucose homeostasis. We will also focus on the molecular mechanisms that mediate the effects of exercise to increase glucose uptake in skeletal muscle. It is now well established that there are different proximal signaling pathways that mediate the effects of exercise and insulin on glucose uptake, and these distinct mechanisms are consistent with the ability of exercise to increase glucose uptake in the face of insulin resistance in people with type 2 diabetes. Ongoing research in this area is aimed at defining the precise mechanism by which exercise increases glucose uptake and insulin sensitivity and the types of exercise necessary for these important health benefits.

ABBREVIATIONS

ADP adenosine diphosphate
AICAR aminoimidazole-4-carboxamide ribonucleoside
AMP adenosine monophosphate
AMPK AMP-activated protein kinase
AS160 Akt substrate of 160 kDa
ATP adenosine triphosphate
CaMKII Ca^{2+}/calmodulin-dependent protein kinase II
GLUT4 glucose transporter type 4
LKB1 liver kinase B1
MIRKO muscle-specific insulin receptor knockout mice
PAS phospho-Akt-substrate
Pi inorganic phosphate
Rab ras homologous from brain
SNARK sucrose nonfermenting AMPK-related kinase

1. INTRODUCTION

The unique ability of humans to perform endurance running has likely contributed to the evolution of *Homo sapiens* from other primates.[1] High levels of physical activity were required in order to evade predators as well as to obtain food. To maintain these high levels of physical activity, the working skeletal muscles require increased substrates for generation of adenosine triphosphate (ATP). A major substrate for the working muscles is carbohydrates, with one source being in the muscle itself in the form of glycogen, and another source glucose coming from the blood. The breakdown of glycogen from the muscle (glycogenolysis) and the regulation of glucose uptake into the muscle from the blood are highly regulated processes, and in this chapter, current knowledge on these functions will be discussed.

Since carbohydrate utilization promotes human survival, genes and traits regulating carbohydrate metabolism during exercise and energy storage have been selected throughout evolution.[2] However, current lifestyles are predominantly sedentary, which coupled with the intake of excessive amounts of carbohydrates, has led to metabolic diseases such as type 2 diabetes. On the other hand, exercise has beneficial effects on carbohydrate metabolism, and as a result, exercise is a well-established tool to prevent and combat type 2 diabetes. The molecular mechanisms that mediate the effects of exercise to increase skeletal muscle glucose uptake and increase insulin sensitivity in healthy people and people with type 2 diabetes will also be discussed.

2. CARBOHYDRATE UTILIZATION DURING REST AND EXERCISE

At rest, the energy used by the human body is predominantly derived from the oxidation of carbohydrates and fats. Blood glucose, plasma-free fatty acids, muscle glycogen, and intramuscular triglycerides are major substrate sources for energy production in skeletal muscles.[3,4] The contribution of proteins to the pool of usable energy is very limited, as amino acid oxidation is usually strictly adjusted to the intake of amino acids.

At rest, ingestion of carbohydrates results in insulin release from the pancreas, and the ensuing increase in plasma insulin concentrations has a myriad of metabolic effects. One important effect of insulin is to promote glucose transport into skeletal muscle. Insulin also suppresses fatty acid release from adipose tissue while increasing fat storage by activation of lipoprotein lipase.[5,6] Intake of physiologically normal carbohydrate levels has no impact on adipose tissue levels via *de novo* lipogenesis,[7] suggesting that the human body can accommodate intake of relatively large amounts of carbohydrates without a need to store carbohydrates as fat.

The contraction of skeletal muscles during physical exercise results in an increased energy demand for the muscle. The challenge for the working muscle is to increase production of ATP, and several cellular processes function to meet this need. Accordingly, metabolic pathways that oxidize both carbohydrates and fat need to be activated simultaneously.[3,4] Intensity, duration, and type of exercise determine the mechanisms through which this extra energy is supplied.

The enzyme ATPase facilitates the breakdown of ATP to ADP + inorganic phosphate (Pi) to generate energy for rapid use; however, only a small amount of ATP is present within the muscle cells.[8] An additional but even smaller source of stored energy is creatine phosphate, which can be resynthesized to ATP by the enzyme creatine kinase to replenish depleted ATP levels. Thus, the major sources of energy during exercise are carbohydrates and fats. Sources of carbohydrates for the muscle include blood glucose, muscle glycogen, and liver glycogen.[9]

Glucose and glycogen are converted to glucose-6-phosphate before they can be used to generate energy. One fate of glucose-6-phosphate is conversion to lactic acid, which results in the formation of three molecules of ATP per glycogen molecule or two molecules of ATP per glucose molecule (anaerobic glycolysis). The ATP generated by anaerobic glycolysis is not large enough to sustain continued muscle activity for long durations. With submaximal exercise, oxygen uptake increases, and within several minutes, a steady state is reached. This steady state indicates that the aerobic processes are supplying the majority of energy required by the contracting muscles. Aerobic generation of ATP from the glucose molecule is many times more efficient than the anaerobic reaction of glycolysis. During the aerobic reaction of glycolysis, glycogen is converted to pyruvic acid, which is then converted to acetyl-CoA and utilized for ATP production in the Krebs cycle within the mitochondria. Although the primary fuels contributing to oxidative metabolism during exercise are fats and carbohydrates, under extreme conditions amino acids can also be used as source of substrate oxidation.[9]

In the fasted state and during low intensity exercise, the bulk of energy required by the muscle is provided by oxidation of free fatty acids that are predominantly derived from the plasma.[10] When exercise increases to a moderate level of intensity (60–70% VO_2 peak), the source of fatty acids for oxidation also includes intramuscular triglyceride. Although both sources of fatty acids contribute to the energy needs of the muscle, even when combined they are not sufficient to meet the energy demand. Therefore, during moderate intensity exercise about half of the total energy derived is from oxidation of carbohydrates, coming from both muscle glycogen and blood glucose.[11] During high-intensity exercise, the contribution of plasma fatty acid oxidation becomes even less and carbohydrate oxidation provides roughly two-thirds of the total energy need. Carbohydrate metabolism is the preferred source of fuel under these conditions because the rate of ATP production is two times higher than fatty acids.[9]

3. MUSCLE GLYCOGEN

As noted above, glycogen is an essential fuel for energy production in the contracting skeletal muscles. Glycogen is a branched polymer of glucose with a mixture of α-1,4 and α-1,6 linkages between glucose units. The liver has the highest concentration of stored glycogen; however, skeletal muscle, as a result of its total weight, is the largest reserve of stored glycogen in the body. Intramuscular glycogen is associated with several organelles including the sarcolemma, sarcoplasmic reticulum, mitochondria, and myofibrils.[3,12] Granules of glycogen, or glycosomes, are also physically associated with several proteins including glycogen phosphorylase, phosphorylase kinase, glycogen synthase, glycogenin, protein phosphatases, and adenosine monophosphate (AMP)-activated protein kinase (AMPK).[3,12] The synthesis of glycogen involves multiples enzymes, and glycogen synthase is the rate-limiting enzyme. The breakdown of glycogen (glycogenolysis) is also controlled by a multienzyme system, and this will be discussed in more detail below.

Glycogen utilization is rapidly initiated at the onset of exercise and increases exponentially with exercise intensity.[13] Regulation of glycogenolysis is very sensitive to the metabolic rate of skeletal muscle during exercise.[14,15] Glycogen phosphorylase is the enzyme responsible for the rate-limiting step during muscle glycogenolysis.[3,16] At rest, glycogen phosphorylase exists primarily in the inactive b form, whereas with the onset of exercise phosphorylase kinase phosphorylates the b form to the active a form.[3] Phosphorylase kinase activation results from elevated calcium levels and binding of epinephrine to β-adrenergic receptors on the sarcolemma. Activation of phosphorylase kinase by stimulation of β-adrenergic receptors on the sarcolemma is mediated by cyclic AMP. Elevated epinephrine levels increase glycogen phosphorylase activity and glycogenolysis in perfused rat hind limbs and glycogenolysis in humans during moderate exercise.[16,17]

Muscle glycogenolysis does not always correlate tightly with levels of phosphorylase kinase a.[18] This suggested that posttranslational factors enhance the glycogenolytic rate during various intensities of exercise. Indeed, AMP and adenosine diphosphate (ADP) levels, and increased levels of Pi can all allosterically regulate activity of the a and b forms of phosphorylase kinase.[3] With increase exercise duration, there is a decrease in glycogen availability in parallel with decreased phosphorylase activity, while there is increased availability of other substrates for oxidation, such as plasma glucose and free fatty acids.

Muscle fiber type can also be a factor in determining the regulation of muscle glycogenolysis. During moderate intensity exercise, muscle glycogenolysis occurs predominantly in type I muscle fibers. As exercise duration increases or if exercise intensity increases, type I fibers become depleted and increasing amounts of glycogen are degraded in type II muscle fibers. Thus, as exercise intensity increases, recruitment of type II fibers increases accordingly. With short-term exercise at intensities approaching and exceeding VO_{2max}, glycogenolysis occurs in all fibers, but the highest rates are in type II fibers.[3]

Once muscle glycogen is depleted or near depleted, fatigue sets in, and exercise capacity is compromised.[14,15] Although duration and intensity of exercise play a role in regulating glycogen breakdown in muscle, diet history and training status also regulate muscle glycogenolysis during exercise. In general, increased carbohydrate intake is associated with greater muscle glycogen utilization, whereas increased fat intake results in decreased muscle glycogen utilization during exercise.[3] This attenuation of muscle glycogenolysis during exercise following the intake of a high-fat diet appears to be dependent on metabolic adaptations resulting from the high-fat diet and independent of muscle glycogen availability, which was similar at the onset of exercise.[19] Following exercise, glycogen synthase is activated and muscle glycogen concentrations are increased in the resting muscle.[20,21] Despite this increase in resting muscle glycogen levels, muscle glycogenolysis is decreased during dynamic exercise following short-term endurance training.[22] This decrease in muscle glycogenolysis is contributing to the well described increases in muscle oxidative capacity.[22]

4. GLUCOSE TRANSPORT

The other major source of carbohydrate during exercise is circulating blood glucose. Blood glucose concentrations during exercise are controlled by a precise regulatory mechanism, and the source of the circulating glucose is primarily the liver. In the resting state, food consumption also regulates blood glucose concentrations, and the removal of glucose from the circulation in response to both food consumption and physical exercise is a critical factor for the maintenance of normal glycemia in humans.

The transport of glucose into skeletal muscle is essential for tissue homeostasis, and under normal physiologic conditions, the transport process is rate limiting for glucose utilization.[23] Transport occurs by facilitated diffusion, and there is an increase in the maximal velocity of transport without an appreciable change in the substrate concentration at which glucose transport is half maximal.[24] The transport of glucose utilizes specific carrier proteins called glucose transporters, which are a family of structurally related proteins that are expressed in a tissue-specific manner.[25] In skeletal muscle from rodents and humans, GLUT4 is the major isoform expressed, whereas expression of the GLUT1, GLUT5, and GLUT12 isoforms is much lower.[26–28] Studies where there was genetic ablation of GLUT4 in the skeletal muscles of mice reveal that GLUT4 is necessary for normal rates of basal, insulin, and exercise-stimulated glucose transport.[29,30]

The mechanism by which exercise increases glucose transport via the GLUT4 transporter has been an area of intense investigation for many years. Likewise, there has been great interest in understanding the mechanism for the effects of insulin on glucose transport. Early studies using subcellular fractionation of skeletal muscle tissue,[31,32] and more recently work using *in vivo* confocal microscopy,[33,34] have clearly established that both exercise and insulin increase glucose transport in skeletal muscle through the translocation of GLUT4 from an intracellular compartment to the sarcolemma and transverse tubules. The GLUT4 translocation process is very complex, involving numerous cellular processes. In skeletal muscle, the movement of transporters occurs by the exocytosis, trafficking, docking, and fusion of GLUT4-containing storage compartment or "vesicles" into the cell-surface membranes. Our knowledge of the composition, specificity, and trafficking of GLUT4 vesicles has increased in recent years, although they are still not fully understood.[35] There is good evidence that multiple soluble *N*-ethylmaleimide attachment protein receptor (SNARE) proteins regulate the docking and fusion of GLUT4-containing vesicles. With stimulation such as exercise, muscle contraction, or insulin, the vesicle-associated SNARE proteins (v-SNARE), including vesicle-associated membrane protein-2 (VAMP-2), bind to the target-membrane SNARE proteins (t-SNARE), which include syntaxin 4 and SNAP23. This complex is thought to facilitate the fusion of GLUT4-containing vesicles into the cell-surface membrane. In studies with syntaxin 4 heterozygous-knockout (KO) mice, syntaxin 4 has been shown to be a major molecule responsible for the regulation of insulin-stimulated GLUT4 redistribution and glucose transport in skeletal muscle.[36] The roles of the SNARE proteins in exercise-stimulated GLUT4 translocation are less well understood, although VAMP2 has been shown to translocate to the cell surface in response to exercise.[37]

When skeletal muscles are stimulated simultaneously by contraction and insulin treatments, there are additive or partially additive effects on glucose transport.[24,38] Consistent with these findings, the combination of exercise and insulin can have additive effects on GLUT4 translocation to the sarcolemma.[38] These data support the concept that there are different mechanisms leading to the stimulation of muscle glucose transport by exercise and insulin.[39]

5. EXERCISE SIGNALS REGULATING GLUCOSE TRANSPORT

The intracellular signaling proteins that regulate the increase in GLUT4 translocation and glucose transport in skeletal muscle with exercise have also been an area of intensive investigation during the last 10 years. Since insulin and exercise both stimulate GLUT4 translocation, it has been hypothesized that there may be similar signaling proteins involved in the translocation process. Insulin signaling involves the rapid phosphorylation of the insulin receptor, insulin receptor substrate-1/2 (IRS-1/2) on tyrosine residues, and the activation of phosphatidylinositol 3-kinase (PI3-K).[40,41] In contrast, exercise does not result in tyrosine phosphorylation of the insulin receptor and IRS-1, and there is no increase in PI3-K activity.[42,43] Additional evidence that exercise can increase glucose transport in the absence of insulin signaling comes from a study investigating mice that lack insulin receptors in skeletal muscle (muscle-specific insulin receptor KO mice; MIRKO).[44,45] While these mice have blunted insulin-stimulated glucose transport,[45] they have normal exercise-stimulated glucose transport.[44] Taken together, these studies reveal that insulin and exercise mediate GLUT4 translocation in skeletal muscle through different proximal signaling mechanisms.

It is well known that a single bout of exercise activates multiple signaling pathways[46–48]; however, the precise signaling mechanism that mediates exercise-stimulated glucose transport is still not fully understood. Muscle contractile activity results in numerous alterations within the muscle fibers including changes in energy status (i.e., increased AMP/ATP), increases in intracellular Ca^{2+} concentration, increased reactive oxygen species, and stretching of the muscle fibers. These modifications can activate various signaling cascades, some of which have been implicated in exercise-stimulated glucose transport[49,50] (Fig. 1).

5.1 AMPK and LKB1

AMPK is a heterotrimeric protein composed of a catalytic α-subunit and regulatory β- and γ-subunits. The α- and β-subunits each exist in two isoforms (α1, α2 and β1, β2), and the γ-subunit exists in three isoforms

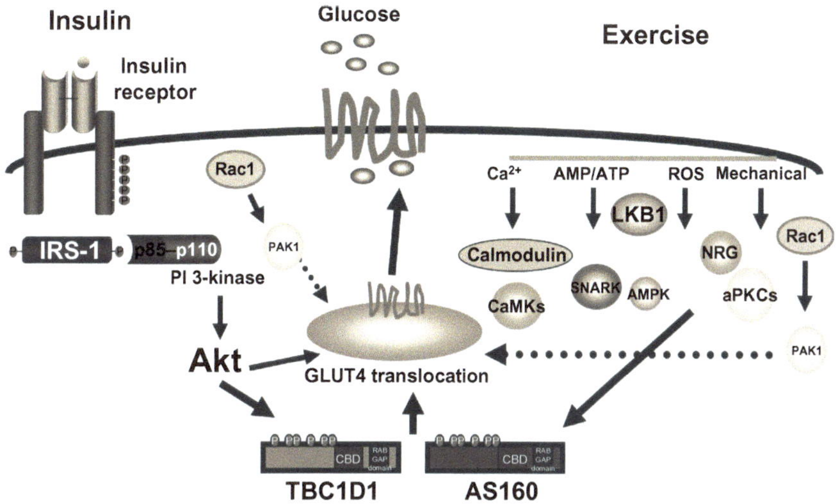

Figure 1 *Exercise and insulin regulation of glucose transport.* Proposed model for the signaling pathways mediating exercise- and insulin-induced skeletal muscle glucose transport. Insulin is initiated by binding to its cell service receptor leading to a cascade of phosphorylation reactions involving IRS-1, PI 3-kinase, and Akt among other proteins. Exercise works through a proximal signaling mechanism from that is distinct form that of insulin and is less well defined. It is likely that the proximal exercise signaling mechanism has redundancy as number of stimuli have been implicated in this process including changes in intracellular Ca^{2+}, the AMP:ATP ratio, generation of reactive oxygen species, and mechanical stresses. The insulin and exercise signaling pathways are thought to converge at the level of the Rab GAP proteins TBC1D1 and AS160, which allow for the release of the GLUT4-containing vesicles from intracellular stores, translocation to the transverse tubules and sarcolemma, and an increase in glucose uptake. *Adapted from Ref. 50.* (See the color plate.)

(γ1, γ2, and γ3). AMPK is activated by phosphorylation by one or more upstream kinases, including LKB1.[52–54]

AMPK and LKB1 have been widely studied for their potential role in exercise-stimulated glucose transport.[55] The initial evidence for a role of AMPK in exercise-stimulated glucose transport came from studies using the AMP-analog, 5-aminoimidazole-4-carboxamide ribonucleoside (AICAR).[56,57] These studies showed that AICAR increases glucose transport in skeletal muscle,[56,57] and similar to muscle contraction, the effects of AICAR are additive with insulin and PI 3-kinase-independent.[56,58] Some studies have shown that mice overexpressing a dominant negative AMPK α2 construct in muscle or α1 and α2 KO mice have impaired exercise-stimulated glucose uptake.[24,59–63] In contrast, other studies using mouse models with ablated AMPK activity demonstrate that inhibition of AMPK has little or no effect on exercise-induced glucose uptake,[62,64,65] or exercise-stimulated glucose uptake *in vivo*,[66] suggesting redundancy in the system. Therefore, it is still controversial whether AMPK is necessary for exercise-stimulated glucose uptake.

The role of LKB1 in exercise-stimulated glucose transport is also not clear. Mice with KO of LKB1 specifically in skeletal muscle have been shown to have a severe blunting of contraction-stimulated glucose transport.[51,67] This decrease in glucose transport could be due to decreased activation of AMPK and one or more of the AMPK-related kinases that are substrates of LKB1. One possible LKB1 substrate that may regulate exercise-stimulate glucose transport is the sucrose-nonfermenting AMPK-related kinase (SNARK). Decreased SNARK activity in skeletal muscle was shown to decrease exercise-stimulated glucose transport.[68]

While contraction-stimulated glucose transport was shown to be impaired in LKB-1 KO mice[51,67] and with decreased SNARK activity,[68] another recent study showed that glucose uptake during treadmill running was similar, if not higher, in LKB-1 KO mice compared to wild-type controls.[69] In yet another study, muscle-specific deletion of LKB1 only partially inhibited exercise-stimulated glucose transport.[51] These data suggest that while AMPK, SNARK, and LKB1 may be important in the regulation of exercise-stimulated glucose uptake, this system must have a high degree of redundancy, and it is likely that there are several overlapping signaling systems that can control exercise-stimulated glucose transport in skeletal muscle. This theory is consistent with the importance of carbohydrate utilization during exercise for survival.

5.2 Ca^{2+}/Calmodulin-Dependent Protein Kinases

Skeletal muscle contractile activity requires an increase in intracellular Ca^{2+} concentrations, and some studies have indicated that Ca^{2+}/calmodulin signaling and Ca^{2+}/calmodulin-dependent protein kinases are critical signals mediating exercise-stimulated glucose transport in skeletal muscle. Incubation of rat skeletal muscle with the Ca^{2+}/calmodulin inhibitor KN-93 decreased contraction-stimulated glucose transport.[70] KN-93 also inhibited exercise-induced CaMKII phosphorylation in the absence of AMPK inhibition, suggesting that CaMKs regulate glucose transport independently of AMPK signaling.[70,71] These studies also showed that overexpressing a constitutively active CaMKKα in mouse skeletal muscle increased AMPK

Thr-172 phosphorylation and skeletal muscle glucose uptake.[71] Electroporation of a specific CaMKII inhibitor into mouse tibialis anterior muscle reduced exercise-stimulated glucose uptake by 30%.[72] However, a separate study found that increases in Ca^{2+} concentration in muscle caused very little increase in glucose uptake when the contractile response of the muscle was impaired.[73] These data point to an indirect effect of Ca^{2+} on muscle glucose uptake, and the study of calcium signaling in the regulation of exercise-stimulated glucose transport needs further investigation.

5.3 Downstream Signals Mediating Exercise-Stimulated Glucose Transport

The signaling proteins downstream in the exercise and insulin signaling pathways have been proposed to converge at the Rab GAP proteins Akt substrate of 160 kDa (AS160/TBC1D4) and Tre-2/USP6, BUB2, cdc16 domain family member 1 (TBC1D1). AS160 and TBC1D1 are linked to GLUT4 translocation via the Rab (ras homologous from brain) proteins. Rab proteins are members of the Ras small GTPases superfamily[74] and have been shown to be involved in many membrane-trafficking events. Active Rabs recruit various effector proteins that are involved in vesicle budding, tethering, and fusion.[49,74,75] In addition to the well-established roles of the Rab proteins, there is evidence that the Rho family GTPase Rac1 is involved in both insulin- and exercise-stimulated GLUT4 translocation.[76,77] Mice deficient in Rac1 (Rac1 KO) have decreased insulin-stimulated GLUT4 translocation,[71,76] and Rac1 inhibition decreased contraction-stimulated glucose uptake in mouse skeletal muscle.[77]

5.4 AS160 and TBC1D1

AS160 was initially demonstrated to regulate insulin-stimulated GLUT4 translocation in 3T3LI adipocytes.[78–80] AS160 has numerous phosphorylation sites, and Rab GAP activity is controlled by phosphorylation. The best-studied phosphorylation sites are a group of six distinct sites that were identified as substrates for Akt. These are collectively referred to as phospho-Akt-substrate (PAS) sites and both insulin and exercise increase AS160 PAS phosphorylation in skeletal muscle.[78,81,82] Prolonged exercise in humans[82–84] and rats,[78] as well as AICAR, are also known to cause AS160 PAS phosphorylation. Therefore, in addition to Akt, AMPK has been shown to phosphorylate AS160.[81] Mutation of four PAS sites significantly inhibits both insulin- and exercise-induced glucose uptake.[85] AS160 also contains a calmodulin-binding domain, and mutation of this domain inhibits exercise-, but not insulin-stimulated glucose uptake.[86] These data show that both phosphorylation and calmodulin binding on AS160 are involved in the regulation of exercise-stimulated glucose uptake. These data also suggest that while AS160 may serve as a point of convergence for both insulin- and exercise-dependent signaling in the regulation of glucose uptake, other proteins may be involved in this regulation of glucose uptake.

TBC1D1 is another potential molecular link among signaling pathways converging on GLUT4 translocation in skeletal muscle.[78,81,87–91] TBC1D1 and AS160 share 47% overall identity and have several comparable structural features. TBC1D1 was first identified in adipocytes in culture but has only very limited expression in this tissue. In contrast, TBC1D1 is highly expressed in skeletal muscle.[89] Insulin increases TBC1D1 PAS phosphorylation in skeletal muscle[90,92,93] but, unlike AS160, TBC1D1 can regulate insulin-stimulated glucose transport through a PAS-independent mechanism.[92] Mutations of TBC1D1 differentially regulate insulin- and exercise-stimulated glucose transport in skeletal muscle.[92,93] Thus, TBC1D1 regulates both insulin- and exercise-stimulated glucose transport in muscle, but through distinct phosphorylation sites. Taken together, these data demonstrate that AS160 and TBC1D1 are a point of convergence for the regulation of GLUT4 translocation for insulin- and exercise-stimulated glucose transport in skeletal muscle.

6. INCREASES IN INSULIN SENSITIVITY FOR GLUCOSE TRANSPORT AFTER EXERCISE

The effects of an acute bout of exercise on glucose transport are relatively short-lived, returning to baseline typically in ~30–40 min. However, once the acute effects of exercise *per se* have disappeared, there is a period characterized by an increased effectiveness of insulin to stimulate glucose transport.[94,95] This increase in postexercise insulin sensitivity has been observed up to 48 h after exercise in humans.[96] The mechanisms for increased insulin sensitivity are not known. Although decreased muscle glycogen concentrations may play a part in exercise-induced increases in insulin sensitivity, the increased insulin action can occur even after full glycogen repletion.[94] The signaling mechanisms mediating the postexercise increase in insulin sensitivity are also not known, but similar to the acute effects of exercise on glucose transport, are not thought to be due to increased activity of the insulin receptor or IRS-1.[94,95,97] However, we and others have data suggesting that there is enhanced IRS-2 tyrosine phosphorylation,[98,99] Akt phosphorylation[44,100,101] and activity,[44] Atk substrate of 160 kDa (AS160) phosphorylation,[100] and expression of cytoplasmic SHP2[101] in the postexercise state.

7. EXERCISE TRAINING: IMPACT ON HEALTHY PEOPLE AND PEOPLE WITH TYPE 2 DIABETES

Regular physical activity leads to numerous adaptations in skeletal muscle which allow the muscle to more efficiently generate ATP and become more resistant to fatigue.[102] In regards to carbohydrate metabolism, some of the key adaptations that occur in skeletal muscle with exercise training include enhanced glucose uptake and increased expression of GLUT4.[103,104] Trained muscles are also characterized by increased concentrations of glycogen, which is an important factor in the decreased rates of fatigue with prolonged exercise. Exercise training causes muscle fiber type transformation to a more oxidative and perhaps slow phenotype,[105–107] and an increase in mitochondrial activity and content.[108–110] In addition, exercise training can increase insulin sensitivity and improve overall glucose homeostasis,[111–113] which are of particular importance for individuals with metabolic diseases such as type 2 diabetes.

Type 2 diabetes arises from a combination of genetic susceptibility and environmental factors including physical inactivity and poor nutrition.[114] Thus, type 2 diabetes typically develops as individuals become more obese and less active, leading to insulin resistance, impaired glucose tolerance, and eventually, the onset of full blown type 2 diabetes. While type 2 diabetes is a multifactorial disease, it is a disease of altered carbohydrate metabolism on many levels. In people with type 2 diabetes, insulin levels are normal or high, but tissues such as liver, skeletal muscle, and adipose tissue become resistant to insulin. The pancreas compensates by producing large amounts of insulin, but this stress can eventually lead to pancreatic failure and the need for exogenous insulin treatment. The hyperinsulinemic state can result in impaired glucose transport into the liver, skeletal muscle, and adipose tissue.[115] While type 2 diabetes is usually adult-onset, the number of children and adolescents afflicted by this disease is dramatically increasing. In fact, there are currently 23.6 million people in the United States, which reflects approximately 8% of the population that have diabetes, a number that has doubled over the last 15 years and is continuing to increase at epidemic rates.[116]

Although these statistics are discouraging, the good news is that regular physical exercise can delay or prevent the onset of type 2 diabetes.[117–120] Studies using randomized trials have found that lifestyle interventions, which included ~150 min of physical activity per week, combined with diet-induced weight loss, reduced the risk of type 2 diabetes by 58% in an at-risk population.[91,117] Exercise interventions, independent of diet, have also been shown to be effective for the prevention and the progression of type 2 diabetes.[118] Exercise training in people with type 2 diabetes can improve blood glucose concentrations, body weight, lipids, blood pressure, cardiovascular disease, mortality, and overall quality of life.[121–127] The Look AHEAD study has demonstrated that combined weight loss and physical activity in people with type 2 diabetes causes modest weight loss of approximately 6%, improved glycolated hemoglobin, improved mobility, and improved kidney function but no improvement in cardiovascular disease over a 10-year period.[121,123,124,126] However, since the level of fitness was only assessed through year 4 of the study, conclusions on the effects of fitness level on cardiovascular disease cannot be made.[121,123,124,126] Increasing physical activity in adults with type 2 diabetes has been shown to result in partial or complete remission of type 2 diabetes in 11.5% of subjects within the first year of intervention, and an additional 7% had partial or complete remission of type 2 diabetes after 4 years of exercise intervention.[122] Taken together, all of these data show that the effects of exercise on carbohydrate metabolism have profound effects on metabolic health, and this knowledge is important as we work to address the epidemic of type 2 diabetes.

ACKNOWLEDGMENTS

Work in the author's laboratory was supported by National Institutes of Health Grants R01-AR42238, R01-DK101043, R01-DK099511 (to L.J.G.), and 5P30 DK36836 (Joslin Diabetes Center, DRC). J.D.M. is supported by a mentor-based fellowship from the American Diabetes Association (to L.J.G.). K.I.S. was been supported by an American College of Sports Medicine Research Endowment Grant and Mary K. Iacocca Fellowship at the Joslin Diabetes Center and is currently supported by National Institutes of Health K01-DK105109.

REFERENCES

1. Bramble DM, Lieberman DE. Endurance running and the evolution of Homo. *Nature*. 2004;432:345–352.
2. Chakravarthy MV, Booth FW. Eating, exercise, and "thrifty" genotypes: connecting the dots toward an evolutionary understanding of modern chronic diseases. *J Appl Physiol*. 2004;96:3–10.

3. Hargreaves M. The metabolic systems; carbohydrate metabolism. In: Farrell PA, Joyner MJ, Caizozzo VJ, eds. *Advanced Exercise Physiology*. Philadelphia, PA: Lippincott Williams and Wilkins; 2012:3–391.
4. Spriet LL. The metabolic systems; lipid metabolism. In: Farrell PA, Joyner MJ, Caizozzo VJ, eds. *Advanced Exercise Physiology*. Philadelphia, PA: Lippincott Williams and Wilkins; 2012:392–407.
5. Sidossis LS, Stuart CA, Shulman GI, Lopaschuk GD, Wolfe RR. Glucose plus insulin regulate fat oxidation by controlling the rate of fatty acid entry into the mitochondria. *J Clin Invest*. 1996;98:2244–2250.
6. Wolfe RR. Metabolic interactions between glucose and fatty acids in humans. *Am J Clin Nutr*. 1998;67:519S–526S.
7. Acheson KJ, Schutz Y, Bessard T, Anantharaman K, Flatt JP, Jequier E. Glycogen storage capacity and de novo lipogenesis during massive carbohydrate overfeeding in man. *Am J Clin Nutr*. 1988;48:240–247.
8. Maughan RJ. *Sports Nutrition. Handbook of Sports Medicine and Science*. London: Blackwell Science; 2002.
9. McArdle W, Katch FKV. *Essentials of exercise physiology*. Philadelphia, PA: Lippincott Williams & Wilkins; 2000.
10. van Baak MA. Physical activity and energy balance. *Public Health Nutr*. 1999;2:335–339.
11. Romijn JA, Coyle EF, Sidossis LS, et al. Regulation of endogenous fat and carbohydrate metabolism in relation to exercise intensity and duration. *Am J Physiol*. 1993;265:E380–E391.
12. Graham TE. Glycogen: an overview of possible regulatory roles of the proteins associated with the granule. *Appl Physiol Nutr Metab*. 2009;34:488–492.
13. van Loon LJ, Greenhaff PL, Constantin-Teodosiu D, Saris WH, Wagenmakers AJ. The effects of increasing exercise intensity on muscle fuel utilisation in humans. *J Physiol*. 2001;536:295–304.
14. Gollnick PD, Piehl K, Saltin B. Selective glycogen depletion pattern in human muscle fibres after exercise of varying intensity and at varying pedalling rates. *J Physiol*. 1974;241:45–57.
15. Hultman E, Harris RC. Carbohydrate metabolism. In: Poortmans JR, ed. *Principles of Exercise Biochemistry*. Basel: Karger; 1988:78–119.
16. Richter EA, Ruderman NB, Gavras H, Belur ER, Galbo H. Muscle glycogenolysis during exercise: dual control by epinephrine and contractions. *Am J Physiol*. 1982;242:E25–E32.
17. Watt MJ, Howlett KF, Febbraio MA, Spriet LL, Hargreaves M. Adrenaline increases skeletal muscle glycogenolysis, pyruvate dehydrogenase activation and carbohydrate oxidation during moderate exercise in humans. *J Physiol*. 2001;534:269–278.
18. Howlett RA, Parolin ML, Dyck DJ, et al. Regulation of skeletal muscle glycogen phosphorylase and PDH at varying exercise power outputs. *Am J Physiol*. 1998;275:R418–R425.
19. Burke LM, Angus DJ, Cox GR, et al. Effect of fat adaptation and carbohydrate restoration on metabolism and performance during prolonged cycling. *J Appl Physiol*. 2000;89:2413–2421.
20. Bergstrom J, Hermansen L, Hultman E, Saltin B. Diet, muscle glycogen and physical performance. *Acta Physiol Scand*. 1967;71:140–150.
21. Bergstrom J, Hultman E. Muscle glycogen synthesis after exercise: an enhancing factor localized to the muscle cells in man. *Nature*. 1966;210:309–310.
22. Chesley A, Heigenhauser GJ, Spriet LL. Regulation of muscle glycogen phosphorylase activity following short-term endurance training. *Am J Physiol*. 1996;270:E328–E335.

23. Kubo K, Foley JE. Rate-limiting steps for insulin-mediated glucose uptake into perfused rat hindlimb. *Am J Physiol.* 1986;250:E100–E102.
24. Nesher R, Karl IE, Kipnis DM. Dissociation of effects of insulin and contraction on glucose transport in rat epitrochlearis muscle. *Am J Physiol.* 1985;249:C226–C232.
25. Bell GI, Burant CF, Takeda J, Gould GW. Structure and function of mammalian facilitative sugar transporters. *J Biol Chem.* 1993;268:19161–19164.
26. Kayano T, Burant CF, Fukumoto H, et al. Human facilitative glucose transporters. Isolation, functional characterization, and gene localization of cDNAs encoding an isoform (GLUT5) expressed in small intestine, kidney, muscle, and adipose tissue and an unusual glucose transporter pseudogene-like sequence (GLUT6). *J Biol Chem.* 1990;265:13276–13282.
27. Klip A, Paquet MR. Glucose transport and glucose transporters in muscle and their metabolic regulation. *Diabetes Care.* 1990;13:228–243.
28. Rogers S, Macheda ML, Docherty SE, et al. Identification of a novel glucose transporter-like protein-GLUT-12. *Am J Physiol Endocrinol Metab.* 2002;282: E733–E738.
29. Katz EB, Stenbit AE, Hatton K, DePinho R, Charron MJ. Cardiac and adipose tissue abnormalities but not diabetes in mice deficient in GLUT4. *Nature.* 1995;377:151–155.
30. Zisman A, Peroni OD, Abel ED, et al. Targeted disruption of the glucose transporter 4 selectively in muscle causes insulin resistance and glucose intolerance. *Nat Med.* 2000;6:924–928.
31. Hirshman MF, Wallberg-Henriksson H, Wardzala LJ, Horton ED, Horton ES. Acute exercise increases the number of plasma membrane glucose transporters in rat skeletal muscle. *FEBS Lett.* 1988;238:235–239.
32. Hirshman MF, Goodyear LJ, Wardzala LJ, Horton ED, Horton ES. Identification of an intracelular pool of glucose transporters from basal and insulin-stimulated rat skeletal muscle. *J Biol Chem.* 1990;265:987–991.
33. Lauritzen HP, Galbo H, Brandauer J, Goodyear LJ, Ploug T. Large GLUT4 vesicles are stationary while locally and reversibly depleted during transient insulin stimulation of skeletal muscle of living mice: imaging analysis of GLUT4-enhanced green fluorescent protein vesicle dynamics. *Diabetes.* 2008;57:315–324.
34. Lauritzen HP, Galbo H, Toyoda T, Goodyear LJ. Kinetics of contraction-induced GLUT4 translocation in skeletal muscle fibers from living mice. *Diabetes.* 2010;59:2134–2144.
35. Leto D, Saltiel AR. Regulation of glucose transport by insulin: traffic control of GLUT4. *Nat Rev Mol Cell Biol.* 2012;13:383–396.
36. Yang C, Coker KJ, Kim JK, et al. Syntaxin 4 heterozygous knockout mice develop muscle insulin resistance. *J Clin Invest.* 2001;107:1311–1318.
37. Kristiansen S, Hargreaves M, Richter EA. Exercise-induced increase in glucose transport, GLUT-4, and VAMP-2 in plasma membrane from human muscle. *Am J Physiol.* 1996;270:E197–E201.
38. Wallberg-Henriksson H, Constable SH, Young DA, Holloszy JO. Glucose transport into rat skeletal muscle: interaction between exercise and insulin. *J Appl Physiol.* 1988;65:909–913.
39. Hayashi T, Wojtaszewski JF, Goodyear LJ. Exercise regulation of glucose transport in skeletal muscle. *Am J Physiol.* 1997;273:E1039–E1051.
40. Folli F, Saad MJ, Backer JM, Kahn CR. Insulin stimulation of phosphatidylinositol 3-kinase activity and association with insulin receptor substrate 1 in liver and muscle of the intact rat. *J Biol Chem.* 1992;267:22171–22177.
41. Goodyear LJ, Giorgino F, Balon TW, Condorelli G, Smith RJ. Effects of contractile activity on tyrosine phosphoproteins and PI 3-kinase activity in rat skeletal muscle. *Am J Physiol.* 1995;268:E987–E995.

42. Greengard P, Valtorta F, Czernik AJ, Benfenati F. Synaptic vesicle phosphoproteins and regulation of synaptic function. *Science*. 1993;259:780–785.
43. Treadway JL, James DE, Burcel E, Ruderman NB. Effect of exercise on insulin receptor binding and kinase activity in skeletal muscle. *Am J Physiol*. 1989;256: E138–E144.
44. Wojtaszewski JF, Higaki Y, Hirshman MF, et al. Exercise modulates postreceptor insulin signaling and glucose transport in muscle-specific insulin receptor knockout mice. *J Clin Invest*. 1999;104:1257–1264.
45. Bruning JC, Michael MD, Winnay JN, et al. A muscle-specific insulin receptor knockout exhibits features of the metabolic syndrome of NIDDM without altering glucose tolerance. *Mol Cell*. 1998;2:559–569.
46. Jessen N, Goodyear LJ. Contraction signaling to glucose transport in skeletal muscle. *J Appl Physiol*. 2005;99:330–337.
47. Kramer HF, Goodyear LJ. Exercise, MAPK, and NF-kappaB signaling in skeletal muscle. *J Appl Physiol*. 2007;103:388–395.
48. Lessard SJ, Rivas DA, Alves-Wagner AB, et al. Resistance to aerobic exercise training causes metabolic dysfunction and reveals novel exercise-regulated signaling networks. *Diabetes*. 2013;62:2717–2727.
49. Richter EA, Hargreaves M. Exercise, GLUT4, and skeletal muscle glucose uptake. *Physiol Rev*. 2013;93:933–1017.
50. Rockl KS, Witczak CA, Goodyear LJ. Signaling mechanisms in skeletal muscle: acute responses and chronic adaptations to exercise. *IUBMB Life*. 2008;60:145–153.
51. Koh HJ, Arnolds DE, Fujii N, et al. Skeletal muscle-selective knockout of LKB1 increases insulin sensitivity, improves glucose homeostasis, and decreases TRB3. *Mol Cell Biol*. 2006;26:8217–8227.
52. Hardie DG. Energy sensing by the AMP-activated protein kinase and its effects on muscle metabolism. *Proc Nutr Soc*. 2011;70:92–99.
53. Hardie DG, Carling D, Carlson M. The AMP-activated/SNF1 protein kinase subfamily: metabolic sensors of the eukaryotic cell? *Annu Rev Biochem*. 1998;67:821–855.
54. Kemp BE, Mitchelhill KI, Stapleton D, Michell BJ, Chen ZP, Witters LA. Dealing with energy demand: the AMP-activated protein kinase. *Trends Biochem Sci*. 1999;24:22–25.
55. Hardie DG. AMPK: positive and negative regulation, and its role in whole-body energy homeostasis. *Curr Opin Cell Biol*. 2014;33C:1–7.
56. Hayashi T, Hirshman MF, Kurth EJ, Winder WW, Goodyear LJ. Evidence for 5' AMP-activated protein kinase mediation of the effect of muscle contraction on glucose transport. *Diabetes*. 1998;47:1369–1373.
57. Winder WW, Holmes BF, Rubink DS, Jensen EB, Chen M, Holloszy JO. Activation of AMP-activated protein kinase increases mitochondrial enzymes in skeletal muscle. *J Appl Physiol*. 2000;88:2219–2226.
58. Merrill GF, Kurth EJ, Hardie DG, Winder WW. AICA riboside increases AMP-activated protein kinase, fatty acid oxidation, and glucose uptake in rat muscle. *Am J Physiol*. 1997;273:E1107–E1112.
59. Abbott MJ, Bogachus LD, Turcotte LP. AMPKalpha2 deficiency uncovers time dependency in the regulation of contraction-induced palmitate and glucose uptake in mouse muscle. *J Appl Physiol*. 2011;111:125–134.
60. Jensen TE, Rose AJ, Jorgensen SB, et al. Possible CaMKK-dependent regulation of AMPK phosphorylation and glucose uptake at the onset of mild tetanic skeletal muscle contraction. *Am J Physiol Endocrinol Metab*. 2007;292:E1308–E1317.
61. Jensen TE, Schjerling P, Viollet B, Wojtaszewski JF, Richter EA. AMPK alpha1 activation is required for stimulation of glucose uptake by twitch contraction, but not by H2O2, in mouse skeletal muscle. *PLoS One*. 2008;3:e2102.

62. Jorgensen SB, Viollet B, Andreelli F, et al. Knockout of the alpha2 but not alpha1 5'-AMP-activated protein kinase isoform abolishes 5-aminoimidazole-4-carboxamide-1-beta-4-ribofuranosidebut not contraction-induced glucose uptake in skeletal muscle. *J Biol Chem*. 2004;279:1070–1079.
63. Lefort N, St Amand E, Morasse S, Cote CH, Marette A. The alpha-subunit of AMPK is essential for submaximal contraction-mediated glucose transport in skeletal muscle in vitro. *Am J Physiol Endocrinol Metab*. 2008;295:E1447–E1454.
64. Fujii N, Hirshman MF, Kane EM, et al. AMP-activated protein kinase alpha2 activity is not essential for contraction- and hyperosmolarity-induced glucose transport in skeletal muscle. *J Biol Chem*. 2005;280:39033–39041.
65. Mu J, Brozinick Jr JT, Valladares O, Bucan M, Birnbaum MJ. A role for AMP-activated protein kinase in contraction- and hypoxia-regulated glucose transport in skeletal muscle. *Mol Cell*. 2001;7:1085–1094.
66. Maarbjerg SJ, Jorgensen SB, Rose AJ, et al. Genetic impairment of AMPKalpha2 signaling does not reduce muscle glucose uptake during treadmill exercise in mice. *Am J Physiol Endocrinol Metab*. 2009;297:E924–E934.
67. Sakamoto K, McCarthy A, Smith D, et al. Deficiency of LKB1 in skeletal muscle prevents AMPK activation and glucose uptake during contraction. *EMBO J*. 2005;24:1810–1820.
68. Koh HJ, Toyoda T, Fujii N, et al. Sucrose nonfermenting AMPK-related kinase (SNARK) mediates contraction-stimulated glucose transport in mouse skeletal muscle. *Proc Natl Acad Sci U S A*. 2010;107:15541–15546.
69. Jeppesen J, Maarbjerg SJ, Jordy AB, et al. LKB1 regulates lipid oxidation during exercise independently of AMPK. *Diabetes*. 2013;62:1490–1499.
70. Wright DC, Hucker KA, Holloszy JO, Han DH. Ca2+ and AMPK both mediate stimulation of glucose transport by muscle contractions. *Diabetes*. 2004;53:330–335.
71. Witczak CA, Fujii N, Hirshman MF, Goodyear LJ. Ca2+/calmodulin-dependent protein kinase kinase-alpha regulates skeletal muscle glucose uptake independent of AMP-activated protein kinase and Akt activation. *Diabetes*. 2007;56:1403–1409.
72. Witczak CA, Jessen N, Warro DM, et al. CaMKII regulates contraction- but not insulin-induced glucose uptake in mouse skeletal muscle. *Am J Physiol Endocrinol Metab* 2010;298:E1150–E1160.
73. Jensen TE, Sylow L, Rose AJ, et al. Contraction-stimulated glucose transport in muscle is controlled by AMPK and mechanical stress but not sarcoplasmatic reticulum Ca(2+) release. *Mol Metab*. 2014;3:742–753.
74. Wennerberg K, Rossman KL, Der CJ. The Ras superfamily at a glance. *J Cell Sci*. 2005;118:843–846.
75. Kaddai V, Marchand-Brustel Y, Cormont M. Rab proteins in endocytosis and Glut4 trafficking. *Acta Physiol (Oxf)*. 2008;192:75–88.
76. Sylow L, Jensen TE, Kleinert M, et al. Rac1 signaling is required for insulin-stimulated glucose uptake and is dysregulated in insulin-resistant murine and human skeletal muscle. *Diabetes*. 2013;62:1865–1875.
77. Sylow L, Jensen TE, Kleinert M, et al. Rac1 is a novel regulator of contraction-stimulated glucose uptake in skeletal muscle. *Diabetes*. 2013;62:1139–1151.
78. Bruss MD, Arias EB, Lienhard GE, Cartee GD. Increased phosphorylation of Akt substrate of 160 kDa (AS160) in rat skeletal muscle in response to insulin or contractile activity. *Diabetes*. 2005;54:41–50.
79. Sano H, Kane S, Sano E, et al. Insulin-stimulated phosphorylation of a Rab GTPase-activating protein regulates GLUT4 translocation. *J Biol Chem*. 2003;278:14599–14602.
80. Thong FS, Dugani CB, Klip A. Turning signals on and off: GLUT4 traffic in the insulin-signaling highway. *Physiology*. 2005;20:271–284.

81. Kramer HF, Witczak CA, Fujii N, et al. Distinct signals regulate AS160 phosphorylation in response to insulin, AICAR, and contraction in mouse skeletal muscle. *Diabetes*. 2006;55:2067–2076.
82. Treebak JT, Birk JB, Rose AJ, Kiens B, Richter EA, Wojtaszewski JF. AS160 phosphorylation is associated with activation of alpha2beta2gamma1- but not alpha2beta2gamma3-AMPK trimeric complex in skeletal muscle during exercise in humans. *Am J Physiol Endocrinol Metab*. 2007;292:E715–E722.
83. Deshmukh A, Coffey VG, Zhong Z, Chibalin AV, Hawley JA, Zierath JR. Exercise-induced phosphorylation of the novel Akt substrates AS160 and filamin A in human skeletal muscle. *Diabetes*. 2006;55:1776–1782.
84. Sriwijitkamol A, Coletta DK, Wajcberg E, et al. Effect of acute exercise on AMPK signaling in skeletal muscle of subjects with type 2 diabetes: a time-course and dose–response study. *Diabetes*. 2007;56:836–848.
85. Kramer HF, Witczak CA, Taylor EB, Fujii N, Hirshman MF, Goodyear LJ. AS160 regulates insulin- and contraction-stimulated glucose uptake in mouse skeletal muscle. *J Biol Chem*. 2006;281:31478–31485.
86. Kramer HF, Taylor EB, Witczak CA, Fujii N, Hirshman MF, Goodyear LJ. Calmodulin-binding domain of AS160 regulates contraction- but not insulin-stimulated glucose uptake in skeletal muscle. *Diabetes*. 2007;56:2854–2862.
87. Kane S, Sano H, Liu SC, et al. A method to identify serine kinase substrates. Akt phosphorylates a novel adipocyte protein with a Rab GTPase-activating protein (GAP) domain. *J Biol Chem*. 2002;277:22115–22118.
88. Karlsson HK, Zierath JR, Kane S, Krook A, Lienhard GE, Wallberg-Henriksson H. Insulin-stimulated phosphorylation of the Akt substrate AS160 is impaired in skeletal muscle of type 2 diabetic subjects. *Diabetes*. 2005;54:1692–1697.
89. Roach WG, Chavez JA, Miinea CP, Lienhard GE. Substrate specificity and effect on GLUT4 translocation of the Rab GTPase-activating protein Tbc1d1. *Biochem J*. 2007;403:353–358.
90. Taylor EB, An D, Kramer HF, et al. Discovery of TBC1D1 as an insulin-, AICAR-, and contraction-stimulated signaling nexus in mouse skeletal muscle. *J Biol Chem*. 2008;283:9787–9796.
91. Treebak JT, Glund S, Deshmukh A, et al. AMPK-mediated AS160 phosphorylation in skeletal muscle is dependent on AMPK catalytic and regulatory subunits. *Diabetes*. 2006;55:2051–2058.
92. An D, Toyoda T, Taylor EB, et al. TBC1D1 regulates insulin- and contraction-induced glucose transport in mouse skeletal muscle. *Diabetes*. 2010;59:1358–1365.
93. Vichaiwong K, Purohit S, An D, et al. Contraction regulates site-specific phosphorylation of TBC1D1 in skeletal muscle. *Biochem J*. 2010;431:311–320.
94. Cartee GD, Young DA, Sleeper MD, Zierath J, Wallberg-Henriksson H, Holloszy JO. Prolonged increase in insulin-stimulated glucose transport in muscle after exercise. *Am J Physiol*. 1989;256:E494–E499.
95. Wojtaszewski JF, Hansen BF, Kiens B, Richter EA. Insulin signaling in human skeletal muscle: time course and effect of exercise. *Diabetes*. 1997;46(11):1775–1781.
96. Mikines KJ, Sonne B, Farrell PA, Tronier B, Galbo H. Effect of physical exercise on sensitivity and responsiveness to insulin in humans. *Am J Physiol*. 1988;254:E248–E259.
97. Wojtaszewski JF, Hansen BF, Gade J, et al. Insulin signaling and insulin sensitivity after exercise in human skeletal muscle. *Diabetes*. 2000;49:325–331.
98. Howlett KF, Sakamoto K, Hirshman MF, et al. Insulin signaling after exercise in insulin receptor substrate-2-deficient mice. *Diabetes*. 2002;51:479–483.
99. Howlett KF, Sakamoto K, Yu H, Goodyear LJ, Hargreaves M. Insulin-stimulated insulin receptor substrate-2-associated phosphatidylinositol 3-kinase activity is enhanced in human skeletal muscle after exercise. *Metabolism*. 2006;55(8):1046–1052.

100. Arias EB, Kim J, Funai K, Cartee GD. Prior exercise increases phosphorylation of Akt substrate of 160 kDa (AS160) in rat skeletal muscle. *Am J Physiol Endocrinol Metab.* 2007;292(4):E1191–E1200.
101. Wadley GD, Konstantopoulos N, Macaulay L, et al. Increased insulin-stimulated Akt pSer473 and cytosolic SHP2 protein abundance in human skeletal muscle following acute exercise and short-term training. *J Appl Physiol.* 2007;102(4):1624–1631.
102. Tonkonogi M, Sahlin K. Physical exercise and mitochondrial function in human skeletal muscle. *Exerc Sport Sci Rev.* 2002;30(3):129–137.
103. James DE, Kraegen EW, Chisholm DJ. Effects of exercise training on in vivo insulin action in individual tissues of the rat. *J Clin Invest.* 1985;76:657–666.
104. Goodyear LJ, Hirshman MF, Valyou PM, Horton ES. Glucose transporter number, function, and subcellular distribution in rat skeletal muscle after exercise training. *Diabetes.* 1992;41:1091–1099.
105. Andersen P, Henriksson J. Training induced changes in the subgroups of human type II skeletal muscle fibres. *Acta Physiol Scand.* 1977;99(1):123–125.
106. Fitts RH. Effects of regular exercise training on skeletal muscle contractile function. *Am J Phys Med Rehabil.* 2003;82(4):320–331.
107. Röckl KS, Hirshman MF, Brandauer J, Fujii N, Witters LA, Goodyear LJ. Skeletal muscle adaptation to exercise training: AMP-activated protein kinase mediates muscle fiber type shift. *Diabetes.* 2007;56(8):2062–2069.
108. Holloszy JO. Biochemical adaptations in muscle: effects of exercise on mitochondrial oxygen uptake and respiratory activity in skeletal muscle. *J Biol Chem.* 1967;242:2278–2282.
109. Gollnick PD, Armstrong RB, Saltin B, Saubert CW, Sembrowich WL, Shepherd RE. Effect of training on enzyme activity and fiber composition of human skeletal muscle. *J Appl Physiol.* 1973;34(1):107–111.
110. Holloszy JO, Booth FW. Biochemical adaptations to endurance exercise in muscle. *Annu Rev Plant Physiol Plant Mol Biol.* 1976;38:273–291.
111. Bjorntorp P, Fahlen M, Grimby G, et al. Carbohydrate and lipid metabolism in middle-aged physically well trained men. *Metabolism.* 1972;21:1037–1044.
112. Ruderman NB, Ganda OP, Johansen K. The effect of physical training on glucose tolerance and plasma lipids in maturity-onset diabetes. *Diabetes.* 1979;28:89–92.
113. Ostergard T, Andersen JL, Nyholm B, et al. Impact of exercise training on insulin sensitivity, physical fitness, and muscle oxidative capacity in first-degree relatives of type 2 diabetic patients. *Am J Physiol Endocrinol Metab.* 2006;290(5):E998–E1005.
114. Hu FB, Manson JE, Stampfer MJ, et al. Diet, lifestyle, and the risk of type 2 diabetes mellitus in women. *N Engl J Med.* 2001;345:790–797.
115. DeFronzo RA, Tripathy D. Skeletal muscle insulin resistance is the primary defect in type 2 diabetes. *Diabetes Care.* 2009;32(Suppl 2):S157–S163.
116. National Task Force on the Prevention, Treatment of Obesity. Overweight, obesity, and health risk. *Arch Intern Med.* 2000;160:898–904.
117. Pan XR, Li GW, Hu YH, et al. Effects of diet and exercise in preventing NIDDM in people with impaired glucose tolerance. The Da Qing IGT and Diabetes Study. *Diabetes Care.* 1997;20(4):537–544.
118. Colberg SR, Sigal RJ, Fernhall B, et al. Exercise and type 2 diabetes: the American College of Sports Medicine and the American Diabetes Association: joint position statement executive summary. *Diabetes Care.* 2010;33:2692–2696.
119. Coyle EF, Coggan AR, Hemmert MK, Ivy JL. Muscle glycogen utilization during prolonged strenuous exercise when fed carbohydrate. *J Appl Physiol.* 1986;61:165–172.
120. Kinney JM, Tucker HN. *Energy Metabolism. Tissue Determinants and Cellular Corollaries.* New York, NY: Raven Press; 1992.

121. Espeland MA, Glick HA, Bertoni A, et al. Impact of an intensive lifestyle intervention on use and cost of medical services among overweight and obese adults with type 2 diabetes: the action for health in diabetes. *Diabetes Care*. 2014;37(9):2548–2556.
122. Gregg EW, Chen H, Wagenknecht LE, et al. Association of an intensive lifestyle intervention with remission of type 2 diabetes. *JAMA*. 2012;308(23):2489–2496.
123. Look AHEAD Research Group. Effect of a long-term behavioural weight loss intervention on nephropathy in overweight or obese adults with type 2 diabetes: a secondary analysis of the Look AHEAD randomised clinical trial. *Lancet Diabetes Endocrinol.* 2014;2(10):801–809.
124. Wing RR, Bolin P, Brancati FL, et al. Cardiovascular effects of intensive lifestyle intervention in type 2 diabetes. *N Engl J Med*. 2013;369(2):145–154.
125. Kelley DE, He J, Menshikova EV, Ritov VB. Dysfunction of mitochondria in human skeletal muscle in type 2 diabetes. *Diabetes*. 2002;51(10):2944–2950.
126. Rejeski WJ, Bray GA, Chen SH, et al. Aging and physical function in type 2 diabetes: 8 years of an intensive lifestyle intervention. *J Gerontol A Biol Sci Med Sci.* 2015;70(3):343–351.
127. Wing RR, Lang W, Wadden TA, et al. Benefits of modest weight loss in improving cardiovascular risk factors in overweight and obese individuals with type 2 diabetes. *Diabetes Care*. 2011;34(7):1481–1486.

CHAPTER THREE

Exercise and Regulation of Lipid Metabolism

Robert C. Noland[1]
Pennington Biomedical Research Center, Louisiana State University, Baton Rouge, Louisiana, USA
[1]Corresponding author: e-mail address: robert.noland@pbrc.edu

Contents

Abstract

The increased prevalence of hyperlipidemia, hypertriglyceridemia, hypercholesterolemia, and fatty liver disease has provided increasingly negative connotations toward lipids. However, it is important to remember that lipids are essential components supporting life. Lipids are a class of molecules defined by their inherent insolubility in water. In biological systems, lipids are either hydrophobic (containing only polar groups) or amphipathic (possess polar and nonpolar groups). These characteristics lend lipids to be highly diverse with a multitude of functions including hormone and membrane synthesis, involvement in numerous signaling cascades, as well as serving as a source of metabolic fuel supporting energy production. Exercise can induce changes

Progress in Molecular Biology and Translational Science, Volume 135
ISSN 1877-1173
http://dx.doi.org/10.1016/bs.pmbts.2015.06.017

in the lipid composition of membranes that effect fluidity and cellular function, as well as modify the cellular and circulating environment of lipids that regulate signaling cascades. The purpose of this chapter is to focus on lipid utilization as metabolic fuel in response to acute and chronic exercise training. Lipids utilized as an energy source during exercise include circulating fatty acids bound to albumin, triglycerides stored in very-low-density lipoprotein, and intramuscular triglyceride stores. Dynamic changes in these lipid pools during and after exercise are discussed, as well as key factors that may be responsible for regulating changes in fat oxidation in response to varying exercise conditions.

1. CROSSOVER CONCEPT

In 1994, Brooks and Mercier proposed the "crossover concept" as a unified theory to describe the dynamic changes in substrate metabolism that occur during endurance exercise across a range of exercise intensities, nutritional status, gender, age, and prior training status.[1] The central tenets to this model are (1) the primary fuel source at rest is lipid; (2) exercise intensity is the primary factor that dictates fuel selection during exercise; (3) as exercise intensity increases beyond 45–65% of VO_2max, glycogen and glucose become the primary fuels, whereas lipid utilization decreases; (4) endurance training, dietary modification, and energy supply can modify substrate selection during exercise, but these play a secondary role to the impact of exercise intensity; and (5) comparisons made among subjects of differing age, gender, training status, or activities requiring vastly different metabolic rates require normalization of data to total energy flux. This model was developed as a compilation of results from several independent studies and has proven to accurately represent acute exercise substrate metabolism.

2. FAT METABOLISM DURING EXERCISE

2.1 Overview of Whole-Body Fat Oxidation During Exercise

Fat contributes 90% of the substrate for resting energy expenditure in skeletal muscle[2]; however, its resting metabolic activity is relatively low so muscle receives only ~20% of total cardiac output. Alternatively, during maximal exercise intramuscular oxygen consumption and blood flow to skeletal muscle can increase ~30-fold,[3,4] representing up to 80% of total cardiac output. Since maximal exercise can induce a 20-fold increase in whole-body metabolic rate and ATP turnover in contracting skeletal muscle can increase 100-fold,[5] a dramatic increase in blood flow is necessary to deliver substrates to meet heightened energy demand.

Substrate preference during exercise varies with intensity. At low exercise intensities (<40% VO_2max), lipid is the predominant fuel, and as intensity increases the body relies more heavily on carbohydrates. Maximal fat oxidation peaks between 45% and 65% VO_2max, whereupon the balance in fuel preference shifts toward carbohydrate. This point is illustrated in the crossover concept and confirmed in studies measuring fuel utilization during exercise via respiratory exchange ratio (RER), tracer methodology, and $^{13}CO_2$ expiration in the breath which revealed fat oxidation increased from rest to exercise and peaked at moderate-intensity exercise (45–65% VO_2max), but declined at higher intensities.[6] Whole-body fat oxidation can increase two- to threefold during submaximal exercise (~65% VO_2max) and this is largely driven by adaptations in skeletal muscle as lipid utilization can increase 5- to 15-fold in this tissue under comparable exercise conditions.[7,8] As a result, *in vivo* studies indicate skeletal muscle accounts for ~55–75% of whole-body fat utilization during submaximal exercise.[8] It is also important to note that during submaximal exercise (~65% VO_2max), the relative contribution of fat toward energy provision progressively rises as the exercise bout is prolonged beyond 90 min; however, this is not observed in exercise lasting less than 60 min.[8,9] This is believed to contribute to a "glycogen-sparing" effect during prolonged exercise, but more research is required to prove this notion.

2.2 Adipose-Derived Free Fatty Acid Utilization During Exercise

2.2.1 Adipose Tissue Lipolysis and Fatty Acid Utilization by Skeletal Muscle

Lipolysis of adipose tissue is constantly occurring at rest, but ~70% of the liberated fatty acids (FAs) are re-esterified into the adipocyte triglyceride (TG) pool.[10] During the first 10–15 min of exercise, it is not uncommon to observe a decrease in plasma FA concentrations due to enhanced clearance by skeletal muscle, although this is more prominent at high exercise intensities.[11] However, after this point circulating FA gradually increases with time due to heightened exercise-induced β-adrenergic outflow to adipose tissue which invokes an approximately threefold increase in lipolysis and reduces the re-esterification rate to nearly 25% within the first 30 min of moderate-intensity exercise.[10] Heightened sympathetic drive during exercise not only enhances circulation to skeletal muscle but also increases adipose tissue blood flow approximately two- to sevenfold.[12,13] These adaptations ensure sufficient delivery of lipid fuel to exercising muscle. In support, Turcotte *et al.*[14] showed that, during knee extensor exercise, the arterial plasma FA concentration increased significantly over time and this was associated with increased fat uptake and oxidation in skeletal muscle. However, it is important to mention that the percentage of FA removed by skeletal muscle decreased over time, which indicates that mobilization of lipid from adipose tissue exceeds the requirement to meet energy demand. Circulating FA levels during prolonged low- to moderate-intensity exercise can reach approximately twofold higher levels than observed at rest, although this does not generally occur at higher exercise intensities.[11,15] Collectively, these data suggest that lipid availability is not likely to be limiting during moderate-intensity exercise.

Peak fat utilization is achieved during moderate-intensity exercise; however, it is noteworthy that total fat oxidation is reduced at both lower and higher exercise intensities. Romijn *et al.*[11] provided insight into this phenomenon with a seminal study examining adipose tissue lipolysis and FA delivery to skeletal muscle at exercise intensities equivalent to 25%, 65%, and 85% VO_2max. At low intensity, fat was the predominant fuel source and the majority of these lipids were derived from plasma, whereas TG contributed little. Based upon this evidence, it appears that the relative percent contribution of fat to total substrate oxidation is maximal during low-intensity exercise and the limitation in fat utilization is in the energy requirement of the exercise bout itself. In accordance with the crossover concept, peak total fat utilization rates were detected at 65% VO_2max. As one might expect, lipolysis and FA delivery to skeletal muscle increased as a function of exercise intensity up to 65% VO_2max. Interestingly, however, despite increased FA delivery during moderate-intensity exercise, the contribution of plasma FA to total substrate oxidation was similar to that detected at 25% VO_2max. The authors provided a resolution to this apparent discrepancy as they determined that, during submaximal exercise, TG catabolism increased to a level that represented nearly half of the lipid utilized for total fat oxidation. When exercise intensity was increased further to 85% VO_2max, TG stores and plasma-derived FA contributed nearly equally to total fat oxidation; however, there was an overall decrease in the contribution of fat to total substrate utilization to a point where it represented $<50\%$ of the fuel required to meet the energy demand of high-intensity exercise. While the decrease in fat oxidation at 85% VO_2max could not be explained by alterations in adipose tissue lipolysis, there was a $\sim 50\%$ reduction in plasma FA levels suggesting lipid re-esterification in adipose tissue increased during high-intensity exercise.[10,11] Interestingly, infusion of intralipid for 2 h at rest (7.4 U/kg/h) and during exercise at 85% VO_2max (15.4 U/mg/h) in order to match the circulating FA concentrations observed during exercise at 65% VO_2max enhances FA uptake and utilization; however, neither reaches levels similar to those observed at 65% VO_2max.[16] Overall, these findings suggest FA delivery may slightly limit fat oxidation at higher intensities; however, the molecular mechanisms inherent to skeletal muscle lipid uptake/metabolism may provide a stronger explanation for the decreased fat oxidation rates at higher workloads.

2.2.2 Fatty Acid Transporters

While much variability exists, it appears that 60–100% of the FA transported into muscle during exercise is oxidized and that the amount oxidized approaches 100% as exercise intensity increases.[17] This is seemingly at odds with evidence showing fat oxidation decreases at high intensity. To reconcile this discrepancy, it seems likely that the decrease in fat oxidation at high intensity is due to decreased FA uptake into muscle. Historically, FA was believed to traverse the cellular membrane via passive diffusion, but discovery of FA transport proteins supports the concept of facilitated lipid diffusion. As depicted in Fig. 1, three proteins involved in cellular FA transport have been discovered: FA transporter, cluster of differentiation 36 (FAT/CD36); a plasma membrane-associated fatty acid-binding protein (FABP-PM); and a class of fatty acid transport proteins (FATP1–6). Fatp members are self-supporting transporters, whereas Fat/cd36 and Fabp-pm may work in a cooperative manner.[18] These transporters account for ~70% of basal lipid transport and are highly regulated.[19] The quantitatively large proportion of FA uptake mediated by these transporters, in combination with the fact that they are highly regulated, supports the notion that exercise-mediated regulation of these transporters modulates fat oxidation during exercise.

Fat/cd36 and Fabp-pm are most abundant in highly oxidative type I muscle fibers.[20] Training-induced adaptations of FAT/CD36 are inconsistent as total protein content is either unchanged[21] or increased[22] in response to moderate-intensity exercise programs. Alternatively, high-intensity exercise training increases FAT/CD36 protein.[23] FABP-PM protein is consistently higher in skeletal muscle from trained humans.[17] Exercise-induced alterations in FABP-PM regulation in humans may also be related to exercise intensity as moderate-intensity exercise does not alter protein

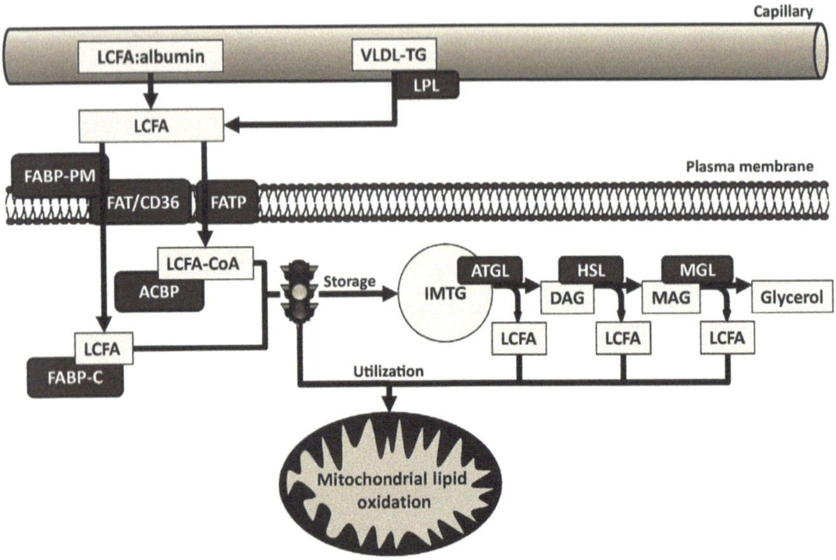

Figure 1 Cellular fatty acid uptake and intracellular lipid handling in skeletal muscle. During moderate-intensity exercise, delivery of lipid (LCFA:albumin and VLDL-TG) through the circulatory system to skeletal muscle increases. In response, upregulation of fatty acid transporters at the plasma membrane during exercise increases fatty acid uptake into skeletal muscle. Additionally, exercise increases lipolysis of intramuscular triglycerides. Collectively, these actions significantly increase fatty acid delivery to the mitochondria for use as metabolic fuel during moderate-intensity endurance exercise. *Abbreviations*: LCFA, long-chain fatty acid; VLDL-TG, very-low-density lipoprotein triglyceride; FABP-PM, plasma membrane fatty acid-binding protein; FAT/CD36, fatty acid translocase, cluster of differentiation 36; FATP, fatty acid transport protein; FABP-C, cytosolic fatty acid-binding protein; ACBP, acyl-CoA-binding protein; IMTG, intramuscular triglyceride; ATGL, adipose triglyceride lipase; HSL, hormone-sensitive lipase; MGL, monoacylglycerol lipase; DAG, diacylglycerol; MAG, monoacylglycerol. (See the color plate.)

expression,[21,22] whereas high-intensity exercise increases protein content.[23,24] While evidence indicates exercise training generally induces expression of these transporters, it is important to note that these changes are often subtle and are not consistently reported. While the aforementioned studies suggest FAT/CD36 and FABP-PM may be subtly responsive to a chronic training stimulus, most research indicates these transporters are primarily regulated in response to acute exercise. Early studies showed FA uptake increased 50–75% after contraction and correlated with Fat/cd36 content.[18,25] In the basal state, Fat/cd36 and Fabp-pm are found in an intracellular vesicular pool and respond to acute exercise by translocating to the plasma membrane and this localization is associated with increased FA uptake.[17] These data suggest FAT/CD36 and FABP-PM likely regulate FA uptake during exercise; however, it is currently unknown whether the differences in fat oxidation at varying exercise intensities can be explained by differential plasmalemmal localization of these proteins.

Fatp1 and Fatp4 are found in skeletal muscle and are most highly expressed in type I, oxidative muscle fibers.[26] Interestingly, Fatp1 and Fatp4 have enzymatic activity that converts long-chain FAs (LCFA) to acyl-CoA thioesters.[27] Overexpression of Fatp1 and Fatp4 increases FA uptake and oxidation; however, the transport capacity of Fatp4 may be greater in muscle.[26] A recent study indicated 8 weeks of variable-intensity training (45–80% VO_2peak) decreased FATP1, whereas an increase in FATP4 was observed which was associated with enhanced fat oxidation.[28] Insulin induces Fatp1 translocation to the plasma membrane[29]; however, whether exercise modulates the plasmalemmal localization of either FATP1 or FATP4 is unknown. Further research is required to uncover the importance of FATP1 and FATP4 in regulating lipid metabolism in response to exercise. As progress is made in this area, it should be remembered that the acyl-CoA synthetase activity of both Fatp1 and Fatp4 requires ATP and CoA.[30,31] It is possible that during exercise, the availability of one or both of these cofactors may become limiting; therefore, FATP1 and FATP4 activity may be dictated by more than either protein expression or localization during exercise.

Once lipids enter the cell, they are bound to a cytosolic fatty acid-binding protein (FABP-C), which reduces cytosolic FA concentrations. This protects against lipotoxicity and reduces the concentration gradient, thereby facilitating greater FA uptake. There are at least seven FABP isoforms expressed in various tissues and all are implicated in intracellular lipid trafficking.[32] Overexpression of FABPs increases FA import, yet knockouts of specific Fabp isoforms do not generally exhibit a gross phenotype due to compensatory upregulation of other Fabp isoforms. With respect to skeletal muscle, two primary cytosolic lipid carriers have been examined: FABP-C and acyl-CoA-binding protein (ACBP) (Fig. 1). Both FABP-C and ACBP mediate intracellular FA transport.[17] *Fabp-c* ablation in mice reduced FA uptake ~45–50% in skeletal muscle.[33] Despite the obvious importance of these proteins in facilitating lipid transport, neither FABP-C nor ACBP was altered in response to exercise training.[17] Overall, there appears to be little need to expand the intracellular lipid transporter system because it may already be sufficient to tolerate the increased lipid uptake during exercise.

2.3 Very-Low-Density Lipoprotein Triglyceride Utilization During Exercise

The majority of circulating lipids are transported within lipoproteins. Lipoproteins carry varying amounts of proteins, cholesterol, phospholipids, and TGs and are classified based on their density. Circulating lipoproteins provide ~10% of total lipid oxidation under normal dietary conditions; however, this increases to ~25% in response to a high-fat diet.[7] This is primarily accounted for by triglycerides carried within very-low-density lipoproteins (VLDL-TG). TGs are removed from VLDLs by tissue-specific lipoprotein lipases (LPLs). LPL is bound to the capillary endothelium and hydrolyzes TGs, thus liberating FA for uptake into the surrounding tissue (Fig. 1). Tissues with high LPL activity include adipose tissue, skeletal muscle, and heart; however, the regulation of this enzyme is tissue specific. In the fed state, LPL activity is highest in adipose tissue, thereby facilitating storage. Alternatively, in states of energy demand, such as fasting and endurance exercise, LPL activity in adipose tissue is diminished, while skeletal muscle Lpl activity increases.[34,35] Since exercise training also increases capillarization and oxidative capacity of skeletal muscle, upregulation of LPL activity in muscle could substantially enhance FA delivery via VLDL-TG hydrolysis during exercise. In this light, VLDL-TG degradation is significantly greater in trained versus untrained muscle and VLDL-TG turnover increases during moderate-intensity exercise.[36,37] VLDL-TG also provides fuel in the post-exercise period as the fractional catabolic rate of VLDL-TG is significantly higher during recovery.[37] This effect can be prolonged as significantly elevated VLDL-TG clearance has been found up to 14 h after exercise.[38] Globally, findings suggest VLDL-TG contributes lipid substrates to help meet the excess energy demands during and after exercise; however, the contribution of VLDL-TG is likely less than that obtained from circulating FA and intramuscular triglyceride (IMTG) stores.

2.4 IMTG Utilization During Exercise

2.4.1 Contribution of IMTG to Exercise Substrate Metabolism

IMTGs are estimated to represent 1–2% of whole-body lipid stores; however, since they are stored in close proximity with mitochondria and have high energy density, they provide a readily accessible fuel reserve during exercise. IMTG content correlates with the oxidative potential of the muscle; thus, oxidative type I fibers can contain threefold greater IMTG than type II fibers.[17] Endurance-trained individuals have elevated IMTG: a phenomenon termed the Athlete's Paradox.[39] This phenomenon can occur in response to as little as 2–8 weeks of endurance training.[17] Heightened IMTG storage in trained muscle appears to translate to greater IMTG utilization during exercise than untrained individuals.[40]

While circulating lipids are an important fuel source, only ~55–65% of the total FA utilized during submaximal exercise originate from the circulating lipid pool.[17] With this in mind, IMTGs can contribute significantly to fat oxidation during exercise as they provide up to 35–50% of the lipid substrate to support total fat oxidation during exercise at various durations and intensities.[17] The quantitative importance of IMTG is likely highest during prolonged (>90 min) moderate-intensity exercise as they can provide ~25% of maximal total energy (~50% of total FA used), but they contribute less during either low- or high-intensity exercise.[11,41]

Studies examining the response of IMTG content to exercise are widespread, but often report different findings. The discrepancies may be related to the methodology used to analyze IMTG content pre- and postexercise and have been discussed in great detail elsewhere.[17,40] Regardless, the general consensus is that IMTGs are utilized during endurance and resistance exercise. With respect to endurance exercise, IMTG stores diminish during prolonged low- to moderate-intensity exercise, but do not change at high intensity.[17,40] The latter findings may help explain the decrease in fat oxidation relative to increased exercise intensity as IMTG hydrolysis may decline during high-intensity exercise. Collectively, these studies reveal that IMTG lipolysis can serve as a significant source of metabolic fuel during exercise.

2.4.2 Skeletal Muscle Lipases

IMTG lipolytic rates during exercise can be 10-fold higher than re-esterification rates.[42] As depicted in Fig. 1, FAs are mobilized from IMTG through the hydrolytic action of a series of lipase enzymes: adipose triglyceride lipase (ATGL), hormone-sensitive lipase (HSL), and monoacylglycerol lipase (MGL). Since Atgl and Hsl account for 98% of the total lipase activity in skeletal muscle,[43] the discussion herein will focus on the activity of these enzymes as they relate to IMTG utilization during and after exercise.

HSL has historically been considered the primary lipase responsible for IMTG hydrolysis; however, recent evidence indicates ATGL is critical for complete IMTG catabolism. ATGL has 10-fold higher substrate specificity for TGs than diacylglycerides (DAGs), indicating this enzyme catalyzes the initial step of IMTG hydrolysis.[44] Within skeletal muscle, ATGL is expressed almost exclusively in oxidative type I fibers, which likely explains why IMTG stores are reduced after exercise up to ~60%, predominantly in type I fibers.[9,45] Eight weeks of endurance training induced a twofold increase in skeletal muscle ATGL protein which was associated with decreased IMTG,[46] thereby suggesting exercise training enhanced ATGL activity. ATGL is regulated via protein–protein interactions with comparative gene identification 58 and various perilipin proteins. While investigation of whether exercise induces alterations in these protein–protein interactions is in its infancy, several studies are yielding exciting insight into mechanisms responsible for the exercise-induced increase in ATGL activity.[47–53]

HSL has greater specificity for DAGs (10-fold) and cholesteryl (5-fold) than TGs or monoacylglycerides.[17] Moreover, *Hsl*-deficient mice exhibit DAG accrual suggesting this enzyme has a primary role in DAG hydrolysis *in vivo*.[54] Hsl is highest in oxidative type I fibers[55] and exercise consistently elevates enzymatic activity in skeletal muscle.[17,40] Increased HSL activity does not appear to be due to alterations in protein content as endurance training has no effect on HSL expression.[56,57] Instead, exercise primarily regulates Hsl via posttranslational modification of five different serine phosphorylation sites.[58] The potential regulation of this enzyme by various signaling cascades offers the advantage of allowing for rapid changes in enzymatic activity. This has been confirmed in skeletal muscle as HSL activity increased within 1 min of initiating an exercise bout and enzymatic activity continues to climb in low- and moderate-intensity exercise until 60 min; however, beyond this period HSL activity declines.[59] The rapid activation of HSL appears to be even more robust at high-intensity exercise (90% VO_2max), but this is short lived as lipolytic activity begins to decline within 10 min.

2.5 Regulation of Mitochondrial Fatty Acid Oxidation

While adipose tissue lipolysis, cellular FA uptake, and IMTG catabolism can certainly impact skeletal muscle fat oxidation during exercise, mechanisms regulating mitochondrial fat oxidation itself are equally likely to dictate exercise-induced adaptations in substrate utilization. In support, Watt *et al.*[60] showed long-chain acyl-CoAs in skeletal muscle accumulated during moderate-intensity exercise and increased with time, indicating FA availability was sufficient. Additionally, Kiens *et al.*[61] found FA content within skeletal muscle increased after high-intensity exercise despite decreased whole-body fat oxidation, leading the authors to conclude that the decline in fat oxidation was primarily due to diminished mitochondrial fat oxidation rather than insufficient substrate provision (delivery, uptake, or lipolysis). These findings provide compelling evidence supporting the notion that mechanisms regulating mitochondrial lipid utilization play an important role in determining fuel selection during exercise.

Maximal fat oxidation rates can vary approximately fivefold during exercise[62] and flux through the tricarboxylic acid (TCA) cycle and electron transport chain (ETC) can increase 70–100-fold during strenuous exercise[4] indicating the excess energy demand during exercise can be substantial. To accommodate increased energy demand during repeated bouts of exercise, muscle adapts by increasing mitochondrial density. Exercise-induced adaptations in mitochondrial capacity provide trained individuals with a more robust energetic capacity to meet the demands of an exercise bout by increasing the potential for mitochondrial ATP production, TCA cycle activity, as well as mobilization, transport, and oxidation of fat.[3] These adaptations not only contribute to increased exercise capacity but also provide a system that is more efficient at utilizing lipid-derived substrates. While an in-depth discussion of exercise-induced mitochondrial adaptations can be found in a separate chapter within this book, this section will focus on concepts that may modify or regulate mitochondrial fat oxidation in response to exercise (Fig. 2).

2.5.1 Regulation of Long-Chain Acyl-CoA Synthetase

There are five known isoforms of long-chain acyl-CoA synthetase (ACSL1–5). Acsl1 appears to be the predominant isoform in skeletal muscle as muscle-specific *Acsl1*-deficient mice exhibited a 91% reduction in Acsl activity, which induced a 60–85% reduction in fat oxidation.[63] Muscle-specific *Acsl1* deficiency also leads to limited exercise capacity as these mice ran only 48% as far as controls. In humans, 10 days of exercise training at 75% VO_2peak increased skeletal muscle Acsl activity, resulting in increased fat oxidation.[64] These data suggest that ACSL activity could regulate fat oxidation during exercise; however, identification of mechanisms regulating ACSL activity in this context is lacking. With this in mind, it is likely that ACSL activity is not limiting at moderate-intensity exercise as evidence indicates FA levels progressively decline, while acyl-CoAs accumulate over time.[60,61] Alternatively, it seems reasonable to predict that the decrease in fat oxidation during high-intensity exercise may be partly attributable to limitations in ACSL activity. ACSL enzymes require free CoA to convert FA to

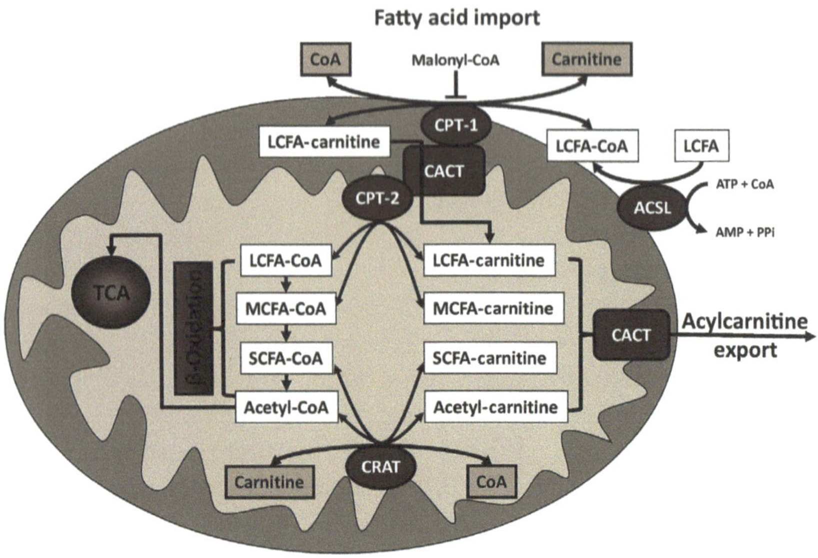

Figure 2 Role of the carnitine shuttle in mitochondrial import and export of fatty acids. Carnitine palmitoyltransferase-1 (CPT-1) converts long-chain acyl-CoA (LCFA-CoA) into long-chain acylcarnitine (LCFA-carnitine), which subsequently enters the mitochondria through carnitine–acylcarnitine translocase (CACT). Within the mitochondrial matrix, carnitine palmitoyltransferase-2 (CPT-2) converts LCFA-carnitine back into LCFA-CoA, which then enters the β-oxidation pathway for use as metabolic fuel. The ability of carnitine to facilitate mitochondrial fatty acid import is essential and supports lipid metabolism during exercise. In the event that mitochondrial lipid entry overwhelms the requirement for metabolic fuel, excess acyl-CoAs generated within the β-oxidative pathway can be converted into acylcarnitine esters by CPT-2 or carnitine acetyltransferase (CRAT). The resultant acylcarnitines can then be exported from the mitochondrial matrix. The ability of carnitine to facilitate the export of excess acyl moieties from the mitochondrial matrix likely preserves optimal mitochondrial performance during exercise. *Abbreviations*: ACSL, long-chain acyl-CoA synthetase; PPi, inorganic pyrophosphate. (See the color plate.)

acyl-CoA (Fig. 2), yet free CoA levels can decrease ~30–45% after exercise at 75–90% VO_2max.[65] Unfortunately, it is difficult to discern whether the cytosolic pool of free CoA that would be required for ACSL activity is altered because the majority of cellular CoA is located within the mitochondrial matrix.[65] Additionally, ACSL enzymes consume two ATP equivalents to produce acyl-CoA (Fig. 2); thus, it is also possible that ATP levels at the outer mitochondrial membrane that would be available in close proximity to ACSL may be limited during high-intensity exercise. Finally, Acsl1 is regulated by phosphorylation and acetylation[66]; however, whether these sites are altered in response to exercise is currently unknown.

2.5.2 Regulation of Carnitine Palmitoyltransferase-1—Malonyl-CoA

As shown in Fig. 2, mitochondrial FA entry is a highly regulated process. First, LCFA must be converted to acyl-CoA by ACSL. As neither LCFAs nor acyl-CoAs can permeate the inner mitochondrial membrane, acyl-CoA must be transferred to carnitine palmitoyltransferase-1 (CPT-1). CPT-1 is associated with the outer mitochondrial membrane and is the rate-limiting enzyme in fat oxidation. This enzyme catalyzes the exchange of CoA bound to LCFAs for carnitine within the mitochondrial inter-membrane space, thus forming an acylcarnitine. Acylcarnitine derivatives can enter the mitochondria via carnitine/acylcarnitine translocase (CACT). Once within the mitochondrial matrix, carnitine is removed from the acyl moiety and exchanged for CoA by a reversible reaction catalyzed by CPT-2, thus providing an acyl-CoA that can enter β-oxidation.

When mitochondrial substrate supply and cellular energy status are high, malonyl-CoA is produced. Malonyl-CoA is a potent CPT-1 inhibitor (Fig. 2); thus, several studies have aimed to determine the relevance of this regulatory pathway during exercise. Malonyl-CoA levels decrease during exercise in rodents, which was postulated to contribute to enhanced fat oxidation during exercise.[67] Roepstorff *et al.*[68] also found reduced malonyl-CoA levels in humans at the onset of moderate-intensity exercise (60% VO_2peak); however, the authors concluded this effect was too subtle to be the primary factor that fine-tunes fat oxidation during exercise. Further doubt regarding the importance of malonyl-CoA to regulate fat oxidation during exercise in humans was found when (1) malonyl-CoA content did not change in muscle in response to moderate-intensity exercise, despite increased whole-body fat oxidation, and (2) decreases in whole-body fat oxidation during high-intensity exercise could not be explained by alterations in malonyl-CoA levels.[69,70] Globally, these findings suggest that while alterations in malonyl-CoA content have the potential to regulate CPT-1, the changes during exercise are likely to be too subtle to have a profound effect on overall mitochondrial fat oxidation.

2.5.3 Regulation of CPT-1—Carnitine Provision

Since CPT-1 activity requires carnitine (Fig. 2), it is equally possible that free carnitine availability could impact mitochondrial FA uptake during exercise. Carnitine content reflects the FA oxidative capacity of the tissue (i.e., oxidative tissues have greater carnitine reserves). Carnitine uptake into skeletal muscle occurs against a large concentration gradient which results in a 50–100-fold greater concentration within muscle than in the circulation.

In combination with its large mass, estimates indicate that nearly 95% of the total carnitine pool within the body lies within skeletal muscle.[71] Free carnitine does not change during low-intensity exercise (<40% VO_2max); however, beyond this point levels progressively diminish with increasing intensity and the decrease can be approximately two- to fivefold after heavy exercise.[6,72] Since carnitine is necessary to facilitate mitochondrial FA uptake, these robust decreases in free carnitine levels led researchers to predict that free carnitine availability may become limiting as workload increases, thereby providing a potential explanation for the differences in fat oxidation observed at varying exercise intensity.

At rest, free carnitine is nearly four- to sixfold more abundant than acylcarnitines in skeletal muscle.[73] During moderate-intensity exercise (50–65% VO_2max), free carnitine levels decline, but still account for greater than half of the total carnitine pool (free carnitine + acylcarnitines).[73] Evidence that carnitine may be limiting was found as provision of supplemental carnitine for 28 days enhanced fat oxidation during moderate-intensity exercise (66% VO_2max) in humans.[74] Furthermore, 24 weeks of carnitine supplementation significantly increased free carnitine content in skeletal muscle which was associated with greater preservation of glycogen stores following 50% VO_2max exercise, suggesting an increased capacity to facilitate fat oxidation.[73] While these studies lend credence to the notion that supplemental carnitine can enhance fat oxidation during moderate-intensity exercise, these findings are not universally supported as separate studies providing supplemental carnitine either acutely or for up to 4 weeks did not enhance fat oxidation.[75–78] Currently, it is speculated that enhancing the duration of carnitine supplementation may facilitate fat oxidation during moderate-intensity exercise. However, a requirement for extended carnitine supplementation would suggest that the effects would likely be due to chronic remodeling of metabolic pathways, rather than acute changes in carnitine status that would modulate substrate selection during submaximal exercise. Further studies in this area should shed light on this controversy.

While significant controversy remains regarding the importance of carnitine during submaximal exercise, the role it plays during high-intensity exercise is somewhat clearer. The decrease in skeletal muscle free carnitine during high-intensity exercise is substantial. Indeed, at exercise intensities exceeding 75% VO_2max, free carnitine represents <25% of the total carnitine pool (free carnitine + acylcarnitines).[72,73] Clearly, a substantial portion of carnitine is sequestered into the acylcarnitine pool during high-intensity exercise, which may represent enough of a drop in free carnitine to impact substrate metabolism. Acetylcarnitine represents ~90% of the total acylcarnitine pool in skeletal muscle and increases in this metabolite far outweigh changes in any other acylcarnitine species.[41,70,72,73] Moreover, decline in free carnitine due to excess sequestration in the form of acetylcarnitine was associated with decreasing fat oxidation in humans when exercising at 75% VO_2max.[41] An intramitochondrial enzyme, carnitine acetyltransferase (CRAT), utilizes free carnitine to convert acetyl-CoA to acetylcarnitine (Fig. 2). Using genetic manipulation strategies, it was determined that Crat activity favors glucose oxidation, while limiting fat utilization.[79,80] Furthermore, through the action of Crat, carnitine is known to stimulate pyruvate dehydrogenase (PDH) activity in isolated mitochondria and tissue homogenates.[79,80] This concept is supported *in vivo* as provision of supplemental carnitine generally increases PDH activity and diminishes lactate accumulation, thus augmenting carbohydrate utilization during high-intensity exercise.[41,73] Collectively, these findings indicate that one of the primary mechanisms through which carnitine facilitates glucose utilization during high-intensity exercise is via activation of CRAT, which facilitates removal of excess acetyl-CoA from the mitochondrial matrix, thus relieving PDH inhibition.

Through the action of CRAT, intramitochondrial carnitine can modulate the acetyl-CoA pool. With this in mind, acetyl-CoA also serves as a substrate for acetylation reactions and a recent study found 388 lysine acetylation sites on 195 mitochondrial proteins, many of which are enzymes in the TCA cycle and ETC.[81] Alterations in acetylation status can alter substrate selection and hyperacetylation of mitochondrial proteins appears to limit exercise capacity.[82,83] While it is currently unknown as to whether acetylation of mitochondrial proteins is altered during high-intensity exercise, this is certainly an area worthy of pursuit.

Regardless of the mechanism(s) through which carnitine modulates substrate metabolism during exercise, the substantial decrease in free carnitine during high-intensity exercise appears to limit fat oxidation. Moreover, this seems to be a preferential response in order to enhance glucose metabolism, which offers an interesting paradox regarding the role of carnitine during exercise at varying intensities. Specifically, this cofactor is essential for CPT-1 activity and seems to facilitate fat oxidation at low- to moderate-intensity exercise; however, during high-intensity exercise, carnitine is preferentially utilized by CRAT to enhance glucose utilization via removing excess acetyl-CoA from the mitochondrial matrix. Mechanisms explaining how varying exercise intensities can induce a shift in either the enzymatic preference of carnitine (CRAT vs. CPT-1) or the subcellular distribution of the carnitine pool (i.e., intramitochondrial to facilitate CRAT vs. extramitochondrial to promote CPT-1) that promote this switch in enzymatic preference are currently unknown.

3. POSTEXERCISE LIPID METABOLISM

3.1 Lipid Dynamics During Exercise

Adipose tissue lipolysis and FA release into circulation can increase approximately fivefold during moderate-intensity exercise.[10,11] This exercise-induced lipolytic drive can persist 3–6 h into recovery[13,84] and the resultant increase in plasma FA concentrations can last 12–24 h after exercise.[85] Additionally, blood flow remains elevated during the postexercise recovery phase.[12,13] These changes are positively associated with energy expenditure during exercise, suggesting the postexercise increase in circulating FA is greatest in response to exercise where high energy demand is incurred.[85]

IMTG levels continue to decrease after exercise cessation if a low-fat diet is consumed.[86,87] This effect can be quite prolonged as IMTG content can be significantly lower than resting values 6–30 h after exercise.[86] In combination with increased adipose tissue lipolysis, the heightened IMTG breakdown during recovery likely provides excess lipids to the mitochondria. Indeed, during the recovery phase heightened oxidation of both circulating FA and IMTG pools is observed, which facilitates replenishment of muscle glycogen.[3] Muscle glycogen begins to recover immediately, whereas IMTG restoration begins 2–4 h after exercise cessation.[88,89] Dietary factors certainly impact restoration of these intramuscular energy reserves. For instance, if a high-carbohydrate, low-fat diet (<20–25% by energy) is consumed after exercise, muscle glycogen is restored quickly, but it can take 3–7 days before IMTGs return to basal levels.[86,87] Alternatively, when moderate- (35–40%) to high- (55–70%) fat diets are consumed after exercise, IMTG levels return to baseline within 12–24 h.[90]

3.2 Excess Postexercise Oxygen Consumption

Following acute exercise, oxygen consumption (VO_2) can remain elevated above basal levels for several hours. Hill and colleagues developed the oxygen debt hypothesis to describe this phenomenon.[91–93] This hypothesis postulated that an individual incurs an oxygen deficit during exercise that is associated with lactate formation. To extend this notion, Hill *et al.* predicted that heightened postexercise VO_2 represented an oxygen debt that was incurred to facilitate lactate removal. This concept was modified by Margaria *et al.*[94] who found a curvilinear decline in VO_2 postexercise and proposed that oxygen debt should be described in terms of an initial, fast (alactacid) and secondary, slow (lactacid) phase. Unfortunately, this "lactate-centric" model has proven to be too simplistic as it is now known that a variety of additional components contribute to elevated VO_2 following exercise. As such, Gaesser and Brooks[95] suggested using the more generic phrase "excess post-exercise oxygen consumption" (EPOC) to describe this phenomenon as it does not ascribe causality. Within the scope of this umbrella term, it is now known that mechanisms involved in the early phase of EPOC can be attributed not only to lactate removal but also include the support of heightened respiration to replenish oxygen reserves, restore ATP/creatine phosphate levels, normalize body temperature, and restore fluid balance.[96] Alternatively, mechanisms contributing to the secondary, prolonged phase of EPOC are less well understood. While there are many potential mechanisms, one trait that is commonly found during EPOC is a heavy reliance on FA as metabolic fuel.

Substrate preference shifts from carbohydrate to lipid in the recovery phase following exhaustive aerobic exercise. During EPOC, fat oxidation can remain at ~25% of levels reported during exercise and represents greater than 60% of oxidative metabolism during recovery.[97] Logically, this substrate shift facilitates replenishment of glycogen stores; however, the energetic demand needed for glycogen synthesis likely contributes little to EPOC. A portion of EPOC may be explained by the fact that the energetic efficiency of oxygen-linked ATP production is lower with fat (~4.1 mol ATP/mol oxygen) than glucose (~6.3 mol ATP/mol oxygen); therefore, relatively more lipid is required to replenish energy stores. This substrate shift is estimated to account for 10–15% of the EPOC observed after exhaustive submaximal exercise.[98]

The evidence above suggests postexercise lipid catabolism occurs in excess of that required to simply restore energy and glycogen reserves. Several studies now indicate that a large portion of EPOC may be driven by increased triglyceride/fatty acid (TG/FA) cycling.[96] As mentioned earlier, lipolysis is heightened during postexercise recovery and circulating FA levels continually rise during this period, suggesting that the liberated FA is in excess of what is required to meet energy demand. To avoid potentially lipotoxic FA levels, excess lipids must be re-esterified into the TG pool: a process known as TG/FA cycling. TG/FA cycling is considered a futile cycle since TG stores undergo lipolysis which liberates FA, but the fate of the FA is simply to be re-esterified back into the TG pool (Fig. 3). The re-esterification process is initiated by ACSL enzymes, which convert FA to acyl-CoA at the expense of two ATP equivalents. As such, TG/FA cycling can be an energy-expensive process and it has been predicted that the energy cost of this futile cycle can account for a significant portion of EPOC.[96]

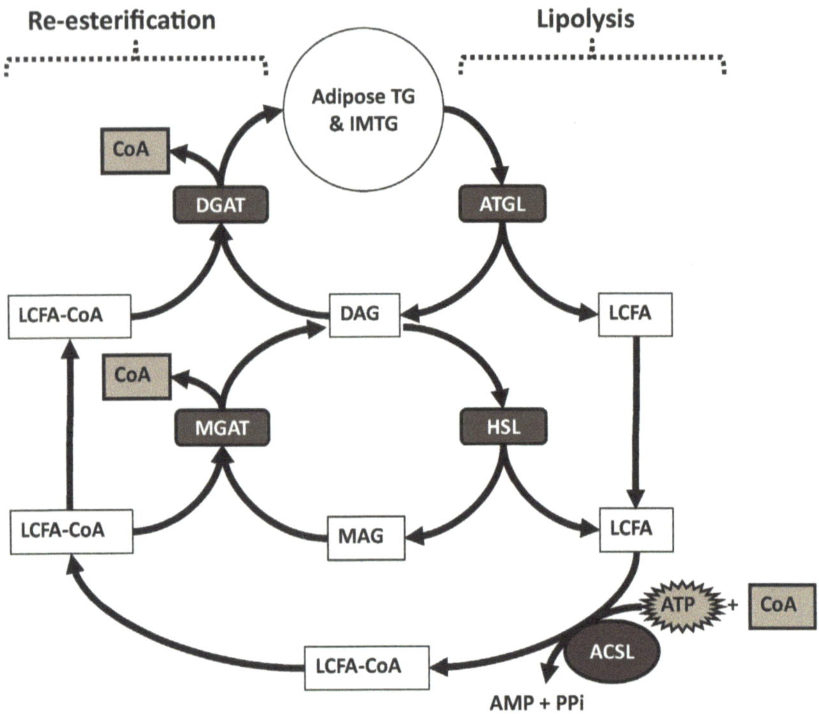

Figure 3 Energy cost of triglyceride/fatty acid (TG/FA) cycling. During exercise, adipose triglyceride lipase (ATGL) and hormone-sensitive lipase (HSL) activity are heightened, which increases mobilization of lipids via lipolysis of triglyceride stores (adipose tissue and IMTG). Heightened lipolysis can be maintained for prolonged periods in the post-exercise recovery phase when energy demand drops substantially. Since the mobilized long-chain fatty acids (LCFA) are not utilized as rapidly during the recovery period, they undergo re-esterification. Together, this phenomenon is referred to as triglyceride/fatty acid (TG/FA) cycling. Re-esterification of lipids into the TG pool requires conversion of LCFA to LCFA-CoA via the action of long-chain acyl-CoA synthetase (ACSL). This reaction consumes two ATP equivalents; therefore, TG/FA cycling can be considered an energy-dependent futile cycle. *Abbreviations*: IMTG, intramuscular triglyceride; DAG, diacylglycerol; MAG, monoacylglycerol; MGAT, monoacylglycerol acyltransferase; DGAT, diacylglycerol acyltransferase. (See the color plate.)

When an EPOC response is elicited, energy expenditure can be elevated for 12–24 h after exercise cessation. However, the duration and magnitude of EPOC are highly variable. Some of the primary factors that contribute to this variability include the duration, intensity, and modality of exercise. These matters will be discussed below.

3.2.1 EPOC and Exercise Duration

Most studies examining the relationship between EPOC and exercise duration suggest that postexercise energy expenditure increases in response to extended exercise duration.[96] Bahr[98] described a positive linear relationship between exercise duration and EPOC. Moreover, despite an overall increase in postexercise energy expenditure, the relative substrate selection during this period remains similar to shorter duration exercise and still favors lipids. Interestingly, however, there appears to be a threshold exercise intensity to elicit this response as studies using exercise protocols <50% VO_2max do not consistently evoke EPOC, regardless of duration.[99,100]

3.2.2 EPOC and Exercise Intensity

The most consistent and dynamic modulator of EPOC is exercise intensity as the vast majority of literature indicates that heightened exercise intensity not only increases the magnitude of EPOC but can also extend the duration.[96] The strength of this relationship is perhaps best depicted by Bahr[98] who showed that as exercise intensity increases, a positive curvilinear increase in EPOC is detected. It appears that an exercise intensity >50% VO_2max is where EPOC begins to increase exponentially. This exercise intensity threshold (~50–60% VO_2max) coincides with the point wherein the crossover principle predicts that the balance of whole-body fuel metabolism begins to shift from lipid toward carbohydrate. As such, it appears that a portion of the lipid metabolism that is sacrificed during high-intensity exercise is recovered during the exaggerated EPOC phase.

3.2.3 EPOC and Exercise Modality

The aforementioned studies show a significant EPOC response following moderate- and high-intensity aerobic exercise. Interval training generally induces greater EPOC magnitude and duration than continuous aerobic exercise.[96] Likewise, resistance exercise also induces a greater EPOC response than aerobic exercise, with the greatest response coinciding with the highest intensity of resistance training.[96] However, it must be mentioned that within the interval and resistance training modalities, the increase in lipid metabolism that occurs during EPOC would not be sufficient to compensate for the decrease in lipid oxidation incurred during the high-intensity phases of the exercise bout itself. Therefore, while these modalities do seem to incur greater benefit in caloric expenditure during the EPOC phase, total fat oxidation may actually be lower within these exercise modalities.

While enhanced fat oxidation and energy expenditure during EPOC offer a potentially intriguing model to maximize fat utilization, it is important to note that the amount of lipid catabolism that occurs during EPOC is likely minimal compared to that which would be spent during exercise. Estimates indicate the energy expenditure during EPOC may be negligible compared to that which occurs during exercise; thus, EPOC is unlikely to contribute substantially to overall energy balance and weight loss.[96] As such, if the goal of an exercise program is to maximize lipid catabolism (i.e., fat loss), it is likely best to partake in moderate-intensity exercise (45–60% VO_2max) that focuses on burning fat during the exercise bout, rather than trying to maximize lipid utilization during the EPOC phase.

4. DIETARY FACTORS INFLUENCING EXERCISE FAT METABOLISM

There is a vast market supported by individuals seeking nutritional supplements that aid sport performance, and to provide a thorough review of this field is beyond the scope of this chapter. However, it is pertinent to provide a brief overview of how common dietary practices influence lipid metabolism during exercise. Generally speaking, these dietary practices take advantage of the fact that biological systems adapt to carbohydrate-enriched diets by enhancing pathways relevant to glucose metabolism, whereas lipid-enriched diets upregulate fat metabolism pathways. Due to the ability of these dietary practices to alter the pathways that dictate utilization of the primary two substrates that provide metabolic fuel during exercise, there is significant interest in how each impacts sport performance.

4.1 Carbohydrate Loading

In 1939, Christensen and Hansen reported athletes placed on a high-carbohydrate diet for 7 days performed better during submaximal exercise than those that were placed on a high-fat diet for a similar time frame.[101] In the 1960s, Bergstrom and colleagues not only showed that a positive association existed between muscle glycogen and submaximal exercise capacity but also described that a brief period of carbohydrate restriction followed by a few days of high carbohydrate loading maximized preexercise muscle glycogen and endurance capacity.[102,103] These findings led to a practice that is commonly referred to as carbohydrate loading. A traditional carbohydrate loading regimen includes an intense training period for ~2 days that depletes muscle glycogen, followed by a 3–4-day period immediately prior to competition where the athlete consumes a carbohydrate-enriched diet (~60–70% total energy from carbohydrate). This practice "supercompensates" the skeletal muscle with glycogen stores in order to delay the fatigue commonly associated with glycogen depletion. While carbohydrate loading has been frequently shown to improve performance of endurance exercise bouts lasting >90 min, this dietary practice is minimally effective in altering performance during either moderate-intensity exercise lasting <60–90 min or high-intensity exercise.[104] Interestingly, the type of carbohydrate consumed (glucose vs. glucose polymer vs. fructose) does not appear to alter glycogen storage or utilization during exercise[105]; however, there is evidence indicating that fructose inhibits fat oxidation to a greater extent than glucose during and after exercise.[106]

4.2 Ketogenic Diet (Low-Carbohydrate, High-Fat)

Adipose tissue and intracellular lipid stores overwhelmingly represent the most abundant fuel reserves in the body that are available to support the energy demands incurred during exercise. In fact, for all practical purposes, these endogenous lipids represent a nearly inexhaustible source of metabolic energy. As such, there is substantial interest in developing strategies to maximize lipid utilization during exercise as a strategy to improve performance. One such strategy that is receiving interest is consumption of a low-carbohydrate, high-fat diet: commonly referred to as a ketogenic diet. Due to the high lipid content, coupled with a limited supply of carbohydrate, ketogenic diets naturally drive enhanced FA utilization and ketogenesis. According to Volek *et al.*,[107] prolonged ketogenic diet consumption results in a coordinated reprogramming of metabolic systems throughout the body that ensures optimal supply and utilization of lipid when faced with a low-carbohydrate environment: a phenomenon known as ketoadaptation. Moreover, the ketoadaptation principle suggests adopting this dietary principle allows humans to adapt to utilize lipid as the primary fuel during submaximal exercise, while simultaneously reducing reliance on glycogen stores.[107]

In agreement with the principles above, it is perhaps predictable that humans consuming lipid-enriched diets for as little as 3–5 days or up to 4–7 weeks exhibited greater whole-body fat oxidation during exercise than individuals consuming a carbohydrate-rich diet.[6] Indeed, several studies indicate that acute provision of excess lipid prior to or during exercise results in use of FAs as the primary metabolic fuel during moderate-intensity exhaustive exercise, whereas carbohydrate use is slowed.[106–113] However, it is important to note that exercise performance in response to short-term exposure to a ketogenic diet (<6 days) often declines, despite substantially enhancing fat oxidation.[108–111] Alternatively, mounting evidence indicates that prolonged (>4 weeks) consumption of a low-carbohydrate, ketogenic diet not only enhances fat utilization but can either maintain or improve exercise performance.[113–116] One such study performed by Zajac *et al.*[113] compared the effects of a 4-week ketogenic diet (70% fat, 15% protein, and 15% carbohydrate) versus a mixed meal diet (30% fat, 20% protein, and 50% carbohydrate) on endurance exercise cycling performance. The authors found that the ketogenic diet increased energy expenditure (higher VO_2) and fat utilization (reduced RER and greater clearance of plasma lipid) during moderate-intensity exercise. Interestingly, as exercise intensity increased, the ketogenic diet actually reduced the lactate threshold, thus strengthening the contention that ketoadaptation limits reliance on glucose during exercise.

Work from Dr. Stephen Phinney's laboratory has extended our understanding of the impact of the ketogenic diet on exercise outcomes by performing studies using a diet that has minimal carbohydrate.[114–116] Subjects in these studies received a 4-week diet modeled after that of the Inuit population who consume roughly 85% fat and 15% protein, with minimal carbohydrate (participants in Dr. Phinney's studies received <10 g carbohydrate per day). At the end of the 4-week dietary intervention, RQ values indicated that FAs were nearly the sole source of metabolic fuel during exercise. Interestingly, despite the fact that keto-adapted individuals had preexercise muscle glycogen levels that were ~50% lower than baseline, their time to exhaustion during a submaximal (65% VO_2max) endurance exercise session was identical to baseline values. These data suggest that the metabolic adaptations in response to a ketogenic diet are sufficient to at least maintain endurance capacity during exhaustive submaximal exercise, while simultaneously maximizing lipid utilization. Equally interesting are findings indicating keto-adapted individuals exhibit reduced inflammatory markers and muscle damage after an exercise bout,[113,114,117,118] indicating ketogenic diets may significantly facilitate postexercise recovery.

Based upon the above information, it seems clear that prolonged consumption of ketogenic diets maximizes fat oxidation, while at least maintaining exercise performance during submaximal exercise. Indeed, current evidence supports the notion that a ketogenic diet may be most beneficial in ultra-endurance athletic endeavors performed at moderate intensity as it would maximize fat utilization, while limiting the reliance on frequent carbohydrate ingestion.[107] Additionally, ketogenic diets are highly effective at improving body composition, especially when combined with resistance training,[107,119] while preserving strength and power performance. Alternatively, it is important to note that low-glycemic, ketogenic diets are not optimal for high-intensity exercise.[106,113] This is largely thought to stem from the fact that the rate of ATP breakdown during high-intensity (>85% VO_2max) endurance exercise is too fast to be adequately matched by ATP production from FAs; therefore, glucose is required to efficiently maintain this balance. Also, lipolysis is reduced during high-intensity exercise, thus limiting the availability of the primary fuel reserve available to those consuming a ketogenic diet.

4.3 Dietary Considerations in the Postexercise Period

Exercise training in the fasted state induces metabolic remodeling to increase FA oxidative capacity; however, these adaptations are abolished when a carbohydrate-rich diet is consumed prior to or during exercise.[106] Similarly, increases in whole-body fat oxidation that are typically observed 2 h after completion of a moderate-intensity exercise bout are abolished in individuals that consume carbohydrates immediately after exercise.[6] These findings indicate that dietary choices not only modulate substrate utilization during exercise but are also likely to impact postexercise metabolism and the molecular adaptations incurred in response to exercise training. In this light, while the EPOC phase may not be sufficient to burn vast amounts of calories, it may be an essential period during which metabolic remodeling occurs to facilitate more efficient lipid utilization in future exercise bouts. As such, careful consideration regarding nutrition must be taken in order to maximize the desired training effects because the aforementioned findings suggest that nutritional choices made pre-, during, and postexercise can override some of the training-induced adaptations in gene transcription related to substrate metabolism. The practical relevance of these findings can be readily described using the dietary models mentioned above. For instance, carbohydrate loading maximizes tissue glycogen stores and has proven to be successful in terms of improving performance during prolonged exhaustive exercise. However, if the primary goal of an individual is to maximize fat utilization (i.e., lose weight), rather than improve performance, then it may be prudent to limit carbohydrate consumption prior to, during, and perhaps even immediately postexercise in order to allow for optimal adaptations of the lipid metabolism pathways in the body to occur.

5. MOLECULAR PROGRAMMING OF LIPID METABOLISM

As mentioned earlier, FA exposure drives reprogramming of metabolic pathways that favor lipid metabolism. Peroxisome proliferator-activated receptors (PPARs) are key transcription factors that are activated by endogenous lipid ligands to induce these responses. There are three commonly described PPAR isoforms: PPAR-alpha (PPARA), PPAR-delta (PPARD), and PPAR-gamma (PPARG). Mechanistically, each PPAR heterodimerizes with the retinoid X receptor (RXR) and the resultant receptor complex then binds to a PPAR response element (*PPRE*) within the DNA in order to modulate transcription (Fig. 4).[120] Discussion of the role of each of these transcription factors in relation to exercise will be discussed below.

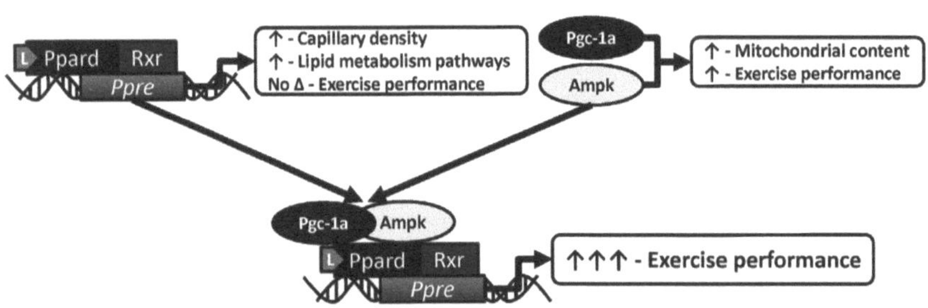

Figure 4 Synergistic coactivation of Ppard by Pgc-1a and Ampk to enhance lipid utilization and exercise performance. Ligand (L)-activated Ppard heterodimerizes with the retinoid X receptor (Rxr) and the resultant complex translocates to the nucleus where it binds to a Ppar response element (*Ppre*) within the DNA. This increases capillary density and upregulates lipid metabolism pathways, but has no effect on exercise performance. Pgc-1a and Ampk activation induce mitochondrial biogenesis, which modestly enhances exercise performance. During exercise, Pgc-1a and Ampk bind to the Ppard:Rxr heterodimer complex resulting in a synergistic improvement in exercise performance. *Abbreviations*: L, ligand—synthetic agonist or endogenous fatty acid; Ppard, peroxisome proliferator-activated receptor delta; Pgc-1a, Ppar-gamma coactivator-1alpha; Ampk, AMP-activated protein kinase alpha; △, change. (See the color plate.)

5.1 PPAR-Alpha

PPARA is most abundant in highly oxidative tissues and regulates pathways in a manner that enhances FA oxidation.[121,122] Studies in liver and heart indicate that the absence of Ppara decreases mitochondrial fat oxidation rates ~70–80%.[123–125] Exercise training has been shown to increase PPARA protein expression in skeletal muscle,[126,127] which led researchers to predict that this transcriptional regulator would induce metabolic remodeling that improves exercise performance. In support, Muoio *et al.*[128] found that *Ppara*-deficient mice exhausted earlier during an acute endurance exercise bout than control mice. Interestingly, however, traditional PPARA target genes (*Pdk4* and *Ucp3*) and fat oxidation rates in skeletal muscle were largely unaltered in *Ppara*-deficient mice. Also, following an acute exhaustive endurance exercise bout, intramuscular glycogen stores were only reduced 50%. In contrast, hepatic glycogen stores were almost entirely depleted and traditional PPARA target genes were robustly reduced in both liver and heart. Data from this study also showed serum glucose, lactate, and ketone levels were reduced in *Ppara*-deficient mice after exercise, whereas circulating FA levels increased. Collectively, these data indicate that PPARA may impact exercise capacity through modulation of cardiac and hepatic lipid metabolism, but the effect on skeletal muscle is minimal.

5.2 PPAR-Delta

Ppard is ubiquitously expressed; however, it is the predominant isoform in skeletal muscle, where it is most abundant in oxidative, type I fibers.[122,128–130] In humans, single-nucleotide polymorphisms in *PPARD* are associated with limitations in endurance exercise training adaptations.[131] Conversely, both short-term exercise and endurance training increase *PPARD* gene expression,[120] and mice overexpressing this transcription factor have been nicknamed "marathon mice" due to robust increases in endurance capacity.[130,132] In a seminal study performed in the laboratory of Dr. Ronald Evans,[133] mice receiving Ppard agonist (GW501516) treatment for 4 weeks exhibited increased oxidative muscle fibers and had a more robust lipid oxidative capacity; however, this alone did not impact exercise endurance capacity (results summarized in Fig. 4). Alternatively, when Ppard agonist treatment was combined with 4 weeks of endurance exercise training, a synergistic increase in oxidative muscle fibers and running capacity was observed over vehicle-treated, exercise-trained mice. Ehrenborg and Krook elegantly summarized these findings by stating that "although pharmacological activation of PPARD may indeed potentiate the daily training benefits induced by exercise, both for the amateur and elite athlete, it will not supersede the need to actually perform the exercise." While this is partially true, it also must be noted that the AMP-activated protein kinase alpha (Ampk) and the PPAR-gamma coactivator 1-alpha (Pgc-1a) physically interact with Ppard, which dramatically increases basal and ligand-dependent transcription.[133] Ampk and Pgc-1a are independently capable of improving exercise performance[133–135]; however, the authors discovered that activation of these pathways in the presence of heightened Ppard activity induced a synergistic improvement in exercise capacity (Fig. 4).[133,136] Extrapolation of these results to the practical setting suggested AMPK agonists (AICAR) and PPARD agonists (GW501516) could be used as "exercise mimetics." As would be expected, the potential impact of these findings was of immediate interest to the field of sports performance because these compounds were deemed to be capable of improving endurance exercise capacity. As a result, within 1 year of publishing these findings, the World Anti-Doping Agency added both AICAR and GW501516 to the Prohibited List in 2009.[137]

5.3 PPAR-Gamma

There are three transcriptional splice variants of PPARG that produce two distinct proteins: PPARG-1 is ubiquitously expressed, while PPARG-2 is expressed primarily in white adipose tissue and brown adipose tissue. While tissue-specific *Pparg* deletion studies have uncovered a multitude of functions in various organ systems (reviewed in Ref. 138), studies using PPARG agonists (broadly referred to as thiazolidinediones (TZDs)) indicate that the primary role of this transcription factor is to induce adipogenesis. While numerous studies have established the value of PPARG activation to improve insulin sensitivity, few have examined the effects of these compounds on exercise performance. Administration of one of the early TZDs, rosiglitazone, for 4–6 months was shown to either have no effect on VO_2peak when given to patients with cardiovascular disease,[139] or slightly improve exercise capacity in type 2 diabetic humans.[140] Unfortunately, the impact of PPARG activation on exercise performance in healthy humans is unknown. Truthfully, it is unlikely that PPARG activation has potential to serve as an ergogenic target; however, studies in rodents suggest it could have an indirect impact relevant to exercise. Specifically, Pparg agonist (rosiglitazone) administration has been shown to provide an antidepressant effect in rodents during a forced swim test, which improved performance.[141,142] While a forced swim test is designed specifically to test parameters related to depression rather than exercise performance, the fact that rosiglitazone exerted antidepressant action during an exercise activity leads one to speculate that PPARG could play a role in exercise adherence. This contention, however, has not been tested.

6. CONCLUDING REMARKS

Lipids are the predominant fuel source used to support submaximal endurance exercise. Since lipid overwhelmingly represents the largest reserve of stored energy in the body, there is much interest in maximizing lipid utilization during exercise in order to either optimize sport performance or maximize weight loss. Skeletal muscle is the primary consumer of lipid during and immediately after exercise; however, this requires an integrative organ systems effort to deliver lipid to muscle for use as fuel. Evidence reviewed in this chapter emphasizes the importance of mobilization of lipid from adipose tissue (lipolysis-derived free FAs), liver (VLDL-TG), and intramuscular (IMTG) reserves to support the increased energy demand incurred by exercise. Of course, increased presentation of lipid to skeletal muscle does not ensure greater fat utilization. Indeed, simultaneous modulation of several regulatory mechanisms that dictate tissue FA uptake, intracellular lipid delivery, and mitochondrial β-oxidation is required to maximize fat use during exercise. In response to acute exercise, the coordinated activation of the sympathetic nervous system and induction of energy deficit-sensing pathways ensures mitochondrial lipid delivery is sufficient to at least match energy demand. In point of fact, even during exercise bouts where lipid utilization is maximal, FA delivery to contracting muscle occurs far in excess of that which is required to support the energy demands of exercise. Excess FA delivery continues into the postexercise period and contributes to increased energy expenditure (EPOC) via heightened TG/FA cycling. While the contribution of the EPOC phase to overall lipid catabolism may be relatively small in comparison to that which is utilized during an exercise bout, the increased TG/FA cycling during the postexercise period may play an important role in adaptations to exercise training. Specifically, energy-sensing pathways such as AMPK can remain active for at least 30–60 min postexercise,[143,144] which may act synergistically with elevated lipid ligands during this period to induce PPARD-mediated remodeling of lipid metabolism pathways in muscle. The importance of the postexercise period in mediating metabolic adaptation is emphasized by the fact that consumption of carbohydrates either during or immediately after exercise negates remodeling of lipid metabolism pathways. In addition to adaptations in response to exercise, evidence from studies testing the impact of a prolonged (>4 weeks) low-carbohydrate, ketogenic diet prior to submaximal endurance exercise indicates that ketoadaptation can further maximize the lipid-handling capacity of skeletal muscle, without impeding performance. Alternatively, many forms of exercise rely more heavily on carbohydrate than fat; therefore, adopting a ketogenic diet will likely impair performance during exercise involving high-intensity effort. Indeed, dietary choices impact substrate preference during exercise and each individual has different goals in terms of what they desire from engaging in exercise (i.e., maximize performance, lose weight, other). As such, one's dietary choices should be tailored to match their specific goals and any drastic changes in dietary or exercise practices should be carefully considered and implemented with caution.

REFERENCES

1. Brooks GA, Mercier J. Balance of carbohydrate and lipid utilization during exercise: the "crossover" concept. *J Appl Physiol (1985)*. 1994;76(6):2253–2261.
2. Dagenais GR, Tancredi RG, Zierler KL. Free fatty acid oxidation by forearm muscle at rest, and evidence for an intramuscular lipid pool in the human forearm. *J Clin Invest*. 1976;58(2):421–431.
3. Egan B, Zierath JR. Exercise metabolism and the molecular regulation of skeletal muscle adaptation. *Cell Metab*. 2013;17(2):162–184.
4. Gibala MJ, MacLean DA, Graham TE, Saltin B. Tricarboxylic acid cycle intermediate pool size and estimated cycle flux in human muscle during exercise. *Am J Physiol*. Aug 1998;275(2 pt 1):E235–E242.
5. Gaitanos GC, Williams C, Boobis LH, Brooks S. Human muscle metabolism during intermittent maximal exercise. *J Appl Physiol (1985)*. 1993;75(2):712–719.
6. Kiens B, Alsted TJ, Jeppesen J. Factors regulating fat oxidation in human skeletal muscle. *Obes Rev*. Oct 2011;12(10):852–858.
7. Helge JW, Watt PW, Richter EA, Rennie MJ, Kiens B. Fat utilization during exercise: adaptation to a fat-rich diet increases utilization of plasma fatty acids and very low density lipoprotein-triacylglycerol in humans. *J Physiol*. 2001;537(pt 3):1009–1020.
8. Roepstorff C, Steffensen CH, Madsen M, et al. Gender differences in substrate utilization during submaximal exercise in endurance-trained subjects. *Am J Physiol Endocrinol Metab*. 2002;282(2):E435–E447.
9. van Loon LJ, Koopman R, Stegen JH, Wagenmakers AJ, Keizer HA, Saris WH. Intramyocellular lipids form an important substrate source during moderate intensity exercise in endurance-trained males in a fasted state. *J Physiol*. 2003;553(pt 2):611–625.
10. Wolfe RR, Klein S, Carraro F, Weber JM. Role of triglyceride-fatty acid cycle in controlling fat metabolism in humans during and after exercise. *Am J Physiol*. 1990;258(2 pt 1):E382–E389.

11. Romijn JA, Coyle EF, Sidossis LS, et al. Regulation of endogenous fat and carbohydrate metabolism in relation to exercise intensity and duration. *Am J Physiol.* 1993;265(3 pt 1):E380–E391.
12. Bulow J, Madsen J. Adipose tissue blood flow during prolonged, heavy exercise. *Pflugers Arch.* 1976;363(3):231–234.
13. Mulla NA, Simonsen L, Bulow J. Post-exercise adipose tissue and skeletal muscle lipid metabolism in humans: the effects of exercise intensity. *J Physiol.* 2000; 524(pt 3):919–928.
14. Turcotte LP, Richter EA, Kiens B. Increased plasma FFA uptake and oxidation during prolonged exercise in trained vs. untrained humans. *Am J Physiol.* 1992;262(6 pt 1): E791–E799.
15. Kiens B, Essen-Gustavsson B, Christensen NJ, Saltin B. Skeletal muscle substrate utilization during submaximal exercise in man: effect of endurance training. *J Physiol.* 1993;469:459–478.
16. Romijn JA, Coyle EF, Sidossis LS, Zhang XJ, Wolfe RR. Relationship between fatty acid delivery and fatty acid oxidation during strenuous exercise. *J Appl Physiol (1985).* 1995;79(6):1939–1945.
17. Kiens B. Skeletal muscle lipid metabolism in exercise and insulin resistance. *Physiol Rev.* 2006;86(1):205–243.
18. Luiken JJ, Turcotte LP, Bonen A. Protein-mediated palmitate uptake and expression of fatty acid transport proteins in heart giant vesicles. *J Lipid Res.* 1999;40(6): 1007–1016.
19. Luiken JJ, Glatz JF, Bonen A. Fatty acid transport proteins facilitate fatty acid uptake in skeletal muscle. *Can J Appl Physiol.* 2000;25(5):333–352.
20. Bonen A, Dyck DJ, Luiken JJ. Skeletal muscle fatty acid transport and transporters. *Adv Exp Med Biol.* 1998;441:193–205.
21. Bradley NS, Snook LA, Jain SS, Heigenhauser GJ, Bonen A, Spriet LL. Acute endurance exercise increases plasma membrane fatty acid transport proteins in rat and human skeletal muscle. *Am J Physiol Endocrinol Metab.* 2012;302(2):E183–E189.
22. Tunstall RJ, Mehan KA, Wadley GD, et al. Exercise training increases lipid metabolism gene expression in human skeletal muscle. *Am J Physiol Endocrinol Metab.* 2002;283(1): E66–E72.
23. Talanian JL, Galloway SD, Heigenhauser GJ, Bonen A, Spriet LL. Two weeks of high-intensity aerobic interval training increases the capacity for fat oxidation during exercise in women. *J Appl Physiol (1985).* 2007;102(4):1439–1447.
24. Kiens B, Kristiansen S, Jensen P, Richter EA, Turcotte LP. Membrane associated fatty acid binding protein (FABPpm) in human skeletal muscle is increased by endurance training. *Biochem Biophys Res Commun.* 1997;231(2):463–465.
25. Bonen A, Luiken JJ, Arumugam Y, Glatz JF, Tandon NN. Acute regulation of fatty acid uptake involves the cellular redistribution of fatty acid translocase. *J Biol Chem.* 2000;275(19):14501–14508.
26. Nickerson JG, Alkhateeb H, Benton CR, et al. Greater transport efficiencies of the membrane fatty acid transporters FAT/CD36 and FATP4 compared with FABPpm and FATP1 and differential effects on fatty acid esterification and oxidation in rat skeletal muscle. *J Biol Chem.* 2009;284(24):16522–16530.
27. DiRusso CC, Li H, Darwis D, Watkins PA, Berger J, Black PN. Comparative biochemical studies of the murine fatty acid transport proteins (FATP) expressed in yeast. *J Biol Chem.* 2005;280(17):16829–16837.
28. Jeppesen J, Jordy AB, Sjoberg KA, et al. Enhanced fatty acid oxidation and FATP4 protein expression after endurance exercise training in human skeletal muscle. *PLoS One.* 2012;7(1). e29391.

29. Stahl A, Evans JG, Pattel S, Hirsch D, Lodish HF. Insulin causes fatty acid transport protein translocation and enhanced fatty acid uptake in adipocytes. *Dev Cell.* 2002;2(4):477–488.
30. Hall AM, Smith AJ, Bernlohr DA. Characterization of the acyl-CoA synthetase activity of purified murine fatty acid transport protein 1. *J Biol Chem.* 2003;278(44): 43008–43013.
31. Hall AM, Wiczer BM, Herrmann T, Stremmel W, Bernlohr DA. Enzymatic properties of purified murine fatty acid transport protein 4 and analysis of acyl-CoA synthetase activities in tissues from FATP4 null mice. *J Biol Chem.* 2005;280(12):11948–11954.
32. Stremmel W, Pohl L, Ring A, Herrmann T. A new concept of cellular uptake and intracellular trafficking of long-chain fatty acids. *Lipids.* 2001;36(9):981–989.
33. Luiken JJ, Koonen DP, Coumans WA, et al. Long-chain fatty acid uptake by skeletal muscle is impaired in homozygous, but not heterozygous, heart-type-FABP null mice. *Lipids.* 2003;38(4):491–496.
34. Ladu MJ, Kapsas H, Palmer WK. Regulation of lipoprotein lipase in muscle and adipose tissue during exercise. *J Appl Physiol (1985).* 1991;71(2):404–409.
35. Ladu MJ, Kapsas H, Palmer WK. Regulation of lipoprotein lipase in adipose and muscle tissues during fasting. *Am J Physiol.* 1991;260(5 pt 2):R953–R959.
36. Kiens B, Lithell H. Lipoprotein metabolism influenced by training-induced changes in human skeletal muscle. *J Clin Invest.* 1989;83(2):558–564.
37. Morio B, Holmback U, Gore D, Wolfe RR. Increased VLDL-TAG turnover during and after acute moderate-intensity exercise. *Med Sci Sports Exerc.* 2004;36(5):801–806.
38. Malkova D, Evans RD, Frayn KN, Humphreys SM, Jones PR, Hardman AE. Prior exercise and postprandial substrate extraction across the human leg. *Am J Physiol Endocrinol Metab.* 2000;279(5):E1020–E1028.
39. Goodpaster BH, He J, Watkins S, Kelley DE. Skeletal muscle lipid content and insulin resistance: evidence for a paradox in endurance-trained athletes. *J Clin Endocrinol Metab.* 2001;86(12):5755–5761.
40. Watt MJ, Heigenhauser GJ, Spriet LL. Intramuscular triacylglycerol utilization in human skeletal muscle during exercise: is there a controversy? *J Appl Physiol (1985).* 2002;93(4):1185–1195.
41. van Loon LJ, Greenhaff PL, Constantin-Teodosiu D, Saris WH, Wagenmakers AJ. The effects of increasing exercise intensity on muscle fuel utilisation in humans. *J Physiol.* 2001;536(pt 1):295–304.
42. Guo Z, Burguera B, Jensen MD. Kinetics of intramuscular triglyceride fatty acids in exercising humans. *J Appl Physiol (1985).* 2000;89(5):2057–2064.
43. Jordy AB, Kiens B. Regulation of exercise-induced lipid metabolism in skeletal muscle. *Exp Physiol.* 2014;99(12):1586–1592.
44. Zimmermann R, Strauss JG, Haemmerle G, et al. Fat mobilization in adipose tissue is promoted by adipose triglyceride lipase. *Science.* 2004;306(5700):1383–1386.
45. Jocken JW, Smit E, Goossens GH, et al. Adipose triglyceride lipase (ATGL) expression in human skeletal muscle is type I (oxidative) fiber specific. *Histochem Cell Biol.* 2008;129(4):535–538.
46. Alsted TJ, Nybo L, Schweiger M, et al. Adipose triglyceride lipase in human skeletal muscle is upregulated by exercise training. *Am J Physiol Endocrinol Metab.* 2009;296(3): E445–E453.
47. Covington JD, Galgani JE, Moro C, et al. Skeletal muscle perilipin 3 and coatomer proteins are increased following exercise and are associated with fat oxidation. *PLoS One.* 2014;9(3). e91675.
48. Mason RR, Meex RC, Russell AP, Canny BJ, Watt MJ. Cellular localization and associations of the major lipolytic proteins in human skeletal muscle at rest and during exercise. *PLoS One.* 2014;9(7). e103062.

49. Ramos SV, MacPherson RE, Turnbull PC, et al. Higher PLIN5 but not PLIN3 content in isolated skeletal muscle mitochondria following acute in vivo contraction in rat hindlimb. *Physiol Rep*. 2014;2(10). pii: e12154.
50. Macpherson RE, Vandenboom R, Roy BD, Peters SJ. Skeletal muscle PLIN3 and PLIN5 are serine phosphorylated at rest and following lipolysis during adrenergic or contractile stimulation. *Physiol Rep*. 2013;1(4). e00084.
51. MacPherson RE, Ramos SV, Vandenboom R, Roy BD, Peters SJ. Skeletal muscle PLIN proteins, ATGL and CGI-58, interactions at rest and following stimulated contraction. *Am J Physiol Regul Integr Comp Physiol*. 2013;304(8):R644–R650.
52. Louche K, Badin PM, Montastier E, et al. Endurance exercise training up-regulates lipolytic proteins and reduces triglyceride content in skeletal muscle of obese subjects. *J Clin Endocrinol Metab*. 2013;98(12):4863–4871.
53. Peters SJ, Samjoo IA, Devries MC, Stevic I, Robertshaw HA, Tarnopolsky MA. Perilipin family (PLIN) proteins in human skeletal muscle: the effect of sex, obesity, and endurance training. *Appl Physiol Nutr Metab*. 2012;37(4):724–735.
54. Haemmerle G, Zimmermann R, Hayn M, et al. Hormone-sensitive lipase deficiency in mice causes diglyceride accumulation in adipose tissue, muscle, and testis. *J Biol Chem*. 2002;277(7):4806–4815.
55. Langfort J, Ploug T, Ihlemann J, Saldo M, Holm C, Galbo H. Expression of hormone-sensitive lipase and its regulation by adrenaline in skeletal muscle. *Biochem J*. 1999;340(pt 2):459–465.
56. Enevoldsen LH, Stallknecht B, Langfort J, et al. The effect of exercise training on hormone-sensitive lipase in rat intra-abdominal adipose tissue and muscle. *J Physiol*. 2001;536(pt 3):871–877.
57. Helge JW, Biba TO, Galbo H, Gaster M, Donsmark M. Muscle triacylglycerol and hormone-sensitive lipase activity in untrained and trained human muscles. *Eur J Appl Physiol*. 2006;97(5):566–572.
58. Stralfors P, Bjorgell P, Belfrage P. Hormonal regulation of hormone-sensitive lipase in intact adipocytes: identification of phosphorylated sites and effects on the phosphorylation by lipolytic hormones and insulin. *Proc Natl Acad Sci USA*. 1984;81(11):3317–3321.
59. Watt MJ, Spriet LL. Regulation and role of hormone-sensitive lipase activity in human skeletal muscle. *Proc Nutr Soc*. 2004;63(2):315–322.
60. Watt MJ, Heigenhauser GJ, O'Neill M, Spriet LL. Hormone-sensitive lipase activity and fatty acyl-CoA content in human skeletal muscle during prolonged exercise. *J Appl Physiol (1985)*. 2003;95(1):314–321.
61. Kiens B, Roemen TH, van der Vusse GJ. Muscular long-chain fatty acid content during graded exercise in humans. *Am J Physiol*. 1999;276(2 pt 1):E352–E357.
62. Venables MC, Achten J, Jeukendrup AE. Determinants of fat oxidation during exercise in healthy men and women: a cross-sectional study. *J Appl Physiol (1985)*. 2005;98(1):160–167.
63. Li LO, Grevengoed TJ, Paul DS, et al. Compartmentalized acyl-CoA metabolism in skeletal muscle regulates systemic glucose homeostasis. *Diabetes*. 2015;64(1):23–35.
64. Cortright RN, Sandhoff KM, Basilio JL, et al. Skeletal muscle fat oxidation is increased in African-American and white women after 10 days of endurance exercise training. *Obesity (Silver Spring)*. 2006;14(7):1201–1210.
65. Sahlin K, Sallstedt EK, Bishop D, Tonkonogi M. Turning down lipid oxidation during heavy exercise—what is the mechanism? *J Physiol Pharmacol*. 2008;59(suppl 7):19–30.
66. Frahm JL, Li LO, Grevengoed TJ, Coleman RA. Phosphorylation and acetylation of acyl-CoA synthetase-1. *J Proteomics Bioinform*. 2011;4(7):129–137.
67. Winder WW. Malonyl-CoA—regulator of fatty acid oxidation in muscle during exercise. *Exerc Sport Sci Rev*. 1998;26:117–132.

68. Roepstorff C, Halberg N, Hillig T, et al. Malonyl-CoA and carnitine in regulation of fat oxidation in human skeletal muscle during exercise. *Am J Physiol Endocrinol Metab.* 2005;288(1):E133–E142.
69. Odland LM, Heigenhauser GJ, Lopaschuk GD, Spriet LL. Human skeletal muscle malonyl-CoA at rest and during prolonged submaximal exercise. *Am J Physiol.* 1996;270(3 pt 1):E541–E544.
70. Odland LM, Howlett RA, Heigenhauser GJ, Hultman E, Spriet LL. Skeletal muscle malonyl-CoA content at the onset of exercise at varying power outputs in humans. *Am J Physiol.* 1998;274(6 pt 1):E1080–E1085.
71. Brass EP. Pharmacokinetic considerations for the therapeutic use of carnitine in hemodialysis patients. *Clin Ther.* 1995;17(2):176–185. discussion 175.
72. Hiatt WR, Regensteiner JG, Wolfel EE, Ruff L, Brass EP. Carnitine and acylcarnitine metabolism during exercise in humans. Dependence on skeletal muscle metabolic state. *J Clin Invest.* 1989;84(4):1167–1173.
73. Wall BT, Stephens FB, Constantin-Teodosiu D, Marimuthu K, Macdonald IA, Greenhaff PL. Chronic oral ingestion of L-carnitine and carbohydrate increases muscle carnitine content and alters muscle fuel metabolism during exercise in humans. *J Physiol.* 2011;589(pt 4):963–973.
74. Gorostiaga EM, Maurer CA, Eclache JP. Decrease in respiratory quotient during exercise following L-carnitine supplementation. *Int J Sports Med.* 1989;10(3):169–174.
75. Abramowicz WN, Galloway SD. Effects of acute versus chronic L-carnitine L-tartrate supplementation on metabolic responses to steady state exercise in males and females. *Int J Sport Nutr Exerc Metab.* 2005;15(4):386–400.
76. Broad EM, Maughan RJ, Galloway SD. Effects of four weeks L-carnitine L-tartrate ingestion on substrate utilization during prolonged exercise. *Int J Sport Nutr Exerc Metab* 2005;15(6):665–679.
77. Broad EM, Maughan RJ, Galloway SD. Carbohydrate, protein, and fat metabolism during exercise after oral carnitine supplementation in humans. *Int J Sport Nutr Exerc Metab.* 2008;18(6):567–584.
78. Decombaz J, Deriaz O, Acheson K, Gmuender B, Jequier E. Effect of L-carnitine on submaximal exercise metabolism after depletion of muscle glycogen. *Med Sci Sports Exerc.* 1993;25(6):733–740.
79. Muoio DM, Noland RC, Kovalik JP, et al. Muscle-specific deletion of carnitine acetyltransferase compromises glucose tolerance and metabolic flexibility. *Cell Metab.* 2012;15(5):764–777.
80. Noland RC, Koves TR, Seiler SE, et al. Carnitine insufficiency caused by aging and overnutrition compromises mitochondrial performance and metabolic control. *J Biol Chem.* 2009;284(34):22840–22852.
81. Kim SC, Sprung R, Chen Y, et al. Substrate and functional diversity of lysine acetylation revealed by a proteomics survey. *Mol Cell.* 2006;23(4):607–618.
82. Palacios OM, Carmona JJ, Michan S, et al. Diet and exercise signals regulate SIRT3 and activate AMPK and PGC-1alpha in skeletal muscle. *Aging.* 2009;1(9):771–783.
83. Vassilopoulos A, Pennington JD, Andresson T, et al. SIRT3 deacetylates ATP synthase F1 complex proteins in response to nutrient- and exercise-induced stress. *Antioxid Redox Signal.* 2014;21(4):551–564.
84. Bahr R, Hansson P, Sejersted OM. Triglyceride/fatty acid cycling is increased after exercise. *Metabolism.* 1990;39(9):993–999.
85. Magkos F, Mohammed BS, Patterson BW, Mittendorfer B. Free fatty acid kinetics in the late phase of postexercise recovery: importance of resting fatty acid metabolism and exercise-induced energy deficit. *Metabolism.* 2009;58(9):1248–1255.
86. Kiens B, Richter EA. Utilization of skeletal muscle triacylglycerol during postexercise recovery in humans. *Am J Physiol.* 1998;275(2 pt 1):E332–E337.

87. Starling RD, Trappe TA, Parcell AC, Kerr CG, Fink WJ, Costill DL. Effects of diet on muscle triglyceride and endurance performance. *J Appl Physiol (1985)*. 1997;82(4): 1185–1189.
88. Bachmann OP, Dahl DB, Brechtel K, et al. Effects of intravenous and dietary lipid challenge on intramyocellular lipid content and the relation with insulin sensitivity in humans. *Diabetes*. 2001;50(11):2579–2584.
89. Boden G, Lebed B, Schatz M, Homko C, Lemieux S. Effects of acute changes of plasma free fatty acids on intramyocellular fat content and insulin resistance in healthy subjects. *Diabetes*. 2001;50(7):1612–1617.
90. Decombaz J, Fleith M, Hoppeler H, Kreis R, Boesch C. Effect of diet on the replenishment of intramyocellular lipids after exercise. *Eur J Nutr*. 2000; 39(6):244–247.
91. Hill AV, Lupton H. Muscular exercise, lactic acid, and the supply and utilization of oxygen. *Q J Med*. 1923;16:135–171.
92. Hill AV, Long CNH, Lupton H. Muscular exercise, lactic acid, and the supply and utilisation of oxygen: parts I-III. *Proc R Soc Lond (Biol)*. 1924;97:84–138.
93. Hill AV, Long CNH, Lupton H. Muscular exercise, lactic acid, and the supply and utilisation of oxygen: parts VII-VIII. *Proc R Soc Lond (Biol)*. 1924;97:155–176.
94. Margaria R, Edwards HT, Dill DB. The possible mechanisms of contracting and paying the oxygen debt and the role of lactic acid in muscular contraction. *Am J Physiol*. 1933;106(3):689–715.
95. Gaesser GA, Brooks GA. Metabolic bases of excess post-exercise oxygen consumption: a review. *Med Sci Sports Exerc*. 1984;16(1):29–43.
96. Borsheim E, Bahr R. Effect of exercise intensity, duration and mode on post-exercise oxygen consumption. *Sports Med*. 2003;33(14):1037–1060.
97. Horton TJ, Pagliassotti MJ, Hobbs K, Hill JO. Fuel metabolism in men and women during and after long-duration exercise. *J Appl Physiol (1985)*. 1998;85(5): 1823–1832.
98. Bahr R. Excess postexercise oxygen consumption—magnitude, mechanisms and practical implications. *Acta Physiol Scand Suppl*. 1992;605:1–70.
99. Gore CJ, Withers RT. Effect of exercise intensity and duration on postexercise metabolism. *J Appl Physiol (1985)*. 1990;68(6):2362–2368.
100. Gore CJ, Withers RT. The effect of exercise intensity and duration on the oxygen deficit and excess post-exercise oxygen consumption. *Eur J Appl Physiol Occup Physiol*. 1990;60(3):169–174.
101. Christensen EH, Hansen O. Arbeitsfähigkeit und Ernährung. *Skand Arch Physiol*. 1939;81:160–171.
102. Bergstrom J, Hermansen L, Hultman E, Saltin B. Diet, muscle glycogen and physical performance. *Acta Physiol Scand*. 1967;71(2):140–150.
103. Bergstrom J, Hultman E. A study of the glycogen metabolism during exercise in man. *Scand J Clin Lab Invest*. 1967;19(3):218–228.
104. Hawley JA, Schabort EJ, Noakes TD, Dennis SC. Carbohydrate-loading and exercise performance. An update. *Sports Med*. 1997;24(2):73–81.
105. Noakes TD, Lambert EV, Lambert MI, McArthur PS, Myburgh KH, Benade AJ. Carbohydrate ingestion and muscle glycogen depletion during marathon and ultramarathon racing. *Eur J Appl Physiol Occup Physiol*. 1988;57(4):482–489.
106. Gonzalez JT, Stevenson EJ. New perspectives on nutritional interventions to augment lipid utilisation during exercise. *Br J Nutr*. 2012;107(3):339–349.
107. Volek JS, Noakes T, Phinney SD. Rethinking fat as a fuel for endurance exercise. *Eur J Sport Sci*. 2015;15(1):13–20.
108. Burke LM, Hawley JA. Effects of short-term fat adaptation on metabolism and performance of prolonged exercise. *Med Sci Sports Exerc*. 2002;34(9):1492–1498.

109. Burke LM, Hawley JA, Angus DJ, et al. Adaptations to short-term high-fat diet persist during exercise despite high carbohydrate availability. *Med Sci Sports Exerc.* 2002;34(1):83–91.
110. Helge JW. Adaptation to a fat-rich diet: effects on endurance performance in humans. *Sports Med.* 2000;30(5):347–357.
111. Stepto NK, Carey AL, Staudacher HM, Cummings NK, Burke LM, Hawley JA. Effect of short-term fat adaptation on high-intensity training. *Med Sci Sports Exerc.* 2002;34(3):449–455.
112. Yeo WK, Carey AL, Burke L, Spriet LL, Hawley JA. Fat adaptation in well-trained athletes: effects on cell metabolism. *Appl Physiol Nutr Metab.* 2011;36(1):12–22.
113. Zajac A, Poprzecki S, Maszczyk A, Czuba M, Michalczyk M, Zydek G. The effects of a ketogenic diet on exercise metabolism and physical performance in off-road cyclists. *Nutrients.* 2014;6(7):2493–2508.
114. Phinney SD, Bistrian BR, Evans WJ, Gervino E, Blackburn GL. The human metabolic response to chronic ketosis without caloric restriction: preservation of submaximal exercise capability with reduced carbohydrate oxidation. *Metabolism.* 1983;32(8):769–776.
115. Phinney SD, Bistrian BR, Wolfe RR, Blackburn GL. The human metabolic response to chronic ketosis without caloric restriction: physical and biochemical adaptation. *Metabolism.* 1983;32(8):757–768.
116. Phinney SD, Horton ES, Sims EA, Hanson JS, Danforth Jr E, LaGrange BM. Capacity for moderate exercise in obese subjects after adaptation to a hypocaloric, ketogenic diet. *J Clin Invest.* 1980;66(5):1152–1161.
117. Ballard KD, Quann EE, Kupchak BR, et al. Dietary carbohydrate restriction improves insulin sensitivity, blood pressure, microvascular function, and cellular adhesion markers in individuals taking statins. *Nutr Res.* 2013;33(11):905–912.
118. Forsythe CE, Phinney SD, Fernandez ML, et al. Comparison of low fat and low carbohydrate diets on circulating fatty acid composition and markers of inflammation. *Lipids.* 2008;43(1):65–77.
119. Paoli A, Grimaldi K, D'Agostino D, et al. Ketogenic diet does not affect strength performance in elite artistic gymnasts. *J Int Soc Sports Nutr.* 2012;9(1):34.
120. Ehrenborg E, Krook A. Regulation of skeletal muscle physiology and metabolism by peroxisome proliferator-activated receptor delta. *Pharmacol Rev.* 2009;61(3):373–393.
121. Su JL, Simmons CJ, Wisely B, Ellis B, Winegar DA. Monitoring of PPAR alpha protein expression in human tissue by the use of PPAR alpha-specific MAbs. *Hybridoma.* 1998;17(1):47–53.
122. Braissant O, Foufelle F, Scotto C, Dauca M, Wahli W. Differential expression of peroxisome proliferator-activated receptors (PPARs): tissue distribution of PPAR-alpha, -beta, and -gamma in the adult rat. *Endocrinology.* 1996;137(1):354–366.
123. Watanabe K, Fujii H, Takahashi T, et al. Constitutive regulation of cardiac fatty acid metabolism through peroxisome proliferator-activated receptor alpha associated with age-dependent cardiac toxicity. *J Biol Chem.* 2000;275(29):22293–22299.
124. Aoyama T, Peters JM, Iritani N, et al. Altered constitutive expression of fatty acid-metabolizing enzymes in mice lacking the peroxisome proliferator-activated receptor alpha (PPARalpha). *J Biol Chem.* 1998;273(10):5678–5684.
125. Campbell FM, Kozak R, Wagner A, et al. A role for peroxisome proliferator-activated receptor alpha (PPARalpha) in the control of cardiac malonyl-CoA levels: reduced fatty acid oxidation rates and increased glucose oxidation rates in the hearts of mice lacking PPARalpha are associated with higher concentrations of malonyl-CoA and reduced expression of malonyl-CoA decarboxylase. *J Biol Chem.* 2002; 277(6):4098–4103.

126. Cresci S, Wright LD, Spratt JA, Briggs FN, Kelly DP. Activation of a novel metabolic gene regulatory pathway by chronic stimulation of skeletal muscle. *Am J Physiol.* 1996;270(5 pt 1):C1413–C1420.
127. Horowitz JF, Leone TC, Feng W, Kelly DP, Klein S. Effect of endurance training on lipid metabolism in women: a potential role for PPARalpha in the metabolic response to training. *Am J Physiol Endocrinol Metab.* 2000;279(2):E348–E355.
128. Muoio DM, Way JM, Tanner CJ, et al. Peroxisome proliferator-activated receptor-alpha regulates fatty acid utilization in primary human skeletal muscle cells. *Diabetes.* 2002;51(4):901–909.
129. de Lange P, Ragni M, Silvestri E, et al. Combined cDNA array/RT-PCR analysis of gene expression profile in rat gastrocnemius muscle: relation to its adaptive function in energy metabolism during fasting. *FASEB J.* 2004;18(2):350–352.
130. Wang YX, Zhang CL, Yu RT, et al. Regulation of muscle fiber type and running endurance by PPARdelta. *PLoS Biol.* 2004;2(10). e294.
131. Hautala AJ, Leon AS, Skinner JS, Rao DC, Bouchard C, Rankinen T. Peroxisome proliferator-activated receptor-delta polymorphisms are associated with physical performance and plasma lipids: the HERITAGE Family Study. *Am J Physiol Heart Circ Physiol.* 2007;292(5):H2498–H2505.
132. Luquet S, Lopez-Soriano J, Holst D, et al. Peroxisome proliferator-activated receptor delta controls muscle development and oxidative capability. *FASEB J.* 2003;17(15):2299–2301.
133. Narkar VA, Downes M, Yu RT, et al. AMPK and PPARdelta agonists are exercise mimetics. *Cell.* 2008;134(3):405–415.
134. Calvo JA, Daniels TG, Wang X, et al. Muscle-specific expression of PPARgamma coactivator-1alpha improves exercise performance and increases peak oxygen uptake. *J Appl Physiol (1985).* 2008;104(5):1304–1312.
135. Tadaishi M, Miura S, Kai Y, Kano Y, Oishi Y, Ezaki O. Skeletal muscle-specific expression of PGC-1alpha-b, an exercise-responsive isoform, increases exercise capacity and peak oxygen uptake. *PLoS One.* 2011;6(12). e28290.
136. Kleiner S, Nguyen-Tran V, Bare O, Huang X, Spiegelman B, Wu Z. PPAR{delta} agonism activates fatty acid oxidation via PGC-1{alpha} but does not increase mitochondrial gene expression and function. *J Biol Chem.* 2009;284(28):18624–18633.
137. Pokrywka A, Cholbinski P, Kaliszewski P, Kowalczyk K, Konczak D, Zembron-Lacny A. Metabolic modulators of the exercise response: doping control analysis of an agonist of the peroxisome proliferator-activated receptor delta (GW501516) and 5-aminoimidazole-4-carboxamide ribonucleotide (AICAR). *J Physiol Pharmacol.* 2014;65(4):469–476.
138. Gray SL, Dalla Nora E, Vidal-Puig AJ. Mouse models of PPAR-gamma deficiency: dissecting PPAR-gamma's role in metabolic homoeostasis. *Biochem Soc Trans.* 2005;33(pt 5):1053–1058.
139. McGuire DK, Abdullah SM, See R, et al. Randomized comparison of the effects of rosiglitazone vs. placebo on peak integrated cardiovascular performance, cardiac structure, and function. *Eur Heart J.* 2010;31(18):2262–2270.
140. Regensteiner JG, Bauer TA, Reusch JE. Rosiglitazone improves exercise capacity in individuals with type 2 diabetes. *Diabetes Care.* 2005;28(12):2877–2883.
141. Budni J, Lobato KR, Binfare RW, et al. Involvement of PI3K, GSK-3beta and PPARgamma in the antidepressant-like effect of folic acid in the forced swimming test in mice. *J Psychopharmacol.* 2012;26(5):714–723.
142. Eissa Ahmed AA, Al-Rasheed NM. Antidepressant-like effects of rosiglitazone, a PPARgamma agonist, in the rat forced swim and mouse tail suspension tests. *Behav Pharmacol.* 2009;20(7):635–642.

143. Ruderman NB, Park H, Kaushik VK, et al. AMPK as a metabolic switch in rat muscle, liver and adipose tissue after exercise. *Acta Physiol Scand*. 2003;178(4):435–442.
144. Dreyer HC, Fujita S, Cadenas JG, Chinkes DL, Volpi E, Rasmussen BB. Resistance exercise increases AMPK activity and reduces 4E-BP1 phosphorylation and protein synthesis in human skeletal muscle. *J Physiol*. 2006;576(pt 2):613–624.

CHAPTER FOUR

Exercise and Regulation of Protein Metabolism

Philip J. Atherton[1], Bethan E. Phillips, Daniel J. Wilkinson
MRC-ARUK Centre of Excellence for Musculoskeletal Ageing Research, Clinical, Metabolic and Molecular Physiology, University of Nottingham, Royal Derby Hospital Centre, Derby, United Kingdom
[1]Corresponding author: e-mail address: philip.atherton@nottingham.ac.uk

Contents

Abstract

Skeletal muscles exhibit radical changes in physiology and metabolism in response to exercise. While exercise induces highly specific physiological changes, e.g., hypertrophy, associated with weightlifting or oxygen utilization associated with aerobic-type exercises, the foundation of these changes is driven by the summation of exercise-induced alterations in muscle protein metabolism. Practically, any type of exercise stimulates muscle protein turnover, the purpose being both to renew, and also modify, the myocellular composition of proteins in line with adaptations according to the mechanical and metabolic demands imposed. The mechanism(s) by which exercise stimulates protein turnover has been the subset of intense study. These studies have been led by the use of stable isotopically labeled amino acids. Essentially, use of these heavier variants (e.g., ^{13}C AA vs. ^{12}C) coupled to mass spectrometry has enabled study of the dynamic responses of muscle protein turnover to exercise. Using these techniques, it has become patently clear that exercise stimulates muscle protein turnover, i.e., muscle protein synthesis (MPS) and breakdown (MPB). Moreover, intake of specific nutrients (i.e., dietary proteins) potentiates MPS while attenuating MPB, facilitating maintenance of proteostasis and exercise adaptation. The mechanisms driving these protein metabolic responses to exercise include the coordinated activation of mRNA translation pathways (e.g., mechanistic target of rapamycin) and multiple MPB pathways (e.g., autophagy and ubiquitin–proteasome). These processes are triggered by exercise-induced hormone, auto/paracrine-acting growth factors, mechanical transduction, and intramyocellular second

Progress in Molecular Biology and Translational Science, Volume 135
ISSN 1877-1173
http://dx.doi.org/10.1016/bs.pmbts.2015.06.015

messenger pathways. Finally, there remains poor understanding of how distinct exercise modes (e.g., resistance vs. endurance) lead to such distinct adaptations from a protein metabolic and molecular standpoint.

1. INTRODUCTION

Muscle protein metabolism is governed via a dynamic equilibrium between muscle protein synthesis (MPS) and muscle protein breakdown (MPB). Typically, in the postabsorptive (fasted) state, rates of MPB exceed those of MPS (i.e., $0.08–0.1\%\,h^{-1}$ for MPB and $0.03–0.07\%\,h^{-1}$ for MPS[1–3]), engendering negative protein balance (-0.04% to $0.05\%\,h^{-1}$), and hence, an overall loss of muscle proteins. However, this catabolic scenario is reversed following intake of dietary protein, upon which rates of MPS exceed that of MPB, and muscle protein balance becomes positive again through an approximate doubling of MPS.[4] The net result of this cycle being the regaining of muscle proteins lost during the postabsorptive state. There is no better example of muscle plasticity than that which occurs in response to exercise. The mechanical loading of muscle associated with exercise stimulates muscle protein turnover, manifesting as increases in both the synthesis and breakdown of muscle proteins. Moreover, since exercise also increases MPB (more than MPS), this leads to negative net balance[5], i.e., exercise in the fasted state is often catabolic. On the other hand, when dietary proteins are consumed in proximity to exercise (which supply the AA substrates for MPS) increases in MPS persist for longer periods (~24 h[6,7]). In this instance, increases in MPS outweigh those of MPB[5] and net protein balance switches to positive. It is this scenario of positive balance that confers remodeling associated with exercise adaptation, e.g., hypertrophy and mitochondrial biogenesis.

1.1 What Techniques Have Informed Us upon How Exercise Impacts Muscle Protein Metabolism?

How do we study protein metabolism *in vivo*? The use of "stable isotope" tracers permits a dynamic measure of MPS to be determined, and with the advent of sensitive, commercially available analytical instrumentation (e.g., mass spectrometry) is considered the "gold standard" for *in vivo* assessment of protein turnover.[8–11] Typically, these measurements require the introduction of a "heavy" version of the metabolite of interest, which for protein turnover is an AA. This is usually achieved whereby one (or more) of the carbon, hydrogen, or nitrogen in the AA is replaced by its naturally occurring stable isotope equivalent; ^{13}Carbon, Deuterium (^{2}H), or ^{15}Nitrogen, respectively, creating labeled AAs such as 1,2 $^{13}C_2$ leucine, *ring*-D_5-phenylalanine, or ^{15}N-glycine. The body treats these "labeled" AAs in the same way as the "lighter" endogenous metabolite, meaning that through regular tissue and/or blood sampling, the fate of these heavy AAs can be traced allowing rates of utilization and turnover (as well as oxidation) to be quantitatively measured.

Many methods and models have been refined over the years for measurement of protein turnover (for more detailed information on these, we refer the reader to Wolfe and Chinkes[12]). The first and most commonly applied method involves measurement of whole-body (i.e., not "tissue specific") protein turnover. Traditionally, ^{15}N-glycine has been used; however, other AA tracers can also be used.[13] Once the tracer has reached equilibrium calculations of whole-body synthesis and breakdown, based on rates of disappearance and appearance of the tracer from an assumed single pool (through regular urine and blood samples), can be made. Despite providing useful information, this method is not *tissue specific* (i.e., does not tell us about muscle) since it does not provide information on which part of the body is causing any of the measured contributions in protein metabolism. In an attempt to overcome a number of these limitations, the use of arterial–venous (A–V) balance was introduced.[14] Here, both rates of tissue synthesis and breakdown can be determined via monitoring the rate of disappearance of the tracer from the arterial pool (as a proxy of synthesis) or the rate of appearance of the tracer into the venous pool (as a proxy of breakdown), assuming that the AA being studied is not subject to secondary metabolism within the tissue, i.e., assuming that it is exclusively being used for protein synthesis.[14–16] These assumptions can be made stronger through the use of phenylalanine as the tracer and the "3-pool model" where intracellular tracer enrichments are also measured. With measures of breath CO_2 ^{13}C enrichment, AA oxidation can also be simultaneously determined, providing more expansive information on whole-body protein turnover and oxidation. (For detailed information on modeling kinetics, see the following works.[12,15–18]) Yet, despite the considerable popularity and utility of these A–V balance methods, they are subject to influence by surrounding tissues, e.g., skin and bone, when calculating across a limb, which limit (although not to the same extent as whole-body kinetics) extrapolation to the tissue of interest, i.e., to leg skeletal muscle when measuring A–V balance across the femoral A–V sides of a limb. Moreover, with calculation reliance on measures of arterial blood flow and lean leg mass (typically quantified by Doppler ultrasound and dual-energy X-ray absorptiometry, respectively) to model the data, extreme perturbations in flow (e.g., following exercise, nutrition) and inaccurate analysis of the contribution of pure lean tissue mass could mean that acquired data may be over/under estimated.

Instead, direct incorporation- or fractional synthesis rate (FSR) measurements of MPS are considered the "gold standard" in protein metabolism measurements. Here, using continuous, bolus or pulsed (or all three together) tracer infusions, rates of incorporation of an AA tracer into proteins can be determined and, hence, a FSR or absolute synthetic rate (ASR) calculated. By sampling the tissue pool, isolating and hydrolyzing the protein of interest, the amount of tracer AA incorporated into protein is determined,[4,19] which with measurement of enrichment in the precursor pool, or a proxy surrogate, means that FSR of MPS can be accurately calculated (see Fig. 1). Unlike A–V methods, FSR is tissue specific and unaffected by blood flow perturbations. Furthermore, by simply stopping a steady-state tracer infusion, measurement of the decay of the tracer enrichment from the arterial and intracellular pool over time can also give a

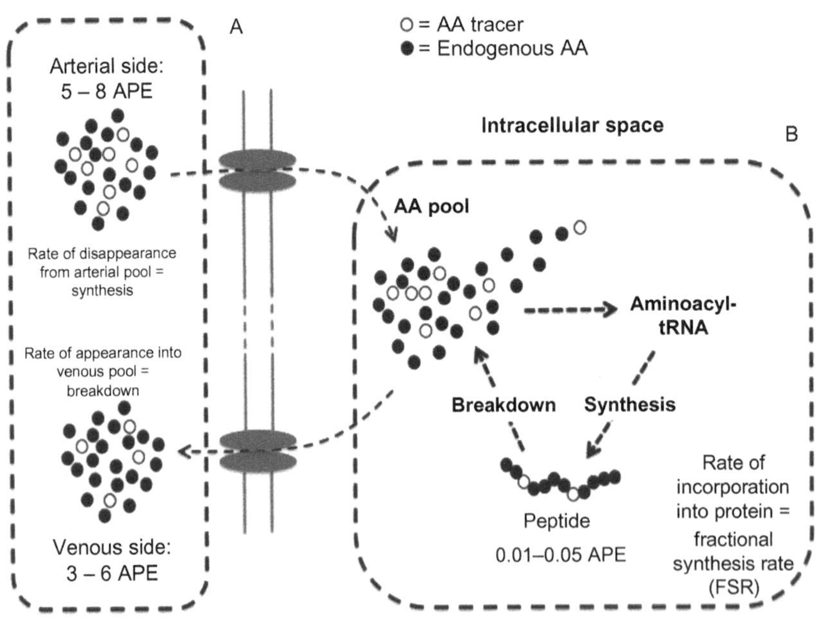

Figure 1 Schematic showing the use of stable isotope tracers for the calculation of protein synthesis and breakdown. (A) represents arterio-venous balance measures, (B) fractional synthetic rate measures. *Abbreviation*: APE, atom percent excess. (See the color plate.)

measurement of fractional breakdown rate.[20,21] However, as with the A–V methods, these so-called gold standard methods also have their limitations. To perform them requires intravenous tracer administration, which restricts measures to acute (<12 h) time periods; there is also the need for invasive venous/arterial cannulation, in addition to repeated biopsies. These aspects restrict these measurements to a laboratory or clinical setting, making their applicability to "normal life" limited.

In contrast, the recent "reintroduction" of the stable isotope tracer deuterium oxide (D_2O) somewhat overcomes some of these limitations associated with traditional AA tracer techniques. The rapid developments over the past decade of multiple D_2O tracer methodologies for use in both acute (hours–days) and chronic (weeks–months) settings with minimally invasive procedures and sampling have been of great interest. Deuterium was one of the first stable isotopes to be successfully chemically isolated.[22] Almost immediately following this isolation, this stable isotope was rapidly implemented for the measurement of cholesterol and protein turnover through the provision of D_2O in the diets of animals and subsequently monitoring the appearance of deuterium in multiple tissue pools.[23–25] Despite the novelty of this technique, it was soon superseded by the isolation and introduction of carbon and nitrogen stable and radiolabeled tracers. However, recent developments in mass spectrometry, in particular the development and commercialization of pyrolysis reactors for H_2 and O measurements, have led to an explosion of interest in the application of D_2O to measure *in vivo* protein metabolism. Indeed with a single bolus[11,26] or continuous daily/weekly ingestion of D_2O,[27–30] it is possible to quantify multiple metabolic pools (i.e., lipids, proteins, fats, nucleic acids) as the rapid equilibrium of D_2O within the body water pool leads to dispersion into multiple tissues and organs. The deuterium can then replace hydrogen from water at stable C–H bonds in metabolites through enzymatic (and nonenzymatic exchange) reactions, with the potential for labeling (and hence measuring of rates of incorporation) of proteins, lipids, DNA, and sugars at a single instance. Indeed, with the ease of oral administration and potential for acute and chronic *in vivo* measures of multiple metabolic pools, it is no wonder the popularity and application of this method are growing rapidly, particularly in the exercise and protein metabolism field.[11,26,27,31–33] Moreover, with its ability to quantify cumulative responses of MPS, this method will prove vital to the study of muscle metabolism over longer periods of exercise training. This should lead to a more holistic understanding of exercise adaptation and muscle metabolism.

2. THE REGULATION OF PROTEIN METABOLISM BY EXERCISE

Physiological adaptations to exercise training depend upon the mode of exercise undertaken and are intricately responsive to the specific demands imposed. The most classical segregation of exercise training modes is that of resistance exercise training (RE-T) and endurance (or aerobic) exercise training (EE-T). Prior chapters have discussed in great detail the metabolism of the other macronutrients (CHO and Lipids) during exercise, particularly with regards to the importance of these as energy substrates. While it is not a primary role of protein/AAs, it is important to highlight their contribution to energy provision before moving forward to discussing aspects of exercise and protein metabolism.

It is generally considered that carbohydrates and fats are the most important substrates for energy provision both under rested and exercised conditions.[34] However, the varied pathways through which both nonessential and essential AAs can be metabolized ensures that under certain conditions protein/AAs can form an important alternate energy substrate. For example, through the process of branched-chain amino acid (BCAA) transamination, BCAAs can provide carbon skeletons for TCA cycle intermediates,[35] while also providing keto acids for decarboxylation in the mitochondria for maintenance of NADH levels,[36] with the nonessential AAs such as alanine capable of reversible transamination to pyruvate,[37] and hence, potential for associated ATP production. The relative role of AAs for energy provision during exercise can be highlighted by the fact that there is a rapid increase in the activity of branched-chain alpha-keto acid dehydrogenase complex (the rate-limiting step for BCAA transamination and oxidation pathways) with the onset of exercise,[38] and this is supported by other studies which found an increase in leucine oxidation rates during exercise.[4,39] Furthermore, the contribution of AAs as an energy substrate seemingly increases as a function of both exercise intensity[39] and duration.[40] Early studies investigating substrate fluxes across organs during exercise found that as exercise progressed in duration there was a greater efflux of alanine from the leg, which was subsequently taken up by the liver; while there was an accompanying increase in the efflux of BCAAs from the liver and uptake by the leg, a pattern indicative for utilization of AAs for energy provision.[40] The need for this increase in AA utilization for energetic purposes during exercise is likely driven by a reduction in the availability of other substrates, as the additional provision of glucose during exercise helps to decrease leucine oxidation rates.[41] Moreover, trained individuals who inherently have more efficient substrate utilization also show a reduced level of leucine oxidation.[42] Though much of this information is derived from studies of moderate- and high-intensity EE, AAs likely also contribute similar energetic benefit during RE; however, this is still relatively poorly understood. Despite protein/AAs providing an evidently important contribution to energy provision, comparative utilization of AAs is believed to be <20% dependent on the type, duration, and intensity of the exercise being performed.[34,43] This is primarily driven by the importance of AAs for maintaining structure, integrity, and function of multiple organ systems.

In the subsequent sections, we will describe protein metabolism in response to both RE and EE regimens before discussing what is known about how these responses are regulated.

2.1 RE-T and Muscle Protein Metabolism

RE-T is not only a pastime of body builders and recreational weightlifters; it also represents the foremost nonpharmacological intervention in offsetting declines in muscle mass with aging and other muscle-wasting conditions.[44] Indeed, despite heavy investment by the pharmaceutical industry into the discovery of promising anabolic agents and neutraceuticals, RE-T remains the most safe and effective way of maintaining and/or building muscle mass (a.k.a. hypertrophy). It is generally accepted that the accumulation of acute post-RE increases in MPS, over a period of RE-T, engenders a sustained situation of greater positive net protein balance (MPS > MPB[11]) culminating in muscle hypertrophy. Indeed, in a comprehensive review, Kumar *et al.* reported that, under fasted conditions, an acute bout of RE increased subsequent rates of MPS in both males and females with little data to contend this.[5] While it is beyond the scope of this chapter to overview hypertrophic responses to RE-T (instead we would refer to the chapter by McGlory and Phillips - Exercise and regulation of muscle hypertrophy), it is important to highlight that cumulative increases in MPS following acute bouts of RE lead to the characteristic increases in muscle fiber size, culminating in increases in whole muscle cross-sectional area,[45,46] that become apparent after just a few weeks RE-T (see later for further discussion and details in relation to muscle protein metabolism[47,48]).

In recent years, a number of studies have sought to establish the most optimal pattern of loading to maximize anabolic responses to RE-T, with investigations focusing upon the impact(s) of (i) intensity, (ii) duration (repetition number), and (iii) mode of contraction. In the period following a bout of RE, MPS increases two- to fourfold above baseline (at rest "mixed" muscle proteins turnover at $\sim 0.05\%\,h^{-1}$ or $1.2\%\,d^{-1}$). In response to a single bout of RE, Kumar *et al.* demonstrated a sigmoidal dose–response to varying exercise intensity, showing that MPS plateaued at intensities >60% one-repetition maximum (1-RM; the maximum amount of resistance that can be overcome for a specific exercise in one effort), even when repetition number was increased at lower intensities (40% 1-RM) to match work volume (*intensity–volume product*).[49] This supports the notion that there is an "intensity ceiling" of 60% 1-RM, above which no further anabolism is conferred. Nonetheless, this does not mean that higher exercise intensities above this ceiling are a prerequisite to stimulate MPS. Work by the Phillips lab showed that a single bout of high-volume low-intensity RE to failure was more effective than low-volume high-intensity exercise with regard to both the amplitude and duration of the subsequent MPS response.[50] This approach did, however, require a greater work volume to be performed by the low-intensity exercise group (30% 1-RM to failure vs. 90% 1-RM to failure). This finding is almost certainly as a consequence of increased type II fiber recruitment and the impact of Henneman's size principle which states that motor units are recruited from smallest to largest in relation to intensity, i.e., while in practice, slow-twitch (type I) fibers are activated before fast fibers, even RE to fatigue at lower loads will eventually activate type II fibers. Interestingly, these acute MPS findings are consistent with recent data from a training study using the same approach, where resulting hypertrophy was identical under high versus low 1-RM training, underlining the notion that lifting bigger weights does not necessarily translate into a bigger anabolic response in the context of skeletal muscle protein metabolism or hypertrophy.[51]

The timecourse of increases in MPS after a single session of RE varies according to an individuals nutritional[52] and training status, but also as a feature of exercise volume and intensity. Typically, immediately after RE, there is a latent period in which MPS remains at postabsorptive values, presumably reflecting a blunting of MPS during exercise due to the suppressive effects of ATP turnover on MPS. In support of this suppression of MPS *during* exercise, a recent study in rodents showed that MPS was suppressed during muscle contraction in manner according to contractile duty cycle.[53] Although there are no equivalent studies assessing MPS *during* exercise in humans, it is likely that MPS is also suppressed during RE in humans. For example, while increases in MPS were observed by 1 h after moderate intensity exercise (6 × 8 repetitions at 75% 1-RM),[49] MPS remained unchanged 3 h after exhaustive eccentric contractions.[7] Thus, the energetic demands of an exercise regime may govern the degree of suppression in MPS and also the latency, before which MPS is "switched-on" to permit remodeling. Nonetheless, other mechanisms for this latency might exist. For example, during and immediately following acute exercise, global responses for mRNA (required as the template for protein production) appear rather chaotic,[54] presumably reflecting their release for the ribosomes in a manner also related to the acute energy crisis within the muscle. Nevertheless, after this period of latency, MPS rises between 45 and 150 min following cessation of RE and is measurably increased out to ~4 h, which in the absence of provision of protein nutrition,[49] engenders negative net muscle protein balance. Indeed, while RE is a potent stimulus for MPS, associated increases in MPB over that of MPS ensures that RE performed in the fasted state is, *de facto*, catabolic.[2] However, when combined with dietary protein, positive balance (which can persist for up to 48 h[6,7] at least in the untrained state) can be restored; hence, permitting the associated accretion of muscle protein over the duration of an RE-T program leading to eventual skeletal muscle hypertrophy.[3]

While much is known about the effects of acute RE upon muscle protein metabolism, due to the intricacies of the measurement techniques (i.e., acute tracer studies being at the beginning and/or end of training rather than temporal measures throughout), much less is known about the influence of chronic RE-T on muscle protein metabolism. Why is this important? We know that responses of muscle protein turnover change throughout RE-T. For instance, sedentary individuals exhibit greater (~50%) MPS responses to a single bout of RE compared to highly trained individuals.[55] Furthermore, using unilateral models of RE-T, trained legs exhibit lower rates of MPS than untrained legs following 8 weeks of training,[56,57] and a dampening of MPS is evident with successive bouts of RE.[11,58] This dampened responsiveness to RE-T in the trained state may be reflective of the well documented yet largely under recognized temporal nature of the hypertrophic response to RE-T. Hypertrophy is not a linear process; one does not simply continue to accrue muscle mass with continuing RE-T. Rather, much of the muscle hypertrophy occurring in response to RE-T occurs in the first few weeks of RE-T,[47,59,60] after which there is a marked plateau or slowing of muscle gains despite progressive training overload.[60] This may explain why slower rates of turnover are evident in trained individuals, i.e., as the rate of hypertrophy slows, the primary driving force behind this hypertrophy, MPS, will do so also. Moreover, it was recently shown that MPS responses to the first bout of RE do not correlate with ensuing hypertrophic responses to RE-T.[61] Why might this be? First, it is likely that the elevation of MPS following unfamiliar RE-T occurs in part due to muscle damage responses to unaccustomed exercise,[62] rather than quantitatively reflecting *bona fide* protein accretion. Second, this could be due to the fact that gross perturbations in muscle protein metabolism in the first few exercise bouts "settle down" with repeated bouts (as evidenced above). Third, even when following established principles of overload (e.g., maintaining relative intensity throughout RE-T), timecourses of hypertrophic adaptations vary from person to person, such that one may not capture the "plateau" in hypertrophy to align to the acute first bout MPS response. Indeed, unless interindividual timecourses of muscle hypertrophy are known, how can hypertrophy be expected to relate to the first bout of RE-T? The D_2O tracer technique (described earlier), which does not rely on information from a single RE-T bout, can help shed new light on the long-term control of protein metabolism in response to exercise.

MPB is less studied in comparison to MPS, primarily due to the less developed and more complex methodologies that are involved in its accurate measurement.[5,15,21] Despite this, there are a number of papers which have highlighted the impact of RE on MPB, which seemingly follow a similar temporal pattern to that of MPS. Postexercise there is a rapid transient increase in MPB of up to ~30–50%, which can last up to 24 h.[2,63] In the rested fasted state, MPB is already significantly higher than MPS creating a net negative balance, and yet despite the considerable stimulation of MPS by RE, the equally as rapid increase in MPB ensures that the muscle remains in a net catabolic state, and it is this MPB effect that is the primary driver behind the net catabolic effect of RE observed in the fasted state.[2,63] However, there does appear to be somewhat of a training adaptation with regards to MPB, as when MPB is measured in the untrained and trained state immediately postexercise, the increase in MPB is dampened in the trained state, indicating that reductions in MPB are an important adaptation associated with RE-T.[55] This is interesting, because as already highlighted, the acute stimulation of MPS is lower in the trained state, which combined with attenuations in MPB postexercise means that trained individuals require less of a stimulation to obtain a positive balance due to the associated adaptations in protein turnover. In the fed state, the impact of RE on MPB is less pronounced than that on MPS, although feeding can lead to associated hyperinsulinemia and an increased abundance of free AAs, which can suppress breakdown in the postabsorptive rested state.[64,65] The evidence of a beneficial effect of postexercise nutrition on suppression of MPB is debatable with the majority of studies showing no/insignificant change(s) in MPB with the addition of nutrition postexercise.[52,66–69] Thus, it seems evident that the main driver behind the positive net balance observed post-acute RE is the influence of nutrition on augmenting the MPS response rather than suppressing the MPB response; however, the cumulative adaptation of chronic RE-T will likely assist in attenuating MPB responses over time.

2.2 EE-T and Muscle Protein Metabolism

In comparison to RE-T, EE-T involves exercise at a submaximal intensity, typically, for prolonged periods (~30 min to several hours). In contrast to RE-T in which muscle hypertrophy is the predominant adaptation, EE-T is associated with enhanced endurance capacity through the induction of shifts in substrate metabolism (e.g., slower use of glycogen), mitochondrial biogenesis, and angiogenesis.[70,71] These physiological adaptations associated with EE-T are stimulated in order to meet the increased demands for oxygen extraction and utilization.[72] Relatively, small amounts of work (as compared to RE-T) have investigated links between EE-T and protein metabolism, principally due to the prevailing "interest" being in muscle hypertrophy associated with RE-T. Nonetheless, akin to RE-T, EE-T is also associated with a stimulation of mixed muscle MPS following different modalities (running, walking, swimming) of EE[73–77] both in males and females. This again follows somewhat of a temporal lag, whereby during the initial 1–1.5 h postexercise, there are minimal increases in mixed muscle MPS, after which MPS increases significantly[78] and can be maintained for up to 24 h (at certain intensities).[79] It does, however, appear, in contrast to RE-T, that increased mixed muscle MPS following EE is predominantly driven by increases in sarcoplasmic and mitochondrial MPS, rather than myofibrillar MPS.[80,81] This makes sense from a physiological viewpoint as increased mitochondrial MPS could potentially reflect increased mitochondrial biogenesis, i.e., increased mitochondrial synthesis was evident ~24 h after a bout of EE.[79] Similarly, it was shown that the synthesis of mitochondrial proteins was preferentially upregulated in response to EE-T, while myofibrillar synthesis was preferentially upregulated in response to RE-T (at least in the trained state[80]). However, there have been suggestions that there may be an effect present in myofibrillar MPS following acute EE, with significant increases in myofibrillar MPS observed between 30 min and 4.5 h postexercise, which can be maintained for up to 24 h when the intensity of the EE is high (60% W_{max}).[79] Yet, the results of this study (compared to other findings) may have been confounded due to differences in the nutritional state of the individuals tested who were in the fasted rather than fed state as with previous studies.

In terms of MPB responses to EE-T, similar to RE-T studies, studies are somewhat limited. It has been shown that during EE-T there is a significant increase in MPB,[73,82] believed to be driven through the release of AAs for energetic purposes.[83] These increases in MPB during exercise are maintained postexercise as measured using both tracer[76] and indirect 3-methylhistidine release techniques,[73] albeit attenuated compared to that during exercise.[82] Clearly, relinquishing AAs and stimulating protein turnover are important processes governing acute metabolic and chronic adaptive responses to exercise of any modality.

3. SIGNAL TRANSDUCTION REGULATING MUSCLE PROTEIN METABOLISM RESPONSES TO EXERCISE

The regulation of protein metabolism involves the activation of cellular pathways in skeletal muscle that transduce signals to the machinery regulating mRNA translation. Of all the signaling pathways associated with exercise and muscle protein metabolism, the mechanistic (formerly mammalian) target of rapamycin complex 1 (MTORC1: composed of the genes *MTOR, RAPTOR, PRAS40, DEPTOR, LST8*, see Fig. 2) is considered a central determinant of muscle protein metabolic responses to exercise.[84] For example, administration of rapamycin, a pharmacological inhibitor of MTORC1, to humans led to the ablation of RE-T-induced increases in MPS,[85] confirming preclinical observations that MTORC1 is required for this response. Activation of MTORC1 is associated with phosphorylation of numerous substrate mRNA "translation factors" including 4E-binding protein (4EBP1), ribosomal protein S6 kinase (P70S6K1), and eukaryotic initiation factors 4 G/A/B (EIF4G/A/B), with ensuing formation of the EIF3F scaffold to promote assembly of a 48S preinitiation complex. In a parallel pathway, activation of the guanine exchange factor, eukaryotic initiation factor 2B (EIF2B) shuttles the initiator tRNA (Met-tRNAi) to the ribosome during formation of the 48S preinitiation complex, promoting "global" protein synthesis and coordinately enhancing translational efficiency. In effect, these events stimulate the polyribosome

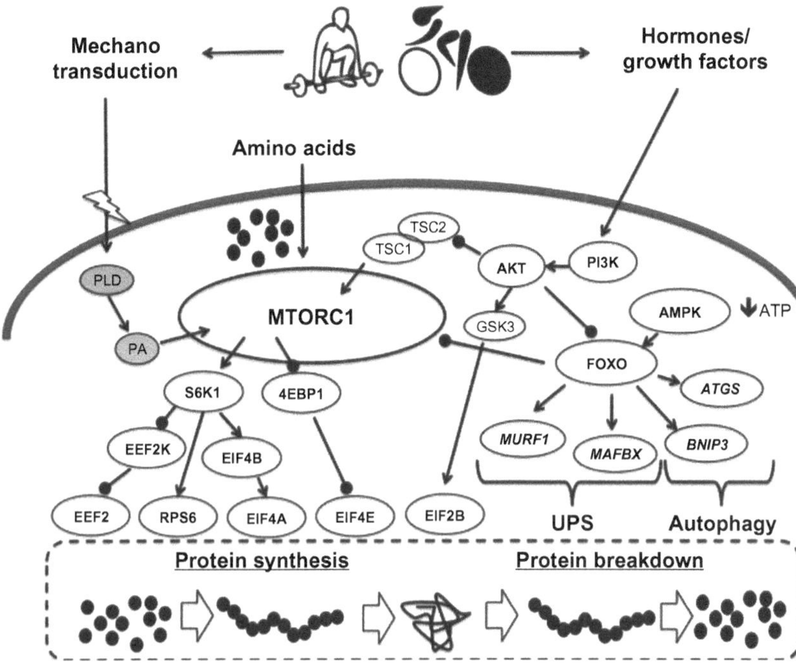

Figure 2 Known intracellular signaling pathways regulating protein synthesis and breakdown. Arrowed lines, activates; circled lines, inhibits. *Abbreviations*: AKT, protein kinase B; AMPK, AMP-activated protein kinase; *ATGS*, autophagy-related genes; BNIP3, BCL-2/E1B-19kD-interacting protein-3; EEF2, eukaryotic elongation factor 2; EEF2K, eukaryotic initiation factor 2 kinase; EIF2B/4A/4B/4E, eukaryotic initiation factor 2B/4A/4B/4E; 4EBP1, eukaryotic initiation factor 4E-binding protein-1; FOXO, forkhead box-containing proteins O; GSK, glycogen synthase kinase; *MAFBX*, muscle atrophy F-box; MTORC1, mechanistic target of rapamycin complex 1; *MURF1*, muscle ring finger protein 1; PLD, phospholipase D; PA, phosphatidic acid; PI3K, phosphoinositide 3-kinase; RPS6, ribosomal protein S6; S6K1, ribosomal protein S6 kinase 1; TSC1/2, tumor suppressor complex 1/2.

formation and promote mRNA translation. For a detailed review of mRNA translational control, we refer the reader to Proud.[86]

Nonetheless, while activation of MTORC1 and its substrates is a common feature of responses to exercise in the regulation of MPS, the proximal mechanisms involved in its activation remain poorly defined. In the most basic sense, "upstream" signal(s) triggered by exercise cause the activation of intracellular signaling pathways (such as MTORC1) responsible for modulating MPS. Yet what triggers these intracellular signaling responses is subject to debate and shrouded in complexity. Exercise triggers numerous physiochemical (i.e., mechanotransduction,[84] endocrine,[87] and auto/paracrine[88]) events which combine to determine muscle protein metabolic responses. So which of these signals is/are most important? Much of the earlier preclinical[89] and cell culture work[90] implicated the muscle (and hepatic) secreted growth factor, insulin-like growth factor-1 (IGF-1) in mediating MPS responses to exercise. This led to a thesis whereby increases in expression and activity of insulin-like growth factor (IGF) and its splice variants bind the IFG-1 receptor on myocytes, which in turn yields activation of a classical insulin-mediated PI3K-protein kinase B (AKT)-MTORC1 signaling pathway, thereby leading to increases in MPS. However, it has since transpired that the muscle canonical IGFr–AKT–MTOR signaling represented a gross oversimplification of the governance of protein metabolism in response to exercise. For example, in a recent study, adult human volunteers performed single bout of RE either in arm muscles alone or concurrently in arm and leg muscles. This protocol engendered either a high (arm and leg exercise) or a low (arm exercise alone) systemic hormone milieu. Yet, despite markedly different systemic concentrations of growth hormone (GH), testosterone, and IGF-1 between the two groups, there was no difference in acute MTORC1-associated signaling or in MPS, or critically in chronic adaptations to training in terms of mass and strength gains.[91,92] Similarly, knockout of the IGF-1 receptor did not impact muscle growth responses to overload in a model of synergist ablation.[93] These data indicate that modulating systemic concentrations of IGF-1, testosterone, or GH (the latter of which acts via inducing IGF-1 expression rather than having any direct effect on MPS in adult humans[94]) within the physiological range does not play a role in adaptive hypertrophy and challenges the classic view of the insulin signaling pathway. However, suppression of testosterone production via use of a gonadotropin releasing hormone analogue in humans ablated muscle growth responses to RE-T,[95] thus suggesting that testosterone remains an integral part of protein metabolic responses to exercise. Nonetheless, the general absence of accord over the role of IGF-1 and other circulating anabolic hormones has led to new hypotheses surrounding the mechanistic basis of protein metabolic responses to exercise with suggestions for it being a more intrinsic process than first thought.

Mechanotransduction is the process of converting mechanical signals that are sensed in response to cellular movement into molecular signals, and numerous candidate "mechanosensors" have been suggested in skeletal muscle. One target recently found was the phospholipase D (PLD) enzyme, which increases production of the lipid second messenger phosphatidic acid in a mechano-sensitive manner. Furthermore, it was demonstrated that phosphatidic acid signaling is upstream of contraction-induced activation of MTOR, since pharmacological inhibition of PLD effectively ablated activation of MTOR in response to muscle contractions.[96] Another possible mechano-sensitive pathway is that of the muscle attachment, or focal adhesion complexes, which represent macromolecular structures situated in the sarcolemma of muscle fibers, that link the extracellular matrix (ECM) to the cytoplasmic cytoskeleton, and consist of a variety of ECM receptors/integrins, intracellular cytoskeletal, and signaling molecules.[97] Interactions of ECM proteins with integrin receptors stimulate intracellular signaling pathways important in cell growth and migration in adult skeletal muscle.[98] Activation of integrin receptors appears to be a common feature of muscle remodeling in response to a variety of conditions including EE-T and muscular dystrophy.[99] Focal adhesion complexes play an integral structural role in the transmission of lateral forces during contraction. Focal adhesion kinase (FAK) is a nonreceptor tyrosine kinase that localizes to focal adhesion complexes and is thought to represents a key component of integrin-mediated signaling.[100] Engagement of integrin receptors induces phosphorylation of FAK at Tyr397, which correlates with its activation,[101] and a growing body of evidence has associated FAK activation with protein metabolic responses to mechanical stress in skeletal muscle.[102] Indeed, expression patterns of FAK have been reported to be load-dependent, i.e., phosphorylation of FAK was lowered following hind-limb suspension in rodents,[103] immobilization in humans[104] and increased in models of overload[105,106] and RE-T in humans.[80] Finally, local overexpression of FAK (FAK pLKO.1 plasmid) in rodents was shown to stimulate muscle hypertrophy[97] suggesting FAK as a genuine mechano-sensitive component of hypertrophy.

One of the greatest challenges associated with understanding how such signals link to protein metabolism and ensuing adaptations is the variance in human responses to exercise. Despite the identification of a numbers of molecular "signals" which are put forward as being "master regulators" of skeletal muscle adaptability (e.g., MTORC1, IGF-1, AMPK and PGC1α),[107] they do not explain the complexity and heterogeneity of exercise adaptation. Similarly, acute increases in MPS after a single bout of RE do not correlate with ensuing hypertrophic responses. A number of acute metabolic studies have highlighted discordance between mainstream signaling paradigms and dynamic measures of protein metabolism[65,108] and adaptation. Furthermore, a recent study using genome-wide profiling of individuals performing a supervised 20-week RE-T program highlighted that a decreased expression of ribosomal RNAs was associated with an increased ability to gain muscle mass,[109] yielding a signature suggesting that *inhibition of* MTORC1 is under some circumstances associated with greater hypertrophic responses following RE-T. Perhaps, this indicates that sufficient MTOR signaling can be coupled with *superior* efficiency for MPS in high-responder individuals (for growth), and hence, there is no need to upregulate the rRNA genes.[109] Perhaps, even more informative was the observation that in the ~25% of subjects that failed to demonstrate muscle growth MTOR activation was clearly indicated in these low responders. Therefore, it clearly cannot be a hallmark of muscle growth, with the canonical signaling of MTOR representing a gross oversimplification of protein metabolic and adaptive responses to exercise training in humans.

The molecular regulation of the apparent exercise-induced stimulation of MPB has also been difficult to ascertain, impacted on by a very limited number of studies, variability in findings across studies,[110,111] and also the fact that multiple proteolytic systems exist.[112] However, there is seemingly a common mechanism independent of exercise mode that increases in MPB are regulated through, ubiquitin–proteasome-mediated degradation.[83] During this process, proteins which are targeted for degradation are identified by the 26S proteasome through the ATP-dependent addition of ubiquitin (Ub) molecules, which is regulated through the three enzymes: Ub-activating enzyme (E1), Ub-conjugating enzyme (E2), and the Ub-ligase enzymes (E3).[113,114] Once four ubiquitin molecules have been conjugated to the protein, it can be identified by the 26S proteasome and degraded to single peptides. To exemplify this, immediately following RE, there is a rapid upregulation of the Ub-ligases *MURF1* and *MAFBx/ATROGIN* 1 mRNA, alongside their upstream transcription factor FOXO1.[115,116] This upregulation follows a temporal pattern, with *MURF1* reaching a peak at ~1–2 h postexercise and returning to baseline levels within 8 h.[117] This molecular regulation does, however, seem to be differentially affected by contraction mode, with increases in mRNA expression only seen in response to concentric contraction. When performing eccentric contractions, there is substantial divergence,[118] with increases in the expression of the structural components of the ubiquitin–proteasome system: *PSMA1*, *ubiquitin splice forms I* and *II*, in addition to *MURF2* and *3*.[116] This difference in molecular regulation between the two contraction modes is likely dependent on the increased levels of damage and remodeling required for eccentric compared to concentric contraction. Following 10-weeks RE-T, the response of the ubiquitin–proteasome system postexercise is somewhat dampened,[118] reflecting that of the measures of MPB following training, which seem to be dampened also.[55] This likely reflects the fact that much of the muscular remodeling due to RE-T occurs in the early phases of the training, and therefore after 10 weeks much of the adaptation has already occurred and there will be less need for MPB to take place. While RE-T seems to primarily upregulate ubiquitin–proteasome pathways, there is evidence that proteolytic responses to EE-T may be mediated through a combination of autophagy–lysosomal and ubiquitin–proteasomal pathways.[83] Following moderate EE-T, both *MURF1* and *ATROGIN-1* mRNA expression levels are increased suggesting a role for the ubiquitin–proteasomal pathway in regulating postexercise proteolysis.[119] However, in addition, following a bout of ultra-endurance running mRNA expression levels of key autophagy genes (*ATG4B, ATG12, Cathepsin L, GABARAPL1, LC3B, BNIP3 and BNIP31*) were significantly upregulated,[120] alongside increased activity of the proteasome B2 subunit and increased MURF1 protein levels,[121] suggesting a combined roll of both systems in the regulation of proteolysis following EE. However, the type of endurance exercise performed in these latter two studies is not typical of most EE regimes. The high energy demand and metabolic stress placed on the individual during this type of exercise likely contributes to considerable upregulation in autophagy-related degradation (perhaps through an AMPK-dependent mechanism). For example, when EE was performed by groups of rats which were either in a fasted (e.g., energy deprived state) or in a fed state, they showed vastly different levels of activation of autophagy, with greater levels observed in the fasted rather than fed states.[122] As with RE, expression profiles after EE follow a temporal pattern peaking again between 1 and 2 h, but being maintained for up to 24 h postexercise unlike RE.[117] Furthermore, responses to training do not seem to reflect that which is observed with RE-T either, with EE-T eliciting an upregulation of ubiquitin–proteasomal pathways to a greater extent after 10 weeks than RE-T.[118] Again, this highlights the discordance in the regulation of protein metabolism between these two different types of exercise.

4. CONCLUSIONS

Using tools such as isotopic tracers in combination with modern molecular biology techniques has led to an explosion of studies, and hence, a greater understanding into the control of exercise and protein metabolism. From what has been discussed in this chapter, findings from a multitude of studies indicate exercise specificity in regulating protein metabolism, in-keeping with the "expected" adaptations required for the exercise type being undertaken. That is, muscle growth associated with RE-T requires increases in myofibrillar protein synthesis and associated myofibrillar deposition, while the prevailing mitochondrial biogenesis associated with EE-T requires preferential stimulation of mitochondrial protein synthesis for associated mitochondrial deposition. Nonetheless, there still remains a great deal more to be done before we fully understand how exercise-specific adaptation, heterogeneity in adaptive potential, and signal integration can be understood.

REFERENCES

1. Smith GI, Villareal DT, Mittendorfer B. Measurement of human mixed muscle protein fractional synthesis rate depends on the choice of amino acid tracer. *Am J Physiol Endocrinol Metab*. 2011;293(3):E666–E671.
2. Phillips SM, Tipton KD, Aarsland A, Wolf SE, Wolfe RR. Mixed muscle protein synthesis and breakdown after resistance exercise in humans. *Am J Physiol*. 1997;273(1 pt 1):E99–E107.
3. Atherton PJ, Smith K. Muscle protein synthesis in response to nutrition and exercise. *J Physiol*. 2012;590(pt 5):1049–1057.
4. Rennie MJ, Edwards RH, Halliday D, Matthews DE, Wolman SL, Millward DJ. Muscle protein synthesis measured by stable isotope techniques in man: the effects of feeding and fasting. *Clin Sci (Lond)*. 1982;63(6):519–523.
5. Kumar V, Atherton P, Smith K, Rennie MJ. Human muscle protein synthesis and breakdown during and after exercise. *J Appl Physiol*. 2009;106(6):2026–2039.
6. Miller BF, Olesen JL, Hansen M, et al. Coordinated collagen and muscle protein synthesis in human patella tendon and quadriceps muscle after exercise. *J Physiol*. 2005;567(pt 3):1021–1033.
7. Cuthbertson DJ, Babraj J, Smith K, et al. Anabolic signaling and protein synthesis in human skeletal muscle after dynamic shortening or lengthening exercise. *Am J Physiol Endocrinol Metab*. 2006;290(4):E731–E738.
8. Smith K, Rennie MJ. The measurement of tissue protein turnover. *Baillieres Clin Endocrinol Metab*. 1996;10(4):469–495.
9. Smith K, Reynolds N, Downie S, Patel A, Rennie MJ. Effects of flooding amino acids on incorporation of labeled amino acids into human muscle protein. *Am J Physiol*. 1998;275(1 pt 1):E73–E78.
10. Toffolo G, Albright R, Joyner M, Dietz N, Cobelli C, Nair KS. Model to assess muscle protein turnover: domain of validity using amino acyl-tRNA vs. surrogate measures of precursor pool. *Am J Physiol Endocrinol Metab*. 2003;285(5):E1142–E1149.
11. Wilkinson DJ, Franchi MV, Brook MS, et al. A validation of the application of D(2)O stable isotope tracer techniques for monitoring day-to-day changes in muscle protein subfraction synthesis in humans. *Am J Physiol Endocrinol Metab*. 2014;306(5): E571–E579.
12. Wolfe RR, Chinkes DL. *Isotope Tracers in Metabolic Research: Principles and Practice of Kinetic Analysis*. 2nd ed. Hoboken, NJ: John Wiley & Sons, Inc.; 2005.
13. Picou D, Taylor-Roberts T. The measurement of total protein synthesis and catabolism and nitrogen turnover in infants in different nutritional states and receiving different amounts of dietary protein. *Clin Sci*. 1969;36(2):283–296.
14. Biolo G, Chinkes D, Zhang XJ, Wolfe RR. Harry M. Vars Research Award. A new model to determine in vivo the relationship between amino acid transmembrane transport and protein kinetics in muscle. *JPEN J Parenter Enteral Nutr*. 1992;16(4):305–315.

15. Biolo G, Fleming RY, Maggi SP, Wolfe RR. Transmembrane transport and intracellular kinetics of amino acids in human skeletal muscle. *Am J Physiol.* 1995;268(1 pt 1): E75–E84.
16. Biolo G, Gastaldelli A, Zhang XJ, Wolfe RR. Protein synthesis and breakdown in skin and muscle: a leg model of amino acid kinetics. *Am J Physiol.* 1994;267(3 pt 1): E467–E474.
17. Barrett EJ, Revkin JH, Young LH, Zaret BL, Jacob R, Gelfand RA. An isotopic method for measurement of muscle protein synthesis and degradation in vivo. *Biochem J.* 1987;245(1):223–228.
18. Katsanos CS, Chinkes DL, Sheffield-Moore M, Aarsland A, Kobayashi H, Wolfe RR. Method for the determination of the arteriovenous muscle protein balance during non-steady-state blood and muscle amino acid concentrations. *Am J Physiol Endocrinol Metab.* 2005;289(6):E1064–E1070.
19. Rennie MJ, Smith K, Watt PW. Measurement of human tissue protein synthesis: an optimal approach. *Am J Physiol.* 1994;266(3 pt 1):E298–E307.
20. Zhang X-J, Chinkes DL, Wolfe RR. Measurement of muscle protein fractional synthesis and breakdown rates from a pulse tracer injection. *Am J Physiol Endocrinol Metab.* 2002;283(4):E753–E764.
21. Zhang XJ, Chinkes DL, Sakurai Y, Wolfe RR. An isotopic method for measurement of muscle protein fractional breakdown rate in vivo. *Am J Physiol.* 1996;270(5 pt 1): E759–E767.
22. Urey HC, Brickwedde F, Murphy G. A hydrogen isotope of mass 2. *Phys Rev.* 1932;39:164.
23. Schoenheimer R, Rittenberg D. Deuterium as an indicator in the study of intermediary metabolism. *Science.* 1935;82(2120):156–157.
24. Rittenberg D, Schoenheimer R. Deuterium as an indicator in the study of intermediary metabolism: XI. Further studies on the biological uptake of deuterium into organic substances, with special reference to fat and cholesterol formation. *J Biol Chem.* 1937;121:235–253.
25. Schoenheimer R, Rittenberg D. The application of isotopes to the study of intermediary metabolism. *Science.* 1938;87(2254):221–226.
26. MacDonald AJ, Small AC, Greig CA, et al. A novel oral tracer procedure for measurement of habitual myofibrillar protein synthesis. *Rapid Commun Mass Spectrom.* 2013;27(15):1769–1777.
27. Robinson MM, Turner SM, Hellerstein MK, Hamilton KL, Miller BF. Long-term synthesis rates of skeletal muscle DNA and protein are higher during aerobic training in older humans than in sedentary young subjects but are not altered by protein supplementation. *FASEB J.* 2011;25(9):3240–3249.
28. Dufner DA, Bederman IR, Brunengraber DZ, et al. Using ^2H_2O to study the influence of feeding on protein synthesis: effect of isotope equilibration in vivo vs. in cell culture. *Am J Physiol Endocrinol Metab.* 2005;288(6):E1277–E1283.
29. Wang D, Liem DA, Lau E, et al. Characterization of human plasma proteome dynamics using deuterium oxide. *Proteomics Clin Appl.* 2014;8(7–8):610–619.
30. Miller BF, Wolff CA, Peelor FF, Shipman PD, Hamilton KL. Modeling the contribution of individual proteins to mixed skeletal muscle protein synthetic rates over increasing periods of label incorporation. *J Appl Physiol.* 2015;118:655–661. http://dx.doi.org/10.1152/japplphysiol.00987.2014.
31. Busch R, Kim Y-K, Neese RA, et al. Measurement of protein turnover rates by heavy water labeling of nonessential amino acids. *Biochim Biophys Acta.* 2006;1760(5): 730–744.
32. Gasier HG, Fluckey JD, Previs SF, Wiggs MP, Riechman SE. Acute resistance exercise augments integrative myofibrillar protein synthesis. *Metabolism.* 2012;61(2):153–156.

33. Previs SF, Fatica R, Chandramouli V, Alexander JC, Brunengraber H, Landau BR. Quantifying rates of protein synthesis in humans by use of ^2H_2O: application to patients with end-stage renal disease. *Am J Physiol Endocrinol Metab.* 2004;286(4):E665–E672.
34. Jeukendrup A. Modulation of carbohydrate and fat utilization by diet, exercise and environment. *Biochem Soc Trans.* 2003;31(pt 6):1270–1273.
35. Wagenmakers AJ. Muscle amino acid metabolism at rest and during exercise: role in human physiology and metabolism. *Exerc Sport Sci Rev.* 1998;26:287–314.
36. Grattagliano I, Lauterburg BH, Palasciano G, Portincasa P. 13C-breath tests for clinical investigation of liver mitochondrial function. *Eur J Clin Invest.* 2010;40(9):843–850.
37. Hider RC, Meade L. The conversion of (1- 14 C)pyruvate into (14 C)alanine in rat skeletal muscle: its relevance to the effect of insulin on protein synthesis. *Biochem J.* 1972;128(1):165–167.
38. Wagenmakers AJM, Brookes JH, Coakley JH, Reilly T, Edwards RHT. Exercise-induced activation of the branched-chain 2-oxo acid dehydrogenase in human muscle. *Eur J Appl Physiol Occup Physiol.* 1989;59(3):159–167.
39. Millward DJ, Davies CT, Halliday D, Wolman SL, Matthews D, Rennie M. Effect of exercise on protein metabolism in humans as explored with stable isotopes. Available at, *Fed Proc.* 1982;41(10):2686–2691. Accessed February 25, 2015, http://www.ncbi.nlm.nih.gov/pubmed/6809497.
40. Ahlborg G, Felig P, Hagenfeldt L, Hendler R, Wahren J. Substrate turnover during prolonged exercise in man. Splanchnic and leg metabolism of glucose, free fatty acids, and amino acids. *J Clin Invest.* 1974;53(4):1080–1090.
41. Bowtell JL, Leese GP, Smith K, et al. Effect of oral glucose on leucine turnover in human subjects at rest and during exercise at two levels of dietary protein. *J Physiol.* 2000;525(1):271–281.
42. Gaine PC, Viesselman CT, Pikosky MA, et al. Aerobic exercise training decreases leucine oxidation at rest in healthy adults. *J Nutr.* 2005;135(5):1088–1092.
43. Rennie MJ, Bohé J, Smith K, Wackerhage H, Greenhaff P. Branched-chain amino acids as fuels and anabolic signals in human muscle. *J Nutr.* 2006;136(suppl 1):264S–268S.
44. Walker DK, Dickinson JM, Timmerman KL, et al. Exercise, amino acids, and aging in the control of human muscle protein synthesis. *Med Sci Sports Exerc.* 2011;43(12):2249–2258. http://dx.doi.org/10.1249/MSS.0b013e318223b037.
45. Ahtiainen JP, Pakarinen A, Alen M, Kraemer WJ, Häkkinen K. Muscle hypertrophy, hormonal adaptations and strength development during strength training in strength-trained and untrained men. *Eur J Appl Physiol.* 2003;89(6):555–563.
46. Hulmi JJ, Tannerstedt J, Selänne H, Kainulainen H, Kovanen V, Mero AA. Resistance exercise with whey protein ingestion affects mTOR signaling pathway and myostatin in men. *J Appl Physiol.* 2009;106(5):1720–1729.
47. Seynnes OR, de Boer M, Narici MV. Early skeletal muscle hypertrophy and architectural changes in response to high-intensity resistance training. *J Appl Physiol.* 2007;102(1):368–373.
48. Blazevich AJ, Cannavan D, Coleman DR, Horne S. Influence of concentric and eccentric resistance training on architectural adaptation in human quadriceps muscles. *J Appl Physiol.* 2007;103(5):1565–1575.
49. Kumar V, Selby A, Rankin D, et al. Age-related differences in the dose-response relationship of muscle protein synthesis to resistance exercise in young and old men. *J Physiol.* 2009;587(pt 1):211–217.
50. Burd NA, West DWD, Staples AW, et al. Low-load high volume resistance exercise stimulates muscle protein synthesis more than high-load low volume resistance exercise in young men. *PLoS One.* 2010;5(8):e12033.
51. Mitchell CJ, Churchward-Venne TA, West DWD, et al. Resistance exercise load does not determine training-mediated hypertrophic gains in young men. *J Appl Physiol.* 2012;113(1):71–77.

52. Tipton KD, Ferrando AA, Phillips SM, Doyle D, Wolfe RR. Postexercise net protein synthesis in human muscle from orally administered amino acids. *Am J Physiol.* 1999;276(4 pt 1):E628–E634.
53. Rose AJ, Alsted TJ, Jensen TE, et al. A Ca(2+)-calmodulin-eEF2K-eEF2 signalling cascade, but not AMPK, contributes to the suppression of skeletal muscle protein synthesis during contractions. *J Physiol.* 2009;587(pt 7):1547–1563.
54. Keller P, Vollaard N, Babraj J, Ball D, Sewell DA, Timmons JA. Using systems biology to define the essential biological networks responsible for adaptation to endurance exercise training. *Biochem Soc Trans.* 2007;35(pt 5):1306–1309.
55. Phillips SM, Tipton KD, Ferrando AA, Wolfe RR. Resistance training reduces the acute exercise-induced increase in muscle protein turnover. *Am J Physiol.* 1999;276(1 pt 1):E118–E124.
56. Phillips SM, Parise G, Roy BD, Tipton KD, Wolfe RR, Tamopolsky MA. Resistance-training-induced adaptations in skeletal muscle protein turnover in the fed state. *Can J Physiol Pharmacol.* 2002;80(11):1045–1053.
57. Kim PL, Staron RS, Phillips SM. Fasted-state skeletal muscle protein synthesis after resistance exercise is altered with training. *J Physiol.* 2005;568(pt 1):283–290.
58. Tang JE, Perco JG, Moore DR, Wilkinson SB, Phillips SM. Resistance training alters the response of fed state mixed muscle protein synthesis in young men. *Am J Physiol Regul Integr Comp Physiol.* 2008;294(1):R172–R178.
59. Tesch PA, Ekberg A, Lindquist DM, Trieschmann JT. Muscle hypertrophy following 5-week resistance training using a non-gravity-dependent exercise system. *Acta Physiol Scand.* 2004;180(1):89–98.
60. DeFreitas JM, Beck TW, Stock MS, Dillon MA, Kasishke PR. An examination of the time course of training-induced skeletal muscle hypertrophy. *Eur J Appl Physiol.* 2011;111(11):2785–2790.
61. Mitchell CJ, Churchward-Venne TA, Parise G, et al. Acute post-exercise myofibrillar protein synthesis is not correlated with resistance training-induced muscle hypertrophy in young men. *PLoS One.* 2014;9(2):e89431.
62. Murton AJ, Billeter R, Stephens FB, et al. Transient transcriptional events in human skeletal muscle at the outset of concentric resistance exercise training. *J Appl Physiol.* 2014;116(1):113–125.
63. Biolo G, Maggi SP, Williams BD, Tipton KD, Wolfe RR. Increased rates of muscle protein turnover and amino acid transport after resistance exercise in humans. *Am J Physiol.* 1995;268(3 pt 1):E514–E520.
64. Wilkes EA, Selby AL, Atherton PJ, et al. Blunting of insulin inhibition of proteolysis in legs of older subjects may contribute to age-related sarcopenia. *Am J Clin Nutr.* 2009;90(5):1343–1350.
65. Greenhaff PL, Karagounis LG, Peirce N, et al. Disassociation between the effects of amino acids and insulin on signaling, ubiquitin ligases, and protein turnover in human muscle. *Am J Physiol Endocrinol Metab.* 2008;295(3):E595–E604.
66. Biolo G, Tipton KD, Klein S, Wolfe RR. An abundant supply of amino acids enhances the metabolic effect of exercise on muscle protein. *Am J Physiol.* 1997;273(1 pt 1): E122–E129.
67. Rasmussen BB, Tipton KD, Miller SL, Wolf SE, Wolfe RR. An oral essential amino acid-carbohydrate supplement enhances muscle protein anabolism after resistance exercise. *J Appl Physiol.* 2000;88(2):386–392.
68. Borsheim E, Tipton KD, Wolf SE, Wolfe RR. Essential amino acids and muscle protein recovery from resistance exercise. *Am J Physiol Endocrinol Metab.* 2002;283(4): E648–E657.
69. Glynn EL, Fry CS, Drummond MJ, et al. Muscle protein breakdown has a minor role in the protein anabolic response to essential amino acid and carbohydrate intake following resistance exercise. *Am J Physiol Regul Integr Comp Physiol.* 2010;299(2):R533–R540.

70. Holloszy JO. Biochemical adaptations in muscle. Effects of exercise on mitochondrial oxygen uptake and respiratory enzyme activity in skeletal muscle. *J Biol Chem.* 1967;242(9):2278–2282.
71. Holloszy JO, Coyle EF. Adaptations of skeletal muscle to endurance exercise and their metabolic consequences. *J Appl Physiol.* 1984;56(4):831–838.
72. Jones AM, Carter H. Oxygen uptake-work rate relationship during two consecutive ramp exercise tests. *Int J Sports Med.* 2004;25(6):415–420.
73. Carraro F, Stuart CA, Hartl WH, Rosenblatt J, Wolfe RR. Effect of exercise and recovery on muscle protein synthesis in human subjects. *Am J Physiol Endocrinol Metab.* 1990;259(4):E470–E476.
74. Harber MP, Crane JD, Dickinson JM, et al. Protein synthesis and the expression of growth-related genes are altered by running in human vastus lateralis and soleus muscles. *Am J Physiol Regul Integr Comp Physiol.* 2009;296(3):R708–R714.
75. Harber MP, Konopka AR, Jemiolo B, Trappe SW, Trappe TA, Reidy PT. Muscle protein synthesis and gene expression during recovery from aerobic exercise in the fasted and fed states. *Am J Physiol Regul Integr Comp Physiol.* 2010;299(5): R1254–R1262.
76. Sheffield-Moore M, Yeckel CW, Volpi E, et al. Postexercise protein metabolism in older and younger men following moderate-intensity aerobic exercise. *Am J Physiol Endocrinol Metab.* 2004;287(3):E513–E522.
77. Tipton KD, Ferrando AA, Williams BD, Wolfe RR. Muscle protein metabolism in female swimmers after a combination of resistance and endurance exercise. *J Appl Physiol.* 1996;81(5):2034–2038.
78. Mascher H, Ekblom B, Rooyackers O, Blomstrand E. Enhanced rates of muscle protein synthesis and elevated mTOR signalling following endurance exercise in human subjects. *Acta Physiol (Oxf).* 2011;202:175–184.
79. Di Donato DM, West DWD, Churchward-Venne TA, Breen L, Baker SK, Phillips SM. Influence of aerobic exercise intensity on myofibrillar and mitochondrial protein synthesis in young men during early and late postexercise recovery. *Am J Physiol Endocrinol Metab.* 2014;306(9):E1025–E1032.
80. Wilkinson SB, Phillips SM, Atherton PJ, et al. Differential effects of resistance and endurance exercise in the fed state on signalling molecule phosphorylation and protein synthesis in human muscle. *J Physiol.* 2008;586(pt 15):3701–3717.
81. Donges CE, Burd NA, Duffield R, et al. Concurrent resistance and aerobic exercise stimulates both myofibrillar and mitochondrial protein synthesis in sedentary middle-aged men. *J Appl Physiol.* 2012;112(12):1992–2001.
82. Andersen G, Ørngreen MC, Preisler N, et al. Protein-carbohydrate supplements improve muscle protein balance in muscular dystrophy patients after endurance exercise: a placebo-controlled crossover study. *Am J Physiol Regul Integr Comp Physiol.* 2015;308(2):R123–R130.
83. Pasiakos SM, Carbone JW. Assessment of skeletal muscle proteolysis and the regulatory response to nutrition and exercise. *IUBMB Life.* 2014;66(7):478–484.
84. Goodman CA, Frey JW, Mabrey DM, et al. The role of skeletal muscle mTOR in the regulation of mechanical load-induced growth. *J Physiol.* 2011;589(pt 22):5485–5501.
85. Drummond MJ, Fry CS, Glynn EL, et al. Rapamycin administration in humans blocks the contraction-induced increase in skeletal muscle protein synthesis. *J Physiol.* 2009;587(pt 7):1535–1546.
86. Proud CG. mTORC1 signalling and mRNA translation. *Biochem Soc Trans.* 2009;37(pt 1):227–231.
87. Hansen S, Kvorning T, Kjaer M, Sjøgaard G. The effect of short-term strength training on human skeletal muscle: the importance of physiologically elevated hormone levels. *Scand J Med Sci Sports.* 2001;11(6):347–354.

88. Bamman MM, Shipp JR, Jiang J, et al. Mechanical load increases muscle IGF-I and androgen receptor mRNA concentrations in humans. *Am J Physiol Endocrinol Metab.* 2001;280(3):E383–E390.
89. Bodine SC, Stitt TN, Gonzalez M, et al. Akt/mTOR pathway is a crucial regulator of skeletal muscle hypertrophy and can prevent muscle atrophy in vivo. *Nat Cell Biol.* 2001;3(11):1014–1019.
90. Rommel C, Bodine SC, Clarke BA, et al. Mediation of IGF-1-induced skeletal myotube hypertrophy by PI(3)K/Akt/mTOR and PI(3)K/Akt/GSK3 pathways. *Nat Cell Biol.* 2001;3(11):1009–1013.
91. West DWD, Kujbida GW, Moore DR, et al. Resistance exercise-induced increases in putative anabolic hormones do not enhance muscle protein synthesis or intracellular signalling in young men. *J Physiol.* 2009;587(pt 21):5239–5247.
92. West DWD, Burd NA, Tang JE, et al. Elevations in ostensibly anabolic hormones with resistance exercise enhance neither training-induced muscle hypertrophy nor strength of the elbow flexors. *J Appl Physiol.* 2010;108(1):60–67.
93. Spangenburg EE, Le Roith D, Ward CW, Bodine SC. A functional insulin-like growth factor receptor is not necessary for load-induced skeletal muscle hypertrophy. *J Physiol.* 2008;586(1):283–291.
94. Doessing S, Heinemeier KM, Holm L, et al. Growth hormone stimulates the collagen synthesis in human tendon and skeletal muscle without affecting myofibrillar protein synthesis. *J Physiol.* 2010;588(pt 2):341–351.
95. Kvorning T, Andersen M, Brixen K, Madsen K. Suppression of endogenous testosterone production attenuates the response to strength training: a randomized, placebo-controlled, and blinded intervention study. *Am J Physiol Endocrinol Metab.* 2006;291(6):E1325–E1332.
96. O'Neil TK, Duffy LR, Frey JW, Hornberger TA. The role of phosphoinositide 3-kinase and phosphatidic acid in the regulation of mammalian target of rapamycin following eccentric contractions. *J Physiol.* 2009;587(pt 14):3691–3701.
97. Klossner S, Durieux A-C, Freyssenet D, Flueck M. Mechano-transduction to muscle protein synthesis is modulated by FAK. *Eur J Appl Physiol.* 2009;106(3):389–398.
98. Schlaepfer DD, Hauck CR, Sieg DJ. Signaling through focal adhesion kinase. *Prog Biophys Mol Biol.* 1999;71(3–4):435–478.
99. Timmons JA, Larsson O, Jansson E, et al. Human muscle gene expression responses to endurance training provide a novel perspective on Duchenne muscular dystrophy. *FASEB J.* 2005;19(7):750–760.
100. Cary LA, Guan JL. Focal adhesion kinase in integrin-mediated signaling. *Front Biosci.* 1999;4:D102–D113.
101. Calalb MB, Polte TR, Hanks SK. Tyrosine phosphorylation of focal adhesion kinase at sites in the catalytic domain regulates kinase activity: a role for Src family kinases. *Mol Cell Biol.* 1995;15(2):954–963.
102. Durieux AC, Desplanches D, Freyssenet D, Fluck M. Mechanotransduction in striated muscle via focal adhesion kinase. *Biochem Soc Trans.* 2007;35(pt 5):1312–1313.
103. Goodman CA, Mabrey DM, Frey JW, et al. Novel insights into the regulation of skeletal muscle protein synthesis as revealed by a new nonradioactive in vivo technique. *FASEB J.* 2011;25(3):1028–1039.
104. De Boer MD, Selby A, Atherton P, et al. The temporal responses of protein synthesis, gene expression and cell signalling in human quadriceps muscle and patellar tendon to disuse. *J Physiol.* 2007;585(pt 1):241–251.
105. Flück M, Carson JA, Gordon SE, Ziemiecki A, Booth FW. Focal adhesion proteins FAK and paxillin increase in hypertrophied skeletal muscle. *Am J Physiol.* 1999;277(1 pt 1):C152–C162.

106. Gordon SE, Flück M, Booth FW. Selected contribution: skeletal muscle focal adhesion kinase, paxillin, and serum response factor are loading dependent. *J Appl Physiol* 2001;90(3):1174–1183. discussion 1165.
107. Coffey VG, Hawley JA. The molecular bases of training adaptation. *Sports Med.* 2007;37(9):737–763.
108. Atherton PJ, Etheridge T, Watt PW, et al. Muscle full effect after oral protein: time-dependent concordance and discordance between human muscle protein synthesis and mTORC1 signaling. *Am J Clin Nutr.* 2010;92(5):1080–1088.
109. Phillips BE, Williams JP, Gustafsson T, et al. Molecular networks of human muscle adaptation to exercise and age. *PLoS Genet.* 2013;9(3):e1003389.
110. Murton AJ, Greenhaff PL. Resistance exercise and the mechanisms of muscle mass regulation in humans: acute effects on muscle protein turnover and the gaps in our understanding of chronic resistance exercise training adaptation. *Int J Biochem Cell Biol.* 2013;45(10):2209–2214.
111. Murton AJ, Constantin D, Greenhaff PL. The involvement of the ubiquitin proteasome system in human skeletal muscle remodelling and atrophy. *Biochim Biophys Acta.* 2008;1782(12):730–743.
112. Clague MJ, Urbé S. Ubiquitin: same molecule, different degradation pathways. *Cell.* 2010;143(5):682–685.
113. Passmore LA, Barford D. Getting into position: the catalytic mechanisms of protein ubiquitylation. *Biochem J.* 2004;379(pt 3):513–525.
114. Hoeller D, Hecker C-M, Wagner S, Rogov V, Dötsch V, Dikic I. E3-independent monoubiquitination of ubiquitin-binding proteins. *Mol Cell.* 2007;26(6):891–898.
115. Deldicque L, Atherton P, Patel R, et al. Effects of resistance exercise with and without creatine supplementation on gene expression and cell signaling in human skeletal muscle. *J Appl Physiol.* 2008;104(2):371–378.
116. Nedergaard A, Vissing K, Overgaard K, Kjaer M, Schjerling P. Expression patterns of atrogenic and ubiquitin proteasome component genes with exercise: effect of different loading patterns and repeated exercise bouts. *J Appl Physiol.* 2007;103(5):1513–1522.
117. Louis E, Raue U, Yang Y, Jemiolo B, Trappe S. Time course of proteolytic, cytokine, and myostatin gene expression after acute exercise in human skeletal muscle. *J Appl Physiol.* 2007;103(5):1744–1751.
118. Stefanetti RJ, Lamon S, Wallace M, Vendelbo MH, Russell AP, Vissing K. Regulation of ubiquitin proteasome pathway molecular markers in response to endurance and resistance exercise and training. *Pflugers Arch.* 2015;467:1523–1537. http://dx.doi.org/10.1007/s00424-014-1587-y.
119. Pasiakos SM, McClung HL, McClung JP, et al. Molecular responses to moderate endurance exercise in skeletal muscle. *Int J Sport Nutr Exerc Metab.* 2010;20(4):282–290.
120. Jamart C, Benoit N, Raymackers J-M, Kim HJ, Kim CK, Francaux M. Autophagy-related and autophagy-regulatory genes are induced in human muscle after ultraendurance exercise. *Eur J Appl Physiol.* 2012;112(8):3173–3177.
121. Jamart C, Francaux M, Millet GY, Deldicque L, Frère D, Féasson L. Modulation of autophagy and ubiquitin-proteasome pathways during ultra-endurance running. *J Appl Physiol.* 2012;112(9):1529–1537.
122. Jamart C, Naslain D, Gilson H, Francaux M. Higher activation of autophagy in skeletal muscle of mice during endurance exercise in the fasted state. *Am J Physiol Endocrinol Metab.* 2013;305(8):E964–E974.

CHAPTER FIVE

Exercise and the Regulation of Mitochondrial Turnover

David A. Hood[1], Liam D. Tryon, Anna Vainshtein, Jonathan Memme, Chris Chen, Marion Pauly, Matthew J. Crilly, Heather Carter
Muscle Health Research Centre, School of Kinesiology and Health Science, York University, Toronto, Ontario, Canada
[1]Corresponding author: e-mail address: dhood@yorku.ca

Contents

Abstract

Exercise is a well-known stimulus for the expansion of the mitochondrial pool within skeletal muscle. Mitochondria have a remarkable ability to remodel their networks and can respond to an array of signaling stimuli following contractile activity to adapt to the metabolic demands of the tissue, synthesizing proteins to expand the mitochondrial reticulum. In addition, when they become dysfunctional, these organelles can be recycled by a specialized intracellular system. The signals regulating this mitochondrial life cycle of synthesis and degradation during exercise are still an area of great research interest. As mitochondrial turnover has valuable consequences in physical performance, in addition to metabolic health, disease, and aging, consideration of the signals which

Progress in Molecular Biology and Translational Science, Volume 135
ISSN 1877-1173
http://dx.doi.org/10.1016/bs.pmbts.2015.07.007

control this cycle is vital. This review focuses on the regulation of mitochondrial turnover in skeletal muscle and summarizes our current understanding of the impact that exercise has in modulating this process.

1. INTRODUCTION

Exercise is a powerful metabolic stress. When performed repeatedly over the course of several weeks, exercise leads to adaptations in skeletal muscle which allow it to meet the increased metabolic demand. One of the most dramatic examples of this phenotypic adaptation is the increase in mitochondrial content within muscle, which is coincident with a reduction in fatigability and an improvement in endurance performance. In contrast to the positive effects that exercise brings about to increase mitochondria, chronic muscle disuse results in the opposite change. Chronic physical inactivity leads to decrements in organelle content and function within muscle, poor performance, and an increase in apoptotic susceptibility. The metabolic derangements associated with inactivity and the loss of mitochondria include a greater storage, rather than oxidation of lipids, and a tendency for obesity and insulin resistance. Aging also brings about phenotypic changes in mitochondria within muscle, attributed to both inactivity and inherent, aging-induced changes in organelle synthesis and degradation pathways. Thus, opposing changes in the mitochondrial network within muscle cells induced by chronic exercise, aging, or disuse are now recognized to have implications for a broad range of health issues. Notably, regular exercise can counteract many of aging- and inactivity-induced detrimental effects observed on metabolism associated with defective mitochondria, and research efforts must continue to seek the molecular underpinnings of how this is brought about.

Our understanding of the structure, function, and turnover of mitochondria has increased tremendously over the last two decades, thanks to an emerging appreciation for the diversity of functions in which the organelle partakes. Long known for their vital roles in cellular energy production, mitochondria are now established participants in calcium (Ca^{2+}) handling, cellular signaling, and organelle-mediated apoptosis. In addition, mitochondrial dysfunction, brought about by either nuclear or mitochondrial DNA (mtDNA) mutations, can lead to a wide variety of pathophysiological conditions affecting a number of organ systems, notably skeletal muscle, the heart, or the brain. Thus, both basic and clinical scientists have become keenly interested in how mitochondrial function and dysfunction contribute to cellular health and disease.

2. OVERVIEW OF MITOCHONDRIAL TURNOVER

The steady-state mitochondrial content of muscle at any time, along with the quality of the organelle pool, is a product of complex pathways of synthesis (biogenesis) and degradation (mitophagy). The synthesis of a new, higher level of functional organelles in response to exercise is the cumulative result of a series of events that begins with the very first bout of exercise in a training program. As discussed in detail below, each acute exercise bout initiates a new cascade of signaling events involving the activation of protein enzymes that modify the activity of downstream proteins such as transcription factors (TFs) that impact the expression of *nu*clear *g*enes *e*ncoding *m*itochondrial *p*roteins (NuGEMPs). The result is an increase in the level of mRNAs which encode precursor proteins destined for mitochondrial localization. Once translated in the cytoplasm, the resulting nuclear-encoded proteins interact with protein chaperones and are delivered to the mitochondria. These mitochondrial-destined proteins are subsequently imported into the different compartments, such as the matrix space or the inner or outer membrane. A subgroup of these proteins include TFs, such as mitochondrial transcription factor A (Tfam), that act directly on mtDNA to increase the mRNA expression of mitochondrial gene products and mtDNA copy number. mtDNA is critical because it encodes 13 vital proteins involved in the mitochondrial respiratory chain. Thus, mitochondrial biogenesis requires the coordinated cooperation of both the nuclear and the mitochondrial genomes to produce an organelle that is functional in providing cellular adenosine triphosphate (ATP).

In contrast to the process of biogenesis, regulation of mitochondrial content and quality is also exerted at the level of mitophagy, the selective degradation of mitochondria by the autophagosome–lysosome system. As discussed below, mitophagy is activated by similar signaling mechanisms such as biogenesis, suggesting that a coordinated regulation of these pathways exists. Rates of mitophagy are enhanced during physiological conditions of increased reactive oxygen species (ROS) production, or when the mitochondrial membrane potential decreases. When this occurs, the affected dysfunctional portion of the organelle reticulum undergoes fission to separate it from the remaining "healthy" mitochondrial network, and the organelle fragment is targeted for degradation. Considerably less is known about the regulation of mitophagy, in comparison to biogenesis, particularly in skeletal muscle, and during exercise.

3. MITOCHONDRIAL MORPHOLOGY AND CHANGES WITH TRAINING

Skeletal muscle contains a heterogeneous pool of mitochondria, distinguished by biochemical composition, function, and subcellular location.[1] Mitochondria may be concentrated under the sarcolemma, termed the subsarcolemmal (SS) mitochondria, or localized in proximity to the contractile apparatus, regarded as the intermyofibrillar (IMF) mitochondria. However, the distinction among subpopulations is not absolute, and there is likely a degree of continuity between them,[2] as mitochondria form a dynamic reticulum,[3] the extensiveness of which is correlated to the oxidative capacity of the tissue.[4] This organelle network is continuously remodeled by fusion and fission events which add to or subtract from it, respectively. The ability of mitochondria to physically interact and alter their morphology enables quality control processes to take place,[5] serves to preserve mtDNA integrity, and influences their respiratory capacity and ROS production. Furthermore, as muscle fibers can be quite large in size, expansion of the mitochondrial reticulum favors the rapid delivery of oxygen to the mitochondrion for consumption and ATP production.

In mammalian cells, mitofusin-1 (Mfn1) and -2 (Mfn2), in addition to optic atrophy factor 1 (Opa1), are mitochondrial membrane proteins regulating the fusion process (Fig. 1). They are essential factors contributing to the coordinated fusion of the outer and inner membranes of the mitochondria, respectively. Fusion processes promote the generation of networks with continuous membranes and a matrix lumen which will facilitate the mobility of mitochondrial proteins, lipids, and mtDNA. Conversely, mitochondrial membrane fission is mediated by dynamin-related protein 1 (Drp1), in association with fission protein 1 (Fis1). Drp1 function requires recruitment to the mitochondria from the cytosol, where it forms an oligomeric structure triggering mitochondrial fission. These division events can be localized to specifically damaged areas of the reticulum, ensuring the selective removal of damaged regions of the network.[5]

Pioneering work by Gollnick and King provided the first suggestion that endurance training has the capability to alter mitochondrial ultrastructure.[6] Endurance training also increases the extensiveness of the mitochondrial

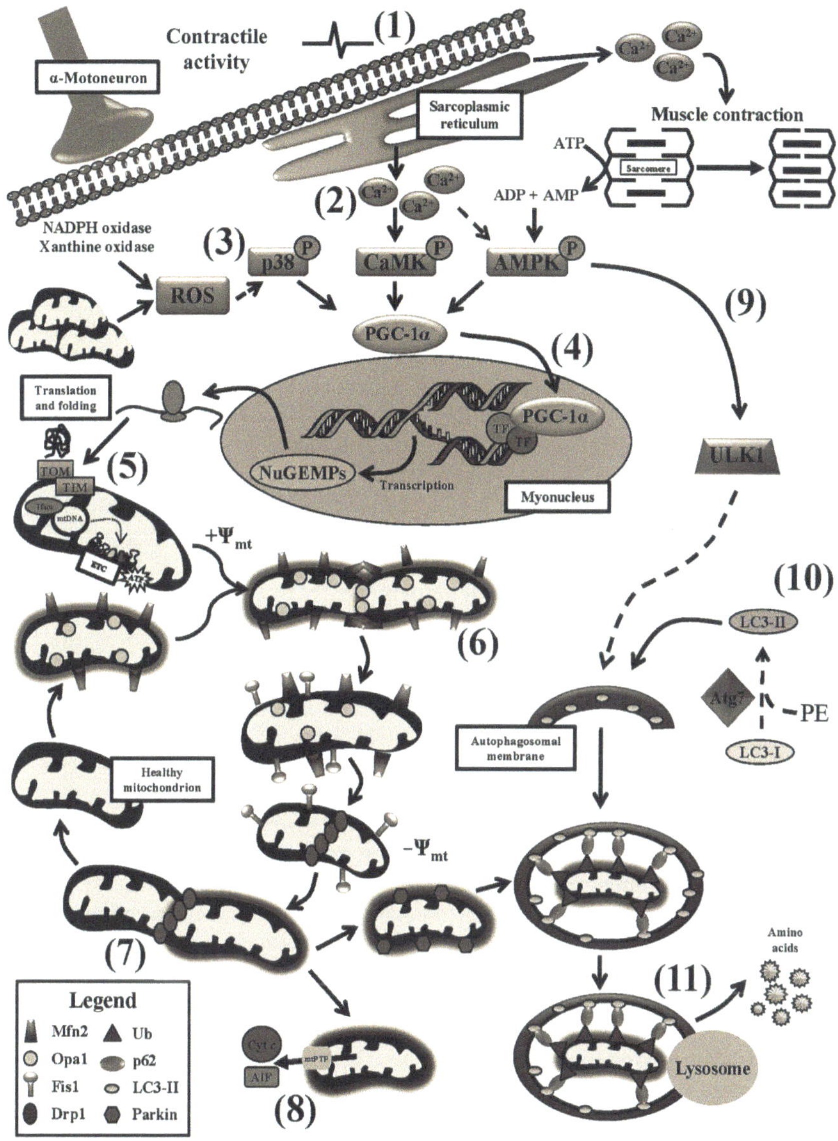

Figure 1 Effect of exercise and training on mitochondrial turnover. During muscle contraction, action potentials propagate down α-motoneurons which innervate muscle fibers (1). Electrical signals are transmitted along the sarcolemma of skeletal muscle and are coupled with the release of Ca^{2+} from the sarcoplasmic reticulum (2). Increases in intracellular Ca^{2+} levels allow for muscle contraction to occur, while also activating Ca^{2+}-sensitive signaling pathways. Contractile activity also consumes cellular ATP, causing a decrease in ATP/ADP ratio and an increase in the formation of AMP, which can activate AMP-activated protein kinase (AMPK). Reactive oxygen species (ROS)

(Continued)

reticulum.[7] Although it is well established that exercise increases mitochondrial content, the regulatory mechanisms which promote these exercise-induced adaptations have only recently been investigated. Recent work utilizing a model of chronic contractile activity showed that Opa1 and Mfn2 were significantly increased, with concomitant reductions in Drp1, providing a molecular basis for the observed changes in mitochondrial ultrastructure.[8] Additionally, the mRNA expression of mitochondrial fusion factors Mfn1 and Mfn2 also increases during the recovery period from a single

Figure 1—Cont'd production from the mitochondrial electron transport chain (ETC) and other intracellular sources are also enhanced during muscle activity, likely leading to the phosphorylation of kinases such as p38 MAPK (3). These signal transduction pathways target transcription factors (TFs), as well as the transcriptional coactivator peroxisome proliferator-activated receptor γ coactivator-1α (PGC-1α), which stimulates mitochondrial biogenesis (4), among other beneficial adaptations to skeletal muscle. PGC-1α is especially critical in augmenting the transcription of NuGEMPs, which need to be translated and imported into the mitochondrion. Additionally, other nuclear-encoded factors such as mitochondrial transcription factor A (Tfam) must be imported into the mitochondrion, where they can interact with mitochondrial DNA (mtDNA) to assist the expression of mtDNA-encoded subunits of the ETC (5). This coordinated expression of both nuclear and mitochondrial genomes is key to the expansion of the mitochondrial pool. Chronic muscle activity also increases the ratio of fusion-to-fission proteins, promoting the fusion of mitochondria to form a reticular network (6). Essential mammalian skeletal muscle fusion proteins involved in these processes include Mfn2 and the inner mitochondrial membrane protein Opa1, which are necessary for fusion of the outer and inner mitochondrial membranes, respectively. Under conditions of cellular stress, mitochondrial fission can occur, allowing for the isolation of dysfunctional components of the organelle (7). Mitochondrial fission is executed by the formation of a Drp1 oligomer and can occur due to a reduction in mitochondrial membrane potential (Ψ_{mt}) to a portion of the organelle. Mitochondrial fission precedes apoptosis, and exercise training can reduce the susceptibility of both skeletal and cardiac muscle to this process, likely by reducing release of proapoptotic factors, such as cytochrome *c* and apoptosis-inducing factor (AIF) from the mitochondrion (8). Mitochondrial fission is also coupled with mitophagy, the specific degradation, and recycling of dysfunctional mitochondria through the autophagosome–lysosomal system (9). This process is thought to occur through AMPK/ULK1 signaling, in cooperation with the activation of other mitochondrial-specific kinases and ubiquitin ligases, such as Parkin. The activation of this pathway is also crucial for the formation of the autophagosomal membrane, which occurs through the Atg7-mediated lipidation of LC3-I with phosphatidylethanolamine (PE), to form LC3-II (10). Specific interactions between ubiquitinated proteins (Ub) on the mitochondrion and autophagosomal membrane are facilitated by p62, ensuring the precise engulfment of the malfunctioning component of the mitochondrial network. The autophagosome subsequently fuses with a lysosome, and its cargo is degraded and recycled within the cell (11). (See the color plate.)

bout of aerobic exercise.[9] Thus, the development of a reticular network likely begins after a single bout of exercise. Mice that were subjected to an acute exercise bout exhibited an increase in membrane interactions between skeletal muscle mitochondria in both the SS and IMF subpopulations, prior to any increase in the expression of mitochondrial fusion proteins.[10] Taken together, these data suggest that signaling to instigate mitochondrial remodeling appears to be one of the earliest events which occur in response to acute exercise. Subsequently, chronic muscle use shifts the expression of fission and fusion machinery to favor enhanced fusion, setting the stage for an intracellular environment more capable of matching the increased metabolic demands of skeletal muscle during training.

4. EXERCISE-INDUCED SIGNALING: A ROLE FOR AMPK

Mitochondrial biogenesis which occurs in response to an exercise training program originates from a variety of intracellular signaling events generated by muscular contraction (Fig. 1). Among these signals is the increase in ATP turnover within muscle cells during exercise, which consequently increases the ADP/ATP ratio of cellular ADP to ATP,[11] and simultaneously the formation of AMP. The elevated levels of AMP allosterically activate the heterotrimeric metabolic energy sensor, AMP-activated protein kinase (AMPK). Activation of this enzyme has been mechanistically linked to an enhanced expression of factors favoring oxidative metabolism, including peroxisome proliferator-activated receptor-γ coactivator-1α (PGC-1α), uncoupling protein 3 (UCP3), cytochrome *c*, succinate dehydrogenase, and citrate synthase.[12–15] These effects appear to be mediated by AMPK regulation of PGC-1α at both the gene and protein levels. Genetically or pharmacologically enhancing AMPK activation increases both the protein and mRNA levels of PGC-1α,[16] as well as augments the binding of TFs to regions in the PGC-1α promoter.[15] Furthermore, AMPK activation increases the phosphorylation of PGC-1α protein, which appears vital to the induction of PGC-1α-regulated genes, as well as the self-regulatory activity of PGC-1α on its own promoter.[13]

Recent studies have also noted that activity of specific subunits of the AMPK heterotrimer is also required for the basal expression of mitochondrial markers. For example, knockout (KO) of AMPK β-subunit(s) results in reduced mitochondrial content and muscle function, as well as impaired endurance performance,[17] emphasizing the necessity of this protein for basal and exercise metabolism.

5. EXERCISE-INDUCED SIGNALING: A ROLE FOR CA^{2+}

An increase in the concentration of cytosolic Ca^{2+} in skeletal muscle occurs during contractile activity (Fig. 1). This elevation of cytosolic Ca^{2+} also occurs at a concentration capable of augmenting oxidative phosphorylation in skeletal muscle mitochondria, as well as activating a number of calcium-regulated enzymes. Increased levels of cytosolic Ca^{2+} have been linked with signaling to mitochondrial biogenesis by *in vitro* experiments utilizing pharmacological agents which mimic the effects of physiological Ca^{2+} release. These studies have implicated cytosolic Ca^{2+} in stimulating the expression of a number of mitochondrial genes,[18,19] effects which appear to be mediated, in part, through enhanced signaling by calcium/calmodulin-dependent protein kinase (CaMK), protein kinase C (PKC), AMPK, and mitogen-activated protein kinases (MAPKs).[18–20]

Contractile activity-induced signaling by these Ca^{2+}-regulated kinases appears to converge on the PGC-1α promoter,[20,21] stimulating an increase in PGC-1α mRNA and protein levels.[20] While the role of PGC-1α as a downstream effector of Ca^{2+} in the context of mitochondrial biogenesis has been described, recent work has also highlighted the idea that PGC-1α may also function upstream of Ca^{2+} signaling in skeletal muscle, as it controls the expression of variety of proteins which modulate Ca^{2+} handling.[22] This provides a link for earlier findings, which have described contractile activity as a potent stimulus for inducing alterations in the expression of Ca^{2+} handling proteins.[23]

6. EXERCISE-INDUCED SIGNALING: A ROLE FOR P38 MAPK

Contractile activity also increases the phosphorylation and activation of p38 MAPK in skeletal muscle. p38 activation is sensitive to changes in ROS, inflammatory cytokines, and insulin which all may be modulated with exercise (Fig. 1). Interestingly, thyroid hormone (T_3) treatment, a well-established model for mitochondrial biogenesis, has also been documented to enhance p38 phosphorylation in skeletal muscle.[24] This may suggest a common mechanism of p38 activation which is important for muscle mitochondrial biogenesis, since both exercise and T_3 are known to result in organelle synthesis.

Akimoto *et al.*[25] first demonstrated that an acute exercise bout was capable of increasing p38 phosphorylation, which was later followed by an increase in PGC-1α mRNA content. The increase in PGC-1α transcript levels was likely accomplished through increased gene transcription. Indeed, PGC-1α promoter activity was increased with overexpression of p38. Furthermore, both promoter activity and PGC-1α protein activity are attenuated during contractile activity if p38 is inhibited.[20] The observed increase in PGC-1α promoter activity may be achieved through the ability of p38 to phosphorylate the TFs, myocyte enhancer factor-2 (MEF2), and activating transcription factor-2 (ATF2), which both bind to promoter elements within the upstream regulatory region of the PGC-1α gene.[26] Therefore, these data suggest that p38 is important for the increase in PGC-1α gene transcription in skeletal muscle with exercise.

p38 has also been shown to directly phosphorylate PGC-1α protein on multiple residues.[27] This posttranslational modification of PGC-1α increases the stability of the protein and alleviates normal coactivator repression, resulting in enhanced PGC-1α activity when compared to its dephosphorylated form. Collectively, these observations place p38 as an important signaling kinase for exercise-induced mitochondrial biogenesis through the activation of PGC-1α at multiple molecular levels.

7. EXERCISE-INDUCED SIGNALING: ACTIVATION OF PGC-1α

Contractile activity-induced signaling from the aforementioned cellular changes acts on a variety of TFs linked with mitochondrial biogenesis, such as nuclear respiratory factor-1/2 (NRF-1/2), early growth response gene-1 (Egr-1), specificity protein-1 (Sp1), c-fos, c-myc, cAMP response element binding protein (CREB), and upstream stimulating factor (USF1). The expression of these factors is upregulated and they translocate to myonuclei to stimulate transcription of mitochondrial genes.[28–31]

These signals are linked to the transcription of genes associated with mitochondrial content and oxidative phosphorylation by a member of a family of transcriptional coactivators, PGC-1α (Fig. 1). In skeletal muscle, PGC-1α is regarded as the most significant regulator of mitochondrial biogenesis and function.[32,33] As it is a transcriptional coactivator, it lacks the capacity to bind to nuclear DNA directly. Instead, it enhances the activity of TFs, such as NRF-1/2 and others,[32,34] by the recruitment of histone-modifying factors and direct interactions with the transcription initiation machinery.[35] The activation of these TFs is coupled with the transcriptional control of other genes involved in mitochondrial function and biogenesis. These include subunits of protein complexes in the electron transport chain (ETC) and factors involved in their assembly, mtDNA transcription and replication machinery, and mitochondrial protein import.[36] PGC-1α also coactivates NRF-1 on the Tfam promoter,[32] a required factor for the expression of the mtDNA, emphasizing its role in the coordinated expression of mitochondrial genes from both nuclear and mitochondrial genomes. It is not surprising then that the expression of PGC-1α is tissue dependent, with the highest levels of this protein existing in the most oxidative tissues, namely cardiac and skeletal muscle.[37] Furthermore, PGC-1α expression can be induced in muscle by contractile activity both *in vitro* and *in vivo*,[31,37] an effect which is preceded by PGC-1α translocation to the nucleus.[30] Taken together, it is clear that PGC-1α sits at the crux of the control of mitochondrial content and function.

Skeletal muscle-specific overexpression of this protein is sufficient to enhance basal mitochondrial content and endurance performance.[38,39] Similarly, whole-body PGC-1α KO mice display compromised basal mitochondrial function and content,[40] while skeletal muscle-specific ablation of PGC-1α is associated with impaired endurance performance, likely owing to abnormalities in mitochondrial structure and function, in addition to a reduced expression of metabolic genes,[41] highlighting the constitutive role of this coactivator in the maintenance of the size and quality of the mitochondrial pool. However, the absolute necessity of PGC-1α in exercise-induced adaptations is still up for debate. Changes in mitochondrial gene expression to an exercise training program are unaffected in a whole-body PGC-1α KO,[42] suggesting that this protein may not be crucial for exercise-induced adaptations to skeletal muscle, and that other factors may be at play in causing these alterations. Furthermore, knockdown of PGC-1α in myotubes revealed that some mitochondrial proteins require PGC-1α for contractile activity-induced organelle synthesis, while others do not.[43] In stark contrast to these findings, skeletal muscle-specific KO of PGC-1α attenuated the expression of mitochondrial proteins, but not a shift in fiber-type, following an endurance training program.[44] These data perhaps imply a specific role for PGC-1α in the adaptations of skeletal muscle to exercise. While the role for PGC-1α in skeletal muscle in an unstressed state is quite clear cut, the function of this coactivator in a state of energetic imbalance is still disputed.

8. AGING AND MUSCLE MITOCHONDRIA

Mitochondrial content and function in skeletal muscle decline with age. Recent reviews by Nair and colleagues[45,46] have summarized the factors which contribute to changes in mitochondrial function in aging muscle. A decline in organelle content is supported by many studies which report reduced Krebs' cycle and ETC enzyme activities, protein markers, and mtDNA content. Electron micrograph evidence of diminished IMF mitochondrial size and a reduced thickness of the SS mitochondrial layer[8] also exist. The fragmented mitochondria evident in aging skeletal muscle are likely a result of altered ratios in the expression of fusion and fission regulatory proteins which govern mitochondrial morphology. Reports from muscle of aged rodents and humans have observed that the balance of these regulatory factors is skewed toward favoring greater rates of fission, compared to fusion, within aged muscle. Mitochondrial respiration and maximal ATP production rates are also impaired with aging, possibly a consequence of reduced mitochondrial protein synthesis,[47] or increased uncoupling of oxygen consumption to ATP synthesis.[48] It is important to recognize that many of these decrements remain even when physical activity levels between young and old subjects are carefully matched,[47,49] suggesting true age-related deficits in mitochondrial function. However, this conclusion remains controversial.[45,50] The level of physical activity of the individual is certainly one of the most important determinants of organelle function in aging muscle.

As noted above, PGC-1α is an important regulator of mitochondrial content in a variety of tissues including muscle. PGC-1α mRNA and protein content are reduced in both slow- and fast-twitch muscles with age,[51] suggesting that reductions in mitochondrial function or content could be attributable to the loss of this coactivator. When PGC-1α was overexpressed in muscle of both young (6 months) and aged (22 months) mice,[39] PGC-1α prevented atrophy and retained mitochondrial content and function. Leick *et al.*[52] also showed that the absence of PGC-1α was necessary to extend the benefits of exercise training into older age. Markers of mitochondrial content were decreased with age as expected, and this was prevented with endurance exercise training. However, training was incapable of preventing the mitochondrial decline in animals that lacked PGC-1α, indicating that PGC-1α is necessary to stimulate the beneficial effects of exercise on mitochondrial content during the aging process. Thus, PGC-1α is important for mitochondrial biogenesis during the aging process.

The cause of the decline in mitochondrial content and function in skeletal muscle with age remain controversial. Research has shown that the protein import pathway, a route employed for the incorporation of new proteins into the organelle reticulum, is unaffected with age.[53] mtDNA deletions and point mutations occur with increasing incidence with age, but appear to occur in later life, after the onset of the decline in mitochondrial function.[54] In contrast, the decline in PGC-1α and a reduction in its transcriptional activity may be one of the most compelling reasons for the decrease in organelle content with age.

Exercise is an important therapeutic intervention to ameliorate this decline and restore organelle function with age. It is well known that exercise is a potent inducer of mitochondrial biogenesis in younger individuals, and many studies report that aging muscle adapts to exercise as well. However, is aging muscle equally adaptive to an exercise regimen as muscle from younger individuals? In assessing this question, it is important to provide equivalent relative workloads to both young and older subjects. In addition, many studies have investigated mitochondrial adaptations in older humans; however, they have not consistently employed the use of a younger group for comparison. This makes it difficult to conclude on the degree of mitochondrial adaptation in young compared to older individuals. Nonetheless, many studies using both cross-sectional and longitudinal designs indicate that mitochondrial concentration can increase in both older men and women with exposure to exercise. Studies which compare active older adults to sedentary counterparts show that the active older groups have preserved mitochondrial content and function, higher PGC-1α expression, and a greater capacity to defend against oxidative stress.

Rodent models of exercise have provided considerable insight into the molecular regulation of mitochondrial biogenesis in aging muscle and afford the possibility of strict control over the absolute training workload. Several studies have utilized the chronic contractile activity model[55] to study the effects of "exercise" on mitochondrial adaptation in aging muscle. Walters *et al.*[56] found that chronic contractile activity-invoked exercise of the flexor digitorum longus in young and aged rodents increased citrate synthase activity; however, the rate of increase in the aged animals was attenuated at the onset of the exercise protocol. Yet, by 90 days, young and aged animals had equivalent levels of this mitochondrial enzyme marker. Work from our laboratory using a short-term chronic contractile activity protocol (7 days) of the rat tibialis anterior muscle indicated reduced mitochondrial biogenesis in old muscle as compared to young muscle.[53] This blunted response in mitochondrial proliferation was attributable to reduced elevations of PGC-1α and Tfam, in addition to lack of alterations in protein import machinery components in aged muscle. These data illustrate the potential corrective nature of chronic exercise in ameliorating organelle dysfunction, but also suggest that the kinetics of mitochondrial adaptations in old muscle are delayed in response to an exercise regimen. This may be a result of the fact that aged muscle is less capable of activating upstream kinases, such as AMP kinase, p38 MAPK, or CaMK in response to exercise.[57] Interestingly, this attenuated signaling, leading to a reduced mRNA response, has also been repeatedly demonstrated in response to resistance exercise protocols in old, compared to young subjects.[58,59] Thus, the reduced activation of important kinases regulating mitochondrial biogenesis may be partly responsible for the delayed and diminished adaptation of mitochondria to exercise in senescent muscle.

In summary, these data suggest that aged muscle is capable of increasing mitochondrial content in response to exercise, but that the rate of onset at which this increase takes place may be reduced. However, it is clear that exercise is an important therapeutic intervention to repair dysfunctional mitochondria and improve aerobic energy provision in aging muscle.

9. ALTERNATIVE EXERCISE PROGRAMS: HIGH-INTENSITY INTERVAL TRAINING

Endurance training has been known for many years to stimulate mitochondrial biogenesis, an effect which contributes to an increased aerobic endurance capacity in trained subjects.[11,60] However, endurance training requires a notable time commitment, in terms of both the frequency and duration of exercise sessions, in order to elicit positive results. In order to circumvent these challenges, alternative training programs to induce similar outcomes have been explored.

In addition to exercise duration and frequency, exercise intensity is a central determinant of the cellular and biochemical alterations which occur in response to a training program. As different muscle fiber types are recruited in a hierarchical manner determined by the intensity of exercise employed, it would appear that an intensity of exercise sufficient enough to recruit the greatest number of fiber types might be the most useful in facilitating whole muscle adaptations. Dudley *et al.* provided one of the first pieces of evidence, suggesting that exercise intensity is directly related to the magnitude with which skeletal muscle mitochondrial enzyme content is augmented.[61] This work also put forward the idea that a variety of high-intensity interval training (HIIT) programs may be capable of yielding increases in skeletal muscle oxidative capacity similar to that of a continuous endurance training program. This has since been corroborated by evidence in human skeletal muscle, as both endurance and interval training approaches provide similar increases in the expression of oxidative genes[62] and mitochondrial content markers.[63,64] The increase in mitochondrial proteins is likely due to the increase in the phosphorylation of p38, AMPK, and p53, which can be evoked by a single bout of HIIT,[65,66] and can mediate the translocation of PGC-1α protein to the nucleus. This translocation has been associated with increases in the mRNA and protein levels of a variety of mitochondrial markers, such as citrate synthase and cytochrome *c* oxidase subunit IV during the recovery phase.[67–69] These changes appear to be a product of the accruing effects of repeated mitochondrial mRNA "bursts" which follow each individual training bout.[70]

In addition to promoting mitochondrial biogenesis, it has been suggested that interval training programs can foster the remodeling of the mitochondrial network. An increase in the protein expression of the mitochondrial fusion protein Mfn1 as well as the fission proteins Fis1 and Drp1 as a result of a HIIT program has been documented.[70] This may have implications for mitochondrial function, as skeletal muscle mitochondria isolated from individuals who have undergone a HIIT program display an increase in the maximal ADP-stimulated respiration rate and respiratory control ratio.[71]

In sum, when contrasting these two approaches to increasing skeletal muscle mitochondrial content at this point in time, the HIIT routine appears to afford a notable reduction in the time invested in exercise, perhaps providing an attractive substitute for continuous training regimens. Whether this exercise regime is suitable for all populations remains under debate.

10. EFFECT OF TRAINING ON mtDNA AND mtDNA DISEASES

Mitochondria are unique organelles as they possess their own genetic material. Human mtDNA is organized as circular and double-stranded DNA that encodes 2 ribosomal RNAs, 22 transfer RNAs, and 13 polypeptide subunits of the respiratory chain complexes found across the inner mitochondrial membrane. However, mtDNA contributes to comprise only a small fraction of the mitochondrial proteome. Thus, the proper assembly of a functional mitochondrion requires the coordinated expression of both mitochondrial and nuclear genomes. Improper transcription and/or translation of either genome can lead to organelle dysfunction. Furthermore, mtDNA is particularly susceptible to ROS-induced damage due in part to its proximity to the ETC.[72] Although the cell has measures in place allowing it to repair mtDNA, these mechanisms are not as effective as those found in the nucleus. Thus, mtDNA mutations which arise from oxidative damage can blunt the expression of mtDNA which is vital for the proper functioning of the ETC. As a result, increases in mitochondrial dysfunction can perpetuate a vicious cycle of mtDNA damage and ROS production.

Pathological mtDNA mutations can be distributed in a heteroplasmic fashion, meaning that the proportion of mutant to wild-type mtDNA copies within a mitochondrion can vary. The level of heteroplasmy can range from 99:1 to 1:99, with the severity of symptoms depending on the extent of the pathology. Symptoms are tissue specific but mutations that directly impair mitochondrial function often affect multiple tissues such as muscle and neurons, which rely heavily on mitochondria for energy. Mitochondrial diseases that specifically affect the neuromuscular system are termed mitochondrial myopathies and include Kearns–Sayre syndrome, myoclonic epilepsy and ragged red fibers, neuropathy, ataxia, retinitis pigmentosa, and Leber's hereditary optic neuropathy.[73]

In healthy individuals, exercise-induced mitochondrial biogenesis occurs concomitant with an increase mtDNA copy number,[74] which parallels the increase in skeletal muscle oxidative capacity.[75] This effect is likely mediated, in part, by an increase in Tfam expression and import into the mitochondrion[76,77] (Fig. 1). However, with mtDNA mutations, ATP production levels are often diminished as a result of defective ETC complexes. In skeletal muscle, a decrease in ATP levels can directly impact muscle contraction and performance. Individuals with cytochrome *b* gene mtDNA mutations suffer exercise intolerance as the predominant clinical feature, with symptoms of fatigue and higher than normal levels of lactic acid.[78] Furthermore, mitochondrial myopathies with high mtDNA heteroplasmy correlate to lower whole-body oxygen consumption and lower exercise capacity. Interestingly, aerobic exercise is an intervention capable of stimulating mitochondrial biogenesis by increasing the expression of respiratory chain enzymes and oxidative capacity in mitochondrial myopathy patients; however, it does not improve the mtDNA mutation levels observed in these patients.[79,80] Resistance exercise training may also offer an alternative therapeutic approach through the activation and recruitment of muscle satellite cells to mature myofibers, thereby reducing the proportion of mutant mtDNA.[81] While considerable progress has been made, more work is needed using larger sample sizes and different training protocols to resolve our understanding of exercise training as a therapeutic option for improving the quality of life for mitochondrial myopathy patients.

11. EXERCISE AND TRAINING ON ROS PRODUCTION AND ANTIOXIDANT ENZYMES

Exercise and contractile activity have the capacity to encourage the production of moderate amounts of ROS in muscle[82,83] (Fig. 1). This occurs due to the improper donation of electrons from the ETC to oxygen during oxidative phosphorylation. The electrons which escape, termed an "electron leak," form reactive superoxide, which is then rapidly transformed into H_2O_2 by cellular antioxidant enzymes, such as the superoxide dismutases. While electron leakage from the ETC has historically been suggested to be the predominant source of cellular ROS, current experimental evidence suggests that NADPH oxidase is likely the major ROS-generating source in contracting muscle.[84] Although high levels of ROS production can be detrimental to cellular components, physiological levels are important regulators of cell signaling pathways, and consequently the regulation of gene expression.[85]

During basal state 4 respiration, mitochondrial ROS production is elevated.[86] ROS formation is reduced under the ADP-stimulated state 3 respiration, similar to what would be observed during aerobic exercise, likely due to more efficient electron transfer between ETC complexes. However, this ROS production appears to have a role in the induction of mitochondrial biogenesis following contractile activity. *In vitro* experiments have identified that exogenous ROS treatment increases PGC-1α expression in muscle cells, through both AMPK-dependent and -independent activation of the PGC-1α promoter.[87] Furthermore, the human PGC-1α promoter contains a variety of binding sites for other ROS-sensitive TFs, including Sp1, CREB, ATF2, nuclear factor-κB, p53, and MEF2,[15] supporting the hypothesis that moderate alterations in ROS during exercise may regulate the activity of the PGC-1α promoter through these or other factors. Additionally, inhibition of the cytosolic ROS source xanthine oxidase during exercise attenuates PGC-1α expression and signaling for mitochondrial biogenesis.[88] These data highlight a shared signaling pathway between these metabolic byproducts and the resulting increase in PGC-1α expression, providing a mechanism for the increase in mitochondrial mass[89] and elongation of the mitochondrial network[90] with ROS treatment.

To control cellular ROS levels, skeletal muscle contains a system of enzymatic and nonenzymatic antioxidants that are strategically located in many compartments of the cell. Enhancement of this antioxidant defense system is one of the many adaptive responses that occur with exercise training. This response is associated with increased expression and activity of nuclear erythroid 2 p45-related factor 2 (Nrf2), a TF that serves as the central regulator of antioxidants and detoxification enzymes.[91] Although limited work analyzing the effect of exercise on the expression and regulation of Nrf2 has been done, targeting this factor and its regulatory axis is an attractive area of research.

12. EXERCISE AND THE PROTEIN IMPORT PATHWAY

The mammalian mitochondrial proteome contains over 1000 proteins.[92] Of these, only 13 are encoded by mtDNA, thereby obligating the existence of an elaborate transport system for the import of nuclear-derived proteins from the cytosol into mitochondria. Defects in the import pathway are lethal in some organisms and lead to disease in humans.[93] Since mitochondria have several compartments, specific pathways have evolved to deliver proteins to the matrix, inner membrane, outer membrane, and intermembrane space. Proteins are synthesized in the cytosol with either internal or N-terminal mitochondrial targeting sequences (MTSs) which interact with specific chaperones to unfold and direct the precursor to the translocase of the outer membrane (Tom) receptor complex (Fig. 1). Cytosolic chaperones include heat-shock proteins 70 and 90 (Hsp70 and Hsp90) and mitochondrial import-stimulating factor (MSF). Precursors are then transferred to Tom40 and its accessory proteins which form an aqueous channel through which the precursor protein passes to be sorted to the outer membrane, the inner membrane, or the translocase of the inner membrane (Tim complex). In a fashion similar to the Tom complex, inner membrane import channel proteins Tim17 and Tim23 bind the precursor protein and form a pore through which the precursor can travel. On the inner face of the inner membrane, precursor entry to the matrix is facilitated by the ratchet-like action of mitochondrial Hsp70 (mtHsp70). Once inside, the N-terminal MTS of the precursor is cleaved by a mitochondrial processing peptidase to form the mature protein, and it is then folded into its active conformation by Hsp60 and chaperonin 10 kDa (cpn10). The import process requires both energy in the form of ATP to assist in cytosolic protein unfolding, as well as an intact membrane potential to help pull the positively charged presequence into the matrix.

The protein import process adapts to conditions of muscle use and disuse and determines, in part, the mitochondrial content within the cell. In response to chronic exercise, the cellular content of a number of protein import machinery components is increased, including the cytosolic chaperones MSF and Hsp70, the intramitochondrial proteins mtHsp70, Hsp60, and cpn10, as well as subunits of both the Tom and Tim complexes.[94,95] Coincident with these changes are parallel contractile activity-induced increases in the rate of import into the matrix and outer membrane.[76,94,95] Tom20 appears to be particularly important in mediating the increase in protein import, since experiments in muscle cells in which this outer membrane protein was artificially over- or underexpressed led to parallel concomitant changes in the rates of protein import into the matrix.[96] Included in the accelerated pathway of import into the matrix is Tfam, the TF that mediates mtDNA transcription and replication. Skeletal muscle contractile activity increases Tfam import, leading to greater Tfam–mtDNA binding, as well as augmented mtDNA gene products.[76] Recent work has shown that tumor suppressor p53 also plays a critical role in this Tfam–mtDNA complex, as the ablation of p53 completely attenuates the expression of mtDNA following acute exercise.[97]

The physiological importance of the adaptation of protein import to exercise is that the capacity for import is increased, producing mitochondria that are more sensitive to small changes in precursor protein concentration. Thus, at any production rate of cytosolic precursor proteins, a higher rate of protein import would occur, a situation that would be advantageous for mitochondrial biogenesis. This would be of particular benefit during conditions of impaired mitochondrial protein import, a situation which could arise during chronic muscle disuse or disease. Indeed, we now have experimental evidence that endurance exercise training can rescue the protein import defect produced in mitochondria by the absence of the outer membrane proteins Bax and Bak.[98]

In summary, the protein import pathway is a complex protein targeting and transfer system which is vital for organelle biogenesis, given the limited coding capacity of mtDNA. This capacity of this pathway is malleable; it decreases during muscle disuse and adapts in a positive manner to contractile activity. In this way, it is an important determinant of mitochondrial content in muscle during alterations in level of physical activity.

13. EFFECT OF EXERCISE ON MITOCHONDRIALLY MEDIATED APOPTOSIS

Apoptosis is the process of programmed cell death which plays an important role in cell function and homeostasis, but that is also capable of contributing to the pathogenesis of physiological aging and skeletal muscle disuse.[99] This process can be induced through various mechanisms, most commonly converging on the caspase proteases, which cleave select nuclear and cytosolic targets leading to DNA fragmentation. Apoptosis can be signaled extrinsically via binding of external ligands to cell death receptors on the outer surface of the cell membrane, or by intrinsic stimuli, resulting in DNA fragmentation contributing to cell death. One of the most notable intrinsic apoptotic pathways is mediated by the mitochondrion (Fig. 1). These organelles contain many proapoptotic factors, including cytochrome *c*, apoptosis-inducing factor (AIF), and endonuclease G that can be released following the opening of the *m*itochondrial *p*ermeability *t*ransition *p*ore (mtPTP) and the *m*itochondrial *a*poptosis-inducing *c*hannel (MAC). A variety of cellular conditions, including excessive ROS production, can cause mitochondrial permeabilization and the release of these proapoptotic factors, inducing DNA fragmentation and apoptosis.[100]

Research has indicated that exercise training can reduce the susceptibility of skeletal muscle to apoptosis. Initial studies have shown that following an endurance training protocol, an increase in the expression of Bcl-2 and mitochondrial superoxide dismutase (MnSOD) occurred, along with a diminished expression of the proapoptotic factor Bax. Chronic contractile activity also revealed a reduction in the release of proapoptotic factors from isolated mitochondria of young animals[101] and reduced DNA fragmentation in older animals.[53] Furthermore, endurance training also resulted in a decreased Bax:Bcl2 ratio, as well as AIF levels in both soleus and cardiac muscle.[102]

This protective effect of endurance exercise on skeletal muscle apoptotic susceptibility may also be relevant in several animal models of human disease. Obese Zucker rats typically display elevated levels of apoptotic factors in cardiac muscle. Interestingly, exercise training can reduce the expression of these factors to levels near that of a nonobese control.[103] More recently, it has been established that hypertension is associated with enhanced skeletal muscle myopathy and apoptosis. Endurance training of the spontaneously hypertensive rat led to a reduction in fragmented nuclei and cytochrome *c* release into the cytosol, while increasing Bcl-2 and MnSOD expression.[104]

Taken together, exercise training appears to be a feasible approach to diminish mitochondrially mediated apoptosis, both basally and under stressed conditions, in turn yielding a more favorable muscle phenotype.

14. AUTOPHAGY AND MITOPHAGY WITH EXERCISE

The cellular remodeling which occurs in response to contractile activity necessitates the turnover of dysfunctional organelles, as well as oxidized and damaged proteins. Macroautophagy (hereafter referred to as autophagy) is a cellular recycling mechanism that is characterized by the encapsulation of damaged organelles and protein aggregates in doubled membrane vesicles known as autophagosomes, and their subsequent delivery to and degradation within the lysosome (Fig. 1). Enhanced lysosomal degradation with strenuous exercise was observed in skeletal muscle over three decades ago.[105] Since then, an acute bout of exercise has been demonstrated to induce autophagy in various organs and tissues including skeletal muscle of humans and rodents.[106–110]

Exercise-induced autophagy involves the phosphorylation of Bcl-2, an antiapoptotic and antiautophagic protein. This allows for the release of Beclin1 from a complex with Bcl-2, consequently allowing for the formation of autophagosomes. It has been thought that exercise may require autophagy for the breakdown of fuel sources to be utilized during an exercise bout. Mice overexpressing a mutated Bcl-2 incapable of being phosphorylated (AAA knock-in, $Bcl2^{AAA}$) exhibit normal basal autophagy, but are deficient in stimulus-induced autophagy, and have an impaired endurance capacity, suggesting that autophagy may be required for exercise performance. However, given that the inability to initiate autophagy in this experimental model is not specific to any one tissue, teasing out the requirement for autophagy specifically within skeletal muscle as a determinant of exercise performance is essential. Intriguingly, skeletal muscle-specific ablation of Atg7 ($Atg7^{-/-}$), an indispensable protein involved in autophagy, does not result in alterations in endurance performance.[111] In fact, these mice have an improved metabolic profile and are protected from obesity, findings which were attributed to the mitochondrial stress and the release of the mitokine FGF21 in $Atg7^{-/-}$ mice.[112] These contrasting findings bring into question the requirement for autophagy as an energy source for muscle contraction. However, repeated bouts of running exercise reveal a progressive drop in performance as well as diminished mitochondrial membrane potential in muscle-specific $Atg7^{-/-}$ mice, suggesting that autophagy may be critical for organelle turnover postexercise. This autophagy activation is, in part, ROS dependent, as treatment with general or mitochondria-specific ROS scavengers attenuates autophagy induction in wild-type mice.[111] Moreover, aerobic exercise enhances mitochondrial LC3II, p62, and ubiquitination following exercise, suggesting the induction of mitochondrial-specific autophagy with exercise.[110] These findings indicate that autophagy is vital for organelle turnover postexercise, thus contributing to exercise-induced adaptations.

An increase in the expression of autophagy genes and proteins has also been noted in humans following ultra-endurance exercise, which consisted of 24–28 h of treadmill running.[109] While limited data have been compiled with regard to the autophagic processes in human skeletal muscle following acute or chronic endurance training, this is certainly an avenue of research which would be attractive to explore.

Autophagy is also essential for exercise-induced benefits, including protection against high-fat diet (HFD)-induced metabolic impairments. $Bcl2^{AAA}$ mice are more susceptible to HFD-induced obesity and fail to exhibit exercise training-mediated protection against HFD-induced disturbance in glucose tolerance. These findings point to the involvement of autophagy in metabolic renovation with repeated bouts of physical activity.[107] Moreover, voluntary exercise training results in increased basal autophagy and mitophagy protein expression.[113] However, this increase is compromised in mice that are heterozygous for Atg6 ($Atg6^{-/+}$). $Atg6^{-/+}$ mice also fail to induce skeletal muscle mitochondrial biogenesis and angiogenesis in response to an exercise training program, along with an inability to improve their endurance capacity. These results point to a role for basal autophagy and/or mitophagy as a requirement for exercise training-induced skeletal muscle adaptations and improvement of physical performance.[113]

There is also evidence to suggest that regular exercise may be able to restore conditions where autophagy may be deficient, or ineffective. Indeed, long-term exercise training was also demonstrated to reactivate autophagy flux in skeletal muscle of animals treated with the lysosomal inhibitor chloroquine.[114] Furthermore, lifelong exercise, in combination with caloric restriction, has been found to improve the reduction in autophagy observed with aging, as well as to dampen the age-related increase in oxidative damage and apoptosis.[115] This suggests that exercise may serve as a potential form of therapy for myopathies characterized by impaired autophagy. However, exercise should be prescribed with caution, as in certain autophagy-deficient conditions, such as those involving a mutation in the extracellular matrix protein collagen VI, the myopathy may be exacerbated.[106]

Although the early events leading to autophagy activation with exercise have not been thoroughly examined, several possibilities exist. Muscle contraction represents a form of energetic imbalance, much like nutrient deprivation or starvation, all of which are well characterized as activators of autophagy. The increased metabolic demands of contracting muscle, along with elevated ROS production, activate the cellular stress responders such as AMPK, SIRT1, and p38, which may lead to the induction of autophagic machinery. In particular, activation of AMPK has been definitively linked with the control of mitophagy.[116] An acute bout of exercise also leads to reduced Akt phosphorylation, which would inhibit mTOR signaling and thereby also promote autophagy.[106,107]

Another possible mechanism involved in the induction of autophagy may be the unfolded protein response (UPR), which is activated in response to sarcoplasmic reticulum stress induced by contractile activity.[117,118] Certain UPR factors have been implicated in the activation of autophagy in various cell types.[119–121] Interestingly, PGC-1α was shown to contribute to UPR induction with exercise acting through the UPR ATF6, and mice lacking ATF6 are exercise intolerant.[117]

Therefore, although several studies indicate that autophagy is activated during an acute bout of exercise, the mechanisms underlying this activation remain to be elucidated. It is also becoming increasingly evident that autophagy is involved in chronic physical activity-induced adaptations through increased organelle turnover. However, the kinetics of autophagy activation during and following an exercise bout still require further exploration.

15. CONCLUSIONS

Exercise training-induced adaptations are wide ranging and can positively impact a variety of organ systems. In particular, alterations to skeletal muscle, such as the increase in mitochondrial content and improvement in the health of the mitochondrial pool, are keys to the enhanced metabolic capacity of the organ. These changes are a product of the signals which originate from shifts in the myocellular environment as a result of exercise, including the consumption of ATP, the production of moderate levels of ROS, and the intracellular release of Ca^{2+}, among others. Concomitant with an increase in the quantity of mitochondria is the expansion of the organelle network and augmented antioxidant capacity. In addition, exercise results in an enhanced removal of dysfunctional mitochondria. Altogether, these alterations are beneficial to muscle in that they afford the organ the ability to function more efficiently and effectively, and are vital for improvements in endurance capacity and well-being, particularly during the aging process. Thus, continued attention to the understanding of the molecular basis of theses mitochondrial adaptations is essential.

ACKNOWLEDGMENTS

Research in our laboratory is supported by the Natural Science and Engineering Research Council of Canada (NSERC), the Canada Research Chairs program, and the Canadian Institutes for Health Research (CIHR). D.A.H. is the holder of a Canada Research Chair in Cell Physiology.

REFERENCES

1. Cogswell AM, Stevens RJ, Hood DA. Properties of skeletal muscle mitochondria isolated from subsarcolemmal and intermyofibrillar regions. *Am J Physiol.* 1993;264: C383–C389.
2. Picard M, White K, Turnbull DM. Mitochondrial morphology, topology, and membrane interactions in skeletal muscle: a quantitative three-dimensional electron microscopy study. *J Appl Physiol (1985).* 2013;114(2):161–171.
3. Kirkwood SP, Munn EA, Brooks GA. Mitochondrial reticulum in limb skeletal muscle. *Am J Physiol.* 1986;20:C395–C402.
4. Ogata T, Yamasaki Y. Ultra-high-resolution scanning electron micrscopy of mitochondria and sarcoplasmic reticulum arrangement in human red, white and intermediate muscle fibers. *Anat Rec.* 1997;248(2):1997.
5. Twig G, Shirihai OS. The interplay between mitochondrial dynamics and mitophagy. *Antioxid Redox Signal.* 2011;14(10):1939–1951.
6. Gollnick PD, King DW. Effect of exercise and training of rat skeletal muscle on mitochondria. *Am J Physiol.* 1969;216(6):1502–1509.
7. Kirkwood SP, Packer L, Brooks GA. Effects of endurance training on a mitochondrial reticulum in limb skeletal muscle. *Arch Biochem Biophys.* 1987;255(1):80–88.
8. Iqbal S, Ostojic O, Singh K, Joseph AM, Hood DA. Expression of mitochondrial fission and fusion regulatory proteins in skeletal muscle during chronic use and disuse. *Muscle Nerve.* 2013;48(6):963–970.
9. Cartoni R, Leger B, Hock MB, et al. Mitofusins 1/2 and ERRα expression are increased in human skeletal muscle after physical exercise. *J Physiol.* 2005;567:349–358.
10. Picard M, Gentil BJ, McManus MJ, et al. Acute exercise remodels mitochondrial membrane interactions in mouse skeletal muscle. *J Appl Physiol (1985).* 2013;115(10):1562–1571.
11. Hood DA. Invited Review: contractile activity-induced mitochondrial biogenesis in skeletal muscle. *J Appl Physiol (1985).* 2001;90(3):1137–1157.
12. Bergeron R, Ren JM, Cadman KS, et al. Chronic activation of AMP kinase results in NRF-1 activation and mitochondrial biogenesis. *Am J Physiol Endocrinol Metab.* 2001;281:E1340–E1346.
13. Jäger S, Handschin C, St-Pierre J, Spiegelman BM. AMP-activated protein kinase (AMPK) action in skeletal muscle via direct phosphorylation of PGC-1alpha. *Proc Natl Acad Sci USA.* 2007;104(29):12017–12022.
14. Winder WW, Holmes BF, Rubink DS, Jensen EB, Chen M, Holloszy JO. Activation of AMP-activated protein kinase increases mitochondrial enzymes in skeletal muscle. *J Appl Physiol (1985).* 2000;88(6):2219–2226.
15. Irrcher I, Ljubicic V, Kirwan AF, Hood DA. AMP-activated protein kinase-regulated activation of the PGC-1alpha promoter in skeletal muscle cells. *PLoS One.* 2008;3(10): e3614.
16. Garcia-Roves PM, Osler ME, Holmström MH, Zierath JR. Gain-of-function R225Q mutation in AMP-activated protein kinase gamma3 subunit increases mitochondrial biogenesis in glycolytic skeletal muscle. *J Biol Chem.* 2008;283(51):35724–35734.

17. O'Neill HM, Maarbjerg SJ, Crane JD, et al. AMPK-activated protein kinase (AMPK) β12 muscle null mice reveal an essential role for AMPK in maintaining mitochondrial content and glucose uptake during exercise. *Proc Natl Acad Sci USA*. 2011;108(38): 16092–16097.
18. Freyssenet D, Di Carlo M, Hood DA. Calcium-dependent regulation of cytochrome c gene expression in skeletal muscle cells: identification of a protein kinase c-dependent pathway. *J Biol Chem*. 1999;274(14):9305–9311.
19. Ojuka EO, Jones TE, Han DH, Chen M, Holloszy JO. Raising Ca2+ in L6 myotubes mimics effects of exercise on mitochondrial biogenesis in muscle. *FASEB J*. 2003;17:675–681.
20. Zhang Y, Uguccioni G, Ljubicic V, et al. Multiple signaling pathways regulate contractile activity-mediated PGC-1α gene expression and activity in skeletal muscle cells. *Physiol Rep*. 2014;2(5):1–12.
21. Handschin C, Rhee J, Lin J, Tarr PT, Spiegelman BM. An autoregulatory loop controls peroxisome proliferator-activated receptor γ coactivator 1α expression in muscle. *Proc Natl Acad Sci USA*. 2003;100(12):7111–7116.
22. Summermatter S, Thurnheer R, Santos G, et al. Remodeling of calcium handling in skeletal muscle through PGC-1α: impact on force, fatigability, and fiber type. *Am J Physiol Cell Physiol*. 2012;302(1):C88–C99.
23. Ohlendieck K, Frömming GR, Murray BE, et al. Effects of chronic low-frequency stimulation on Ca^{2+}-regulatory membrane proteins in rabbit fast muscle. *Pflugers Arch*. 1999;438:700–708.
24. Irrcher I, Walkinshaw DR, Sheehan TE, Hood DA. Thyroid hormone (T3) rapidly activates p38 and AMPK in skeletal muscle in vivo. *J Appl Physiol (1985)*. 2008;104(1):178–185.
25. Akimoto T, Pohnert SC, Li P, et al. Exercise stimulates Pgc-1alpha transcription in skeletal muscle through activation of the p38 MAPK pathway. *J Biol Chem*. 2005;280(20):19587–19593.
26. Yan Z, Li P, Akimoto T. Transcriptional control of the Pgc-1alpha gene in skeletal muscle in vivo. *Exerc Sport Sci Rev*. 2007;35(3):97–101.
27. Puigserver P, Spiegelman BM. Peroxisome proliferator-activated receptor-gamma coactivator 1alpha (PGC-1alpha): transcriptional coactivator and metabolic regulator. *Endocr Rev*. 2003;24(1):78–90.
28. Irrcher I, Hood DA. Regulation of Egr-1, SRF, and Sp1 mRNA expression in contracting skeletal muscle cells. *J Appl Physiol (1985)*. 2004;97:2207–2213.
29. Xia Y, Buja LM, Scarpulla RC, McMillin JB. Electrical stimulation of neonatal cardiomyocytes results in the sequential activation of nuclear genes governing mitochondrial proliferation and differentiation. *Proc Natl Acad Sci USA*. 1997;94: 11399–11404.
30. Wright DC, Han DH, Garcia-Roves PM, Geiger PC, Jones TE, Holloszy JO. Exercise-induced mitochondrial biogenesis begins before the increase in muscle PGC-1alpha expression. *J Biol Chem*. 2007;282(1):194–199.
31. Pilegaard H, Saltin B, Neufer PD. Exercise induces transient transcriptional activation of the PGC-1 gene in human skeletal muscle. *J Physiol*. 2003;546(3):851–858.
32. Wu Z, Puigserver P, Andersson U, et al. Mechanisms controlling mitochondrial biogenesis and respiration through the thermogenic coactivator PGC-1. *Cell*. 1999;98(1):115–124.
33. Vega RB, Huss JM, Kelly DP. The coactivator PGC-1 cooperates with peroxisome proliferator-activated receptor alpha in transcriptional control of nuclear genes encoding mitochondrial fatty acid oxidation enzymes. *Mol Cell Biol*. 2000;20(5): 1868–1876.

34. Scarpulla RC. Transcriptional paradigms in mammalian mitochondrial biogenesis and function. *Physiol Rev*. 2008;88:611–638.
35. Puigserver P, Adelmant G, Wu Z, et al. Activation of PPAR coactivator-1 through transcription factor docking. *Science*. 1999;286:1368–1371.
36. Kelly DP, Scarpulla RC. Transcriptional regulatory circuits controlling mitochondrial biogenesis and function. *Genes Dev*. 2004;18(4):357–368.
37. Irrcher I, Adhihetty PJ, Sheehan T, Joseph AM, Hood DA. PPARγ coactivator-1α expression during thyroid hormone- and contractile activity-induced mitochondrial adaptations. *Am J Physiol Cell Physiol*. 2003;284:C1669–C1677.
38. Calvo JA, Daniels TG, Wang X, et al. Muscle-specific expression of PPARgamma coactivator-1alpha improves exercise performance and increases peak oxygen uptake. *J Appl Physiol (1985)*. 2008;104(5):1304–1312.
39. Wenz T, Rossi SG, Rotundo RL, Spiegelman BM, Moraes CT. Increased muscle PGC-1alpha expression protects from sarcopenia and metabolic disease during aging. *Proc Natl Acad Sci USA*. 2009;106(48):20405–20410.
40. Adhihetty PJ, Uguccioni G, Leick L, Hidalgo J, Pilegaard H, Hood DA. The role of PGC-1alpha on mitochondrial function and apoptotic susceptibility in muscle. *Am J Physiol Cell Physiol*. 2009;297(1):C217–C225.
41. Handschin C, Chin S, Li P, et al. Skeletal muscle fiber-type switching, exercise intolerance, and myopathy in PGC-1alpha muscle-specific knock-out animals. *J Biol Chem*. 2007;282(41):30014–30021.
42. Leick L, Wojtaszewski JFP, Johansen ST, et al. PGC-1alpha is not mandatory for exercise- and training-induced adaptive gene responses in mouse skeletal muscle. *Am J Physiol Endocrinol Metab*. 2008;294(2):E463–E474.
43. Uguccioni G, Hood DA. The importance of PGC-1α in contractile activity-induced mitochondrial adaptations. *Am J Physiol Endocrinol Metab*. 2011;300:E361–E371.
44. Geng T, Li P, Okutsu M, et al. PGC-1alpha plays a functional role in exercise-induced mitochondrial biogenesis and angiogenesis but not fiber-type transformation in mouse skeletal muscle. *Am J Physiol Cell Physiol*. 2010;298(3):C572–C579.
45. Johnson ML, Robinson MM, Nair KS. Skeletal muscle aging and the mitochondrion. *Trends Endocrinol Metab*. 2013;24(5):247–256.
46. Hebert SL, Lanza IR, Nair KS. Mitochondrial DNA alterations and reduced mitochondrial function in aging. *Mech Ageing Dev*. 2010;131(7–8):451–462.
47. Rooyackers OE, Adey DB, Ades PA, Nair KS. Effect of age on in vivo rates of mitochondrial protein synthesis in human skeletal muscle. *Proc Natl Acad Sci USA*. 1996;93(26):15364–15369.
48. Conley KE, Jubrias SA, Esselman PC. Oxidative capacity and ageing in human muscle. *J Physiol*. 2000;526:203–210.
49. Ghosh S, Lertwattanarak R, Lefort N, et al. Reduction in reactive oxygen species production by mitochondria from elderly subjects with normal and impaired glucose tolerance. *Diabetes*. 2011;60(8):2051–2060.
50. Kent-Braun JA, Ng AV. Skeletal muscle oxidative capacity in young and older women and men. *J Appl Physiol (1985)*. 2000;89(3):1072–1078.
51. Chabi B, Ljubicic V, Menzies KJ, Huang JH, Saleem A, Hood DA. Mitochondrial function and apoptotic susceptibility in aging skeletal muscle. *Aging Cell*. 2008;7(1):2–12.
52. Leick L, Lyngby SS, Wojtaszewski JFP, Pilegaard H. PGC-1alpha is required for training-induced prevention of age-associated decline in mitochondrial enzymes in mouse skeletal muscle. *Exp Gerontol*. 2010;45(5):336–342.
53. Ljubicic V, Joseph AM, Adhihetty PJ, et al. Molecular basis for an attenuated mitochondrial adaptive plasticity in aged skeletal muscle. *Aging (Albany, NY)*. 2009;1(9):818–831.

54. Conley KE, Marcinek DJ, Villarin J. Mitochondrial dysfunction and age. *Curr Opin Clin Nutr Metab Care*. 2007;10(6):688–692.
55. Ljubicic V, Adhihetty PJ, Hood DA. Application of animal models: chronic electrical stimulation-induced contractile activity. *Can J Appl Physiol*. 2005;30(5):625–643.
56. Walters TJ, Sweeney HL, Farrar RP. Influence of electrical stimulation on a fast-twitch muscle in aging rats. *J Appl Physiol (1985)*. 1991;71(5):1921–1928.
57. Ljubicic V, Hood DA. Diminished contraction-induced intracellular signaling towards mitochondrial biogenesis in aged skeletal muscle. *Aging Cell*. 2009;8(4):394–404.
58. Funai K, Parkington JD, Carambula S, Fielding RA. Age-associated decrease in contraction-induced activation of downstream targets of Akt/mTor signaling in skeletal muscle. *Am J Physiol Regul Integr Comp Physiol*. 2006;290(4):R1080–R1086.
59. Fry CS, Drummond MJ, Glynn EL, et al. Aging impairs contraction-induced human skeletal muscle mTORC1 signaling and protein synthesis. *Skelet Muscle*. 2011;1(1):11.
60. Holloszy JO. Biochemical adaptations in muscle. Effects of exercise on mitochondrial oxygen uptake and respiratory enzyme activity in skeletal muscle. *J Biol Chem*. 1967;242(9):2278–2282.
61. Dudley GA, Abraham WM, Terjung RL. Influence of exercise intensity and duration on biochemical adaptations in skeletal muscle. *J Appl Physiol Respir Exerc Physiol*. 1982;53(4):844–850.
62. Wang L, Psilander N, Tonkonogi M, Ding S, Sahlin K. Similar expression of oxidative genes after interval and continuous exercise. *Med Sci Sports Exerc*. 2009;41(12):2136–2144.
63. Burgomaster KA, Howarth KR, Phillips SM, et al. Similar metabolic adaptations during exercise after low volume sprint interval and traditional endurance training in humans. *J Physiol*. 2008;586(1):151–160.
64. Scribbans TD, Edgett BA, Vorobej K, et al. Fibre-specific responses to endurance and low volume high intensity interval training: striking similarities in acute and chronic adaptation. *PLoS One*. 2014;9(6):e98119.
65. Gibala MJ, McGee SL, Garnham AP, Howlett KF, Snow RJ, Hargreaves M. Brief intense interval exercise activates AMPK and p38 MAPK signaling and increases the expression of PGC-1α in human skeletal muscle. *J Appl Physiol (1985)*. 2009;106(3):929–934.
66. Bartlett JD, Hwa Joo C, Jeong TS, et al. Matched work high-intensity interval and continuous running induce similar increases in PGC-1α mRNA, AMPK, p38, and p53 phosphorylation in human skeletal muscle. *J Appl Physiol (1985)*. 2012;112(7):1135–1143.
67. Little JP, Safdar A, Bishop D, Tarnopolsky MA, Gibala MJ. An acute bout of high-intensity interval training increases the nuclear abundance of PGC-1α and activates mitochondrial biogenesis in human skeletal muscle. *Am J Physiol Regul Integr Comp Physiol*. 2011;300(6):R1303–R1310.
68. Gurd BJ, Perry CGR, Heigenhauser GJF, Spriet LL, Bonen A. High-intensity interval training increases SIRT1 activity in human skeletal muscle. *Appl Physiol Nutr Metab*. 2010;35(3):350–357.
69. Little JP, Safdar A, Cermak N, Tarnopolsky MA, Gibala MJ. Acute endurance exercise increases the nuclear abundance of PGC-1alpha in trained human skeletal muscle. *Am J Physiol Regul Integr Comp Physiol*. 2010;298(4):R912–R917.
70. Perry CGR, Lally J, Holloway GP, Heigenhauser GJF, Bonen A, Spriet LL. Repeated transient mRNA bursts precede increases in transcriptional and mitochondrial proteins during training in human skeletal muscle. *J Physiol*. 2010;588(Pt 23):4795–4810.
71. Daussin FN, Zoll J, Dufour SP, et al. Effect of interval versus continuous training on cardiorespiratory and mitochondrial functions: relationship to aerobic performance

improvements in sedentary subjects. *Am J Physiol Regul Integr Comp Physiol.* 2008;295(1):R264–R272.
72. Yakes FM, Van Houten B. Mitochondrial DNA damage is more extensive and persists longer than nuclear DNA damage in human cells following oxidative stress. *Proc Natl Acad Sci USA.* 1997;94(2):514–519.
73. Tarnopolsky MA, Raha S. Mitochondrial myopathies: diagnosis, exercise intolerance, and treatment options. *Med Sci Sports Exerc.* 2005;37(12):2086–2093.
74. Puntschart A, Claassen H, Jostarndt K, Hoppeler H, Billeter R. mRNAs of enzymes involved in energy metabolism and mtDNA are increased in endurance-trained athletes. *Am J Physiol.* 1995;269(3 Pt 1):C619–C625.
75. Williams RS. Mitochondrial gene expression in mammalian striated muscle. Evidence that variation in gene dosage is the major regulatory event. *J Biol Chem.* 1986;261(26): 12390–12394.
76. Gordon JW, Rungi AA, Inagaki H, Hood DA. Effects of contractile activity on mitochondrial transcription factor A expression in skeletal muscle. *J Appl Physiol (1985).* 2001;90:389–396.
77. Bengtsson J, Gustafsson T, Widegren U, Jansson E, Sundberg CJ. Mitochondrial transcription factor A and respiratory complex IV increase in response to exercise training in humans. *Pflugers Arch.* 2001;443(1):61–66.
78. Andreu AL, Hanna MG, Reichmann H, et al. Exercise intolerance due to mutations in the cytochrome b gene of mitochondrial DNA. *N Engl J Med.* 1999;341(14): 1037–1044.
79. Jeppesen TD, Dunø M, Schwartz M, et al. Short- and long-term effects of endurance training in patients with mitochondrial myopathy. *Eur J Neurol.* 2009;16(12): 1336–1339.
80. Taivassalo T, Gardner JL, Taylor RW, et al. Endurance training and detraining in mitochondrial myopathies due to single large-scale mtDNA deletions. *Brain.* 2006;129(Pt 12):3391–3401.
81. Murphy JL, Blakely EL, Schaefer AM, et al. Resistance training in patients with single, large-scale deletions of mitochondrial DNA. *Brain.* 2008;131(Pt 11):2832–2840.
82. Davies KJA, Quintanilhat AT, Brooks GA, Packer L. Free radicals and tissue damage produced by exercise. *Biochem Biophys Res Commun.* 1982;107(4):1198–1205.
83. McArdle F, Pattwell DM, Vasilaki A, McArdle A, Jackson MJ. Intracellular generation of reactive oxygen species by contracting skeletal muscle cells. *Free Radic Biol Med.* 2005;39(5):651–657.
84. Sakellariou GK, Jackson MJ, Vasilaki A. Redefining the major contributors to superoxide production in contracting skeletal muscle. The role of NAD(P)H oxidases. *Free Radic Res.* 2014;48(January):12–29.
85. Powers SK, Talbert EE, Adhihetty PJ. Reactive oxygen and nitrogen species as intracellular signals in skeletal muscle. *J Physiol.* 2011;589(Pt 9):2129–2138.
86. Adhihetty PJ, Ljubicic V, Menzies KJ, Hood DA. Differential susceptibility of subsarcolemmal and intermyofibrillar mitochondria to apoptotic stimuli. *Am J Physiol Cell Physiol.* 2005;3:C994–C1001.
87. Irrcher I, Ljubicic V, Hood DA. Interactions between ROS and AMP kinase activity in the regulation of PGC-1α transcription in skeletal muscle cells. *Am J Physiol Cell Physiol.* 2009;296:C116–C123.
88. Kang C, O'Moore KM, Dickman JR, Ji LL. Exercise activation of muscle peroxisome proliferator-activated receptor-gamma coactivator-1alpha signaling is redox sensitive. *Free Radic Biol Med.* 2009;47:1394–1400.
89. Lee HC, Yin PH, Chi CW, Wei YH. Increase in mitochondrial mass in human fibroblasts under oxidative stress during replicative cell senescence. *J Biomed Sci.* 2002;9: 517–526.

90. Koopman WJH, Verkaart S, Visch HJ, et al. Inhibition of complex I of the electron transport chain causes O^{2-}-mediated mitochondrial outgrowth. *Am J Physiol Cell Physiol*. 2005;288:C1440–C1450.
91. Muthusamy VR, Kannan S, Sadhaasivam K, et al. Acute exercise stress activates Nrf2/ARE signaling and promotes antioxidant mechanisms in the myocardium. *Free Radic Biol Med*. 2012;52(2):366–376.
92. Pagliarini DJ, Calvo SE, Chang B, et al. A mitochondrial protein compendium elucidates complex I disease biology. *Cell*. 2008;134(1):112–123.
93. Harbauer AB, Zahedi RP, Sickmann A, Pfanner N, Meisinger C. The protein import machinery of mitochondria—a regulatory hub in metabolism, stress, and disease. *Cell Metab*. 2014;19(3):357–372.
94. Takahashi M, Chesley A, Freyssenet D, Hood DA. Contractile activity-induced adaptations in the mitochondrial protein import system. *Am J Physiol*. 1998;274(43): C1380–C1387.
95. Joseph AM, Hood DA. Plasticity of TOM complex assembly in skeletal muscle mitochondria in response to chronic contractile activity. *Mitochondrion*. 2012;12(2): 305–312.
96. Grey JY, Connor MK, Gordon JW, Yano M, Mori M, Hood DA. Tom20-mediated mitochondrial protein import in muscle cells during differentiation. *Am J Physiol Cell Physiol*. 2000;279:C1393–C1400.
97. Saleem A, Hood DA. Acute exercise induces tumour suppressor protein p53 translocation to the mitochondria and promotes a p53-Tfam-mitochondrial DNA complex in skeletal muscle. *J Physiol*. 2013;591(Pt 14):3625–3636.
98. Zhang Y, Iqbal S, O'Leary MFN, et al. Altered mitochondrial morphology and defective protein import reveal novel roles for Bax and/or Bak in skeletal muscle. *Am J Physiol Cell Physiol*. 2013;305(5):C502–C511.
99. Adhihetty PJ, O'Leary MFN, Hood DA. Mitochondria in skeletal muscle: adaptable rheostats of apoptotic susceptibility. *Exerc Sport Sci Rev*. 2008;36(3):116–121.
100. Kroemer G, Galluzzi L, Brenner C. Mitochondrial membrane permeabilization in cell death. *Physiol Rev*. 2007;87:99–163.
101. Adhihetty PJ, Ljubicic V, Hood DA. Effect of chronic contractile activity on SS and IMF mitochondrial apoptotic susceptibility in skeletal muscle. *Am J Physiol Endocrinol Metab*. 2007;292:E748–E755.
102. Vainshtein A, Kazak L, Hood DA. Effects of endurance training on apoptotic susceptibility in striated muscle. *J Appl Physiol (1985)*. 2011;110(6):1638–1645.
103. Peterson JM, Bryner RW, Sindler A, Frisbee JC, Alway SE. Mitochondrial apoptotic signaling is elevated in cardiac but not skeletal muscle in the obese Zucker rat and is reduced with aerobic exercise. *J Appl Physiol (1985)*. 2008;105(6):1934–1943.
104. Mcmillan EM, Graham DA, Rush JWE, Quadrilatero J. Decreased DNA fragmentation and apoptotic signaling in soleus muscle of hypertensive rats following 6 weeks of treadmill training. *J Appl Physiol (1985)*. 2012;113(7):1048–1057.
105. Salminen A, Vihko V. Autophagic response to strenuous exercise in mouse skeletal muscle fibers. *Virchows Arch B Cell Pathol Incl Mol Pathol*. 1984;45(1):97–106.
106. Grumati P, Coletto L, Schiavinato A, et al. Physical exercise stimulates autophagy in normal skeletal muscles but is detrimental for collagen VI-deficient muscles. *Autophagy*. 2014;7(12):1415–1423.
107. He C, Bassik MC, Moresi V, et al. Exercise-induced BCL2-regulated autophagy is required for muscle glucose homeostasis. *Nature*. 2012;481(7382):511–515.
108. Jamart C, Benoit N, Raymackers J-MM, Kim HJ, Kim CK, Francaux M. Autophagy-related and autophagy-regulatory genes are induced in human muscle after ultraendurance exercise. *Eur J Appl Physiol*. 2012;112(8):3173–3177.

109. Jamart C, Francaux M, Millet GY, Deldicque L, Frère D, Féasson L. Modulation of autophagy and ubiquitin-proteasome pathways during ultra-endurance running. *J Appl Physiol (1985)*. 2012;112(9):1529–1537.
110. Saleem A, Carter HN, Hood DA. p53 is necessary for the adaptive changes in the cellular milieu subsequent to an acute bout of endurance exercise. *Am J Physiol Cell Physiol*. 2014;306(3):C241–C249.
111. LoVerso F, Carnio S, Vainshtein A, Sandri M. Autophagy is not required to sustain exercise and PRKAA1/AMPK activity but is important to prevent mitochondrial damage during physical activity. *Autophagy*. 2014;10:1883–1894.
112. Kim KH, Jeong YT, Oh H, et al. Autophagy deficiency leads to protection from obesity and insulin resistance by inducing Fgf21 as a mitokine. *Nat Med*. 2013;19(1):83–92.
113. Lira VA, Okutsu M, Zhang M, et al. Autophagy is required for exercise training-induced skeletal muscle adaptation and improvement of physical performance. *FASEB J*. 2013;27(10):4184–4193.
114. Jiang D, Chen K, Lu X, Gao HJ, Qin ZH, Lin F. Exercise ameliorates the detrimental effect of chloroquine on skeletal muscles in mice via restoring autophagy flux. *Acta Pharmacol Sin*. 2014;35(1):135–142.
115. Wohlgemuth SE, Seo AY, Marzetti E, Lees HA, Leeuwenburgh C. Skeletal muscle autophagy and apoptosis during aging: effects of calorie restriction and life-long exercise. *Exp Gerontol*. 2010;45(2):138–148.
116. Egan DF, Shackelford DB, Mihaylova MM, et al. Phosphorylation of ULK1 (hATG1) by AMP-activated protein kinase connects energy sensing to mitophagy. *Science*. 2011;331(6016):456–461.
117. Wu J, Ruas JL, Estall JL, et al. The unfolded protein response mediates adaptation to exercise in skeletal muscle through a PGC-1α/ATF6α complex. *Cell Metab*. 2011;13(2):160–169.
118. Kim HJ, Jamart C, Deldicque L, et al. Endoplasmic reticulum stress markers and ubiquitin–proteasome pathway activity in response to a 200-km run. *Med Sci Sports Exerc*. 2011;43(1):18–25.
119. Li J, Ni M, Lee B, Barron E, Hinton DR, Lee AS. The unfolded protein response regulator GRP78/BiP is required for endoplasmic reticulum integrity and stress-induced autophagy in mammalian cells. *Cell Death Differ*. 2008;15(9):1460–1471.
120. Yu Z, Wang AM, Adachi H, et al. Macroautophagy is regulated by the UPR-mediator CHOP and accentuates the phenotype of SBMA mice. *PLoS Genet*. 2011;7(10):e1002321.
121. Ngo JK, Pomatto LCD, Davies KJA. Upregulation of the mitochondrial Lon Protease allows adaptation to acute oxidative stress but dysregulation is associated with chronic stress, disease, and aging. *Redox Biol*. 2013;1(1):258–264.

CHAPTER SIX

Endurance Exercise and the Regulation of Skeletal Muscle Metabolism

Frank W. Booth[*,†,‡,§,1], **Gregory N. Ruegsegger**[*], **Ryan G. Toedebusch**[*], **Zhen Yan**[¶,||,#,**]

[*]Department of Biomedical Sciences, University of Missouri, Columbia, Missouri, USA
[†]Department of Nutrition and Exercise Physiology, University of Missouri, Columbia, Missouri, USA
[‡]Department of Medical Pharmacology and Physiology, University of Missouri, Columbia, Missouri, USA
[§]Dalton Cardiovascular Research Center, University of Missouri, Columbia, Missouri, USA
[¶]Department of Medicine, University of Virginia, Charlottesville, Virginia, USA
[||]Department of Pharmacology, University of Virginia, Charlottesville, Virginia, USA
[#]Department of Molecular Physiology and Biological Physics, University of Virginia, Charlottesville, Virginia, USA
[**]Center for Skeletal Muscle Research, University of Virginia, Charlottesville, Virginia, USA
[1]Corresponding author: e-mail address: boothf@missouri.edu

Contents

Abstract

Almost a half century ago, regular endurance exercise was shown to improve the capacity of skeletal muscle to oxidize substrates to produce ATP for muscle work. Since then, adaptations in skeletal muscle mRNA level were shown to happen with a single bout of exercise. Protein changes occur within days if daily endurance exercise continues. Some of the mRNA and protein changes cause increases in mitochondrial concentrations. One mitochondrial adaptation that occurs is an increase in fatty acid oxidation at a given absolute, submaximal workload. Mechanisms have been described as to how endurance training increases mitochondria. Importantly, *Pgc-1α* is a master regulator of mitochondrial biogenesis by increasing many mitochondrial proteins. However, not all

Progress in Molecular Biology and Translational Science, Volume 135
ISSN 1877-1173
http://dx.doi.org/10.1016/bs.pmbts.2015.07.016

adaptations to endurance training are associated with increased mitochondrial concentrations. Recent evidence suggests that the energetic demands of muscle contraction are by themselves stronger controllers of body weight and glucose control than is muscle mitochondrial content. Endurance exercise has also been shown to regulate the processes of mitochondrial fusion and fission. Mitophagy removes damaged mitochondria, a process that maintains mitochondrial quality. Skeletal muscle fibers are composed of different phenotypes, which are based on concentrations of mitochondria and various myosin heavy chain protein isoforms. Endurance training at physiological levels increases type IIa fiber type with increased mitochondria and type IIa myosin heavy chain. Endurance training also improves capacity of skeletal muscle blood flow. Endurance athletes possess enlarged arteries, which may also exhibit decreased wall thickness. VEGF is required for endurance training-induced increases in capillary–muscle fiber ratio and capillary density.

1. INTRODUCTION

Regular exercise not only enhances exercise capacity but also brings about beneficial health outcomes.[1,2] Skeletal muscle adaptation with remodeling of the mitochondrial network, the contractile apparatus, and the vasculatures underlies the functional improvement and many of the health benefits.[3,4] These adaptations for improved physical performance are analogous to an improvement of power and efficiency of a reciprocating engine by modifications of the cylinders, mechanical connections, and fuel supplies. In fact, central to the improvement of skeletal muscle contractile and metabolic functions is mitochondrial remodeling with improvement of mitochondrial quantity and quality. Extensive research has led to an improved understanding of the underlying molecular mechanisms that regulate mitochondrial biogenesis, the process by which new mitochondria are formed.[5] These efforts have culminated with the discovery of peroxisome proliferator-activated receptor-γ coactivator-1α (*Pgc-1α*) gene and the unveiling of its function in exercise-induced mitochondrial biogenesis in skeletal muscle.[6] It is now appreciated that mitochondrial biogenesis is not the only regulatory event that is required for the improvement of mitochondrial network in skeletal muscle following exercise training.[7] Mitochondria, like many other organelles in mammalian cells, are subject to damages due to the exposure to toxins, aging, and other insults. Therefore, it is imperative that muscle cells have means not only to generate new mitochondria, but also maintain the healthy ones and remove the damaged/dysfunctional ones. The regulation of mitochondrial life cycle ultimately determines the overall quantity, and most importantly, quality and function of mitochondrial network in skeletal muscle, which are the determinants of metabolic function and physical performance. This dynamic process allows for a program trading in old, dysfunctional mitochondria for new, healthy mitochondria.

2. MITOCHONDRIAL BIOGENESIS

Mitochondria play a crucial role in muscle contraction by generating ATP (via oxidative phosphorylation). Mitochondria also have important function in the regulation of calcium level, cell signaling, and apoptosis. The seminal study by Holloszy in 1967 provided the first direct evidence of exercise training-induced mitochondrial biogenesis in skeletal muscle, in which a strenuous program of treadmill running in rats led to significant increases of mitochondrial protein and enzyme activities.[8] It is now well accepted that mitochondrial biogenesis, the process of expansion of preexisting mitochondria, hinges on coordination of nuclear and mitochondrial-encoded gene expression.

The discovery of *Pgc-1α* in brown adipose tissue, originally as an inducible regulator of adaptive thermogenesis, has triggered mounting interest in elucidating the molecular and signaling mechanisms underlying exercise-induced mitochondrial biogenesis in skeletal muscle.[6,9] *Pgc-1α*, a master regulator of mitochondrial biogenesis, functions in the nucleus by controlling the expression of nuclear respiratory factors 1 and 2 (*Nrf-1* and *Nrf-2*), which in turn control the expression of mitochondrial transcription factor A (*Tfam*).[10] The central role of *Pgc-1α* in promoting mitochondrial biogenesis has been demonstrated in studies where increased *Pgc-1α* expression leads to increased expression of mitochondrial enzymes in the electron transport chain, such as ATP synthetase (β-subunit), cytochrome c oxidase subunits, and δ-aminolevulinate synthase.[11,12] Convincing evidence by Schreiber *et al.* showed that overexpression of *Pgc-1α* leads to upregulation of 151 genes, encoding mitochondrial proteins that are involved in various metabolic functions.[13] It is worth noting that the *Pgc-1α* isoform, *Pgc-1β*, although involved in mitochondrial biogenesis and basal oxygen consumption, is not induced by exercise, suggesting specific pathways of mitochondrial biogenesis induced by exercise training.[14]

Particularly relevant to exercise training-induced mitochondrial biogenesis in skeletal muscle are the findings that endurance exercise promotes *Pgc-1α* expression through transcriptional control and activity through post-translational modification. Goto *et al.* were the first to report that *Pgc-1α* mRNA level increased significantly in recruited skeletal muscle of rats that were swim trained compared to the nonexercised control group,[15] which was later observed in animal models and humans by other groups.[12,16,17] Evidence also exists supporting increased Pgc-1α activity by exercise.[18,19] Exciting experimental evidence in supporting *Pgc-1α* in mitochondrial biogenesis in skeletal muscle was obtained in transgenic mice with overexpression of *Pgc-1α* under the control of muscle-specific muscle creatine kinase (*Mck*) promoter, which led to profound changes toward oxidative muscle phenotype with robust mitochondrial biogenesis and resistance to electrically stimulated fatigue.[20] In fact, gene manipulation conditions that induce expression of *Pgc-1α* in skeletal muscle also mimic many effects of exercise on mitochondrial biogenesis.[21,22] More importantly, mice with genetic deletion of the *Pgc-1α* gene have impaired protection that exercise training provides against age-related decline in oxidative function (citrate synthase activity) and antioxidant gene expression.[23] Skeletal muscle-specific *Pgc-1α* knockout mice showed normal exercise training-induced fiber-type transformation, but significantly attenuated mitochondrial biogenesis and angiogenesis in skeletal muscle.[24] Taken together, *Pgc-1α* is necessary for some exercise-induced metabolic adaptations, including mitochondrial biogenesis, in skeletal muscle.

Exercise-induced mitochondrial biogenesis in skeletal muscle is partly controlled by activation of PGC-1α and consequent translocation to the nucleus,[19] which could be detected prior to its increased expression.[18] More recently, it has been shown that a single bout of exercise leads to increased PGC-1α level in the mitochondria,[25] where enhanced interaction between PGC-1α and TFAM at the mitochondrial DNA D-loop has been observed.[26] Therefore, increased PGC-1α activity along with subcellular translocation and interaction with important regulatory factors may also be involved in the coordination of the transcription of the nuclear and mitochondrial encoded mitochondrial genes, as well as the replication of mitochondrial DNA in mitochondrial adaptation in response to exercise training. This exercise-induced activation of PGC-1α is likely mediated through posttranslational modification. It has been shown that sirtuin (silent mating-type information regulation 2 homolog) 1 (SIRT1) deacetylates PGC-1α,[27] and SIRT1 activation is associated with increases in *Pgc-1α* target gene expression.[28] These findings support that SIRT1 activates PGC-1α through deacetylation. Interestingly, this SIRT1-mediated PGC-1α activation depends on prior phosphorylation of PGC-1α by AMPK,[29,30] integrating both contractile and metabolic cues in exercise-induced skeletal muscle adaptation.

As aforementioned, transcriptional control of the *Pgc-1α* gene appears to be one of the important mechanisms by which exercise training induces mitochondrial biogenesis in skeletal muscle.[12,31–34] A signaling–transcription coupling mechanism appears to sense the contractile and metabolic stress cues induced by exercise and promotes *Pgc-1α* gene regulation to induce adaptations in skeletal muscle.[3] Both AMPK[35] and p38γ mitogen-activated protein kinase (MAPK)[36,37] have been shown to be involved in the regulation of *Pgc-1α* in skeletal muscle in response to exercise training. Ampk activation may promote *Pgc-1α* expression[38] through PGC-1α phosphorylation and activation,[35] as Pgc-1α works in an autoregulatory loop.[39] An acute bout of exercise can readily activate AMPK by phosphorylation at Thr172,[40] and long-term exercise training leads to an increase in AMPK protein content.[41,42] Considering the fact that neither acute exercise nor long-term exercise training is capable of increasing Ampkα protein content in skeletal muscle of *Pgc-1α*-deficient mice,[43] it suggests that PGC-1α and AMPK are interdependent in the context of exercise-induced mitochondrial biogenesis in skeletal muscle.

p38 Mapk is a pathway that is readily activated by exercise.[44–46] It has been shown that a single bout of voluntary running in mice is sufficient to activate the p38 MAPK pathway,[36] and p38 MAPK activity is sufficient to promote the *Pgc-1α* promoter activity through MEF2 and ATF2 *in vivo*.[47,48] Molecular genetic studies revealed that muscle-specific deletion of either p38α or p38β MAPK does not affect exercise-induced mitochondrial biogenesis and angiogenesis in skeletal muscle, whereas mice with muscle-specific deletion of the *p38γ Mapk* or *Pgc-1α* gene have attenuated mitochondrial biogenesis and angiogenesis.[24,37] Therefore, the *p38γ Mapk–Pgc-1α* regulatory axis is functionally important for exercise-induced metabolic adaptations, specifically mitochondrial biogenesis and angiogenesis in skeletal muscle.

Pgc-1α does not appear to influence some beneficial metabolic effects of prolonged exercise, according to the Muoio lab.[49] When comparing mice specifically overexpressing *Pgc-1α* in skeletal muscle to nontransgenic littermates fed a high-fat diet, no differences in glucose tolerance or improvements in weight loss were observed following 10 weeks of voluntary wheel running and 10 weeks of voluntary wheel running with 25% caloric restriction.[49] Similarly, sedentary muscle-specific *Pgc-1α* overexpressing mice were more susceptible to diet-induced glucose tolerance and had decreased insulin action compared to nontransgenic mice. Thus, the authors conclude the energetic demands of muscle contraction induced by exercise are stronger controllers of body weight and glucose control than muscle mitochondrial content. They further suggest that many of the metabolic disease risk-lowering benefits of chronic endurance exercise training are not due to enhanced mitochondrial function. The Muoio group[49] finally concludes that the aforementioned processes are more complex than previously proposed.

3. MITOCHONDRIAL DYNAMICS AND MAINTENANCE

Mitochondria in skeletal muscle form a specialized network with distinct populations (i.e., subsarcolemmal and intermyofibrillar). The intermyofibrillar mitochondria are in proximity to the contractile apparatus, providing immediate high energy for contractile activities, whereas the subsarcolemmal mitochondria are situated in proximity to sarcolemma, responsible for ATP, for maintenance of its membrane electrical potential and metabolism of nutrients from outside the muscle cells.[50] The mitochondrial network in skeletal muscle is a dynamic organelle in terms of physical structure. The joining of nearby individual mitochondria and separation of mitochondria is referred to as fusion and fission, respectively.[51,52] The balance of fusion and fission controls mitochondrial structure and function. Generally speaking, healthy, metabolically active cells tend to have a large interconnected network of mitochondria, while dysfunctional mitochondria are separated from the network and removed by a process of autophagy (mitophagy, a category of autophagy).[52,53] Therefore, the balance between fusion and fission, and the signals responsible for their regulation are of extreme importance to the maintenance of the mitochondrial network.

Mitochondrial dynamics requires proteins involved in fission and fusion.[54] Mitofusins 1 and 2 (MFN1 and MFN2) control fusion of the mitochondrial outer membrane, and MGM1/OPA1 controls fusion of the mitochondrial inner membrane.[54] Dynamin-related protein 1 (DRP1), a GTPase, works together with Fission1 (FIS1) or mitochondrial fission factor (MFF) for fission of the mitochondrial outer membrane.[52,53] Genetic studies in cell culture and tissues have provided convincing evidence that fusion and fission are both critical for maintaining healthy mitochondrial population and cell function.[55–59] Conditions favoring one process over the other may be related to the functional status of the network. For example, mitochondrial fusion appears to be shut down in depolarized mitochondria, favoring fission.[60] Importantly, the depolarization and fission events appear to precede the removal of those mitochondria by mitophagy,[60] providing a possible link from the regulation of mitochondrial dynamics to the removal of damaged portions of the mitochondrial network.

There is clear evidence that exercise regulates both mitochondrial fusion and fission. For example, high intensity interval training in human subjects progressively increased the protein content of MFN1 and FIS1,[61] and a single bout of treadmill running in rats induced increased *Mfn1* and *MfnN2* mRNA.[62] A single bout of cycling exercise in human subjects led to increased *MFN1* and *MFN2* mRNA expression at 24-h postexercise,[63] which appears to follow the induction of *Pgc-1α* and *Errα* mRNA following the same exercise regimen. *In vitro* experiments suggest that *Mfn1* and *Mfn2* transcription is regulated by PGC-1α through ERRα.[63] Therefore, PGC-1α may also play an important role in regulating mitochondrial dynamics induced by exercise training in skeletal muscle. It remains to be certain how exercise precisely regulates mitochondrial fission and fusion at individual mitochondrial level, and what is the relationship of the pre-existing quality of the mitochondrial network region to the responses to exercise regimen.

4. MITOPHAGY

Mitochondrial toxins, free radicals, and other chemical agents may cause damage to part of the mitochondrial network, and functional mitochondria continually mix their content through fusion, while severely damaged mitochondria are prevented from fusing with healthy mitochondria.[60] The primary mechanism for removal of severely damaged mitochondria is thought to be autophagy specifically targeting mitochondria (mitophagy). This clearance process is particularly important for skeletal muscle function as accumulation of damaged mitochondria might ultimately hinder the ability of mitochondria to function properly and impair contractile and metabolic functions. In fact, a recent study has shown high-fat diets resulted in mitochondrial stress and an accumulation of damaged mitochondria in skeletal muscle, while exercise training clearly prevents/reverses these abnormalities.[64]

Autophagy is an evolutionarily conserved process for degradation of damaged organelles and macromolecules and recycle of nutrients by a lysosome-dependent mechanism.[65] From the perspective of biological function, autophagy can be classified as nonselective autophagy (stimulated by starvation to nonselectively degrade cellular components) and selective autophagy (triggered by accumulation of aggregate/misfolded proteins and damaged organelles and specifically targeting these cellular components). Upon activation of a number of autophagy-related kinases, a preautophagosomal structure called phagophore is formed, which elongates and engulfs the target and forms a double-membrane structure known as autophagosome. To complete the autophagy process, the autophagosome has to fuse with lysosome, which contains lysosomal enzymes, to form autophagolysosome. The proteins that are involved in this process are generally called autophagy-related genes (*Atgs*). The autophagic removal of mitochondria is called mitophagy, which is a critical process in the maintenance of mitochondrial quality. In most of cases, mitophagy is a selective process of removal of damaged mitochondria, in which depolarized, dysfunctional mitochondria undergo mitophagy.[66,67]

Mitochondrial degradation through mitophagy is clearly dependent on specific signals present on the damaged, dysfunctional region of the mitochondrial network. The pathway that involves PTEN-induced putative kinase protein 1 (PINK1) and PARKIN is the most studied pathway that relay the signals associated with mitochondrial damage to the induction of mitophagy.[68] Upon depolarization of mitochondrial inner membrane, PINK1 is stabilized and recruits PARKIN from the cytosol onto mitochondrial outer membrane,[68,69] which leads to polyubiquitination of mitochondrial outer membrane proteins and mitophagy.[70,71] Mitochondrial outer membrane proteins BNIP3 and BNIP3L (also known as NIX) are also important for mitophagy.[60,72] Experimental data show that forced expression of BNIP3 in adult skeletal fibers induces massive mitophagy,[73] whereas BNIP3L recruits the autophagy machinery to the mitochondria through its aminoterminal LC3-interacting region (LIR), which binds to light chain 3 (LC3, also known as ATG8).[74] Therefore, BNIP3L is a mitophagy receptor that is physically targeting autophagy machinery to the damaged mitochondria. Another mitophagy receptor is the mitochondrial outer membrane protein FUN14 domain containing 1 (FUNDC1), which has a LIR and is important in recruiting mitophagy machinery through its interaction with LC3.[75] Interestingly, both exercise training and enhanced PGC-1α expression promote BNIP3 protein expression in mouse skeletal muscle,[76] while acute bout exercise does not seem to induce PARKIN and BNIP3 expression under either fed or fasted conditions.[77] The relationship of the above processes that are directly related to damaged, dysfunctional mitochondria under the conditions of exercise training remains to be fully elucidated.

AMPK has been shown to activate ATG1 (also known as ULK1) through posttranslational modification of phosphorylation.[78] Since AMPK is readily activated by exercise in skeletal muscle[79] and ULK1 is required for mitophagy,[80] it is conceivable that exercise-mediated activation of AMPK–ULK1 signaling axis serves as a "top-down" signal, in combination with aforementioned "bottom-up" signals from damaged mitochondria, to promote mitophagy to removed damaged mitochondria with an assurance of great fidelity. A recent descriptive studies in mice strongly support this notion.[81] However, the precise functional role of this signaling axis in exercise-induced autophagy has yet to be explored. It is important to conduct integrative research to reconcile the global exercise signal(s) and the specific mitochondrial signal(s) to fully understand the mitochondrial quality control mechanisms in skeletal muscle, which will provide tremendously important information for future design of exercise programs and pharmacological therapeutics for prevention and treatment of diseases related to impaired mitochondrial function.

5. MULTIPLE TYPES OF EXERCISE ENDURANCE TESTS

Cardiorespiratory endurance, or aerobic endurance is: "the ability of the whole body to sustain prolonged, rhythmic exercise" (quotations are taken from Wilmore and Costill[82] in this paragraph). Two tests of cardiorespiratory endurance—submaximal and maximal—appear in the literature. The test of "*Submaximal* (cardiorespiratory) endurance capacity is more closely related to actual competitive endurance performance and likely is determined by both the individual's VO_2max and his or her lactate threshold." *Maximal* cardiorespiratory endurance capacity (aerobic power) is defined as "the highest rate of oxygen consumption attainable during maximal or exhaustive exercise." For animal studies to mimic human standards for maximal cardiovascular endurance tests, a determination of VO_2max or VO_2peak can be obtained, as described by Britton and Wisloff.[83–85] Maximal endurance capacity tests are ramped by nature, progressively increasing workload every 3–5 min past the lactate threshold until approaching or exceeding maximal workload. When exercise capacity tests are performed without VO_2 determinations as the "gold" standard for proof of maximal cardiorespiratory endurance capacity, the test endpoint relies upon an animal's motivation to avoid electric shock, which is not a surrogate for VO_2max or VO_2peak. Muscular endurance is "skeletal muscle's ability to repeatedly develop and sustain near-maximal or maximal forces."

Cardiorespiratory endurance capacity declines with aging. For example, times for top marathoners increased exponentially with age. For ages

30 → 50 → 70 → 90 years old, pace (min/mile) for men increased from 4:46 → 4:55 → 6:02 → 8:22, respectively, and for women from 5:10 → 5:36 → 7:26 → 11:04, respectively.[86]

6. TYPE I FIBERS ARE RELATED TO HUMAN ENDURANCE

In general, type I (slow-twitch) fibers have a high level of aerobic endurance, associated with better metabolic health at a given level of physical activity.[87,88] Type II fibers have a relatively poor aerobic endurance. In a classic paper in 1972, Gollnick and Saltin[89] showed that sports groups that were "extensively engaged in the endurance work" had the highest enzyme activities of mitochondrial Krebs cycle succinate dehydrogenase and the highest percentage of type I (slow-twitch) fibers within their vastus lateralis and deltoid muscles. The oxidative capacity of both types I and fast type II muscle fibers were higher in trained than untrained men. Mitochondrial concentration (mg/g wet muscle weight) is both inherent and acquired. Thus today, consensus is that endurance-type physical activity in humans can increase skeletal muscle mitochondrial concentration, but not alter the type I to II fiber-type ratio. On the other hand, physical inactivity such as limb immobilization decreases both mitochondrial concentration and the absolute numbers of type II muscle fibers. Whether long-term (years) endurance training in humans can increase type I percentage in skeletal muscle is unknown and continues to be debated. The possibility exists that many years of aerobic endurance training could increase type I fiber percentage,[90,31] but longitudinal human studies in the same subjects from childhood through adolescence are not permitted by most governmental committees for the protection of human subjects.

A skeletal muscle cell (termed fiber) is multinucleated and is the structural contractile unit. The terminology for the phenotype and gene expression of skeletal muscle fibers has been categorized according to their: (1) specific contractile functions of speed (fast or slow) based upon their contractile protein isoform,[91] and (2) density of mitochondria (high or low).[92] The complex history for the terminology of skeletal muscle fiber types is nearing a century old and is described in greater detail by Schiaffino's and Reggiani's review.[93] Historically, Peter *et al.* promoted a binominal nomenclature for fiber-type specializations in 1972 to describe the dual functions of: (1) contraction and (2) energy metabolism.[92] A less informative mixed nomenclature Roman numeral system and Arabic letter is predominant today. Slow oxidative fibers (type I) have the slowest contractile speed and a high mitochondrial density supporting postural/cardiovascular endurance function. The myosin heavy chain protein isoform has a slow contractile speed that uses less ATP per unit of work because of its lower myosin ATPase activity, relative to fast fibers, and is thus more metabolically efficient. The high mitochondrial density allows an oxidative supply of ATP for contractile work. The faster contracting muscle fibers have three common classifications; they are: (1) fast oxidative (type IIa), (2) fast oxidative–glycolytic (type IIx), and (3) fast glycolytic (type IIb). Rodents have all four fiber types, but human limb muscles have no IIb. Each of the four fiber types has its own myosin heavy chains (MHC) gene that is designated by the attachment of the Roman numeral after "myosin heavy chain." Contractile speeds progressively increase in the next order, designated by [Roman numeral for fiber type (human gene name) in italics]; they are: I (*MYH7*) < IIa (*MYH2*) < IIx (*MYH1*) < IIb (*MYH4*). Mitochondrial density is high in type I and IIa and then decreasing as follows I = IIa > IIx > IIb. Major confounders in published papers to the above nomenclature are: (1) MHC protein isoforms within a single fiber can be switched by changes in physical activity and physical inactivity, (2) a small percentage of single muscle fibers can have mixed populations of multiple MHC isoforms, (3) mitochondrial densities are in continuum from type I/IIa through IIb fibers, (4) superphysiological applications (electrical stimulations or gene modifications) cause MHC isoforms to switch beyond limits found in human exercise training, and (5) which fiber type (type I or IIa) has the higher mitochondrial density is species specific.

The current usage of a fiber-type classification based solely on contractile speed has produced misinformation on mitochondrial density as fast oxidative type IIa fiber sometimes being called a glycolytic fiber, instead of being called an oxidative–glycolytic. For example, Holloszy's lab,[31] contends that the common equation of "oxidative muscle fibers" exclusively with type I slow twitch fibers is wrong. He wrote that in rats, type IIa fibers have a considerably higher content of mitochondria and capacity for substrate oxidation than the type I fibers. Further, he points out that in humans, type I fibers have a greater content of mitochondria than type IIa fibers, but this difference largely disappears in response to strenuous aerobic endurance exercise training. Support for the previous contention is the next two quotations from Schiaffino and Reggiani[93]: "Both 2A and 2B fibers have high levels of glycolytic enzymes, in spite of the different oxidative enzyme complement; this led to the classification of slow oxidative (type 1), fast-twitch oxidative glycolytic (2A), and fast-twitch glycolytic fibers (2B)."

Misinformation is perpetuated by calling all type II fibers as "glycolytic," failing to acknowledge that type IIa fibers have higher mitochondrial density than IIx and IIb.

Another common fiber-type misnomer is the extrapolation to physiological aerobic endurance training experiments from gene manipulation experiments in mice,[20] or from 8, 12, or 24 daily chronic stimulation experiments in rats,[94] that show MHC 1 ↔ 2A fiber-type switches. Schiaffino's review article[93] summarizes his expert opinion from a lifetime of work in the field, as follows:

> *Intrinsic differences exist among various fiber types with respect to the activity-dependent type 1 ↔ 2A switch: 1) in soleus secondary generation fibers, the switch can take place easily, e.g., 2A → 1 switch during postnatal development, 1 → 2A switch after short-term inactivity; 2) in soleus primary generation fibers, the 1 → 2A switch requires longer inactivity periods, such as long-term spinal cord injury in humans or spinal cord isolation in rats, or as a result of more drastic experimental conditions, e.g., switch induced by high-frequency stimulation; and 3) finally, as discussed above, this switch is apparently even more difficult for fast muscles as 2A → 1 switch is rare and seen only after very long periods of low-frequency stimulation especially in combination with hypothyroidism.*

Intriguingly, Schiaffino and Reggiani[93] posit that muscle fiber adaptations in response to endurance exercise are restricted to intrinsic ranges within a given fiber type that limit mechanical adaptations. While a slow-to-fast fiber-type switch can occur after long periods of inactivity or injury, adaptations to increased activity (i.e., running, swimming) induce fiber-type switching to a more oxidative fast phenotype (e.g., 2X → 2A in humans), but do not increase switching to type I/slow fibers.[93] Similarly, there appear to be differences in the intrinsic firing patterns of slow and fast motor neurons that restrict fiber-type switching and potential adaptations to endurance exercise. Motor neuron depolarization frequencies (Hz) of firing are fixed and not altered by endurance training or inactivity. Rather, following endurance training, the total activity of a motor neuron (minutes/hours per day) may change, thus fostering the corresponding muscle fibers to transition to a fiber with a greater oxidative capacity (e.g., 2X → 2A transition in humans). Type I fibers can increase their mitochondrial capacity with training. Thus, it has been suggested that motor neurons likely do not change their firing frequency,[93] and thus slow motor neurons cannot change into fast motor neurons and vice versa. For example, given that slow motor neuron firing rate drastically decreases with inactivity, MHC protein decreases to 16% of control when nongravity bearing.[95] In summary, given the restrictive ranges by which fiber-type switching occurs, a disconnection between contractile and metabolic adaptations to endurance exercise appears to be present, such that endurance exercise increases fiber oxidative capacity without associated changes in MHC composition. This notion suggests that the metabolic and contractile differences, which accompany endurance-exercise training, may be controlled by distinct signaling pathways (see Ref. 93).

Interestingly, human and rodent studies have unexpectedly revealed a general trend toward an increased shortening velocity (V_O) (not decreased because of more slow MHC) and a decreased normalized force production (P_O/CSA) at the myofibrillar level with aerobic endurance training (see Ref. 90). Reidy *et al.*[90] had used individual muscle fibers to show that the type or amount of contractile proteins present cannot explain altered contractile force and speed observed with aerobic endurance training.

Long-term (>20 years) "recreational levels" of physical activity (male subjects studied at 67–74 years old) had an implied reinnervation of type 1 muscle fibers as compared to the age-matched sedentary group. Reidy *et al.*[90] wrote: "decades of high-level exercise allowed adaptation to denervation of individual muscle fibers beginning at the age of 50 years by saving otherwise lost muscle fibers through selective recruitment to slow motor units. These effects on size and structure of myofibers may delay functional decline in late aging." They further suggested that long-term physical activity mainly spared the slow motoneurons from age-related lesion/death, thereby increasing the chance that peripheral reinnervation to denervated fibers occurred because of sprouting of slow axons. As slow motor units are recruited first (at the lowest work intensity), prior to fast motor units, with increasing intensity of aerobic endurance types of exercise, they suggested, "... the intensity of recreational training reported is something that the general population may achieve" since walking recruits type I fibers.

While early aerobic endurance studies concentrated on human physical performance, in more recent decades, evidence has appeared to support the statement that a higher percentage of type 1 fibers in skeletal muscle is associated with better metabolic health.[96] For example, maximally insulin-stimulated glucose uptake is greater in rat type I than in type IIa or type IIb skeletal muscle fibers.[97] However, maximal contractile-stimulated glucose uptake was greater in rat type IIa than type I skeletal muscle in the same study.

7. REGULATION OF MYOSIN HEAVY CHAIN COMPOSITION BY AEROBIC ENDURANCE TYPES OF EXERCISE

Major comprehensive reviews are available by Schiaffino and Reggiani in 2011[93] and Yan *et al.*,[3] respectively, for signaling pathways controlling muscle fiber-type-specific gene programs. These reviews provide detailed interpretations for the roles of calcineurin-NFAT, MEF2, HDAC, CaMK, AMPK, PPARβ/δ, and PGC-1α/β signaling in skeletal muscle fiber types. Readers are referred to these reviews.

Here, we restrict ourselves to a few critical comments related to physiological animal models used in translational studies on MHC isoforms in human exercise. First, conversion from type II to type I skeletal muscle fibers in transgenic mice, mouse studies have not been shown to translate to aerobically endurance-trained humans. Holloszy and coworkers[31] wrote, "It is possible that the extreme training programs of professional cyclists, triathletes, etc., that involve many hours of intense exercise daily for years may cause an increase in slow muscle fibers, but this remains to be documented," which would be difficult to obtain because it would require longitudinal biopsies beginning in children. We posit that most ambulatory human physical activity exceeds a threshold for converting from MHCIIa to MHCI. However, publications do find long periods without using slow fibers against gravity results in a loss of slow MHCI. For example, Trappe *et al.*[98] found that a shift from slower contracting type I fibers to a faster contracting type IIa fibers after 4 weeks of human bed rest, confirming other human[99,100] and animal[101,102] studies. The "switch" from type I to type II is not a switch within the same fiber so much as a preferential loss of type 2 fibers so that the absolute amount of MHCI protein per muscle decreases. Thomason, Baldwin, and coworkers[95] showed this. In their experimental design of going from normal control to 56 days of hindlimb suspension to a subsequent 28-day recovery period, slow MHCI protein changed 13.4 → 2.1 → 11.2 mg, while fast MHC changed from 0.6 → 0.115 → 0.1 mg. Remarkably, the major factor in the switch from a slow to fast muscle, and its reversal, was due to the absolute loss, and regain of slow MHC, not due to an increase in fast MHC. The terminology of skeletal muscle fiber types is sometimes misused in fiber switching. For example, all fast fibers are glycolytic, but not all are oxidative. Type IIa fibers are oxidative–glycolytic fibers, whereas types IIx and IIb are glycolytic only.

To reemphasize the concepts so well described by Schiaffino and Reggiani,[93] mentioned above, each muscle fiber type has its own "adaptive range" of possible transformations in response to different activity patterns. Fast muscles have the capacity to adapt mostly in the range 2B ↔ 2X ↔ 2A, due to intrinsic properties that limit their range of possible adaptations under physiological conditions. Genes can be manipulated to modify MHC protein expression outside of physiological limits. An adequate knowledge of fiber types, as it relates to physical activity/inactivity adaptations, is necessary to provide valid interpretations.

8. BLOOD FLOW DURING ENDURANCE EXERCISE AND ITS EFFECT ON METABOLISM

During aerobic endurance exercise, the metabolic demands of skeletal muscle increase dramatically. The ability of the vascular network to selectively increase delivery of blood to working muscle 100-fold,[103] as well as remove metabolic waste products, termed vascular transport capacity,[104–107] is an important rate-limiting factor to exercise. One example is that when adding arm exercise to leg exercise during maximal exercise, the increase in leg blood is attenuated.[108] Another example is that the partial pressure of oxygen of working skeletal muscle at maximal effort is 3–4 mmHg.[109] In this regard, the regulation of skeletal muscle metabolism is directly influenced by the state of the vascular network and the muscles ability to efficiently exchange substrates to and from the vasculature.

Additionally, blood flow capacity differs between fiber types at rest, with acute exercise, and with adaptation to aerobic endurance exercise training. Increases in skeletal muscle blood flow during an acute bout of exercise are related to muscle fiber type and recruitment patterns.[110–112] Type I fibers, which are recruited before type II fibers for submaximal aerobic endurance exercise, are recognized to have increased capillarization, arteriolar density, and endothelium-dependent dilation compared with type II fibers in the rat.[104] Green and coauthors[113] recently reviewed published studies of resistance and conduit vessel adaptation in elite-level athletes who were in the aerobic-endurance trained state. They conclude, "... Sufficient evidence exists to conclude that aerobic endurance athletes possess enlarged arteries, which may also exhibit decreased wall thickness." They mention the obvious limitation for human studies is the *in vivo* assessments of size, structure, and function of different arteries in the live human. Green and coauthors[113] further conclude that limited evidence exists in aerobic-endurance athletes to support the concept of enhanced arterial responsiveness to pharmacological or physiological stimulation. They suggest that shear stress is likely implicated in conduit and resistance artery enlargement with aerobic endurance training. However, the mechanisms may be more complicated as changes in wall thickness may be systemic rather than localized to muscles recruited in training,[113] raising the question as to the nature of the adaptive stimulus in blood vessels in nonrecruited muscle during the aerobic endurance training.

9. ANGIOGENESIS

Mean fiber area/per capillary is type I = II < IIx in human skeletal muscle.[114] The capillary to fiber perimeter exchange index was greatest in rat type I fibers.[115] Capillary density increases with endurance training.[116] Delavar *et al.*[117] interpreted their data to suggest that the presence of VEGF within the myofiber is necessary for endurance (treadmill) training responses in capillarity and oxidative capacity and for improved running speed and endurance. They employed a complex genetically engineered mouse. The mouse had a skeletal myofiber-specific VEGF conditional gene-deletion (skmVEGF −/−). VEGF protein levels in skmVEGF −/− were reduced by >80% in untrained and trained hindlimb skeletal muscles. Increases of 20–25% in VEGF protein happened only in VEGF +/+ mice, but not in skmVEGF −/− mice after 8 weeks of endurance training on a treadmill. The increase in capillary-muscle fiber ratio, capillary density, and citrate synthase activities were almost twice as much in VEGF +/+ as in the skmVEGF −/− mice. The same training program in VEGF +/+ mice increased aerobic endurance by 99%.

10. SUMMARY AND CONCLUSIONS

Endurance exercise requires a continuous supply of energy for a lengthy duration of submaximal (no increase in blood lactate) contractile activity. Such energy demand is met mainly from within skeletal muscle (stored glycogen and fatty acids) and from oxygen, glucose, and fatty acids supplied by blood flow to working skeletal muscle. The energy demand disrupts homeostatic levels of high-energy nucleotides (ATP, ADP, AMP, etc.), glycogen, and various metabolites within the muscle fiber. It has been long known that homeostatic disruption at a given absolute, submaximal workload is lessened with endurance training. A reasonable hypothesis is that feedback from most homeostatic disruptions is sensed. The sensor then signals a change in gene expression (an adaptation), which minimizes the magnitude of succeeding homeostatic disruptions. The current state of the science, in our opinion, is that more is known about adaptations minimizing homeostatic disruptions than the sensors of the disruptions. Further, with passage of time, mechanisms of skeletal muscle adaptation to contraction have become progressively more and more complex. However, complexity should be expected. Skeletal muscle has functions to supply of amino acids to liver for blood glucose during fasting, strength to defend and build defenses, and endurance to travel distances for food. Our current understanding of endurance exercise-induced skeletal muscle adaptations involves orchestrated signaling–transcription coupling, coordinated nuclear-mitochondrial intergenomic communication, and dynamic spatial-temporal cellular and subcellular remodeling of this extremely elegant organ system. These adaptive changes in mitochondrial network, the contractile apparatus, and the vasculatures lead to improvements of physical and metabolic functions, analogous to an improvement of power and efficiency of a reciprocating engine. This will maximize specific survival challenges at the expenses of other functions (liver gluconeogenesis, power, and endurance). Acute responses and chronic adaptive changes in skeletal muscle proteins maximizing endurance exercise are polygenic (complex).

REFERENCES

1. Booth FW, Gordon SE, Carlson CJ, Hamilton MT. Waging war on modern chronic diseases: primary prevention through exercise biology. *J Appl Physiol (1985)*. 2000;88(2):774–787.
2. Haskell WL, Lee IM, Pate RR, et al. Physical activity and public health: updated recommendation for adults from the American College of Sports Medicine and the American Heart Association. *Circulation*. 2007;116(9):1081–1093.
3. Yan Z, Okutsu M, Akhtar YN, Lira VA. Regulation of exercise-induced fiber type transformation, mitochondrial biogenesis, and angiogenesis in skeletal muscle. *J Appl Physiol (1985)*. 2011;110(1):264–274.
4. Thyfault JP, Booth FW. Lack of regular physical exercise or too much inactivity. *Curr Opin Clin Nutr Metab Care*. 2011;14(4):374–378.
5. Lanza IR, Sreekumaran Nair K. Regulation of skeletal muscle mitochondrial function: genes to proteins. *Acta Physiol (Oxf)*. 2010;199(4):529–547.
6. Arany Z. PGC-1 coactivators and skeletal muscle adaptations in health and disease. *Curr Opin Genet Dev*. 2008;18(5):426–434.
7. Yan Z, Lira VA, Greene NP. Exercise training-induced regulation of mitochondrial quality. *Exerc Sport Sci Rev*. 2012;40(3):159–164.
8. Holloszy JO. Biochemical adaptations in muscle. Effects of exercise on mitochondrial oxygen uptake and respiratory enzyme activity in skeletal muscle. *J Biol Chem*. 1967;242(9):2278–2282.

9. Yan Z. Exercise, PGC-1alpha, and metabolic adaptation in skeletal muscle. *Appl Physiol Nutr Metab*. 2009;34(3):424–427.
10. Virbasius JV, Scarpulla RC. Activation of the human mitochondrial transcription factor A gene by nuclear respiratory factors: a potential regulatory link between nuclear and mitochondrial gene expression in organelle biogenesis. *Proc Natl Acad Sci USA*. 1994;91(4):1309–1313.
11. Wu Z, Puigserver P, Andersson U, et al. Mechanisms controlling mitochondrial biogenesis and respiration through the thermogenic coactivator PGC-1. *Cell*. 1999;98(1): 115–124.
12. Baar K, Wende A, Jones T, et al. Adaptations of skeletal muscle to exercise: rapid increase in the transcriptional coactivator PGC-1. *FASEB J*. 2002;16(14): 1879–1886.
13. Schreiber SN, Emter R, Hock MB, et al. The estrogen-related receptor alpha (ERRalpha) functions in PPARgamma coactivator 1alpha (PGC-1alpha)-induced mitochondrial biogenesis. *Proc Natl Acad Sci USA*. 2004;101(17):6472–6477.
14. Meirhaeghe A, Crowley V, Lenaghan C, et al. Characterization of the human, mouse and rat PGC1 beta (peroxisome-proliferator-activated receptor-gamma co-activator 1 beta) gene in vitro and in vivo. *Biochem J*. 2003;373(pt 1):155–165.
15. Goto M, Terada S, Kato M, et al. cDNA cloning and mRNA analysis of PGC-1 in epitrochlearis muscle in swimming-exercised rats. *Biochem Biophys Res Commun*. 2000;274(2):350–354.
16. Terada S, Goto M, Kato M, Kawanaka K, Shimokawa T, Tabata I. Effects of low-intensity prolonged exercise on PGC-1 mRNA expression in rat epitrochlearis muscle. *Biochem Biophys Res Commun*. 2002;296(2):350–354.
17. Norrbom J, Sundberg C, Ameln H, Kraus W, Jansson E, Gustafsson T. PGC-1{alpha} mRNA expression is influenced by metabolic perturbation in exercising human skeletal muscle. *J Appl Physiol*. 2004;96(1):189–194.
18. Wright D, Han D, Garcia-Roves P, Geiger P, Jones T, Holloszy J. Exercise-induced mitochondrial biogenesis begins before the increase in muscle PGC-1alpha expression. *J Biol Chem*. 2007;282(1):194–199.
19. McGee S, Hargreaves M. Exercise and myocyte enhancer factor 2 regulation in human skeletal muscle. *Diabetes*. 2004;53(5):1208–1214.
20. Lin J, Wu H, Tarr PT, et al. Transcriptional co-activator PGC-1 alpha drives the formation of slow-twitch muscle fibres. *Nature*. 2002;418(6899):797–801.
21. Ojuka E, Jones T, Han D, Chen M, Holloszy J. Raising Ca^{2+} in L6 myotubes mimics effects of exercise on mitochondrial biogenesis in muscle. *FASEB J*. 2003;17(6): 675–681.
22. Wu H, Kanatous S, Thurmond F, et al. Regulation of mitochondrial biogenesis in skeletal muscle by CaMK. *Science*. 2002;296(5566):349–352.
23. Leick L, Lyngby SS, Wojtasewski JF, Pilegaard H. PGC-1alpha is required for training-induced prevention of age-associated decline in mitochondrial enzymes in mouse skeletal muscle. *Exp Gerontol*. 2010;45(5):336–342.
24. Geng T, Li P, Okutsu M, et al. PGC-1alpha plays a functional role in exercise-induced mitochondrial biogenesis and angiogenesis but not fiber-type transformation in mouse skeletal muscle. *Am J Physiol Cell Physiol*. 2010;298(3):C572–C579.
25. Safdar A, Little JP, Stokl AJ, Hettinga BP, Akhtar M, Tarnopolsky MA. Exercise increases mitochondrial PGC-1alpha content and promotes nuclear-mitochondrial cross-talk to coordinate mitochondrial biogenesis. *J Biol Chem*. 2011;286(12): 10605–10617.
26. Gleyzer N, Vercauteren K, Scarpulla RC. Control of mitochondrial transcription specificity factors (TFB1M and TFB2M) by nuclear respiratory factors (NRF-1 and NRF-2) and PGC-1 family coactivators. *Mol Cell Biol*. 2005;25(4):1354–1366.

27. Nemoto S, Fergusson MM, Finkel T. SIRT1 functionally interacts with the metabolic regulator and transcriptional coactivator PGC-1{alpha}. *J Biol Chem*. 2005;280(16): 16456–16460.
28. Gurd BJ, Yoshida Y, McFarlan JT, et al. Nuclear SIRT1 activity, but not protein content, regulates mitochondrial biogenesis in rat and human skeletal muscle. *Am J Physiol Regul Integr Comp Physiol*. 2011;301(1):R67–R75.
29. Cantó C, Gerhart-Hines Z, Feige J, et al. AMPK regulates energy expenditure by modulating NAD(+) metabolism and SIRT1 activity. *Nature*. 2009;458(7241):1056–1060.
30. Canto C, Jiang LQ, Deshmukh AS, et al. Interdependence of AMPK and SIRT1 for metabolic adaptation to fasting and exercise in skeletal muscle. *Cell Metab*. 2010;11(3):213–219.
31. Garcia-Roves P, Huss J, Holloszy J. Role of calcineurin in exercise-induced mitochondrial biogenesis. *Am J Physiol Endocrinol Metab*. 2006;290(6):E1172–E1179.
32. Mathai AS, Bonen A, Benton CR, Robinson DL, Graham TE. Rapid exercise-induced changes in PGC-1alpha mRNA and protein in human skeletal muscle. *J Appl Physiol*. 2008;105(4):1098–1105.
33. Röckl KSC, Hirshman MF, Brandauer J, Fujii N, Witters LA, Goodyear LJ. Skeletal muscle adaptation to exercise training: AMP-activated protein kinase mediates muscle fiber type shift. *Diabetes*. 2007;56(8):2062–2069.
34. Deldicque L, Atherton P, Patel R, et al. Effects of resistance exercise with and without creatine supplementation on gene expression and cell signaling in human skeletal muscle. *J Appl Physiol (1985)*. 2008;104(2):371–378.
35. Jäger S, Handschin C, St-Pierre J, Spiegelman BM. AMP-activated protein kinase (AMPK) action in skeletal muscle via direct phosphorylation of PGC-1alpha. *Proc Natl Acad Sci USA*. 2007;104(29):12017–12022.
36. Akimoto T, Pohnert SC, Li P, et al. Exercise stimulates Pgc-1alpha transcription in skeletal muscle through activation of the p38 MAPK pathway. *J Biol Chem*. 2005;280(20):19587–19593.
37. Pogozelski AR, Geng T, Li P, et al. p38gamma mitogen-activated protein kinase is a key regulator in skeletal muscle metabolic adaptation in mice. *PLoS One*. 2009;4(11): e7934.
38. Irrcher I, Ljubicic V, Kirwan AF, Hood DA. AMP-activated protein kinase-regulated activation of the PGC-1alpha promoter in skeletal muscle cells. *PLoS One*. 2008;3(10): e3614.
39. Handschin C, Rhee J, Lin J, Tarr P, Spiegelman B. An autoregulatory loop controls peroxisome proliferator-activated receptor gamma coactivator 1alpha expression in muscle. *Proc Natl Acad Sci USA*. 2003;100(12):7111–7116.
40. Durante PE, Mustard KJ, Park SH, Winder WW, Hardie DG. Effects of endurance training on activity and expression of AMP-activated protein kinase isoforms in rat muscles. *Am J Physiol Endocrinol Metab*. 2002;283(1):E178–E186.
41. Clark SA, Chen ZP, Murphy KT, et al. Intensified exercise training does not alter AMPK signaling in human skeletal muscle. *Am J Physiol Endocrinol Metab*. 2004;286(5):E737–E743.
42. Sriwijitkamol A, Ivy JL, Christ-Roberts C, DeFronzo RA, Mandarino LJ, Musi N. LKB1-AMPK signaling in muscle from obese insulin-resistant Zucker rats and effects of training. *Am J Physiol Endocrinol Metab*. 2006;290(5):E925–E932.
43. Leick L, Wojtaszewski JFP, Johansen ST, et al. PGC-1alpha is not mandatory for exercise- and training-induced adaptive gene responses in mouse skeletal muscle. *Am J Physiol Endocrinol Metab*. 2008;294(2):E463–E474.
44. Boppart M, Asp S, Wojtaszewski J, Fielding R, Mohr T, Goodyear L. Marathon running transiently increases c-Jun NH2-terminal kinase and p38 activities in human skeletal muscle. *J Physiol*. 2000;526(pt 3):663–669.

45. Boppart M, Hirshman M, Sakamoto K, Fielding R, Goodyear L. Static stretch increases c-Jun NH2-terminal kinase activity and p38 phosphorylation in rat skeletal muscle. *Am J Physiol Cell Physiol.* 2001;280(2):C352–C358.
46. Yu M, Blomstrand E, Chibalin A, Krook A, Zierath J. Marathon running increases ERK1/2 and p38 MAP kinase signalling to downstream targets in human skeletal muscle. *J Physiol.* 2001;536(pt 1):273–282.
47. Akimoto T, Li P, Yan Z. Functional interaction of regulatory factors with the Pgc-1alpha promoter in response to exercise by in vivo imaging. *Am J Physiol Cell Physiol* 2008;295(1):C288–C292.
48. Akimoto T, Sorg B, Yan Z. Real-time imaging of peroxisome proliferator-activated receptor-gamma coactivator-1alpha promoter activity in skeletal muscles of living mice. *Am J Physiol Cell Physiol.* 2004;287(3):C790–C796.
49. Wong KE, Mikus CR, Slentz DH, et al. Muscle-specific overexpression of PGC-1alpha does not augment metabolic improvements in response to exercise and caloric restriction. *Diabetes.* 2015;64(5):1532–1543.
50. Cogswell AM, Stevens RJ, Hood DA. Properties of skeletal muscle mitochondria isolated from subsarcolemmal and intermyofibrillar regions. *Am J Physiol.* 1993;264(2 pt 1):C383–C389.
51. Westermann B. Mitochondrial fusion and fission in cell life and death. *Nat Rev Mol Cell Biol.* 2010;11(12):872–884.
52. Chan DC. Fusion and fission: interlinked processes critical for mitochondrial health. *Annu Rev Genet.* 2012;46:265–287.
53. Youle RJ, van der Bliek AM. Mitochondrial fission, fusion, and stress. *Science.* 2012;337(6098):1062–1065.
54. Chan DC. Mitochondria: dynamic organelles in disease, aging, and development. *Cell.* 2006;125(7):1241–1252.
55. Griparic L, van der Wel NN, Orozco IJ, Peters PJ, van der Bliek AM. Loss of the intermembrane space protein Mgm1/OPA1 induces swelling and localized constrictions along the lengths of mitochondria. *J Biol Chem.* 2004;279(18):18792–18798.
56. Papanicolaou KN, Khairallah RJ, Ngoh GA, et al. Mitofusin-2 maintains mitochondrial structure and contributes to stress-induced permeability transition in cardiac myocytes. *Mol Cell Biol.* 2011;31(6):1309–1328.
57. Chen H, Vermulst M, Wang YE, et al. Mitochondrial fusion is required for mtDNA stability in skeletal muscle and tolerance of mtDNA mutations. *Cell.* 2010;141(2): 280–289.
58. Chen H, Detmer SA, Ewald AJ, Griffin EE, Fraser SE, Chan DC. Mitofusins Mfn1 and Mfn2 coordinately regulate mitochondrial fusion and are essential for embryonic development. *J Cell Biol.* 2003;160(2):189–200.
59. Ishihara N, Nomura M, Jofuku A, et al. Mitochondrial fission factor Drp1 is essential for embryonic development and synapse formation in mice. *Nat Cell Biol.* 2009;11(8): 958–966.
60. Twig G, Elorza A, Molina AJ, et al. Fission and selective fusion govern mitochondrial segregation and elimination by autophagy. *EMBO J.* 2008;27(2):433–446.
61. Perry CG, Lally J, Holloway GP, Heigenhauser GJ, Bonen A, Spriet LL. Repeated transient mRNA bursts precede increases in transcriptional and mitochondrial proteins during training in human skeletal muscle. *J Physiol.* 2010;588(pt 23):4795–4810.
62. Ding H, Jiang N, Liu H, et al. Response of mitochondrial fusion and fission protein gene expression to exercise in rat skeletal muscle. *Biochim Biophys Acta.* 2010;1800(3):250–256.
63. Cartoni R, Léger B, Hock MB, et al. Mitofusins 1/2 and ERRalpha expression are increased in human skeletal muscle after physical exercise. *J Physiol (Lond).* 2005;567(pt 1):349–358.

64. Laker RC, Xu P, Ryall KA, et al. A novel MitoTimer reporter gene for mitochondrial content, structure, stress, and damage in vivo. *J Biol Chem.* 2014;289(17): 12005–12015.
65. Mizushima N, Levine B, Cuervo A, Klionsky D. Autophagy fights disease through cellular self-digestion. *Nature.* 2008;451(7182):1069–1075.
66. Kim I, Rodriguez-Enriquez S, Lemasters JJ. Selective degradation of mitochondria by mitophagy. *Arch Biochem Biophys.* 2007;462(2):245–253.
67. Rodriguez-Enriquez S, He L, Lemasters JJ. Role of mitochondrial permeability transition pores in mitochondrial autophagy. *Int J Biochem Cell Biol.* 2004;36(12): 2463–2472.
68. Narendra DP, Jin SM, Tanaka A, et al. PINK1 is selectively stabilized on impaired mitochondria to activate Parkin. *PLoS Biol.* 2010;8(1):e1000298.
69. Narendra D, Tanaka A, Suen DF, Youle RJ. Parkin is recruited selectively to impaired mitochondria and promotes their autophagy. *J Cell Biol.* 2008;183(5):795–803.
70. Chan NC, Salazar AM, Pham AH, et al. Broad activation of the ubiquitin-proteasome system by Parkin is critical for mitophagy. *Hum Mol Genet.* 2011;20(9):1726–1737.
71. Sarraf SA, Raman M, Guarani-Pereira V, et al. Landscape of the PARKIN-dependent ubiquitylome in response to mitochondrial depolarization. *Nature.* 2013;496(7445): 372–376.
72. Schon EA, DiMauro S, Hirano M. Human mitochondrial DNA: roles of inherited and somatic mutations. *Nat Rev Genet.* 2012;13(12):878–890.
73. Romanello V, Guadagnin E, Gomes L, et al. Mitochondrial fission and remodelling contributes to muscle atrophy. *EMBO J.* 2010;29(10):1774–1785.
74. Novak I, Kirkin V, McEwan DG, et al. Nix is a selective autophagy receptor for mitochondrial clearance. *EMBO Rep.* 2010;11(1):45–51.
75. Chen G, Han Z, Feng D, et al. A regulatory signaling loop comprising the PGAM5 phosphatase and CK2 controls receptor-mediated mitophagy. *Mol Cell.* 2014;54(3): 362–377.
76. Lira VA, Okutsu M, Zhang M, et al. Autophagy is required for exercise training-induced skeletal muscle adaptation and improvement of physical performance. *FASEB J.* 2013;27(10):4184–4193.
77. Jamart C, Naslain D, Gilson H, Francaux M. Higher activation of autophagy in skeletal muscle of mice during endurance exercise in the fasted state. *Am J Physiol Endocrinol Metab.* 2013;305(8):E964–E974.
78. Egan DF, Shackelford DB, Mihaylova MM, et al. Phosphorylation of ULK1 (hATG1) by AMP-activated protein kinase connects energy sensing to mitophagy. *Science.* 2011;331(6016):456–461.
79. Winder WW, Holmes BF, Rubink DS, Jensen EB, Chen M, Holloszy JO. Activation of AMP-activated protein kinase increases mitochondrial enzymes in skeletal muscle. *J Appl Physiol.* 2000;88(6):2219–2226.
80. Kundu M, Lindsten T, Yang C-Y, et al. Ulk1 plays a critical role in the autophagic clearance of mitochondria and ribosomes during reticulocyte maturation. *Blood.* 2008;112(4):1493–1502.
81. Pagano AF, Py G, Bernardi H, Candau RB, Sanchez AM. Autophagy and protein turnover signaling in slow-twitch muscle during exercise. *Med Sci Sports Exerc.* 2014;46(7): 1314–1325.
82. Wilmore JH, Costill DL. *Physiology of Sport and Exercise.* 3rd ed. Champion, Illinois: Human Kinetics; 2004.
83. Koch LG, Kemi OJ, Qi N, et al. Intrinsic aerobic capacity sets a divide for aging and longevity. *Circ Res.* 2011;109(10):1162–1172.
84. Koch LG, Britton SL, Wisloff U. A rat model system to study complex disease risks, fitness, aging, and longevity. *Trends Cardiovasc Med.* 2012;22(2):29–34.

85. Hoydal MA, Wisloff U, Kemi OJ, Ellingsen O. Running speed and maximal oxygen uptake in rats and mice: practical implications for exercise training. *Eur J Cardiovasc Prev Rehabil.* 2007;14(6):753–760.
86. Trappe S. Marathon runners: how do they age? *Sports Med.* 2007;37(4–5):302–305.
87. Hickey MS, Carey JO, Azevedo JL, et al. Skeletal muscle fiber composition is related to adiposity and in vitro glucose transport rate in humans. *Am J Physiol.* 1995;268(3 pt 1): E453–E457.
88. Tanner CJ, Barakat HA, Dohm GL, et al. Muscle fiber type is associated with obesity and weight loss. *Am J Physiol Endocrinol Metab.* 2002;282(6):E1191–E1196.
89. Gollnick PD, Armstrong RB, Saubert CW, Piehl K, Saltin B. Enzyme activity and fiber composition in skeletal muscle of untrained and trained men. *J Appl Physiol.* 1972;33(3):312–319.
90. Reidy PT, Hinkley JM, Trappe TA, Trappe SW, Harber MP. Protein composition of endurance trained human skeletal muscle. *Int J Sports Med.* 2014;35(6):476–481.
91. Baldwin KM, Haddad F. Effects of different activity and inactivity paradigms on myosin heavy chain gene expression in striated muscle. *J Appl Physiol (1985).* 2001;90(1): 345–357.
92. Peter JB, Barnard RJ, Edgerton VR, Gillespie CA, Stempel KE. Metabolic profiles of three fiber types of skeletal muscle in guinea pigs and rabbits. *Biochemistry.* 1972;11(14):2627–2633.
93. Schiaffino S, Reggiani C. Fiber types in mammalian skeletal muscles. *Physiol Rev.* 2011;91(4):1447–1531.
94. Pette D, Vrbova G. Adaptation of mammalian skeletal muscle fibers to chronic electrical stimulation. *Rev Physiol Biochem Pharmacol.* 1992;120:115–202.
95. Thomason DB, Herrick RE, Surdyka D, Baldwin KM. Time course of soleus muscle myosin expression during hindlimb suspension and recovery. *J Appl Physiol (1985).* 1987;63(1):130–137.
96. Booth FW, Roberts CK, Laye MJ. Lack of exercise is a major cause of chronic diseases. *Compr Physiol.* 2012;2(2):1143–1211.
97. Henriksen EJ, Bourey RE, Rodnick KJ, Koranyi L, Permutt MA, Holloszy JO. Glucose transporter protein content and glucose transport capacity in rat skeletal muscles. *Am J Physiol.* 1990;259(4 pt 1):E593–E598.
98. Trappe S, Trappe T, Gallagher P, Harber M, Alkner B, Tesch P. Human single muscle fibre function with 84 day bed-rest and resistance exercise. *J Physiol.* 2004;557(pt 2): 501–513.
99. Zhou MY, Klitgaard H, Saltin B, Roy RR, Edgerton VR, Gollnick PD. Myosin heavy chain isoforms of human muscle after short-term spaceflight. *J Appl Physiol (1985).* 1995;78(5):1740–1744.
100. Ohira Y, Yoshinaga T, Ohara M, et al. Myonuclear domain and myosin phenotype in human soleus after bed rest with or without loading. *J Appl Physiol (1985).* 1999;87(5):1776–1785.
101. Caiozzo VJ, Baker MJ, Herrick RE, Tao M, Baldwin KM. Effect of spaceflight on skeletal muscle: mechanical properties and myosin isoform content of a slow muscle. *J Appl Physiol (1985).* 1994;76(4):1764–1773.
102. Caiozzo VJ, Haddad F, Baker MJ, Baldwin KM. Influence of mechanical loading on myosin heavy-chain protein and mRNA isoform expression. *J Appl Physiol (1985).* 1996;80(5):1503–1512.
103. Mortensen SP, Saltin B. Regulation of the skeletal muscle blood flow in humans. *Exp Physiol.* 2014;99:1552–1558.
104. Sexton WL, Laughlin MH. Influence of endurance exercise training on distribution of vascular adaptations in rat skeletal muscle. *Am J Physiol.* 1994;266(2 pt 2):H483–H490.

105. Laughlin MH, Ripperger J. Vascular transport capacity of hindlimb muscles of exercise-trained rats. *J Appl Physiol (1985)*. 1987;62(2):438–443.
106. Sexton WL, Korthuis RJ, Laughlin MH. High-intensity exercise training increases vascular transport capacity of rat hindquarters. *Am J Physiol*. 1988;254(2 pt 2): H274–H278.
107. Sexton WL, Korthuis RJ, Laughlin MH. Vascular flow capacity of hindlimb skeletal muscles in spontaneously hypertensive rats. *J Appl Physiol (1985)*. 1990;69(3): 1073–1079.
108. Boushel R, Saltin B. Ex vivo measures of muscle mitochondrial capacity reveal quantitative limits of oxygen delivery by the circulation during exercise. *Int J Biochem Cell Biol*. 2013;45(1):68–75.
109. Wagner PD. Muscle intracellular oxygenation during exercise: optimization for oxygen transport, metabolism, and adaptive change. *Eur J Appl Physiol*. 2012;112(1):1–8.
110. Laughlin MH, Armstrong RB. Muscular blood flow distribution patterns as a function of running speed in rats. *Am J Physiol*. 1982;243(2):H296–H306.
111. Laughlin MH, Armstrong RB. Rat muscle blood flows as a function of time during prolonged slow treadmill exercise. *Am J Physiol*. 1983;244(6):H814–H824.
112. Laughlin MH, Armstrong RB. Muscle blood flow during locomotory exercise. *Exerc Sport Sci Rev*. 1985;13:95–136.
113. Green DJ, Spence A, Rowley N, Thijssen DH, Naylor LH. Vascular adaptation in athletes: is there an 'athlete's artery'? *Exp Physiol*. 2012;97(3):295–304.
114. Lithell H, Lindgarde F, Hellsing K, et al. Body weight, skeletal muscle morphology, and enzyme activities in relation to fasting serum insulin concentration and glucose tolerance in 48-year-old men. *Diabetes*. 1981;30(1):19–25.
115. Gouspillou G, Sgarioto N, Norris B, et al. The relationship between muscle fiber type-specific PGC-1alpha content and mitochondrial content varies between rodent models and humans. *PLoS One*. 2014;9(8):e103044.
116. Hermansen L, Wachtlova M. Capillary density of skeletal muscle in well-trained and untrained men. *Acta Physiol Scand*. 1970;79(2):16A.
117. Delavar H, Nogueira L, Wagner PD, Hogan MC, Metzger D, Breen EC. Skeletal myofiber VEGF is essential for the exercise training response in adult mice. *Am J Physiol Regul Integr Comp Physiol*. 2014;306(8):R586–R595.

CHAPTER SEVEN

Exercise and the Regulation of Skeletal Muscle Hypertrophy

Chris McGlory, Stuart M. Phillips[1]
Department of Kinesiology, McMaster University, Hamilton, Ontario, Canada
[1]Corresponding author: e-mail address: phillis@mcmaster.ca

Contents

Abstract

Skeletal muscle is a critical organ serving as the primary site for postprandial glucose disposal and the generation of contractile force. The size of human skeletal muscle mass is dependent upon the temporal relationship between changes in muscle protein synthesis (MPS) and muscle protein breakdown. The aim of this chapter is to review our current understanding of how resistance exercise influences protein turnover with a specific emphasis on the molecular factors regulating MPS. We also will discuss recent data relating to the prescription of resistance exercise to maximize skeletal muscle hypertrophy. Finally, we evaluate the impact of age and periods of disuse on the loss of muscle mass and the controversy surround the etiology of muscle disuse atrophy.

1. INTRODUCTION

Exercise *per se* is a potent stimulus for the remodeling and reconditioning of skeletal muscle.[1] Alterations in the load, frequency, and duration of contractile activity are known to induce differential phenotypic adaptations specific to the nature of the stimulus.[1] Resistance exercise, for

Progress in Molecular Biology and Translational Science, Volume 135
ISSN 1877-1173
http://dx.doi.org/10.1016/bs.pmbts.2015.06.018

example, is characterized by repeated bouts of loaded contractions that result in the rapid recruitment of type 2 muscle fibers and stimulation of muscle protein synthesis (MPS).[1,2] When combined with adequate nutrition, resistance exercise improves muscular strength partly through an increase in the radial diameter of individual muscle fibers, a process termed *hypertrophy*.[3] At the outset, it is important to first clarify the nomenclature of the term hypertrophy. The increase in skeletal muscle size with resistance exercise is due to an increase in the cross-sectional area (CSA) of individual muscle fibers. There are, however, documented reports—mainly in rodents—of increases in muscle size associated with increases in the number of individual myofibers within a muscle,[4] which is termed *hyperplasia*. Hyperplasia is proposed to be the result of satellite cell proliferation and new fiber formation, but to date, there is little empirical evidence that hyperplasia occurs in physiological models of human resistance exercise. That is not to say that satellite cells are not important in exercise adaptation. The primary aim of this chapter is to review our current understanding of the mechanisms by which resistance exercise stimulates skeletal muscle *hypertrophy*. We will begin by discussing the biological basis of resistance exercise-induced muscle hypertrophy and a degree of incongruence between current recommendations and recent data relating to prescription of resistance exercise training. We will then examine the variability in the hypertrophic response to resistance training between individuals and how, at the molecular level, this variability and MPS in general are regulated. Specifically, we will discuss the influence of factors intrinsic to skeletal muscle such as the activation of the mechanistic target of rapamycin complex 1 (mTORC1) upon skeletal muscle loading. We will conclude by describing how muscle unloading influences the regulation of muscle mass. For the purposes of concision and relevance to the human model, we will focus on data procured primarily from humans; however, when necessary, studies from rodent and cell-based systems will be incorporated to support relevant points of discussion.

2. RESISTANCE EXERCISE TO ENHANCE SKELETAL MUSCLE MASS

Human skeletal muscle mass is a function of the balance between MPS and muscle protein breakdown (MPB). In healthy nondiseased humans, it is the changes in the rates of MPS and not those of MPB that influence myofiber size. It is for this reason that the focus of the majority of research in recent years has been on factors that ultimately impact upon MPS. Many of these studies have employed stable isotope technology to directly track the incorporation of labeled amino acids into new muscle proteins and into the contractile apparatus of a myofiber. Since the introduction of this technology into the exercise and nutritional sciences, a large body of data has been generated characterizing how various exercise and nutritional manipulations influence muscle protein turnover. In this regard, we now know that the consumption of proteins or amino acids, particularly the essential amino acids, leads to a systemic hyperaminoacidemia stimulating a significant and transient increase in rates of MPS.[5] When a bout of resistance exercise is performed prior to the consumption of amino acids, there is a synergistic impact on rates of MPS over and above that observed with amino acid ingestion alone.[6] In essence, resistance exercise sensitizes skeletal muscle to the anabolic effects of amino acid consumption, an effect that can persist for up to 24–48 h postexercise.[7]

One of the earliest documented reports that resistance exercise enhances muscle function was by De Lorme and Watkins who, in 1948,[8] prescribed "progressive resistance exercise" to treat patients with poliomyelitis. Since this work and that of others demonstrating the efficacy of resistance exercise to enhance skeletal muscle mass,[9] resistance exercise training is now widely used in both clinical and athletic settings to combat diseases associated with the loss of muscle mass (myopenia) and strength (dynapenia) as well as to enhance athletic success and general fitness. The term progression refers to the appropriate manipulation of (i) load per repetition, which is most often prescribed as a percentage of one repetition maximum (1 RM); (ii) number of sets performed; (iii) recovery between sets; (iv) muscle action (concentric vs. eccentric); and (v) total exercise volume (i.e., load × repetitions). Despite receiving significant experimental attention, the optimal resistance exercise-training program incorporating various modifications of the above variables, at least from the perspective of inducing skeletal muscle hypertrophy, remains unknown, and it could be argued that on a population basis it may not even exist.

Current recommendations from the most recent American College of Sport Medicine position stand[10] state that for "...novice and intermediate individuals, it is recommended that moderate loading be used (70–85% 1 RM) for 8–12 repetitions per set for one to three sets per exercise'. Whilst for advanced trainers 'it is recommended that a loading range of 70–100% of 1 RM be used for 1–12 repetitions per set for three to six sets per exercise in a periodized manner such that the majority of training is devoted to 6–12 RM and less training devoted to 1–6 RM." These recommendations and the literature supporting them are surprisingly thin. In a series of studies, it has been demonstrated that repetition load plays a minimal, if not negligible, role in the hypertrophic response to loaded contractions as long as volitional fatigue is achieved.[11] Indeed, in one study recreationally trained participants who performed 30% of their 1 RM until failure, elicited similar increases in MPS 4 h postresistance exercise relative to participants who performed resistance exercise at 90% of their 1 RM. Another interesting finding was that myofibrillar MPS was elevated 24 h postresistance exercise in the 30% 1 RM condition but not the 90% 1 RM condition. Expanding upon these findings, a follow-up study determined whether the similar acute responses (i.e., MPS) translated into longitudinal changes in muscle size.[12] Again, contrary to the existing guidelines, performing resistance exercise to failure with a low repetition load (in this instance 30% 1 RM) produced the same gains in muscle size as compared with performing resistance exercise to failure at a load of 80% of 1 RM. It is important to note that strength gains in that study were greater in the 80% 1 RM group. Thus, from a strength perspective, training at a higher repetition load may be beneficial, particularly for athletes. It also must be acknowledged that exercise regimens characterized by higher exercise volumes (i.e., sets × repetitions × sessions) display favorable gains in muscle growth.[13] So, taken together, these data demonstrate that as long as volitional failure is achieved, and the exerciser performs a comparable amount of effort, repetition load has little influence in the hypertrophic response to resistance exercise at least within the range of 30–80% of 1 RM.[11]

3. HETEROGENEITY IN RESPONSE TO RESISTANCE TRAINING

As previously discussed, an acute bout of resistance exercise coupled with amino acid ingestion results in an increase in the rate of MPS, whereas frequent, episodic bouts of resistance exercise, and amino acid ingestion promote skeletal muscle hypertrophy.[9] The increases in muscle growth with resistance training are often cited as occurring at a linear rate; however, we propose that expressing changes in rates of hypertrophy, as an average rate per day likely obscures some important insight into the process of muscular hypertrophy. In fact, there are numerous reports of large variability in both MPS and gains in muscle size in response to resistance training between individuals.[14,15] The individual variance in gains in muscle mass with resistance training was well demonstrated by Hubal *et al.*[14] in

585 participants with assessment of changes in isometric and dynamic strength as well as changes in muscle CSA following 12 weeks of progressive resistance training of the *biceps brachii*. Intriguingly, in response to the training protocol, 1 RM strength gains ranged from 0% to +250%, whereas changes in muscle size ranged from −2% to +59%. When examining the differences between the sexes, changes in muscle CSA were 2.5% greater in men compared with women, while relative gains in strength were greater in women. There are other studies that also support the notion of extreme individual variability in the hypertrophic response to supervised resistance training (Fig. 1). Indeed, Mitchell *et al.*[12] observed that following 16 weeks of resistance exercise training increases in fat- and bone-free mass ranged from 0 to 9 kg while there was also large variability in the gains in quadriceps volume (443 cm^3) and leg press 1 RM (200 kg). These findings are in agreement with those of Hubal *et al.*,[14] suggesting that a "one size fits all" approach is likely neither appropriate nor effective for promotion of optimal gains with resistance training. Another interesting finding from the study of Mitchell *et al.*[12] was that, in addition to the variable hypertrophic response, the increase in MPS following the first bout of resistance exercise failed to predict the magnitude of gains in muscle mass. So, it appears that not only are there differences in the hypertrophic response *between* individuals following resistance exercise training but the magnitude of the response *within* individual also lacks uniformity. Factors extraneous to the control of the study (i.e., nutritional intake) could be cited as confounding factors, but the marked range in skeletal muscle mass gains among humans is now thought to be biological in nature. However, biological

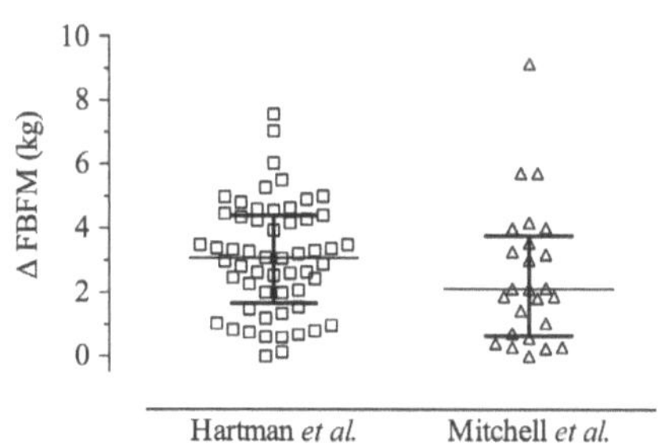

Figure 1 Change in fat- and bone-free mass (FBFM) in response to 10 weeks of resistance exercise training (Mitchell *et al.*[12]) and in response to 12 weeks of resistance training in combination with fat-free milk ($n = 18$), fat-free soy ($n = 19$), and control ($n = 19$) (Hartman *et al.*[16]). Data are expressed as individual delta change (kg), median, and interquartile range.

basis for resistance exercise-induced skeletal muscle hypertrophy remains unknown and is currently a topic of intense debate.[17]

4. THE INFLUENCE OF SYSTEMIC HORMONES

For many years, it has been known that resistance exercise results in a transient (15–30 min) increase in what are purported to be "anabolic" hormones such as free and protein-bound forms of testosterone, insulin-like growth factor 1 (IGF1), and growth hormone.[18] Although there is evidence to show that exogenous administration of supraphysiological doses of testosterone enhances skeletal muscle hypertrophy,[19] there is comparably little support for the efficacy of growth hormone[20] and IGF1[21] to exert the same effects, at least in adult humans. Indeed, it is indisputable that secretion of growth hormone supports bone and muscle growth in children, and supraphysiological doses of growth hormone enhance collagen protein synthesis.[21] However, the administration of growth hormone directly does not enhance rates of skeletal MPS.[20] One argument is that growth hormone-enhanced collagen synthesis may facilitate increased loading of the muscle during resistance exercise, which, subsequently, stimulates greater rates of MPS. However, this hypothesis has yet to be experimentally corroborated. Alternatively, growth hormone is known to promote IGF1 production in the liver. IGF1 also is elevated in a transient manner postexercise[22] and thus has been implicated, in addition to both growth hormone and testosterone, as a mediator of MPS. Data to support this assertion have come predominantly from studies in rodent models. In one such study, blockade of the IGF1 receptor resulted in a reduction in the elevation of contraction-induced rates of MPS,[23] whereas a contrary report demonstrated that deletion of the IGF1 receptor in mice did not influence load-induced skeletal muscle hypertrophy.[24] Furthermore, neither IGF1 receptor phosphorylation nor any other marker of receptor tyrosine kinase activation is increased following resistance exercise in rodents,[25] suggesting that paracrine-related growth factors are not released in sufficient quantities to impact receptor function. Notwithstanding, the fundamental differences between rodents and humans,[26,27] it is also important to point out that whereas insulin can induce phosphorylation of the IGF1 receptor, resistance exercise does not.[25] Moreover, there is now significant data in human models to show that physiological increases in insulin concentrations, corresponding to those seen with ingestion of a meal, play a permissive, and not a stimulatory role in the regulation of MPS.[28,29] In our view, the physiological increases in growth hormone and insulin-like growth factor in response to resistance exercise may be similarly permissive, if they have a role at all, in mediating changes in rates of MPS.

Whereas there is little evidence to support the role of growth hormone in skeletal muscle accretion, the pharmacological administration of testosterone is known to have a direct effect on the accrual of skeletal muscle.[19] Indeed, the suppression of circulating testosterone concentrations—induced by treatment with goserelin—to 10% of circulating concentrations compromises strength gains following 12 weeks of strength training.[30] On the contrary, administrations of 600 mg of testosterone enanthate to healthy young humans increased fat-free mass, muscle size, and muscle strength even in the absence of resistance exercise.[19] Importantly, there are profound and physiologically relevant differences between the pharmacological suppression and pharmacological doses of testosterone and the natural increases observed during the postresistance exercise period.[31] The acute increase in testosterone following a bout of resistance exercise is often ~0.1% to 1% of the dose administered exogenously and acts for comparably shorter durations, depending upon the dosing strategy. A good example of this point is the fact that despite a ~45-fold lower exercise-induced increase in testosterone, women exhibit comparable rises in rates of MPS and hypertrophy as compared to men.[32] In addition, the postexercise rise in testosterone does not exceed that of what is known to occur with diurnal variation.[31]

In a study from West *et al.*,[33] the notion that resistance exercise-induced increases in purportedly anabolic hormones play a significant role in skeletal muscle hypertrophy was examined. In this experiment, 12 healthy males were required to perform a series of unilateral arm curls on separate days for 15 weeks. One arm was exercised alone ("low hormone" condition), whereas participants performed a bout of heavy lower-limb resistance exercise, designed to elicit an increase in circulating "anabolic" hormones, on the days that the other arm performed resistance exercise ("high hormone" condition). Interestingly, although there was a significant and profound increase in testosterone, IGF1, and growth hormone in the "high hormone" condition, there was no significant difference in muscle CSA or strength compared to the "low hormone" condition. If it were true that the acute increase in such hormones were playing a role in adaptations to resistance exercise, then the "high hormone" condition would be expected to produce greater gains in muscle CSA and strength, yet they failed to do so. The interested reader is directed to a debate regarding the role that hormonal changes play in resistance exercise-induced hypertrophy.[31]

5. THE ROLE OF THE mTORC1

A central protein complex in signal transduction, mTORC1, is an evolutionarily conserved threonine/serine protein kinase that is critical for cell survival, cell proliferation, and the synthesis of new muscle proteins (Fig. 2). It is composed of a catalytic subunit, mechanistic target of rapamycin (mTOR) as well as adaptor proteins: regulatory-associated protein of mTOR (RAPTOR), mammalian lethal with SEC13 protein 8 (mLST8), DEP domain-containing mTOR-interacting protein (DEPTOR), and proline-rich Akt substrate of 40 kDa (PRAS40).[34] There is an array of molecular input signals to mTORC1 that can regulate its activity. One such input is the guanosine triphosphatase (GTPase) Ras homolog enriched in brain (RHEB). When guanosine triphosphate (GTP) is bound to RHEB, there is a RHEB-mediated stimulation of mTORC1 assembly

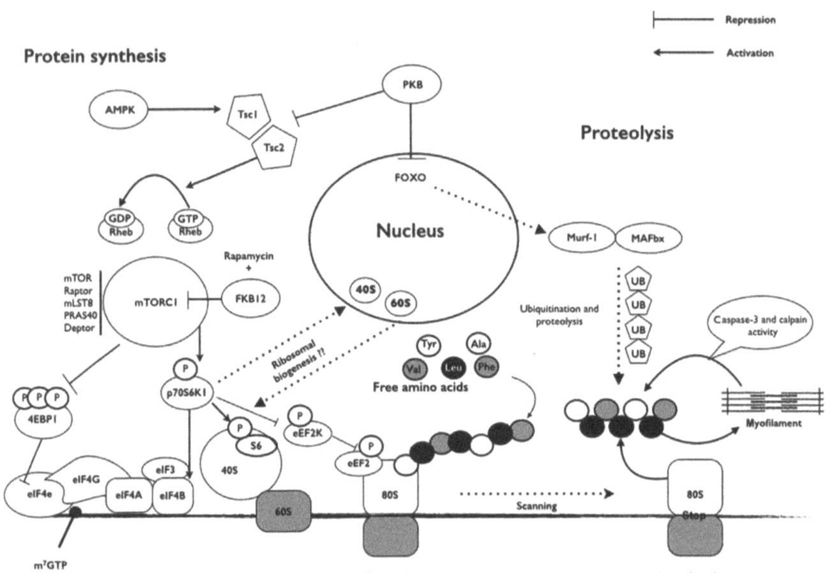

Figure 2 Schematic illustration of the putative molecular factors regulating skeletal muscle protein synthesis and proteolysis. Stimulation of the mechanistic target of rapamycin complex 1 precipitates a series of posttranslational modifications including, but not limited to, the phosphorylation of eukaryotic initiation factors (eIFs), binding proteins, and protein kinases. Upregulation of the ubiquitin proteasome pathway results in the expression of genes encoding for ubiquitin ligases and the subsequent degradation of muscle proteins. Abbreviations and names of genes, mRNAs, and proteins can be found in the text.

and activation.[35] This process is further regulated by an upstream GTPase-activating protein (GAP) called tuberous sclerosis complex 2 (TSC2). TSC2 activates RHEB to drive its GTPase activity, increasing the guanosine diphosphate (GDP)/GTP-bound state of RHEB, and thus reducing the ability to activate mTORC1.[36] In turn, the influence of TSC2 on RHEB can be directly enhanced by adenosine monophosphate-activated protein kinase (AMPK).[37] Moreover, AMPK (an energy stress sensor) also has been shown to inhibit mTORC1 via phosphorylation of DEPTOR.[38] Alternatively, protein kinase B (PKB) (an important downstream protein for receptor tyrosine kinases) phosphorylates TSC2 to inhibit its interaction with RHEB[39] as well as directly inhibiting the function of the mTORC1 inhibitor PRAS40,[40] which exerts a stimulatory effect on mTORC1 activity.

Amino acid sufficiency is signaled to mTORC1 via multiple mechanisms including the lipid kinase human vacuolar sorting protein 34, mitogen-activated protein kinase 3, and the RagGTPases.[35] In essence, mTORC1 acts like a control center integrating inputs on energy stress, hormonal status, and amino acid sufficiency and relaying that information through its downstream targets to control the rate of MPS. mTORC1 targets many downstream substrates including, but not limited to, 70-kDa ribosomal protein S6 kinase 1 (P70S6K1) and eukaryotic initiation factor 4E-binding protein 1 (4EBP1).[41] Both P70S6K1 and 4EBP1 play a crucial role in orchestrating the MPS response to anabolic stimulation. Indeed, P70S6K1 is phosphorylated by mTORC1 on Thr389 to upregulate the opening round of translation initiation via the S6K1 Ally/Ref-like target (SKAR) protein[42] and translational elongation via eukaryotic elongation factor-2 kinase (EEF2K).[43] The function of 4EBP1 is inhibited by mTORC1 (phosphorylation occurs on Thr37/46), which relieves the inhibition on eukaryotic initiation factor 4E (EIF4E). This phosphorylation event enables EIF4E to interact with eukaryotic initiation factor 4G (EIF4G) key to the commencement of protein initiation.[44] It is noteworthy that there are other signaling molecules that either regulate or are targets of, mTORC1, and we acknowledge the existence of another mTOR complex, mechanistic target of rapamycin complex 2 (mTORC2). But the role of mTORC2 and the aforementioned unrelated signaling events are outside the purview of the present discussion, so we refer the interested reader to several informative papers.[45–47]

Given that translation initiation is proposed to be orchestrated by mTORC1-4EBP1/P70S6K1 signaling,[48] it is unsurprising that increased rates of MPS in response to resistance exercise (RE) are often accompanied by elevations in the phosphorylation status of these molecules.[49–53] However, many studies only show associations or correlations between changes in protein phosphorylation (and presumably activation) and rates of MPS, not a direct cause and effect relationship. One of the first investigations to show that mTORC1 was critical for mechanical overload-induced increases in muscle hypertrophy was conducted in rodents by Bodine *et al.*[54] In this classic study, the authors showed that treatment with the highly specific mTORC1 inhibitor, rapamycin, repressed synergist ablation-induced increases in skeletal muscle hypertrophy as well as PKB-mTORC1-P70S6K1 signaling. Other researchers expanded upon these findings in mice, using various rapamycin resistant mutants of mTOR to show that rapamycin also reduced mTORC1 signaling and individual muscle fiber CSA following chronic mechanical loading.[55] However, it is important to note that there are significant interspecies differences between rodents and humans with regards to protein turnover[26] that should be acknowledged when interpreting these data. For example, the work of Goodman *et al.*[55] showed that chronic mechanical loading also induced a degree of hyperplasia, a phenomenon that, to our knowledge, does not occur in human models of resistance exercise training. Nevertheless, what these data show, at least in rodent models, is that loading-induced skeletal muscle hypertrophy occurs predominantly in an mTORC1-dependent manner.

To date, only a few studies in humans have used rapamycin to examine if RE-induced increases in MPS are reliant on mTORC1-mediated signaling.[56,57] By administering rapamycin 2 h prior to a bout of RE, the authors of one study were able to significantly raise blood rapamycin concentrations at the commencement of exercise. Skeletal MPS was determined using the stable isotope technique 1 and 2 h post-RE during which the phosphorylation status of mTOR and associated signaling agents was measured. Rapamycin significantly reduced MPS during the initial 1–2 h post-RE recovery period. Moreover, the reduction in MPS was accompanied by a reduction in $mTOR^{ser2448}$ and $P70S6K1^{thr389}$ phosphorylation at 1 h post-RE. These data show that activation (phosphorylation) of mTOR signaling plays a key role in contraction-induced increases in human skeletal MPS, a finding that corroborates earlier reports in rodents.[54,55] Despite a reduction in MPS, rapamycin treatment did not inhibit phosphorylation of P70S6K1 2 h post-RE nor did it have any impact on 4EBP1 phosphorylation at any time point during the recovery period. The latter finding is not entirely surprising given that 4EBP1 exhibits resistance to the effects of rapamycin,[58] and the dose of rapamycin employed was, in comparison to those as delivered in rodent experiments, relatively low. These data may also suggest that multiple redundant signaling pathways are required to maximize rates of MPS in response to RE. Clearly, more work in human models of exercise is now needed to define the exact molecular mechanisms by which RE stimulates a rise in the rates of human MPS.

A pertinent question is, since mTORC1 is a focal point controlling cell size in response to anabolic stimulation, what signals regulate mTORC1? We now know that mTORC1 activation in response to contraction occurs in a 3-phosphoinositide-dependent protein kinase (PI3K)-PKB-independent manner, and exercise-induced increases in circulating "anabolic" hormones (IGF1 and testosterone) also play little-to-no role. One candidate that has received much attention is focal adhesion kinase (FAK), which is an integrin protein located within the costameres of skeletal muscle and has exhibited mechanical sensitivity in rodent, cell, avian, and human models of loaded contraction.[59–62] The sensitivity of FAK to mechanical loading was highlighted in a study by Fluck *et al.*,[62] who identified an increase in FAK activity following chronic overload of avian skeletal muscle. Furthermore, in a study from our laboratory, 14 days of limb unloading via limb immobilization in humans reduced $FAK^{Tyr576/577}$ phosphorylation,[59] a finding that is consistent with another report in a human immobilization model.[63] While informative, these studies do not provide mechanistic insight into how FAK activity, in response to contractile perturbation, regulates MPS. There is evidence that FAK is required for IGF1-mediated increases in protein synthesis in myotubes via TSC2-mTORC1-P70S6K1 signaling.[64] However, in humans, we know that IGF1 plays a minimal role in RE-induced increases in MPS. Other candidate signaling molecules that have been shown to impact mechanically sensitive mTORC1 activity in rodent models include phospholipase D (PLD)[65] and diacylglycerolkinase zeta,[66] but whether these molecules are responsible for load-induced modulation of mTORC1 and MPS remains a topic of debate. Clearly, more work in humans is needed to identify how mechanical load is transmitted to a biochemical signal that is capable of mediating contraction-induced increases in MPS.

In nearly all of studies of signaling pathway activity, particularly the human studies, the traditional method to assess the activity or functional relevance of a protein in response to exercise and or feeding is to measure its posttranslational phosphorylation status by immunoblotting (Western blotting) or (γ^{-32}P) adenosine triphosphate (ATP)-dependent kinase assays.[67] Emergent data using immunohistochemistry have also provided information relating to the subcellular trafficking of mTOR and this approach has yielded exciting information.[68,69] Indeed, the electrical stimulation of mouse skeletal muscle to elicit eccentric contraction demonstrated that changes in the phosphorylation status of TSC2 within an RxRxx motif resulted in its translocation away from the lysosomal surface.[68] Given that the mTORC1 regulator RHEB is located at the lysosomal surface, and that TSC2 may negatively influence GTP-bound RHEB, these data suggest that resistance exercise may activate mTORC1 by alleviating its inhibition by TSC2. There is also evidence that the provision of amino acids promotes the translocation of mTOR to the lysosome, which may partially account for the synergistic impact of combining amino acid feeding and resistance exercise on MPS. Such a hypothesis awaits experimental confirmation in human models of exercise and feeding but presents an intriguing area of future research nonetheless.

6. TRANSLATIONAL RESPONSES TO RESISTANCE TRAINING

It is proposed that there are two ways in which resistance exercise enhances muscle hypertrophy. Firstly, enhance translational efficiency that is underpinned by an increase in the number of ribosomes binding to a single mRNA transcript, thus resulting in greater protein yield per mRNA. For example, ribosomes are typically spaced approximately 80–100 nucleotides apart, but following stimulation ribosomes can stack much closer, with reports of ribosomes capable of assembling only 27–29 nucleotides apart.[70] The second is enhanced translational capacity, which refers to the upregulation of ribosomal content. The ribosome consists of two subunits, the 40S and the 60S subunit, and is a ribonucleoprotein serving to increase protein content by the decoding of mRNA.[71,72] Both of the subunits are comprised of a variety of ribosomal proteins and it is the increase in these ribosomal proteins and the ribosomal subunits in response to anabolic stimulation that is believed to act in concert with enhanced translational efficiency to promote increases in cell size. Ribosomal biogenesis is a highly controlled and complex biological process regulated in part via mTORC1.[71] Indeed, distinct to many ribosomal protein and growth-related mRNAs is a unique terminal oligopyrimidine tract sequence (5′TOP) and these mRNAs are specifically regulated by mTORC1-P70S6K1 signaling.[73] Given that resistance exercise is a potent stimulator of both mTORC1 and MPS,[74] it has been proposed that resistance exercise enhances rates of MPS in part via upregulation of ribosomal content.[71]

Whereas there is a significant body of research that has focused upon the role of ribosomal biogenesis in cardiac muscle, there is comparably less data available regarding the role of ribosomal biogenesis in skeletal muscle. A common readout of increased ribosomal biogenesis in skeletal muscle is an increase in total RNA content.[55] This assumption is based on the premise that the ribosome is ~30% protein and ~60% RNA and that the majority of RNA is ribosomal RNA (rRNA).[71,75] In this regard, data from rodent models have shown elevations in rRNA content are associated with enhanced mTORC1 signaling.[55,76] Additionally, treatment with rapamycin has been shown to not only prevent P70S6K1 phosphorylation but also inhibit load-induced increases in rRNA content.[55] Such findings would support the emerging thesis that, at least in models where constant load is applied, increases in load-induced MPS are by logical extension facilitated not only by enhanced translational efficiency but also by enhanced translational capacity.

We know that there is an incongruence between resistance exercise-induced increases in rates of MPS and the hypertrophic response to resistance exercise training in humans.[77] Thus, it is not known whether increases in translational capacity seen in rodents are also seen in humans. Potentially debunking this contention, Phillips *et al.*[78] reported that 20 weeks of supervised resistance exercise training actually downregulated the mRNA expression of the translational machinery. In this study,[78] it was shown that the individuals who demonstrated the greatest gains in skeletal muscle mass actually exhibited the biggest reductions in rRNA gene expression. Moreover, those that gained the greatest skeletal muscle mass also displayed a gene signature consistent with that of a rapamycin, in other words indicative of reduced mTORC1 activity. These data also would support other reports of incongruence between acute changes in kinase phosphorylation and skeletal muscle hypertrophy[49,79] raising the question as to the relevance of assessing the phosphorylation/activity status of signaling molecules in the acute postexercise recovery period, at least in the context of skeletal muscle hypertrophy. However, it is important to acknowledge that despite rRNA expression negatively correlating with gains in muscle mass, the participants that responded the greatest to the resistance exercise-training program also tended to have higher levels of rRNA pretraining suggesting that these data are more indicative of enhanced translational efficiency. Thus, more work is clearly necessary in order to establish the role, if any, of increased ribosomal abundance in long-term gains in human skeletal muscle mass in response to resistance exercise training in humans.

7. THE IMPACT OF AGING AND UNLOADING

Aging is associated with the progressive loss of skeletal muscle mass, termed sarcopenia.[80,81] The limited capacity of older adults to mount a "youthful" protein synthetic response following ingestion of smaller doses of amino acids has already been well characterized and is proposed to be a main contributor to the sarcopenic condition.[82] However, this process is confounded by the fact that older people fail to mount a similar protein synthetic response to a bout of resistance exercise,[83] resulting in compromised strength and muscle size gains when compared to their younger counterparts.[84] The "anabolic resistance" of aging skeletal muscle to an acute bout of resistance exercise in older adults has been shown to occur at a range of exercise intensities and is associated with a reduction in mTORC1-P70S6K1-related signaling.[85] Indeed, in older adults, the MPS response with performance of increasing intensities is sigmoidal, with minimal elevations occurring from 20% to 40% 1 RM, followed by a significant increase at 60% 1 RM and then no further enhancement at 75% or 90% 1 RM.[83] This MPS response is also accompanied by lower P70S6K1 and 4EBP1 phosphorylation 1 h postexercise.[83] Expanding upon these findings others have shown that activation of signaling proteins are lower even as long as 6 and 24 h following exercise,[85] thus providing one mechanism of age-related "anabolic resistance." However, the etiology of anabolic resistance is likely multifactorial and complex in nature, the questions of how and why aging results in a lack of sensitivity to the normally potent anabolic stimulus of resistance exercise remain unknown. Discerning the relative contributions of age and physical inactivity toward the development of sarcopenia is challenging as they often are highly associated. For example, simply reducing the number of daily steps performed is known to result in a lower postprandial MPS response to ingestion of a protein-containing meal.[86] It could, we would contend, be argued that both physical inactivity and aging share similar hallmarks of sarcopenia and perhaps even a similar molecular signature; however, it is likely that a molecular aging program likely exists separate from inactivity *per se*. No doubt, experimental work is now needed to pinpoint the exact molecular mechanisms underpinning anabolic resistance in older adults.

Whereas resistance exercise results in a state of skeletal muscle hypertrophy, periods of muscle disuse induce a state of muscle atrophy.[87–90] The mechanisms underpinning disuse muscle atrophy remain a topic of intense debate with some arguing that elevated rates of MPB result in the gradual losses of muscle mass[91] whereas as others suggest that it is a reduction in rates of MPS,[92,93] particularly in response to amino acid ingestion. There are good arguments for both sides of these proposed mechanisms. Indeed, MPB is mediated by numerous processes such as the calcium-dependent calpain system[94,95] and the ubiquitin proteasome pathway.[96] Many readouts suggestive of increased ubiquitin proteasome activity, such as the mRNA expression of muscle atrophy F-box/atrogin-1 (MAFBX) and muscle-specific-RING-finger protein 1 (MURF1), have been shown to be increased during early periods of disuse.[96,97] However, it is not possible for the ubiquitin proteasome pathway to directly target contractile muscle proteins and other mediators of MPB that are required to induce full degradation of intact myofibrils. Moreover, the majority of data to support the assertion that muscle disuse atrophy is mediated by accelerated rates of MPB emanate from rodent models[96] of disuse or human models of disease[98] that do not accurately mimic uncomplicated human disuse atrophy. On the other hand, rates of MPS during muscle unloading are reduced in both the postabsoptive[59] and postprandial state[90] with immobilization. Reductions in rates of MPS that have been calculated to account for a significant proportion, but all, of the decline in muscle mass observed with disuse.[26,93]

Rather than an either/or approach (MPS vs. MPB) as to the primary mechanism responsible for disuse atrophy, there is mounting evidence to suggest that both an increase in rates of MPB and a reduction in rates of MPS act in concert to reduce the size of skeletal muscle during periods of unloading.[87] Indeed, one proposition is that subsequent to immobilization there is an initial rapid but transient increase in rates of MPB that leads to a more sustained and prolonged reduction in rates of MPS. Support for this contention comes from data showing that elevations in the early stage of disuse could be targeting the degradation of translational proteins such as eukaryotic initiation factor 3F.[99,100] The reduction in the content of these proteins subsequently serves to limit the capacity of the translational machinery to mount a sufficient response to anabolic stimulation necessary to sustain positive protein balance. Although to our knowledge, no study has identified if a period of muscle disuse in humans results in the targeted degradation of these proteins and, as such, this hypothesis awaits experimental confirmation.

8. SUMMARY

To conclude, resistance exercise-induced skeletal muscle hypertrophy is a complex process involving many signal transduction pathways. Unfortunately, despite many years of research the biological signal that translates the mechanical tension of resistance exercise (i.e., load) to these signal transduction pathways that subsequently activates the translational machinery, remains elusive. Current developments in mass spectrometry and stable isotope technology as well as use of the novel "omic" techniques will no doubt advance our understanding in the coming years. To date, the majority of research in the field has been generated from rodent models of overload for example using synergist ablation. These rodent models have provided valuable data regarding the mechanisms that regulate the size of skeletal muscle; however, there are fundamental differences between rodents and humans that need to be appreciated when interpreting such data. If we are to significantly enhance our understanding of the factors that control the size of human muscle mass both in response to muscle loading and unloading, more work needs to be conducted in human models of resistance exercise using a range of techniques to complement what we already know.

ACKNOWLEDGMENTS

S.M.P. would like to acknowledge grant support from The Canadian Institutes for Health Research, the Natural Science and Engineering Research Council of Canada, the Canadian Diabetes Association, and the Canadian Foundation for Innovation for support in writing this work. Both C.M. and S.M.P. would like to extend their appreciation to Dr. Lee Hamilton and Amy Hector for providing insightful debate during the preparation of this manuscript.

REFERENCES

1. Egan B, Zierath JR. Exercise metabolism and the molecular regulation of skeletal muscle adaptation. *Cell Metab.* 2013;17(2):162–184.
2. Phillips SM. Physiologic and molecular bases of muscle hypertrophy and atrophy: impact of resistance exercise on human skeletal muscle (protein and exercise dose effects). *Appl Physiol Nutr Metab.* 2009;34(3):403–410.
3. Claassen H, Gerber C, Hoppeler H, Luthi JM, Vock P. Muscle filament spacing and short-term heavy-resistance exercise in humans. *J Physiol.* 1989;409:491–495.
4. Kelley G. Mechanical overload and skeletal muscle fiber hyperplasia: a meta-analysis. *J Appl Physiol (1985).* 1996;81(4):1584–1588.
5. Tipton KD, Ferrando AA, Phillips SM, Doyle Jr D, Wolfe RR. Postexercise net protein synthesis in human muscle from orally administered amino acids. *Am J Physiol.* 1999;276(4 pt 1):E628–E634.

6. Witard OC, Tieland M, Beelen M, Tipton KD, van Loon LJ, Koopman R. Resistance exercise increases postprandial muscle protein synthesis in humans. *Med Sci Sports Exerc.* 2009;41(1):144–154.
7. Burd NA, West DW, Moore DR, et al. Enhanced amino acid sensitivity of myofibrillar protein synthesis persists for up to 24 h after resistance exercise in young men. *J Nutr.* 2011;141(4):568–573.
8. De Lorme TL, Watkins AL. Technics of progressive resistance exercise. *Arch Phys Med Rehabil.* 1948;29(5):263–273.
9. Cermak NM, Res PT, de Groot LC, Saris WH, van Loon LJ. Protein supplementation augments the adaptive response of skeletal muscle to resistance-type exercise training: a meta-analysis. *Am J Clin Nutr.* 2012;96(6):1454–1464.
10. American College of Sports Medicine. American college of sports medicine position stand. Progression models in resistance training for healthy adults. *Med Sci Sports Exerc.* 2009;41(3):687–708.
11. Schoenfeld BJ, Wilson JM, Lowery RP, Krieger JW. Muscular adaptations in low- versus high-load resistance training: a meta-analysis. *Eur J Sport Sci.* 2014;1–10.
12. Mitchell CJ, Churchward-Venne TA, West DWD, et al. Resistance exercise load does not determine training-mediated hypertrophic gains in young men. *J Appl Physiol.* 2012;113(1):71–77.
13. Burd NA, Holwerda AM, Selby KC, et al. Resistance exercise volume affects myofibrillar protein synthesis and anabolic signalling molecule phosphorylation in young men. *J Physiol.* 2010;588(pt 16):3119–3130.
14. Hubal MJ, Gordish-Dressman H, Thompson PD, et al. Variability in muscle size and strength gain after unilateral resistance training. *Med Sci Sports Exerc.* 2005;37(6):964–972.
15. Bamman MM, Petrella JK, Kim JS, Mayhew DL, Cross JM. Cluster analysis tests the importance of myogenic gene expression during myofiber hypertrophy in humans. *J Appl Physiol (1985).* 2007;102(6):2232–2239.
16. Hartman JW, Tang JE, Wilkinson SB, et al. Consumption of fat-free fluid milk after resistance exercise promotes greater lean mass accretion than does consumption of soy or carbohydrate in young, novice, male weightlifters. *Am J Clin Nutr.* 2007;86(2):373–381.
17. Burd NA, Moore DR, Mitchell CJ, Phillips SM. Big claims for big weights but with little evidence. *Eur J Appl Physiol.* 2013;113(1):267–268.
18. Kraemer WJ, Ratamess NA. Hormonal responses and adaptations to resistance exercise and training. *Sports Med.* 2005;35(4):339–361.
19. Bhasin S, Storer TW, Berman N, et al. The effects of supraphysiologic doses of testosterone on muscle size and strength in normal men. *N Engl J Med.* 1996;335(1):1–7.
20. Rennie MJ. Claims for the anabolic effects of growth hormone: a case of the emperor's new clothes? *Br J Sports Med.* 2003;37(2):100–105.
21. Doessing S, Heinemeier KM, Holm L, et al. Growth hormone stimulates the collagen synthesis in human tendon and skeletal muscle without affecting myofibrillar protein synthesis. *J Physiol.* 2010;588(pt 2):341–351.
22. West DW, Kujbida GW, Moore DR, et al. Resistance exercise-induced increases in putative anabolic hormones do not enhance muscle protein synthesis or intracellular signalling in young men. *J Physiol.* 2009;587(pt 21):5239–5247.
23. Fedele MJ, Lang CH, Farrell PA. Immunization against IGF-I prevents increases in protein synthesis in diabetic rats after resistance exercise. *Am J Physiol Endocrinol Metab.* 2001;280(6):E877–E885.
24. Spangenburg EE, Le Roith D, Ward CW, Bodine SC. A functional insulin-like growth factor receptor is not necessary for load-induced skeletal muscle hypertrophy. *J Physiol.* 2008;586(1):283–291.

25. Hamilton DL, Philp A, MacKenzie MG, Baar K. A limited role for PI(3,4,5)P3 regulation in controlling skeletal muscle mass in response to resistance exercise. *PLoS One*. 2010;5(7):e11624.
26. Phillips SM, Glover EI, Rennie MJ. Alterations of protein turnover underlying disuse atrophy in human skeletal muscle. *J Appl Physiol (1985)*. 2009;107(3):645–654.
27. Cunningham ML. A mouse is not a rat is not a human: species differences exist. *Toxicol Sci*. 2002;70(2):157–158.
28. Greenhaff PL, Karagounis LG, Peirce N, et al. Disassociation between the effects of amino acids and insulin on signaling, ubiquitin ligases, and protein turnover in human muscle. *Am J Physiol Endocrinol Metab*. 2008;295(3):E595–E604.
29. Trommelen J, Groen B, Hamer H, de Groot LC, van Loon LJ. Mechanisms in endocrinology: exogenous insulin does not increase muscle protein synthesis rate when administrated systemically: a systematic review. *Eur J Endocrinol*. 2015;173(1): R25–R34.
30. Kvorning T, Andersen M, Brixen K, Madsen K. Suppression of endogenous testosterone production attenuates the response to strength training: a randomized, placebo-controlled, and blinded intervention study. *Am J Physiol Endocrinol Metab*. 2006;291(6):E1325–E1332.
31. Schroeder ET, Villanueva M, West DD, Phillips SM. Are acute post-resistance exercise increases in testosterone, growth hormone, and IGF-1 necessary to stimulate skeletal muscle anabolism and hypertrophy? *Med Sci Sports Exerc*. 2013;45(11): 2044–2051.
32. West DWD, Burd NA, Churchward-Venne TA, et al. Sex-based comparisons of myofibrillar protein synthesis after resistance exercise in the fed state. *J Appl Physiol*. 2012;112(11):1805–1813.
33. West DW, Burd NA, Tang JE, et al. Elevations in ostensibly anabolic hormones with resistance exercise enhance neither training-induced muscle hypertrophy nor strength of the elbow flexors. *J Appl Physiol (1985)*. 2010;108(1):60–67.
34. Foster KG, Fingar DC. Mammalian target of rapamycin (mTOR): conducting the cellular signaling symphony. *J Biol Chem*. 2010;285(19):14071–14077.
35. Laplante M, Sabatini DM. mTOR signaling in growth control and disease. *Cell*. 2012;149(2):274–293.
36. Inoki K, Li Y, Xu T, Guan KL. Rheb GTPase is a direct target of TSC2 GAP activity and regulates mTOR signaling. *Genes Dev*. 2003;17(15):1829–1834.
37. Inoki K, Zhu T, Guan KL. TSC2 mediates cellular energy response to control cell growth and survival. *Cell*. 2003;115(5):577–590.
38. Gwinn DM, Shackelford DB, Egan DF, et al. AMPK phosphorylation of raptor mediates a metabolic checkpoint. *Mol Cell*. 2008;30(2):214–226.
39. Inoki K, Li Y, Zhu T, Wu J, Guan KL. TSC2 is phosphorylated and inhibited by Akt and suppresses mTOR signalling. *Nat Cell Biol*. 2002;4(9):648–657.
40. Vander Haar E, Lee SI, Bandhakavi S, Griffin TJ, Kim DH. Insulin signalling to mTOR mediated by the Akt/PKB substrate PRAS40. *Nat Cell Biol*. 2007; 9(3):316–323.
41. Hershey JW. Translational control in mammalian cells. *Annu Rev Biochem*. 1991;60:717–755.
42. Richardson CJ, Broenstrup M, Fingar DC, et al. SKAR is a specific target of S6 kinase 1 in cell growth control. *Curr Biol*. 2004;14(17):1540–1549.
43. Wang X, Li W, Williams M, Terada N, Alessi DR, Proud CG. Regulation of elongation factor 2 kinase by p90(RSK1) and p70 S6 kinase. *EMBO J*. 2001; 20(16):4370–4379.
44. Gingras AC, Gygi SP, Raught B, et al. Regulation of 4E-BP1 phosphorylation: a novel two-step mechanism. *Genes Dev*. 1999;13(11):1422–1437.

45. Laplante M, Sabatini DM. Regulation of mTORC1 and its impact on gene expression at a glance. *J Cell Sci*. 2013;126(pt 8):1713–1719.
46. Gkogkas C, Sonenberg N, Costa-Mattioli M. Translational control mechanisms in long-lasting synaptic plasticity and memory. *J Biol Chem*. 2010;285(42):31913–31917.
47. Dowling RJ, Topisirovic I, Fonseca BD, Sonenberg N. Dissecting the role of mTOR: lessons from mTOR inhibitors. *Biochim Biophys Acta*. 2010;1804(3):433–439.
48. Laplante M, Sabatini DM. mTOR signaling at a glance. *J Cell Sci*. 2009;122(pt 20):3589–3594.
49. Areta JL, Burke LM, Ross ML, et al. Timing and distribution of protein ingestion during prolonged recovery from resistance exercise alters myofibrillar protein synthesis. *J Physiol*. 2013;591(pt 9):2319–2331.
50. Moore DR, Robinson MJ, Fry JL, et al. Ingested protein dose response of muscle and albumin protein synthesis after resistance exercise in young men. *Am J Clin Nutr*. 2009;89(1):161–168.
51. Churchward-Venne TA, Burd NA, Mitchell CJ, et al. Supplementation of a suboptimal protein dose with leucine or essential amino acids: effects on myofibrillar protein synthesis at rest and following resistance exercise in men. *J Physiol*. 2012;590(pt 11):2751–2765.
52. Mitchell CJ, Churchward-Venne TA, West DW, et al. Resistance exercise load does not determine training-mediated hypertrophic gains in young men. *J Appl Physiol (1985)*. 2012;113(1):71–77.
53. Baar K, Esser K. Phosphorylation of p70(S6k) correlates with increased skeletal muscle mass following resistance exercise. *Am J Physiol*. 1999;276(1 pt 1):C120–C127.
54. Bodine SC, Stitt TN, Gonzalez M, et al. Akt/mTOR pathway is a crucial regulator of skeletal muscle hypertrophy and can prevent muscle atrophy in vivo. *Nat Cell Biol*. 2001;3(11):1014–1019.
55. Goodman CA, Frey JW, Mabrey DM, et al. The role of skeletal muscle mTOR in the regulation of mechanical load-induced growth. *J Physiol*. 2011;589(pt 22):5485–5501.
56. Drummond MJ, Fry CS, Glynn EL, et al. Rapamycin administration in humans blocks the contraction-induced increase in skeletal muscle protein synthesis. *J Physiol Lond*. 2009;587(7):1535–1546.
57. Gundermann DM, Walker DK, Reidy PT, et al. Activation of mTORC1 signaling and protein synthesis in human muscle following blood flow restriction exercise is inhibited by rapamycin. *Am J Physiol Endocrinol Metab*. 2014;306(10):E1198–E1204.
58. Choo AY, Yoon SO, Kim SG, Roux PP, Blenis J. Rapamycin differentially inhibits S6Ks and 4E-BP1 to mediate cell-type-specific repression of mRNA translation. *Proc Natl Acad Sci USA*. 2008;105(45):17414–17419.
59. Glover EI, Phillips SM, Oates BR, et al. Immobilization induces anabolic resistance in human myofibrillar protein synthesis with low and high dose amino acid infusion. *J Physiol*. 2008;586(pt 24):6049–6061.
60. Atherton PJ, Szewczyk NJ, Selby A, et al. Cyclic stretch reduces myofibrillar protein synthesis despite increases in FAK and anabolic signalling in L6 cells. *J Physiol*. 2009;587(pt 14):3719–3727.
61. Gordon SE, Fluck M, Booth FW. Selected contribution: skeletal muscle focal adhesion kinase, paxillin, and serum response factor are loading dependent. *J Appl Physiol (1985)*. 2001;90(3):1174–1183. discussion 1165.
62. Fluck M, Carson JA, Gordon SE, Ziemiecki A, Booth FW. Focal adhesion proteins FAK and paxillin increase in hypertrophied skeletal muscle. *Am J Physiol*. 1999;277(1 pt 1):C152–C162.
63. de Boer MD, Selby A, Atherton P, et al. The temporal responses of protein synthesis, gene expression and cell signalling in human quadriceps muscle and patellar tendon to disuse. *J Physiol*. 2007;585(pt 1):241–251.

64. Crossland H, Kazi AA, Lang CH, et al. Focal adhesion kinase is required for IGF-I-mediated growth of skeletal muscle cells via a TSC2/mTOR/S6K1-associated pathway. *Am J Physiol Endocrinol Metab*. 2013;305(2):E183–E193.
65. Hornberger TA, Chu WK, Mak YW, Hsiung JW, Huang SA, Chien S. The role of phospholipase D and phosphatidic acid in the mechanical activation of mTOR signaling in skeletal muscle. *Proc Natl Acad Sci USA*. 2006;103(12):4741–4746.
66. You JS, Lincoln HC, Kim CR, et al. The role of diacylglycerol kinase zeta and phosphatidic acid in the mechanical activation of mammalian target of rapamycin (mTOR) signaling and skeletal muscle hypertrophy. *J Biol Chem*. 2014;289(3):1551–1563.
67. McGlory C, White A, Treins C, et al. Application of the [gamma-32P] ATP kinase assay to study anabolic signaling in human skeletal muscle. *J Appl Physiol*. 2014;116(5):504–513.
68. Jacobs BL, Goodman CA, Hornberger TA. The mechanical activation of mTOR signaling: an emerging role for late endosome/lysosomal targeting. *J Muscle Res Cell Motil*. 2014;35(1):11–21.
69. Jacobs BL, Goodman CA, Hornberger TA. The mechanical activation of mTOR signaling: an emerging role for late endosome/lysosomal targeting. *J Muscle Res Cell Motil*. 2013;35(1):11–21.
70. Wolin SL, Walter P. Ribosome pausing and stacking during translation of a eukaryotic mRNA. *EMBO J*. 1988;7(11):3559–3569.
71. Chaillou T, Kirby TJ, McCarthy JJ. Ribosome biogenesis: emerging evidence for a central role in the regulation of skeletal muscle mass. *J Cell Physiol*. 2014;229(11):1584–1594.
72. Mahoney SJ, Dempsey JM, Blenis J. Cell signaling in protein synthesis ribosome biogenesis and translation initiation and elongation. *Prog Mol Biol Transl Sci*. 2009;90:53–107.
73. Merrick WC. Eukaryotic protein synthesis: still a mystery. *J Biol Chem*. 2010; 285(28):21197–21201.
74. Drummond MJ, Fry CS, Glynn EL, et al. Rapamycin administration in humans blocks the contraction-induced increase in skeletal muscle protein synthesis. *J Physiol*. 2009;587(pt 7):1535–1546.
75. Moss T, Langlois F, Gagnon-Kugler T, Stefanovsky V. A housekeeper with power of attorney: the rRNA genes in ribosome biogenesis. *Cell Mol Life Sci*. 2007;64(1):29–49.
76. Miyazaki M, McCarthy JJ, Fedele MJ, Esser KA. Early activation of mTORC1 signalling in response to mechanical overload is independent of phosphoinositide 3-kinase/Akt signalling. *J Physiol*. 2011;589(pt 7):1831–1846.
77. Mitchell CJ, Churchward-Venne TA, Parise G, et al. Acute post-exercise myofibrillar protein synthesis is not correlated with resistance training-induced muscle hypertrophy in young men. *PLoS One*. 2014;9(2):e89431.
78. Phillips BE, Williams JP, Gustafsson T, et al. Molecular networks of human muscle adaptation to exercise and age. *PLoS Genet*. 2013;9(3):e1003389.
79. Mitchell CJ, Churchward-Venne TA, Bellamy L, Parise G, Baker SK, Phillips SM. Muscular and systemic correlates of resistance training-induced muscle hypertrophy. *PLoS One*. 2013;8(10):e78636.
80. Janssen I. Evolution of sarcopenia research. *Appl Physiol Nutr Metab*. 2010; 35(5):707–712.
81. Janssen I. The epidemiology of sarcopenia. *Clin Geriatr Med*. 2011;27(3):355–363.
82. Moore DR, Churchward-Venne TA, Witard O, et al. Protein ingestion to stimulate myofibrillar protein synthesis requires greater relative protein intakes in healthy older versus younger men. *J Gerontol A Biol Sci Med Sci*. 2014;70(1):57–62.

83. Kumar V, Selby A, Rankin D, et al. Age-related differences in the dose-response relationship of muscle protein synthesis to resistance exercise in young and old men. *J Physiol.* 2009;587(pt 1):211–217.
84. Drummond MJ, Marcus RL, Lastayo PC. Targeting anabolic impairment in response to resistance exercise in older adults with mobility impairments: potential mechanisms and rehabilitation approaches. *J Aging Res.* 2012;2012:486930.
85. Fry CS, Drummond MJ, Glynn EL, et al. Aging impairs contraction-induced human skeletal muscle mTORC1 signaling and protein synthesis. *Skelet Muscle.* 2011;1(1):11.
86. Breen L, Stokes KA, Churchward-Venne TA, et al. Two weeks of reduced activity decreases leg lean mass and induces "anabolic resistance" of myofibrillar protein synthesis in healthy elderly. *J Clin Endocrinol Metab.* 2013;98(6):2604–2612.
87. Wall BT, Dirks ML, van Loon LJ. Skeletal muscle atrophy during short-term disuse: implications for age-related sarcopenia. *Ageing Res Rev.* 2013;12(4):898–906.
88. Suetta C, Hvid LG, Justesen L, et al. Effects of aging on human skeletal muscle after immobilization and retraining. *J Appl Physiol (1985).* 2009;107(4):1172–1180.
89. Dirks ML, Wall BT, Nilwik R, Weerts DH, Verdijk LB, van Loon LJ. Skeletal muscle disuse atrophy is not attenuated by dietary protein supplementation in healthy older men. *J Nutr.* 2014;144(8):1196–1203.
90. Wall BT, Snijders T, Senden JM, et al. Disuse impairs the muscle protein synthetic response to protein ingestion in healthy men. *J Clin Endocrinol Metab.* 2013;98(12):4872–4881.
91. Reid MB, Judge AR, Bodine SC. CrossTalk opposing view: the dominant mechanism causing disuse muscle atrophy is proteolysis. *J Physiol.* 2014;592(pt 24):5345–5347.
92. Phillips SM, McGlory C. Rebuttal from Stuart M. Phillips and Chris McGlory. *J Physiol.* 2014;592(pt 24):5349.
93. Phillips SM, McGlory C. CrossTalk proposal: the dominant mechanism causing disuse muscle atrophy is decreased protein synthesis. *J Physiol.* 2014;592(pt 24):5341–5343.
94. Bartoli M, Richard I. Calpains in muscle wasting. *Int J Biochem Cell Biol.* 2005;37(10):2115–2133.
95. Salem M, Nath J, Rexroad CE, Killefer J, Yao J. Identification and molecular characterization of the rainbow trout calpains (Capn1 and Capn2): their expression in muscle wasting during starvation. *Comp Biochem Physiol B Biochem Mol Biol.* 2005;140(1):63–71.
96. Bodine SC, Baehr LM. Skeletal muscle atrophy and the E3 ubiquitin ligases MuRF1 and MAFbx/atrogin-1. *Am J Physiol Endocrinol Metab.* 2014;307(6):E469–E484.
97. Murton AJ, Constantin D, Greenhaff PL. The involvement of the ubiquitin proteasome system in human skeletal muscle remodelling and atrophy. *Biochim Biophys Acta.* 2008;1782(12):730–743.
98. Pasini E, Aquilani R, Dioguardi FS, D'Antona G, Gheorghiade M, Taegtmeyer H. Hypercatabolic syndrome: molecular basis and effects of nutritional supplements with amino acids. *Am J Cardiol.* 2008;101(11A):11E–15E.
99. Csibi A, Leibovitch MP, Cornille K, Tintignac LA, Leibovitch SA. MAFbx/atrogin-1 controls the activity of the initiation factor eIF3-f in skeletal muscle atrophy by targeting multiple C-terminal lysines. *J Biol Chem.* 2009;284(7):4413–4421.
100. Csibi A, Cornille K, Leibovitch MP, et al. The translation regulatory subunit eIF3f controls the kinase-dependent mTOR signaling required for muscle differentiation and hypertrophy in mouse. *PLoS One.* 2010;5(2):e8994.

CHAPTER EIGHT

Exercise and the Regulation of Adipose Tissue Metabolism

Thomas Tsiloulis, Matthew J. Watt[1]
Biology of Lipid Metabolism Laboratory, Department of Physiology, Monash University, Clayton, Victoria, Australia
[1]Corresponding author: e-mail address: matthew.watt@monash.edu

Contents

Abstract

Adipose tissue is a major regulator of metabolism in health and disease. The prominent roles of adipose tissue are to sequester fatty acids in times of energy excess and to release fatty acids via the process of lipolysis during times of high-energy demand, such as exercise. The fatty acids released during lipolysis are utilized by skeletal muscle to

Progress in Molecular Biology and Translational Science, Volume 135
ISSN 1877-1173
http://dx.doi.org/10.1016/bs.pmbts.2015.06.016

produce adenosine triphosphate to prevent fatigue during prolonged exercise. Lipolysis is controlled by a complex interplay between neuro-humoral regulators, intracellular signaling networks, phosphorylation events involving protein kinase A, translocation of proteins within the cell, and protein–protein interactions. Herein, we describe in detail the cellular and molecular regulation of lipolysis and how these processes are altered by acute exercise. We also explore the processes that underpin adipocyte adaptation to endurance exercise training, with particular focus on epigenetic modifications, control by microRNAs and mitochondrial adaptations. Finally, we examine recent literature describing how exercise might influence the conversion of traditional white adipose tissue to high energy-consuming "brown-like" adipocytes and the implications that this has on whole-body energy balance.

1. INTRODUCTION

The ability of skeletal muscle to maintain adenosine triphosphate (ATP) production during physical activity requires the input of several metabolic substrates; most notably, carbohydrates and fatty acids provide the bulk of the energy requirements at rest and during most forms of exercise.[1] Fatty acids are predominantly stored as triglycerides within lipid droplets of adipocytes, which make up the vast majority of the adipocyte volume.[2] Without doubt, the major role of adipose tissue is to store and release fatty acids to match energy supply to energy requirements. This is achieved by a complex interplay between neurohumoral regulators, intracellular signaling networks, and protein–protein interactions. Hence, understanding the factors regulating adipose tissue fatty acid metabolism during exercise is important for understanding systemic metabolic control and also has implications for understanding the pathogenesis of obesity (i.e., adipose tissue excess) and its associated comorbidities.

Expanding on our understanding of adipocyte biology is the observation of adipocytes with distinct developmental lineages, morphology, and metabolic characteristics. The presence of significant depots of brown adipose tissue, which is a major site of nonshivering thermogenesis described in most mammals, was recently reconfirmed in humans,[3,4] while other adipocytes known as "beige" or "brite" adipocytes that share features of brown adipocytes have been identified in clusters within traditional white adipose tissue.[5] Understanding the role of these adipocytes in the regulation of metabolism is in its infancy; however, recent studies suggest that regular endurance exercise training may induce transcriptional programs that predict a "browning" of adipose tissue.

The endeavor of this review is to provide an update on the understanding of adipose tissue metabolism, with a particular focus on lipolysis, to outline the regulation of adipose tissue metabolism during acute exercise, and to provide an overview of the molecular changes induced by exercise training that influence adipose tissue metabolism and the programming of adipocyte lineages. Our intention is to focus primarily on human data, with some reference to rodent work and cell-based studies.

2. ADIPOSE TISSUE LOCALIZATION AND COMPOSITION

Adipose tissue is located in several anatomical locations. These include subcutaneous adipose tissue, which is located under the skin and stores ~80% of total body fat with the major stores found in the upper (abdominal, subscapular fat) and lower body (gluteal–femoral fat). Intraabdominal adipose tissues include the visceral adipose tissue located around the digestive organs (mesenteric and omental) and the retroperitoneal depot (kidney), which together account for ~20% of total body fat[6,7] (Fig. 1A). While these depots exhibit heterogeneity with respect to their molecular, morphological, and metabolic profiles,[8] and differences exist between men and women, the metabolic functions of adipose tissue are retained in some capacity at each site.

It is important to note that adipose tissue is made up of many cell types that serve different functions. While adipocytes make up ~83% of adipose tissue volume,[9] they only constitute ~20–40% of the cellular content. In fact, the number of cells contained in the stromal vascular fraction of adipose tissue outnumbers adipocytes by ~3 to 1. These cells include endothelial cells, leukocytes, lymphocytes, and other immune cells (Fig. 1B).

3. ADIPOCYTE METABOLISM AT REST AND DURING EXERCISE

Several metabolic processes including lipolysis, uptake of fatty acids from lipoprotein–triglycerides, *de novo* lipogenesis, and fatty acid oxidation modulate adipocyte fat balance.

3.1 Lipolysis

The primary function of adipocytes is to store fatty acids during times of energy excess and release these fatty acids into the circulation for uptake and oxidation by the major energy-consuming tissues during times of

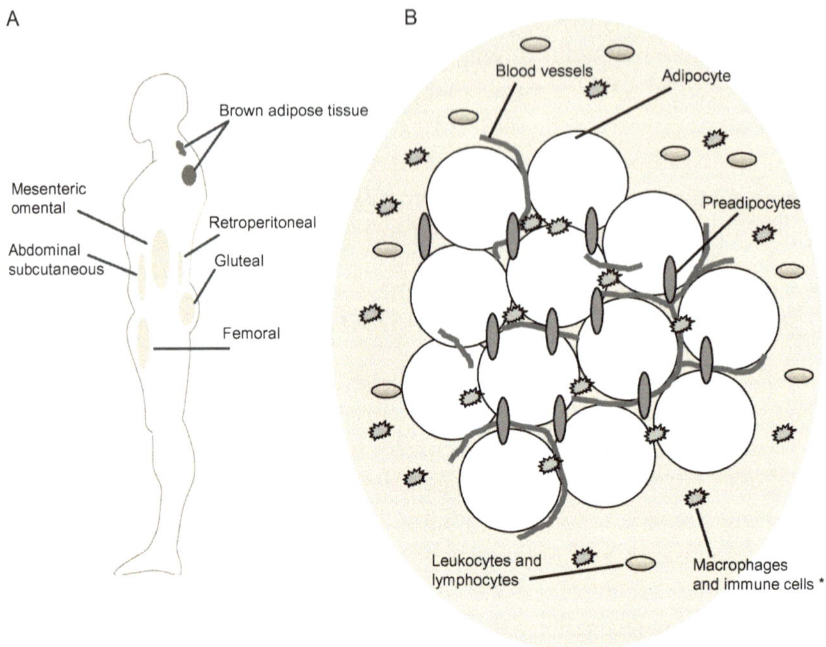

Figure 1 Adipose tissue localization and composition in humans. (A) White and brown adipose tissue is located in various anatomical locations in humans. Subcutaneous adipose tissue is located under the skin and includes the abdominal and gluteal–femoral depots. Visceral adipose tissue is located near the digestive organs and includes the omental and mesenteric adipose tissue depots. (B) The vast majority of the adipose tissue mass is composed of adipocytes. There are many other cell types that constitute adipose tissue. It is highly vascularized and contains a number of immune cells* such as B cells, mast cells, Tregs (T regulatory cells), macrophages, leukocytes, and lymphocytes. (See the color plate.)

increased energy demand, such as exercise. Isotope dilution (e.g., [$^2H_5$]-glycerol tracer) and splanchnic/leg balance techniques have been used to determine lipolysis during exercise. During moderate intensity exercise (40–50% VO_2 max), whole-body adipose tissue lipolysis is increased two- to threefold above resting rates by 30 min and progressively increases with the duration of exercise, reaching rates approximately fivefold higher than rest by 4 h.[10–13] The fatty acid availability from adipose tissue lipolysis matches whole-body fatty acid oxidation rates during the first 90 min of moderate intensity exercise, but there is an "overshoot" of lipolysis late in exercise (>120 min), which contributes to a progressive rise in plasma free fatty acid (FFA). Notably, FFA supply from adipocytes is sufficient to provide all substrate for fatty acid oxidation at rest and during moderate intensity exercise. An often-ignored factor regulating fatty acid release is reesterification. At rest, ~20% of the fatty acids released from adipocytes are reesterified in the adipocytes. The absolute reesterification rate in adipocytes increases approximately twofold during moderate intensity exercise; however, this equates to only ~12% of the lipolytic rate.[13]

Adipose tissue lipolysis increases by ~2.5-fold from rest during low-intensity exercise (25% VO_2 max) but does not increase further at higher exercise intensities, despite marked increases in sympathetic activation.[1,14] Hence, these data indicate that adipocyte lipolysis is sensitive to small increases in β-adrenergic stimulation during exercise. Lipolysis rates decrease in the minutes after low and moderate intensity exercise (25–65% VO_2 max) but remain above preexercise levels for up to 24 h.[1,14,15] The increase in fatty acid availability in the recovery from exercise exceeds fatty acid oxidation by peripheral tissues, which leaves a substantial amount of FFAs for either FFA/triglyceride recycling or uptake by the liver. Furthermore, the increased glycerol availability in the recovery from exercise can act as a gluconeogenic substrate.

Many factors influence lipolysis during exercise. Lipolysis rates are higher in woman compared with men at rest and during moderate intensity exercise (45–65% VO_2 max).[16] Sex differences aside, the anatomical location of adipose tissue is important. Under resting conditions, upper body subcutaneous adipose tissue provides the major proportion of the systemic fatty acids and the lower body fat provides only a small proportion.[17] In studies of isolated adipocytes, both "spontaneous" and maximal $\beta_{2,3}$-adrenergic-stimulated lipolysis are higher in subcutaneous compared with omental-derived adipocytes of lean men.[18] The lipolytic responses of specific depots during exercise and their contribution to whole-body lipolysis are unknown, most likely because of an inability to accurately perform arterio-venous (a-v) balance measures across these tissues. The size of individual adipocytes may also be important, with evidence that large adipocytes have increased lipolytic capacity compared with small adipocytes, which may result from the enrichment of regulatory proteins of lipolysis (see Section 5.1).[19] Finally, environmental (e.g., heat, hypoxia) and nutritional factors (e.g., postabsorptive vs. postprandial states, hydration) impact lipolysis but discussion of these factors is beyond the scope of this review.

3.2 *De Novo* Lipogenesis

De novo lipogenesis is the process by which carbon precursors of acetyl-CoA are synthesized into fatty acids. Lipogenesis is mostly derived from carbohydrates and is a relatively minor contributor to whole-body lipid stores, contributing 1–3% of the total fat balance in humans consuming a typical diet.[20] Lipogenesis increases considerably when individuals are fed a hypercaloric, high-carbohydrate diet, and adipose tissue, rather than the liver, seems to be the major site for lipogenesis under this nutritional load.[21] Thus, in contrast to other species such as mice, *de novo* lipogenesis is for the most part a minor contributor to adipose tissue mass and whole-body energy balance and will not be discussed further.

3.3 Uptake of Fatty Acids from VLDL-Triglyceride

Adipose tissue is important for sequestering triglyceride contained in very low density lipoproteins (VLDLs) at rest. Whether plasma VLDL-TG provides a major substrate for tissues during exercise is controversial, with reports of reduced or unchanged VLDL-TG clearance and oxidation immediately after exercise. Irrespectively, the oxidation of VLDL-TG accounts for only a small fraction (~3%) of total energy expenditure during moderate-intensity exercise.[22] There is a relative hypotriglyceridemia postexercise, which is most likely mediated by reduced secretion rates early after the cessation of exercise and increased whole-body VLDL-TG clearance in the following 6–12 h.[23] The proportional contribution of adipose tissue in postexercise VLDL-TG clearance is unknown.

3.4 Oxidative Metabolism

While adipose tissue typically constitutes a substantial fraction of body mass (10–45%), its contribution to whole-body energy turnover is negligible. Elia[24] provided estimates for the metabolic rates of key organs in humans (in kcal/kg/day) including the liver (200), brain (240), heart (440), kidneys (440), skeletal muscle (13), adipose tissue (4.5), and residual mass (12). When taking mass into consideration,[25] adipose tissue accounts for ~4% of the daily energy expenditure (80 kg male, 15% body fat). Surprisingly, adipose tissue oxygen consumption is not increased above resting levels during exercise and postexercise.[26] It is possible that the oxygen consumption during exercise could be underestimated due to small arterio-venous (a-v) differences when the blood flow is high.

4. NEURAL AND HORMONAL CONTROL OF ADIPOSE TISSUE METABOLISM DURING EXERCISE

There are several neural and hormone regulators of adipose tissue metabolism, which act in synergy to control the rate of fatty acids flux

Table 1 Regulators of Adipocyte Lipolysis in Humans

Activators	**Effect of Exercise**	**Receptor**	**Signaling**
Epinephrine	↑	$\beta_{1/2/3}$-Adrenergic receptor	Adenylyl cyclase → cAMP → PKA
Norepinephrine	↑	α-Adrenergic receptor	—| Adenylyl cyclase inhibition
Glucagon	↑	Gs protein-coupled receptor	Adenylyl cyclase → cAMP → PKA
Cortisol	↑	Glucocorticoid receptor	—| Inhibits insulin signaling cascade
Natriuretic peptides	↑	Natriuretic peptide receptors (A/B)	Guanylyl cyclase → cGMP → PKG
Inhibitors			
Insulin	↓	Insulin receptor	IRS → PI3-K → PKB → PDE-3B —| cAMP
Adenosine	↓	Adenosine receptor (A_1) attached to Gi protein-coupled receptor	Decrease cAMP levels
Fatty acids	↑	–	Mobilization and oxidation

(essentially lipolysis) to meet the metabolic demands of the exercising individual. The regulatory factors and their influences on known lipolytic cascades are summarized in Table 1.

4.1 Catecholamines

Norepinephrine and epinephrine are the most important regulators of adipose tissue metabolism, with adrenaline being most prominent during exercise. The catecholamines modulate lipolysis by acting through lipolysis-inducing β-adrenoceptor (β-AR) subtypes 1–3 (β-AR_{1-3}) and lipolysis-inhibiting α_2-AR subtypes. Briefly, catecholamines bind to the β-ARs (mainly β-AR_{1-2} in humans), which are coupled to stimulatory Gs proteins that activate adenylyl cyclase. This increases intracellular cyclic AMP (cAMP) concentrations and activation of protein kinase A (PKA), which in turn phosphorylates multiple proteins at specific serine residues to increase lipolysis (see Section 5) (Fig. 2).

Plasma epinephrine and norepinephrine increase from resting levels with increasing exercise intensity and also with the duration of steady-state exercise.[27] During epinephrine infusion in resting individuals, the metabolic threshold for lipolysis is ~400–700 pmol/L, a concentration marginally greater than resting epinephrine levels and considerably lower than levels attained during moderate intensity exercise (~2000 pmol/L).[28] The sympathoadrenal response is essential for exercise-induced increases in lipolysis because combined α- and β-adrenergic blockade completely suppresses

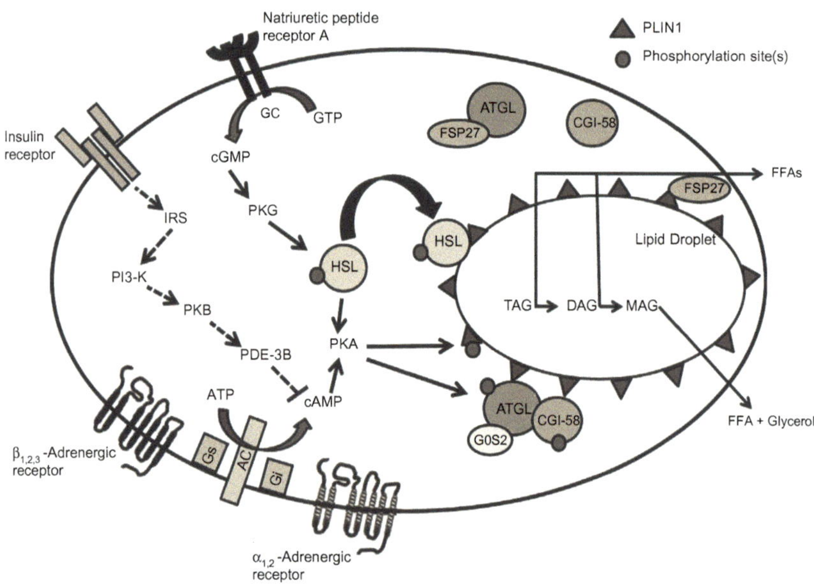

Figure 2 Regulatory control of lipolysis. Lipolytic and antilipolytic hormones act through their receptors to activate signaling pathways. Thereafter, changes in the cellular localization of key lipolytic proteins, their phosphorylation state, and protein–protein interactions dictate the breakdown of triglyceride stored in lipid droplets. The red arrows denote inhibitory pathways, and the blue arrows stimulatory pathways of lipolysis. Abbreviations: AC, adenylyl cyclase; ATP, adenosine triphosphate; ATGL, adipose triglyceride lipase; CGI-58, comparative gene identification 58; cAMP, cyclic adenosine monophosphate; cGMP, cyclic guanine monophosphate; DAG, diacylglycerol; FSP27, fat-specific protein 27; FFA, free fatty acid; G0S2, G0/G1 switch gene 2; GC, guanylyl cyclase; Gi, inhibitory GTP-binding protein; Gs, stimulatory GTP-binding protein; GTP, guanine triphosphate; HSL, hormone-sensitive lipase; IRS, insulin receptor substrate; MAG, monoacylglycerol; PDE-3B, phosphodiesterase 3B; PI3-K, phosphoinositide 3-kinase; PKA, protein kinase A; PKB, protein kinase B; PKG, protein kinase G; TAG, triacylglycerol. (See the color plate.)

the increase in plasma FFA during moderate intensity exercise[29] and the exercise-induced increase in lipolysis is abolished in spinal cord injury patients, who do not have intact somatic afferent and efferent neural activity.[30] Local blockade of all β-AR subtypes with propranolol reduces the exercise-induced increase in adipocyte lipolysis, suggesting that adrenal medulla-derived adrenaline is the main adrenergic agent contributing to exercise-induced lipolysis and that norepinephrine released from sympathetic neurons only contributes a small portion of the lipolytic effect.[31] This response is unique to exercise because other states of stimulated lipolysis, such as fasting and seasonal weight loss, are mediated by direct activation of the SNS innervation of white adipocytes.

4.2 Insulin

Insulin exerts powerful antilipolytic responses and is most important in the postprandial suppression of lipolysis. Insulin stimulates phosphodiesterase 3B, which lowers cAMP levels and reduces PKA activity[32,33] (Fig. 2). The decrease in plasma insulin during exercise appears to be important for the normal exercise-induced increase in plasma FFA because a failure to suppress insulin during exercise results in a twofold increase in plasma FFA.[29]

4.3 Other Hormones

Hormones that bind to stimulatory GTP-binding proteins stimulate the lipolytic cascade as described above.[34] These include glucagon[35,36] and cortisol,[37,38] both of which are increased late in prolonged exercise (Fig. 2). More recent work has identified the natriuretic peptides (NPs), atrial natriuretic peptide (ANP), and B-type natriuretic peptide as regulators of lipolysis. NPs increase during moderate intensity exercise[39] and NP stimulation of the membrane-associated NP guanylyl cyclase receptor of the A subtype increases intracellular cGMP levels,[40,41] which activates cGMP-dependent protein kinase I and, in turn, perilipin 1 (PLIN1) and hormone-sensitive lipase (HSL) phosphorylation, and lipolysis[42] (Fig. 2). Adenosine is an endogenous purine nucleoside that inhibits lipolysis by activation of the A1 adenosine receptor (A_1AR), which decreases adenylyl cyclase activity and inactivates PKA.[43] While adenosine inhibits lipolysis at rest, studies that infused the adenosine receptor antagonist theophylline during exercise reported no change in lipolysis,[44] questioning the importance of adenosine regulation during exercise.

4.4 Blood Flow

Adipose tissue blood flow increases by approximately twofold above rest during prolonged moderate intensity exercise (40–60% $\dot{V}O_2$ max) and is essential for lipolysis.[26] There is a rapid vasoconstriction after exercise, which reduces adipose tissue blood flow to the preexercise level, but interestingly, blood flow increases again in the postexercise period (after ~30 min). This partially accounts for the substantial lipid mobilization from adipose tissue after exercise.

5. CELLULAR AND MOLECULAR CONTROL OF ADIPOSE TISSUE METABOLISM

Lipolysis occurs at the surface of cytoplasmic lipid droplets that are composed of a neutral lipid core containing triglyceride and/or cholesterol esters coated with a phospholipid monolayer.[45] Lipolysis requires the recruitment of lipases and their activators to the lipid droplet, and interactions with lipid droplet-associated proteins for access to the triglyceride substrate (Fig. 2). Herein, we describe the intracellular events controlling adipocyte lipolysis, including the enzymatic regulation by lipases, the protein interactions, and the critical role of phosphorylation.

5.1 Triglyceride Lipases and Other Interacting Proteins

The sequential hydrolysis of triglycerides by specific enzymes results in the liberation of a fatty acid at each step with the generation of diglyceride, monoglyceride, and glycerol (Fig. 1). The three major lipases controlling lipolysis are adipose triglyceride lipase (ATGL), HSL, and monoacylglycerol lipase (MAGL), and each has specific substrate affinities (Fig. 2). Notably, there are >10 other proteins with triglyceride lipase activity, but these contribute only marginally to overall cellular lipolysis in humans.[46]

5.1.1 Adipose Triglyceride Lipase

Three independent laboratories identified ATGL as a major triglyceride in 2004.[47–49] The human gene encodes a 504-amino acid protein with a molecular mass of ~55 kDa.[47] The N-terminal region of ATGL contains ~260 amino acids and contains a "predicted esterase of the α/β-hydrolase" fold domain and a patatin domain.[50] ATGL exhibits high substrate specificity for triglyceride, specifically at the *sn*-2 position of the glycerol backbone, very weak activity against diacylglycerol, and no activity against cholesterol

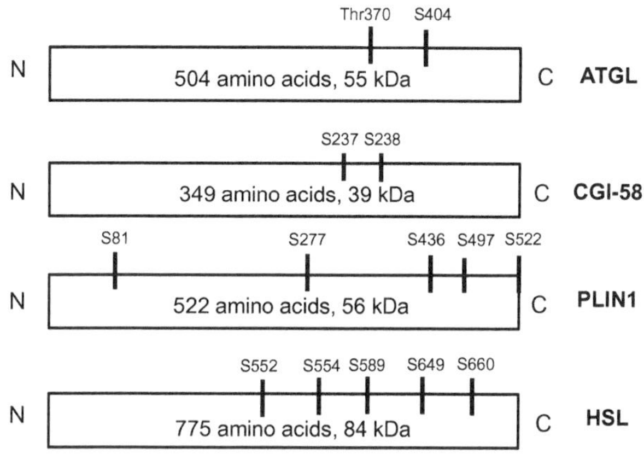

Figure 3 Schematic representation showing the phosphorylation sites of important lipolytic proteins. Phosphorylation of lipolytic proteins is critical in mobilizing fatty acids from lipid droplets. Sites denote human proteins. S, serine; T, threonine. Sequences for ATGL (Q96AD5), CGI-58 (Q8WTS1), PLIN1 (O60240), and HSL (XP_006723281).

or retinyl ester bonds.[47,51] The specificity for the *sn*-2 position of triglyceride is important because the *sn*-1.3 diacylglycerol product is the preferred substrate for subsequent hydrolysis by HSL, thereby facilitating complete triglyceride breakdown. The functional importance of ATGL was demonstrated by studies in ATGL$^{-/-}$ mice, which have reduced basal and catecholamine-stimulated lipolysis and increased adipocyte mass.[52] ATGL is phosphorylated at serine (Ser)404 (Ser-406 in mouse) by PKA, which mildly increases lipolysis,[53] and is phosphorylated by 5′-AMP-activated protein kinase (AMPK) at the same site,[54] although the functional significance of this latter event remains unclear. Phosphorylation of ATGL at threonine (Thr)372 is required for ATGL to localize to the lipid droplet for efficient lipolysis[55] (Fig. 3).

5.1.2 Comparative Gene Identification 58

When coactivated by the protein comparative gene identification 58 (CGI-58), ATGL activity increases 20-fold in recombinant protein studies *in vitro*.[56] The importance of this interaction is highlighted by the observation that ectopic overexpression of CGI-58 or ATGL does not impact triglyceride content in cells, whereas triglyceride storage is markedly reduced when the proteins are coexpressed.[56] Thus, CGI-58 is a critical interacting partner for optimal ATGL function. Phosphorylation of CGI-58 at Ser-237 (Ser-239 in mouse) by PKA is important for regulating its cellular localization to facilitate ATGL coactivation[57] (Fig. 3).

5.1.3 G_0/G_1 Switch Gene 2

The protein product of G_0/G_1 switch gene 2 (G0S2) binds to the C-terminal lipid-binding domain of ATGL, which inhibits ATGL triglyceride hydrolase activity and prevents its localization to the lipid droplet.[58,59] Overexpression of G0S2 in adipocytes reduces both basal and β-adrenergic-stimulated lipolysis while knockdown of G0S2 increases adipocyte lipolysis.[59] G0S2 deletion in mice increases lipolysis and decreases adipose tissue mass.[60] G0S2 is most likely a long-term controller of lipolysis because it regulates ATGL in a dose-dependent manner,[59] and its expression is regulated by nutritional status.[60] Thus, G0S2 is most likely to be important for maintaining homeostatic control of fatty acid flux from adipocytes over hours rather than minutes.

5.1.4 Perilipin 1

PLIN1 is located on the surface of lipid droplets[61] and acts as a "scaffold protein" by facilitating access of lipases to the lipid droplet and orchestrating protein–protein interactions in adipocytes (see below). Phosphorylation of PLIN1 by PKA is required for maximal lipolysis.[61] Murine PLIN1 has six PKA phosphorylation sites,[62] and Ser-492 and Ser-517 are critical for the regulation of lipolysis.[63] Phosphorylation of Ser-517 maximizes activation of ATGL-dependent lipolysis,[64] and PKA-mediated phosphorylation of Ser-492 drives fragmentation and dispersion of lipid droplets, which increases the surface area for lipase binding.[65] Although these results pertain to murine PLIN1, these sites are conserved in human PLIN1 at Ser-81, Ser-436, Ser-497, and Ser-522,[63] suggesting a similar physiological function in humans (Fig. 3).

5.1.5 Fat-Specific Protein 27

Fat-specific protein (FSP) 27 is an adipocyte-specific lipid droplet-associated protein that negatively regulates lipolysis. FSP27 promotes triglyceride storage in two ways: first, PLIN1 interacts with the CIDE-N domain of FSP27 to increase lipid transfer and lipid droplet growth,[66] and second, FSP27 interacts with and controls ATGL activity to regulate both basal and stimulated lipolysis.[67]

5.1.6 Hormone-Sensitive Lipase

HSL exhibits both triglyceride and diglyceride hydrolase activities. The relative hydrolase activity is 11-fold greater against diglyceride than triglyceride *in vitro*[68] and studies in $HSL^{-/-}$ mice report increased diglyceride content in adipocytes,[69] suggesting that diglyceride is the preferred substrate. HSL has a preference for primary ester bonds in the *sn*-3 position of the glycerol backbone,[70] but can also hydrolyse *sn*-1 esters. As such, it hydrolyzes both DAG stereoisomers generated by ATGL.[51,69] HSL activity and cellular localization are influenced by reversible phosphorylation. HSL is phosphorylated by PKA at Ser-552, Ser-649, and Ser-650; by AMPK at Ser-554; and by extracellular-regulated kinase (ERK) at Ser-589[71] (Fig. 3). PKA phosphorylation of HSL results in two- to threefold activation of the enzyme lipolytic activity, promotes translocation of HSL to the lipid droplet,[72] and induces a conformational change that increases the exposed hydrophobic surface area, which facilitates binding of HSL to the lipid substrate.[73] Collectively, these events increase lipolysis during β-adrenergic stimulation. Phosphorylation of HSL by AMPK is proposed to prevent PKA-mediated phosphorylation of HSL and reduce lipolysis,[71] although this remains controversial, while ERK increases HSL activity and lipolysis in culture adipocytes.[74]

5.1.7 Monoacylglycerol Lipase

MAGL is the final step in the lipolysis process. It liberates fatty acid from monoglyceride to produce glycerol. Interestingly, ATGL and HSL account for up to 95% of the adipocyte lipolysis.[75] MAGL is ubiquitously expressed with highest expression levels in adipose tissue, and lack of MAGL impairs lipolysis and is associated with increased monoglyceride levels in adipose tissue.[76]

5.2 Coordinating Lipolysis: Lipid Droplet-Associated Proteins, Protein–Protein Interactions, and Phosphorylation

Under resting conditions, ATGL is mostly localized to endoplasmic reticulum-related membranes in the cytoplasm and a small fraction is complexed with G0S2 on the lipid droplets. PLIN1 binds to CGI-58 and prevents its access to ATGL, thereby suppressing lipolysis (Fig. 2). Upon β-AR stimulation, PKA phosphorylation of PLIN1 releases CGI-58, a process that requires phosphorylation of CGI Ser-237.[57] ATGL translocates to lipid droplets, where it binds and is stimulated by CGI-58 to increase lipolysis. HSL is phosphorylated by PKA and translocates to lipid droplets. Here, it interacts with PLIN1, which facilitates HSL access to the tri- and diglyceride substrates.[77] Prolonged β-adrenergic stimulation downregulates G0S2 protein expression, thereby releasing more ATGL to sustain lipolysis (Fig. 2). Finally, β-adrenergic stimulation is associated with fragmentation of large unilocular lipid droplets, which disperse into smaller lipid droplets. This process increases the lipid droplet surface area for access by lipases to increase lipolysis. Lipid droplet fragmentation is a complex process involving the coordination of phosphorylation-dependent events, including phosphorylation of PLIN1 at Ser-492 (human Ser-497),[65] and changes in the microtubule organization and microfilament structure.

5.3 The Cellular Regulation of Lipolysis During Acute Exercise *In Vivo*

The *in vivo* evidence describing the cellular regulation of lipolysis during exercise is limited and has failed to keep pace with the detailed cell biology findings reported above. Genetic ablation of ATGL and HSL in mice demonstrates that these proteins are important for maintaining FFA availability during exercise,[78] whereas the exercise-induced increase in plasma FFA is normal in *Plin1*$^{-/-}$ mice.[79] ATGL Ser-406 phosphorylation in mice[53] and HSL Ser-563 and Ser-660 phosphorylation in rats[80] and humans[81] are increased during moderate intensity exercise, which corresponds to increased epinephrine and plasma FFA levels. One study, to our knowledge, has examined the protein–protein interactions between some key lipolytic proteins and their intracellular localization during exercise. Exercise increases HSL abundance at the lipid droplet and its interaction with PLIN1, while CGI-58 concomitantly dissociates from PLIN1.[80] Thus, the limited *in vivo* studies support components of the lipolytic model outlined for cells, assuming β-adrenergic stimulation models an "exercise" response.

6. METABOLIC AND MOLECULAR ADAPTATIONS IN ADIPOSE TISSUE WITH EXERCISE TRAINING

6.1 Lipolysis and the Lipolytic Proteins

Endurance exercise training results in adipocyte-specific and circulatory adaptations that impact lipolysis. The capacity for lipolysis appears to increase with training. Adipocytes isolated from marathon runners are more responsive to epinephrine when compared to sedentary controls at pharmacological but not physiological concentrations,[82] demonstrating that training increases the capacity for lipolysis but not necessarily the sensitivity to β-adrenergic stimulation. Similarly, epinephrine-stimulated lipolysis from intact subcutaneous adipose tissue was not different between athletes and untrained men.[83] During low-intensity exercise, trained and sedentary subjects exhibit similar lipolysis rates when exercising at the same absolute workload,[84] suggesting that endurance-trained athletes when compared to untrained individuals exhibit similar lipolytic rates. Data from intervention studies show that plasma fatty acid mobilization during exercise does not increase[85] and can even decrease[86] following several weeks of endurance exercise training. Moreover, whole-body lipolytic sensitivity to physiological catecholamine concentrations is not affected by exercise training.[83,87] While these observations seem at odds with others reporting increases in β-adrenoceptor activity and decreases in the antilipolytic effects of the α_2-adrenoceptor pathway in response to exercise training,[88–91] the lack of a change in lipolysis after training may reflect the combined effects of increased adipose tissue blood flow[84,92] and a suppressed exercise-induced catecholamine release in endurance training individuals. At a molecular level, some, but not all, studies report an increase in the contents of the major lipolytic regulators ATGL, CGI-58, PLIN1, and HSL after endurance exercise training and an increase in the association between ATGL and CGI-58 (Table 2). Notably, these studies have been conducted in rodents and studies in humans are required to examine the molecular responses of lipolytic proteins to endurance exercise training.

6.2 Mitochondrial Biogenesis

Mitochondrial biogenesis is a major adaption of skeletal muscle to exercise training and is induced by a complex interplay between numerous signaling pathways that respond to metabolic, mechanical, and hypoxic stresses that are generated within the myocyte during contraction. Adipocytes are highly plastic and mitochondrial biogenesis can be induced by pharmacological interventions in both isolated cells and free-living animals. Currently, there is a paucity of data describing the effects of exercise on mitochondrial biogenesis and its relevance for adipocyte function. Swim training in mice increases the mRNA/protein levels of key transcriptional regulators of mitochondrial biogenesis and increases mitochondrial DNA content in subcutaneous adipose tissue.[93] In support of these findings, the expression of genes involved in oxidative phosphorylation was increased following 6 months of endurance training, but notably, mitochondrial mass and function were not assessed.[94] We are aware of only one study to directly examine exercise training on adipocyte mitochondrial biogenesis in humans. Camera *et al.*[95] reported no change in adipose tissue citrate synthase activity, mitochondrial volume, or expression of genes that predict increased oxidative capacity after 10 days of endurance training in untrained men. While the

Table 2 Changes in Lipolytic Proteins after Endurance Exercise Training

Protein	Exercise Protocol	Species	Response/Major Finding	References
ATGL	1 h of cycling at 65% VO_2 max	♂ Humans	ATGL, CGI58 mRNA unchanged	89
	90 min running at 30 m/min, 5 days/week for 9 weeks. 5° incline	♂ Wistar rats	ATGL mRNA and protein increased CGI-58 ATGL protein association increased	93
	1 h running at 75–85% peak VO_2. 5 days/week for 8 weeks	♂ Wistar rats	ATGL protein increased 484% (low fat fed only) ATGL protein increased 318% (low fat fed only)	94
	Absolute cumulative running distance of ~10 km. 4 days/week for 6 weeks	♂ LRC and HRC rats	Increased phospho-ATGL (25% in HRC) Increased PLIN1 protein (25% in HRC) Increased CGI-58 (15% in LRC)	95
	1 h running at 25 m/min. 5 days/week for 6 weeks	♂ DIO Wistar rat	ATGL, CGI58 protein increased	92
	1 h continuous swimming. 5 days/week for 8 weeks	♂ Swiss mice	ATGL and CGI58 unchanged	90
	45 min running at 16 m/min. 5 days/week for 5 weeks. Progressive training protocol	♂ C57BL/6 mice	ATGL unchanged	78
	Ad lib wheel cage running (measured revs per day)	♂ Syrian golden hamsters	ATGL protein content increased	91
HSL	20 weeks of vigorous intensity exercise (70–75% heart rate reserve) or walked on treadmill for 3 days/week. HRR = max HR − resting HR obtained from VO_2 max	♀ Human	HSL gene expression increased only with vigorous exercise	96

Table 2 Changes in Lipolytic Proteins after Endurance Exercise Training—cont'd

Protein	Exercise Protocol	Species	Response/Major Finding	References
	Progressive swim training. 6 h/day, 5 day/week for 18 weeks	♀ Wistar rats	HSL protein content doubled in retroperitoneal and unchanged in mesenteric HSL enzyme activity increased	97
	1 h running at 25 m/min. 5 days/week for 6 weeks	♂ DIO Wistar rats	HSL, PLIN1 protein content increased	92
	3 h swimming. 5 days/week for 4 weeks	♂ C57BL/6 mice	No alteration in HSL gene expression	98

The studies shown have followed a prolonged exercise training protocol, with their subsequent lipolytic responses. Abbreviations: AT, adipose tissue; DIO, diet-induced obese; HRC, high running capacity; LD, lipid droplet; LRC, low running capacity.

length of the training program might explain differences between these human studies, the actual requirement for adipocyte mitochondrial biogenesis with exercise training is uncertain because ATP turnover (oxygen uptake) is not actually increased during acute exercise.[26] At present, there is much to be learned about the magnitude of mitochondrial biogenesis in adipocytes, the potential drivers of this process (most likely endocrine related), and even the requirement for biogenesis in sustaining metabolic functions after exercise training.

A major issue faced with examining adipocyte adaptations is the marked heterogeneity of cells in adipose tissue. Hence, a measure of a cellular or molecular response in adipose tissue cannot be regarded as a change in adipocyte function *per se* and caution must be taken when interpreting studies of molecular adaptation in adipose tissue.

6.3 DNA Methylation

Epigenetic signals such as DNA methylation and histone modifications play an important role in regulating gene expression. While the epigenetic pattern is mainly established early in life, recent studies indicate that both acute and chronic physical activities influence DNA methylation in a dose-dependent and gene-specific manner.[99] While this work has almost exclusively focused on effects in skeletal muscle, one study reported the effects of

6 months of endurance exercise training (3 h/week) on adipose tissue DNA methylation in healthy, sedentary men.[100] Exercise increased global DNA methylation, altering ~4% of the individual CG dinucleotide sites examined. Altered mRNA expression occurred in approximately one-third of the gene regions with altered DNA methylation, indicating functional relevance of these changes. Interestingly, exercise training coincided with differential DNA methylation of 39 candidate genes for obesity and type 2 diabetes in adipose tissue, implicating exercise-induced DNA methylation in adipose tissue metabolism and protection against disease.

6.4 miRNA, Adipose Tissue, and Exercise

miRNAs are small noncoding RNAs that alter gene function. They bind to complementary target sites in mRNA, where they regulate gene expression by two general mechanisms: translation repression or cleavage and deadenylation and degradation of target mRNA genes.[101] It has been estimated that over 60% of human protein coding genes are predicted to contain miRNA binding sites within the 3′-untranslated regions (UTRs), and each mRNA may be targeted by many miRNAs.[102] Notably, most putative miRNA binding sites have yet to be experimentally validated.[103] For more in depth information pertaining to miRNAs, refer to the chapter 'Exercise and the role of microRNAs in regulation', written by Aaron P. Russell.

Many miRNAs regulate adipogenesis (both negative and positive)[38]; however, their roles in metabolic regulation are unresolved. Studies in cultured adipocytes have implicated miR-224 in regulating fatty acid metabolism (target: acyl-CoA synthetase long-chain family member 4) and miR-145 in triglyceride lipolysis (targets: HSL, Foxo1, and CGI-58). To our knowledge, there are no studies examining miRNA responses to acute or chronic exercise, or possible effects on adipocyte metabolism in mature adipocytes *in vivo*.

6.5 Evidence That Exercise Training Alters Adipocyte Phenotypes

There are several types of adipocytes in human adipose tissues that arise from different mesenchymal stem cell lineages and possess vastly different functions. The first are white adipocytes, which have been extensively addressed above. The second are brown adipocytes, which arise from mesenchymal precursor cells common to the myogenic cell lineage[104] and are located primarily in the subclavicular and spinal regions of humans (Fig. 1A). The major function of brown adipocytes is to dissipate chemical energy as heat via the actions of uncoupling protein 1 (UCP1) in response to sympathetic stimulation. This process is known as thermogenesis and occurs in response to cold and after food ingestion in rodents. The third type of adipocytes are the "brite" or "beige" adipocytes that are interspersed within white adipose tissue. Their name derives from the observations that these adipocytes are "induced" from precursor cells that are distinct from classical brown adipocytes (although there is evidence of transdifferentiation from white-to-brown adipocytes),[105] share an overlapping gene expression pattern with brown adipocytes, express UCP1, and exhibit thermogenic activity, albeit to a lesser extent than brown adipocytes.[106] This remarkable uncoupling activity and energy dissipation potential, and the observations that brown adipocyte oxidative metabolism contributes to energy expenditure in humans[107] and that this activity appears to be reduced in obesity, have led to intense interest in understanding the processes controlling the "browning" of adipose tissue with a view to designing strategies to combat the obesity epidemic.

How might exercise induce white-to-brown conversion? While noradrenergic stimulation is the predominant mechanism for brown adipocyte activation and "browning" of white adipocytes, the interest from exercise physiologists stemmed from the discovery in mice of irisin, a PGC-1α-dependent and exercise-responsive myokine that stimulates UCP1 expression and a broad program of brown-fat-like development in white adipose tissue.[108] Increases in circulating irisin were reported in humans in response to prolonged, moderate intensity (~1 h at 45–70% VO_2 max) and repeated sprint exercise, although the increase is marginal (~1.2-fold).[109–112] Interestingly, most studies report an increase in *FNDC5* gene expression, from which irisin is transcribed, after acute exercise and chronic exercise training; however, changes in circulating irisin levels are not commonly reported.[113] The lack of consensus in this literature has been ascribed to differences in study designs, analytical techniques for assessing irisin, subject activity levels, age, sex, mass, and disease status (i.e., type 2 diabetes). Aside from irisin, several other secreted factors promote adipocyte "browning" and include cardiac natriuretic peptide (ANP[114]), bone morphogenetic protein 7,[115] fibroblast growth factor 21,[116] adenosine,[117] and several micro RNAs (e.g., miR196a, miR193b-365 cluster). Aside from ANP, which is increased during exercise, there is no conclusive evidence describing whether these "browning" factors respond to acute exercise or regular exercise training.

Studies in rodents mostly demonstrate that exercise has no effect on brown adipocyte mass or function.[113] Whether exercise causes adipocyte "browning" is controversial. There is evidence showing increased *UCP1* expression in select adipose pads (most prominently subcutaneous inguinal white adipose tissue) of exercise-trained compared with sedentary mice.[108,118] In humans, neither 12 weeks of endurance/strength training[110] nor 6 months of aerobic exercise training[94] changed the expression of the brown/brite marker genes *PRDM16*, *TBX1*, *TMEM26*, and *CD137* in subcutaneous adipose tissue. Despite the absence of change in transcriptional markers, *UCP1* expression was increased by 80% after exercise training, hinting at the presence of "brite" adipocytes.[110] This research field is evolving and "browning" may be depot specific and/or occurring in defined clusters within adipose tissue, as is proposed for mice. Immunohistochemical approaches examining "brite" markers across several regions in various adipose tissue depots may help address this issue, as will identifying the distinct resident progenitor cells in various human adipose depots and their sensitivities to putative "browning" factors.

Irrespective of the potential upstream drivers of adipocyte "browning," the physiological relevance of such an "adaptive" response remains unclear. It was postulated that brown adipocyte activation immediately after exercise may facilitate the clearance of excess fatty acids derived from adipocyte lipolysis (see Section 3), although this hypothesis requires testing. Exercise requires the conversion of glucose/fatty acid substrates for ATP production to meet continued ATP demand, and metabolic substrates are required in the recovery from exercise to replenish intracellular fuel stores and fuel repair processes and other cellular functions. Therefore, the benefit of inducing a transcriptional program designed to increase uncoupled respiration, and therefore inefficient utilization of metabolic substrates, seems counterproductive and is incongruent with exercise adaptations in other tissues that promote more efficient ATP generation.

7. CONCLUSIONS AND FUTURE DIRECTIONS

Cell biology studies have provided a wealth of information detailing the molecular and cellular regulation of adipocyte metabolism and adaptation to extracellular stimuli. Over the next few years, we are likely to learn more about the molecular switches controlling adipocyte differentiation/transdifferentiation and the metabolic and endocrine properties of brown/beige adipocytes. Studies in humans have failed to keep pace with the cell biology. While much is known regarding lipid metabolism at rest and during exercise, the hormonal regulators of these processes, and the effects of exercise training on metabolic fluxes, other important adipocyte-related questions remain unanswered in the fields of exercise metabolism and developmental biology. Future studies will need to decipher the cellular and molecular regulation of metabolism in adipocytes *per se*, without the confounding effects of contaminating immune, endothelial, stem, and progenitor cells contained within the adipose tissue. Prominent gaps relate to understanding whether, and how, miRNAs and other noncoding RNAs regulate adipocyte metabolism and the impact of exercise in modulating this regulatory system. Understanding the factors driving the "browning" process of white adipose tissue is a high priority in the field, and exercise may play a role here. Also, epigenetic modification to exercise training is apparent, but the functional outcomes in adipocytes are unknown and may bear relevance on dictating the health of future generations. With the cell biology studies acting as a compass, the remarkable advances of *in vivo* imaging and increased accessibility to powerful molecular screening tools should enable a new wave of research in adipocyte biology to address these fundamental biological questions.

REFERENCES

1. Romijn JA, Coyle EF, Sidossis LS, et al. Regulation of endogenous fat and carbohydrate metabolism in relation to exercise intensity and duration. *Am J Physiol.* 1993;265(3 pt 1):E380–E391.
2. Lass A, Zimmermann R, Oberer M, Zechner R. Lipolysis—a highly regulated multi-enzyme complex mediates the catabolism of cellular fat stores. *Prog Lipid Res.* 2011;50(1):14–27.
3. Cypess AM, Lehman S, Williams G, et al. Identification and importance of brown adipose tissue in adult humans. *N Engl J Med.* 2009;360(15):1509–1517.
4. van Marken Lichtenbelt WD, Vanhommerig JW, Smulders NM, et al. Cold-activated brown adipose tissue in healthy men. *N Engl J Med.* 2009;360(15):1500–1508.
5. Wu J, Bostrom P, Sparks LM, et al. Beige adipocytes are a distinct type of thermogenic fat cell in mouse and human. *Cell.* 2012;150(2):366–376.
6. Neamat-Allah J, Wald D, Husing A, et al. Validation of anthropometric indices of adiposity against whole-body magnetic resonance imaging—a study within the German European prospective investigation into cancer and nutrition (EPIC) cohorts. *PLoS One.* 2014;9(3):e91586.
7. Abate N, Garg A, Peshock RM, Stray-Gundersen J, Grundy SM. Relationships of generalized and regional adiposity to insulin sensitivity in men. *J Clin Invest.* 1995;96(1):88–98.
8. Tchkonia T, Thomou T, Zhu Y, et al. Mechanisms and metabolic implications of regional differences among fat depots. *Cell Metab.* 2013;17(5):644–656.
9. Garrow JS. New approaches to body composition. *Am J Clin Nutr.* 1982;35(5 suppl):1152–1158.
10. Ahlborg G, Felig P, Hagenfeldt L, Hendler R, Wahren J. Substrate turnover during prolonged exercise in man. Splanchnic and leg metabolism of glucose, free fatty acids, and amino acids. *J Clin Invest.* 1974;53(4):1080–1090.

11. Havel RJ, Naimark A, Borchgrevink CF. Turnover rate and oxidation of free fatty acids of blood plasma in man during exercise: studies during continuous infusion of palmitate-1-C14. *J Clin Invest.* 1963;42:1054–1063.
12. Burguera B, Proctor D, Dietz N, Guo Z, Joyner M, Jensen MD. Leg free fatty acid kinetics during exercise in men and women. *Am J Physiol Endocrinol Metab.* 2000;278(1):E113–E117.
13. Wolfe RR, Klein S, Carraro F, Weber JM. Role of triglyceride-fatty acid cycle in controlling fat metabolism in humans during and after exercise. *Am J Physiol.* 1990;258(2 pt 1):E382–E389.
14. Henderson GC, Fattor JA, Horning MA, et al. Lipolysis and fatty acid metabolism in men and women during the postexercise recovery period. *J Physiol.* 2007;584(pt 3): 963–981.
15. Magkos F, Mohammed BS, Patterson BW, Mittendorfer B. Free fatty acid kinetics in the late phase of postexercise recovery: importance of resting fatty acid metabolism and exercise-induced energy deficit. *Metab Clin Exp.* 2009;58(9):1248–1255.
16. Mittendorfer B, Horowitz JF, Klein S. Effect of gender on lipid kinetics during endurance exercise of moderate intensity in untrained subjects. *Am J Physiol Endocrinol Metab.* 2002;283(1):E58–E65.
17. Martin ML, Jensen MD. Effects of body fat distribution on regional lipolysis in obesity. *J Clin Invest.* 1991;88(2):609–613.
18. Hoffstedt J, Arner P, Hellers G, Lonnqvist F. Variation in adrenergic regulation of lipolysis between omental and subcutaneous adipocytes from obese and non-obese men. *J Lipid Res.* 1997;38(4):795–804.
19. Laurencikiene J, Skurk T, Kulyte A, et al. Regulation of lipolysis in small and large fat cells of the same subject. *J Clin Endocrinol Metab.* 2011;96(12):E2045–E2049.
20. McDevitt RM, Bott SJ, Harding M, Coward WA, Bluck LJ, Prentice AM. De novo lipogenesis during controlled overfeeding with sucrose or glucose in lean and obese women. *Am J Clin Nutr.* 2001;74(6):737–746.
21. Aarsland A, Chinkes D, Wolfe RR. Hepatic and whole-body fat synthesis in humans during carbohydrate overfeeding. *Am J Clin Nutr.* 1997;65(6):1774–1782.
22. Sondergaard E, Rahbek I, Sorensen LP, et al. Effects of exercise on VLDL-triglyceride oxidation and turnover. *Am J Physiol Endocrinol Metab.* 2011;300(5):E939–E944.
23. Magkos F, Patterson BW, Mohammed BS, Mittendorfer B. A single 1-h bout of evening exercise increases basal FFA flux without affecting VLDL-triglyceride and VLDL-apolipoprotein B-100 kinetics in untrained lean men. *Am J Physiol Endocrinol Metab.* 2007;292(6):E1568–E1574.
24. Elia M. Organ and tissue contribution to metabolic rate. In: Kinney JM, Tucker HN, eds. *Energy Metabolism: Tissue Determinants and Cellular Corollaries.* New York, NY: Raven Press; 1992:61–80.
25. Gallagher D, Belmonte D, Deurenberg P, et al. Organ-tissue mass measurement allows modeling of REE and metabolically active tissue mass. *Am J Physiol.* 1998;275(2 pt 1): E249–E258.
26. Mulla NA, Simonsen L, Bulow J. Post-exercise adipose tissue and skeletal muscle lipid metabolism in humans: the effects of exercise intensity. *J Physiol.* 2000;524(pt 3): 919–928.
27. Kjaer M, Secher NH, Galbo H. Physical stress and catecholamine release. *Baillieres Clin Endocrinol Metab.* 1987;1(2):279–298.
28. Clutter WE, Bier DM, Shah SD, Cryer PE. Epinephrine plasma metabolic clearance rates and physiologic thresholds for metabolic and hemodynamic actions in man. *J Clin Invest.* 1980;66(1):94–101.

29. Marker JC, Hirsch IB, Smith LJ, Parvin CA, Holloszy JO, Cryer PE. Catecholamines in prevention of hypoglycemia during exercise in humans. *Am J Physiol.* 1991;260(5 pt 1): E705–E712.
30. Kjaer M, Dela F, Sorensen FB, et al. Fatty acid kinetics and carbohydrate metabolism during electrical exercise in spinal cord-injured humans. *Am J Physiol Regul Integr Comp Physiol.* 2001;281(5):R1492–R1498.
31. Arner P, Kriegholm E, Engfeldt P, Bolinder J. Adrenergic regulation of lipolysis in situ at rest and during exercise. *J Clin Invest.* 1990;85(3):893–898.
32. Degerman E, Smith CJ, Tornqvist H, Vasta V, Belfrage P, Manganiello VC. Evidence that insulin and isoprenaline activate the cGMP-inhibited low-Km cAMP phosphodiesterase in rat fat cells by phosphorylation. *Proc Natl Acad Sci USA.* 1990;87(2):533–537.
33. Choi SM, Tucker DF, Gross DN, et al. Insulin regulates adipocyte lipolysis via an Akt-independent signaling pathway. *Mol Cell Biol.* 2010;30(21):5009–5020.
34. Lafontan M, Langin D. Lipolysis and lipid mobilization in human adipose tissue. *Prog Lipid Res.* 2009;48(5):275–297.
35. Kanaley JA, Dall R, Moller N, et al. Acute exposure to GH during exercise stimulates the turnover of free fatty acids in GH-deficient men. *J Appl Physiol.* 2004;96(2): 747–753.
36. Wakelam MJ, Murphy GJ, Hruby VJ, Houslay MD. Activation of two signal-transduction systems in hepatocytes by glucagon. *Nature.* 1986;323(6083):68–71.
37. Djurhuus CB, Gravholt CH, Nielsen S, et al. Effects of cortisol on lipolysis and regional interstitial glycerol levels in humans. *Am J Physiol Endocrinol Metab.* 2002;283(1): E172–E177.
38. Peng Y, Yu S, Li H, Xiang H, Peng J, Jiang S. MicroRNAs: emerging roles in adipogenesis and obesity. *Cell Signal.* 2014;26(9):1888–1896.
39. Moro C, Crampes F, Sengenes C, et al. Atrial natriuretic peptide contributes to physiological control of lipid mobilization in humans. *FASEB J.* 2004;18(7):908–910.
40. Moro C, Galitzky J, Sengenes C, Crampes F, Lafontan M, Berlan M. Functional and pharmacological characterization of the natriuretic peptide-dependent lipolytic pathway in human fat cells. *J Pharmacol Exp Ther.* 2004;308(3):984–992.
41. Sengenes C, Berlan M, De Glisezinski I, Lafontan M, Galitzky J. Natriuretic peptides: a new lipolytic pathway in human adipocytes. *FASEB J.* 2000;14(10):1345–1351.
42. Sengenes C, Bouloumie A, Hauner H, et al. Involvement of a cGMP-dependent pathway in the natriuretic peptide-mediated hormone-sensitive lipase phosphorylation in human adipocytes. *J Biol Chem.* 2003;278(49):48617–48626.
43. Koupenova M, Ravid K. Adenosine, adenosine receptors and their role in glucose homeostasis and lipid metabolism. *J Cell Physiol.* http://dx.doi.org/10.1002/jcp.24352.
44. Raguso CA, Coggan AR, Sidossis LS, Gastaldelli A, Wolfe RR. Effect of theophylline on substrate metabolism during exercise. *Metab:Clin Exp.* 1996;45(9):1153–1160.
45. Khor VK, Shen WJ, Kraemer FB. Lipid droplet metabolism. *Curr Opin Clin Nutr Metab Care.* 2013;16(6):632–637.
46. Steinberg GR, Kemp BE, Watt MJ. Adipocyte triglyceride lipase expression in human obesity. *Am J Physiol Endocrinol Metab.* 2007;293(4):E958–E964.
47. Zimmermann R, Strauss JG, Haemmerle G, et al. Fat mobilization in adipose tissue is promoted by adipose triglyceride lipase. *Science (New York, NY).* 2004;306(5700): 1383–1386.
48. Villena JA, Roy S, Sarkadi-Nagy E, Kim KH, Sul HS. Desnutrin, an adipocyte gene encoding a novel patatin domain-containing protein, is induced by fasting and glucocorticoids: ectopic expression of desnutrin increases triglyceride hydrolysis. *J Biol Chem.* 2004;279(45):47066–47075.

49. Jenkins CM, Mancuso DJ, Yan W, Sims HF, Gibson B, Gross RW. Identification, cloning, expression, and purification of three novel human calcium-independent phospholipase A2 family members possessing triacylglycerol lipase and acylglycerol transacylase activities. *J Biol Chem*. 2004;279(47):48968–48975.
50. Ganal MW, Bonierbale MW, Roeder MS, Park WD, Tanksley SD. Genetic and physical mapping of the patatin genes in potato and tomato. *Mol Gen Genet*. 1991;225(3):501–509.
51. Eichmann TO, Kumari M, Haas JT, et al. Studies on the substrate and stereo/regioselectivity of adipose triglyceride lipase, hormone-sensitive lipase, and diacylglycerol-O-acyltransferases. *J Biol Chem*. 2012;287(49):41446–41457.
52. Haemmerle G, Lass A, Zimmermann R, et al. Defective lipolysis and altered energy metabolism in mice lacking adipose triglyceride lipase. *Science (New York, NY)*. 2006;312(5774):734–737.
53. Pagnon J, Matzaris M, Stark R, et al. Identification and functional characterization of protein kinase A phosphorylation sites in the major lipolytic protein, adipose triglyceride lipase. *Endocrinology*. 2012;153(9):4278–4289.
54. Ahmadian M, Abbott MJ, Tang T, et al. Desnutrin/ATGL is regulated by AMPK and is required for a brown adipose phenotype. *Cell Metab*. 2011;13(6):739–748.
55. Xie X, Langlais P, Zhang X, et al. Identification of a novel phosphorylation site in adipose triglyceride lipase as a regulator of lipid droplet localization. *Am J Physiol Endocrinol Metab*. 2014;306(12):E1449–E1459.
56. Lass A, Zimmermann R, Haemmerle G, et al. Adipose triglyceride lipase-mediated lipolysis of cellular fat stores is activated by CGI-58 and defective in Chanarin–Dorfman syndrome. *Cell Metab*. 2006;3(5):309–319.
57. Sahu-Osen A, Montero-Moran G, Schittmayer M, et al. CGI-58/ABHD5 is phosphorylated on Ser-239 by protein kinase A: control of subcellular localization. *J Lipid Res*. 2014;56(1):109–121.
58. Schweiger M, Paar M, Eder C, et al. G0/G1 switch gene-2 regulates human adipocyte lipolysis by affecting activity and localization of adipose triglyceride lipase. *J Lipid Res*. 2012;53(11):2307–2317.
59. Yang X, Lu X, Lombes M, et al. The G(0)/G(1) switch gene 2 regulates adipose lipolysis through association with adipose triglyceride lipase. *Cell Metab*. 2010;11(3):194–205.
60. Zhang X, Xie X, Heckmann BL, Saarinen AM, Czyzyk TA, Liu J. Targeted disruption of G0/G1 switch gene 2 enhances adipose lipolysis, alters hepatic energy balance, and alleviates high-fat diet-induced liver steatosis. *Diabetes*. 2014;63(3):934–946.
61. Brasaemle DL, Subramanian V, Garcia A, Marcinkiewicz A, Rothenberg A. Perilipin A and the control of triacylglycerol metabolism. *Mol Cell Biochem*. 2009;326(1–2):15–21.
62. Greenberg AS, Egan JJ, Wek SA, Moos Jr MC, Londos C, Kimmel AR. Isolation of cDNAs for perilipins A and B: sequence and expression of lipid droplet-associated proteins of adipocytes. *Proc Natl Acad Sci USA*. 1993;90(24):12035–12039.
63. McDonough PM, Maciejewski-Lenoir D, Hartig SM, et al. Differential phosphorylation of perilipin 1A at the initiation of lipolysis revealed by novel monoclonal antibodies and high content analysis. *PLoS One*. 2013;8(2):e55511.
64. Miyoshi H, Perfield 2nd JW, Souza SC, et al. Control of adipose triglyceride lipase action by serine 517 of perilipin A globally regulates protein kinase A-stimulated lipolysis in adipocytes. *J Biol Chem*. 2007;282(2):996–1002.
65. Marcinkiewicz A, Gauthier D, Garcia A, Brasaemle DL. The phosphorylation of serine 492 of perilipin a directs lipid droplet fragmentation and dispersion. *J Biol Chem*. 2006;281(17):11901–11909.

66. Sun Z, Gong J, Wu H, et al. Perilipin1 promotes unilocular lipid droplet formation through the activation of Fsp27 in adipocytes. *Nat Commun.* 2013;4:1594.
67. Grahn TH, Kaur R, Yin J, et al. Fat-specific protein 27 (FSP27) interacts with adipose triglyceride lipase (ATGL) to regulate lipolysis and insulin sensitivity in human adipocytes. *J Biol Chem.* 2014;289(17):12029–12039.
68. Fredrikson G, Stralfors P, Nilsson NO, Belfrage P. Hormone-sensitive lipase of rat adipose tissue. Purification and some properties. *J Biol Chem.* 1981;256(12):6311–6320.
69. Haemmerle G, Zimmermann R, Hayn M, et al. Hormone-sensitive lipase deficiency in mice causes diglyceride accumulation in adipose tissue, muscle, and testis. *J Biol Chem.* 2002;277(7):4806–4815.
70. Young SG, Zechner R. Biochemistry and pathophysiology of intravascular and intracellular lipolysis. *Genes Dev.* 2013;27(5):459–484.
71. Watt MJ, Steinberg GR. Regulation and function of triacylglycerol lipases in cellular metabolism. *Biochem J.* 2008;414(3):313–325.
72. Egan JJ, Greenberg AS, Chang MK, Wek SA, Moos Jr MC, Londos C. Mechanism of hormone-stimulated lipolysis in adipocytes: translocation of hormone-sensitive lipase to the lipid storage droplet. *Proc Natl Acad Sci USA.* 1992;89(18):8537–8541.
73. Krintel C, Morgelin M, Logan DT, Holm C. Phosphorylation of hormone-sensitive lipase by protein kinase A in vitro promotes an increase in its hydrophobic surface area. *FEBS J.* 2009;276(17):4752–4762.
74. Greenberg AS, Shen WJ, Muliro K, et al. Stimulation of lipolysis and hormone-sensitive lipase via the extracellular signal-regulated kinase pathway. *J Biol Chem.* 2001;276(48):45456–45461.
75. Schweiger M, Schreiber R, Haemmerle G, et al. Adipose triglyceride lipase and hormone-sensitive lipase are the major enzymes in adipose tissue triacylglycerol catabolism. *J Biol Chem.* 2006;281(52):40236–40241.
76. Zechner R, Zimmermann R, Eichmann TO, et al. FAT SIGNALS—lipases and lipolysis in lipid metabolism and signaling. *Cell Metab.* 2012;15(3):279–291.
77. Sztalryd C, Xu G, Dorward H, et al. Perilipin A is essential for the translocation of hormone-sensitive lipase during lipolytic activation. *J Cell Biol.* 2003;161(6):1093–1103.
78. Huijsman E, van de Par C, Economou C, et al. Adipose triacylglycerol lipase deletion alters whole body energy metabolism and impairs exercise performance in mice. *Am J Physiol Endocrinol Metab.* 2009;297(2):E505–E513.
79. Beylot M, Neggazi S, Hamlat N, Langlois D, Forcheron F. Perilipin 1 ablation in mice enhances lipid oxidation during exercise and does not impair exercise performance. *Metab Clin Exp.* 2012;61(3):415–423.
80. Ogasawara J, Nomura S, Rahman N, et al. Hormone-sensitive lipase is critical mediators of acute exercise-induced regulation of lipolysis in rat adipocytes. *Biochem Biophys Res Commun.* 2010;400(1):134–139.
81. Watt MJ, Holmes AG, Pinnamaneni SK, et al. Regulation of HSL serine phosphorylation in skeletal muscle and adipose tissue. *Am J Physiol Endocrinol Metab.* 2006;290(3):E500–E508.
82. Crampes F, Beauville M, Riviere D, Garrigues M. Effect of physical training in humans on the response of isolated fat cells to epinephrine. *J Appl Physiol.* 1986;61(1):25–29.
83. Horowitz JF, Braudy RJ, Martin 3rd WH, Klein S. Endurance exercise training does not alter lipolytic or adipose tissue blood flow sensitivity to epinephrine. *Am J Physiol.* 1999;277(2 pt 1):E325–E331.
84. Klein S, Coyle EF, Wolfe RR. Fat metabolism during low-intensity exercise in endurance-trained and untrained men. *Am J Physiol.* 1994;267(6 pt 1):E934–E940.

85. Horowitz JF, Leone TC, Feng W, Kelly DP, Klein S. Effect of endurance training on lipid metabolism in women: a potential role for PPARalpha in the metabolic response to training. *Am J Physiol Endocrinol Metab*. 2000;279(2):E348–E355.
86. Martin 3rd WH, Dalsky GP, Hurley BF, et al. Effect of endurance training on plasma free fatty acid turnover and oxidation during exercise. *Am J Physiol*. 1993;265(5 pt 1): E708–E714.
87. van Aggel-Leijssen DP, Saris WH, Homan M, van Baak MA. The effect of exercise training on beta-adrenergic stimulation of fat metabolism in obese men. *Int J Obes Relat Metab Disord*. 2001;25(1):16–23.
88. De Glisezinski I, Crampes F, Harant I, et al. Endurance training changes in lipolytic responsiveness of obese adipose tissue. *Am J Physiol*. 1998;275(6 pt 1):E951–E956.
89. Stich V, de Glisezinski I, Crampes F, et al. Activation of antilipolytic alpha(2)-adrenergic receptors by epinephrine during exercise in human adipose tissue. *Am J Physiol*. 1999;277(4 pt 2):R1076–R1083.
90. Stich V, de Glisezinski I, Galitzky J, et al. Endurance training increases the beta-adrenergic lipolytic response in subcutaneous adipose tissue in obese subjects. *Int J Obes Relat Metab Disord*. 1999;23(4):374–381.
91. Richterova B, Stich V, Moro C, et al. Effect of endurance training on adrenergic control of lipolysis in adipose tissue of obese women. *J Clin Endocrinol Metab*. 2004;89(3):1325–1331.
92. Stallknecht B, Larsen JJ, Mikines KJ, Simonsen L, Bulow J, Galbo H. Effect of training on insulin sensitivity of glucose uptake and lipolysis in human adipose tissue. *Am J Physiol Endocrinol Metab*. 2000;279(2):E376–E385.
93. Trevellin E, Scorzeto M, Olivieri M, et al. Exercise training induces mitochondrial biogenesis and glucose uptake in subcutaneous adipose tissue through eNOS-dependent mechanisms. *Diabetes*. 2014;63(8):2800–2811.
94. Ronn T, Volkov P, Tornberg A, et al. Extensive changes in the transcriptional profile of human adipose tissue including genes involved in oxidative phosphorylation after a 6-month exercise intervention. *Acta Physiol (Oxf)*. 2014;211(1):188–200.
95. Camera DM, Anderson MJ, Hawley JA, Carey AL. Short-term endurance training does not alter the oxidative capacity of human subcutaneous adipose tissue. *Eur J Appl Physiol*. 2010;109(2):307–316.
96. You T, Wang X, Yang R, Lyles MF, Gong D, Nicklas BJ. Effect of exercise training intensity on adipose tissue hormone sensitive lipase gene expression in obese women under weight loss. *J Sport Health Sci*. 2012;1(3):184–190.
97. Enevoldsen LH, Stallknecht B, Langfort J, et al. The effect of exercise training on hormone-sensitive lipase in rat intra-abdominal adipose tissue and muscle. *J Physiol*. 2001;536(pt 3):871–877.
98. Higa TS, Bergamo FC, Mazzucatto F, Fonseca-Alaniz MH, Evangelista FS. Physical training prevents body weight gain but does not modify adipose tissue gene expression. *Braz J Med Biol Res*. 2012;45(10):988–994.
99. Barres R, Zierath JR. DNA methylation in metabolic disorders. *Am J Clin Nutr*. 2011;93(4):897S–900S.
100. Ronn T, Volkov P, Davegardh C, et al. A six months exercise intervention influences the genome-wide DNA methylation pattern in human adipose tissue. *PLoS Genet*. 2013;9(6):e1003572.
101. Bartel DP. MicroRNAs: target recognition and regulatory functions. *Cell*. 2009;136(2):215–233.
102. Friedman RC, Farh KK-H, Burge CB, Bartel DP. Most mammalian mRNAs are conserved targets of microRNAs. *Genome Res*. 2009;19(1):92–105.
103. McGregor RA, Choi MS. MicroRNAs in the regulation of adipogenesis and obesity. *Curr Mol Med*. 2011;11(4):304–316.

104. Seale P, Bjork B, Yang W, et al. PRDM16 controls a brown fat/skeletal muscle switch. *Nature*. 2008;454(7207):961–967.
105. Rosenwald M, Perdikari A, Rulicke T, Wolfrum C. Bi-directional interconversion of brite and white adipocytes. *Nat Cell Biol*. 2013;15(6):659–667.
106. Rosen ED, Spiegelman BM. What we talk about when we talk about fat. *Cell*. 2014;156(1–2):20–44.
107. Ouellet V, Labbe SM, Blondin DP, et al. Brown adipose tissue oxidative metabolism contributes to energy expenditure during acute cold exposure in humans. *J Clin Invest*. 2012;122(2):545–552.
108. Bostrom P, Wu J, Jedrychowski MP, et al. A PGC1-alpha-dependent myokine that drives brown-fat-like development of white fat and thermogenesis. *Nature*. 2012;481(7382):463–468.
109. Lee S, Lee HC, Kwon YW, et al. Adenylyl cyclase-associated protein 1 is a receptor for human resistin and mediates inflammatory actions of human monocytes. *Cell Metab*. 2014;19(3):484–497.
110. Norheim F, Langleite TM, Hjorth M, et al. The effects of acute and chronic exercise on PGC-1alpha, irisin and browning of subcutaneous adipose tissue in humans. *FEBS J*. 2014;281(3):739–749.
111. Kurdiova T, Balaz M, Vician M, et al. Effects of obesity, diabetes and exercise on Fndc5 gene expression and irisin release in human skeletal muscle and adipose tissue: in vivo and in vitro studies. *J Physiol*. 2014;592(pt 5):1091–1107.
112. Huh JY, Panagiotou G, Mougios V, et al. FNDC5 and irisin in humans: I. Predictors of circulating concentrations in serum and plasma and II. mRNA expression and circulating concentrations in response to weight loss and exercise. *Metab Clin Exp*. 2012;61(12):1725–1738.
113. Elsen M, Raschke S, Eckel J. Browning of white fat: does irisin play a role in humans? *J Endocrinol*. 2014;222(1):R25–R38.
114. Bordicchia M, Liu D, Amri EZ, et al. Cardiac natriuretic peptides act via p38 MAPK to induce the brown fat thermogenic program in mouse and human adipocytes. *J Clin Invest*. 2012;122(3):1022–1036.
115. Schulz TJ, Huang TL, Tran TT, et al. Identification of inducible brown adipocyte progenitors residing in skeletal muscle and white fat. *Proc Natl Acad Sci USA*. 2011;108(1):143–148.
116. Fisher FM, Kleiner S, Douris N, et al. FGF21 regulates PGC-1alpha and browning of white adipose tissues in adaptive thermogenesis. *Genes Dev*. 2012;26(3):271–281.
117. Gnad T, Scheibler S, von Kugelgen I, et al. Adenosine activates brown adipose tissue and recruits beige adipocytes via A2A receptors. *Nature*. 2014;516(7531):395–399.
118. Wu MV, Bikopoulos G, Hung S, Ceddia RB. Thermogenic capacity is antagonistically regulated in classical brown and white subcutaneous fat depots by high-fat diet and endurance training in rats: impact on whole-body energy expenditure. *J Biol Chem*. 2014;289(49):34129–34140.

CHAPTER NINE

Exercise and the Regulation of Hepatic Metabolism

Elijah Trefts, Ashley S. Williams, David H. Wasserman[1]
Department of Molecular Physiology and Biophysics, Vanderbilt University School of Medicine, Nashville, Tennessee, USA
[1]Corresponding author: e-mail address: david.wasserman@vanderbilt.edu

Contents

Abstract

The accelerated metabolic demands of the working muscle cannot be met without a robust response from the liver. If not for the hepatic response, sustained exercise would be impossible. The liver stores, releases, and recycles potential energy. Exercise would result in hypoglycemia if it were not for the accelerated release of energy as glucose. The energetic demands on the liver are largely met by increased oxidation of fatty acids mobilized from adipose tissue. Adaptations immediately following exercise facilitate the replenishment of glycogen stores. Pancreatic glucagon and insulin responses orchestrate the hepatic response during and immediately following exercise. Like skeletal muscle and other physiological systems, liver adapts to repeated demands of exercise by increasing its capacity to produce energy by oxidizing fat. The ability of regular physical activity to increase fat oxidation is protective and can reverse fatty liver disease. Engaging in regular physical exercise has broad ranging positive health implications including those that improve the metabolic health of the liver.

The liver is a battery, a rechargeable battery at that. It releases stored energy at times of high metabolic demand and replenishes energy stores during the

Progress in Molecular Biology and Translational Science, Volume 135
ISSN 1877-1173
http://dx.doi.org/10.1016/bs.pmbts.2015.07.010

nutrient excess associated with a meal. The liver is a recycler converting metabolites into macronutrients, amino acids into proteins, and transforming potential energy into chemical energy. The liver is a detoxifier removing nitrogenous molecules, hemoglobin, hormones, foreign substances, immunoglobulin, and other compounds from the circulation. The muscle contracts, the adipose tissue stores fat, and the heart pumps blood. The functions of the liver are far too vast to describe with a single dominant process, but all make broad contributions to arterial homeostasis and thereby homeostasis of numerous cell types. Physical exercise poses a unique challenge to the liver as metabolic demands of working muscles require the liver to mobilize energy stores, recycle metabolites, and convert compounds that are toxic in excess to innocuous forms. The focus of this review will be on how the liver adapts to the metabolic demands of physical exercise.

1. LIVER RESPONSE TO ACUTE EXERCISE

1.1 Mobilization of Liver Energy Stores Maintains Glucose Homeostasis

The energy requirements of exercise necessitate a marked increase in glucose uptake by muscle, as well as increased utilization of lipids and muscle glycogen. Energy for working muscle may also be derived from branched chain amino acids. The mechanism for the increase in muscle fuel utilization is discussed in detail previously[1] and elsewhere in this volume. The requirement of glucose uptake for the working muscle is transferred in part to the liver, which must release glucose at a rate that matches the accelerated rate of glucose uptake to maintain glucose homeostasis.[2] There may be deviations from glucose homeostasis during high intensity and/or prolonged exercise. During high-intensity exercise, the stimulus to release glucose from the liver exceeds glucose utilization causing a rise in arterial glucose.[3] In contrast, during prolonged exercise, hypoglycemia may result as liver glycogen nears depletion.[4] Hypoglycemia can cause a more rapid onset of fatigue, perhaps due to neuroglycopenia. In cases of hypoglycemia, fatigue can be delayed by ingestion of glucose or glucose polymers.[5]

Accelerated muscle glucose uptake during exercise is matched by the sum of increased mobilization of hepatic glycogen and gluconeogenesis. Potential energy in the form of glycogen is at its highest concentration in the liver. With the onset of exercise, this energy store is hydrolyzed by activation of glycogen phosphorylase,[6] contributing to the release of glucose from the liver. Moreover, exercise causes the liver to discharge chemical energy with the hydrolysis of adenosine triphosphate (ATP). Adenosine monophosphate (AMP) and adenosine diphosphate (ADP) accumulate,[7,8] resulting in a precipitous fall in liver energy charge (Fig. 1). This discharge of chemical energy with exercise is necessary for gluconeogenesis and reactions that support this pathway (e.g., fatty acid activation, ureagenesis). Studies using 5′-aminoimidazole-4-carboxamide-1-beta-D-ribofuranoside (AICAR) have aided in establishing a role for hepatic adenine nucleotides in the regulation of liver glucose metabolism. AICAR is converted by phosphorylation to an AMP analog (ZMP) inside cells.[9] Infusing AICAR at rates to create liver ZMP concentrations that match liver AMP concentrations evident during exercise creates a potent breakdown of liver glycogen,[9,10]

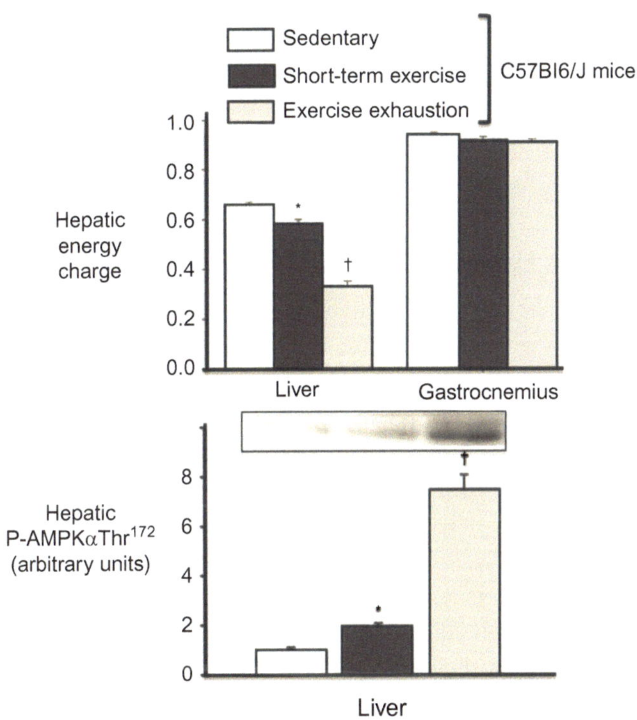

Figure 1 Hepatic energy discharge in response to short term and exhaustive exercise in C57Bl/6J mice. Hepatic energy charge decreases with exercise and becomes critically low with exhaustive exercise (top). Skeletal muscle energy charge is not affected by exercise. The decrease in energy charge is associated with an increase in hepatic AMPK activation (bottom). Energy charge is calculated by the equation (ATP + 0.5ADP)/(ATP + ADP + AMP). Data are mean ± SE. *Significantly different from SED ($P < 0.05$). †Significantly different from SED and ST ($P < 0.05$). *Modified from Camacho* et al.[8] (See the color plate.)

due presumably to increased allosteric activation of glycogen phosphorylase. Thus, the AMP produced with chemical energy discharge may combine with or mediate endocrine stimuli to enhance the release of potential energy by mobilization of hepatic glycogen.

AMP-activated protein kinase (AMPK) serves as a transducer, sensing energy state and activating metabolic pathways in accordance with metabolic needs.[11] Free fatty acids are mobilized from adipose tissue and delivered to the liver as the body transitions into a gluconeogenic mode during prolonged exercise.[12] AMPK activation increases the oxidation of fatty acids delivered to the liver resulting in the production of chemical energy from extrahepatic sources.[13] This process is reflected by increased hepatic oxygen uptake and ketone body output. Elevated fatty acid oxidation also increases the liver mitochondrial redox state, which favors gluconeogenic flux.[12]

Accelerated gluconeogenic flux during exercise requires more than activation of reactions within the liver. The increased flux through gluconeogenesis requires increased gluconeogenic precursor delivery from extrahepatic sites to the liver and transport across the membrane by the liver (Fig. 2).

1.2 Pancreatic Hormones Stimulate Hepatic Glucose Output

Exercise is characterized by complex neural and endocrine responses. If exercise is sustained (>~20 min), a decrease in insulin secretion and increases in glucagon, catecholamines, and cortisol secretion, among other hormones, are observed.[2] The signal for these hormonal and autonomic changes has been difficult to elucidate. An increase in afferent nerve activity originating at the working limb, a deficit in fuel availability, and a neural feed-forward mechanism have all been postulated as possible stimuli.[2] Afferent sensors in the carotid sinus area are required for the full increases in glucagon and norepinephrine during exercise.[14] Surprisingly, denervation of the pancreas does not impair the glucagon and insulin responses to exercise,[15] suggesting that an endocrine or a paracrine factor may be involved. The myokine interleukin 6 (IL-6) is released in response to physical exercise[16] and has been shown to stimulate glucagon release from the pancreatic alpha-cell under stressful conditions.[17] The pancreas is highly sensitive to small changes in blood glucose during exercise.[18–21] In light of this, it is surprising that preventing the exercise-induced increase in muscle glucose uptake by deleting the exercise-sensitive glucose transporter, Glut4, prevents neither the pancreatic hormone response nor the increase in

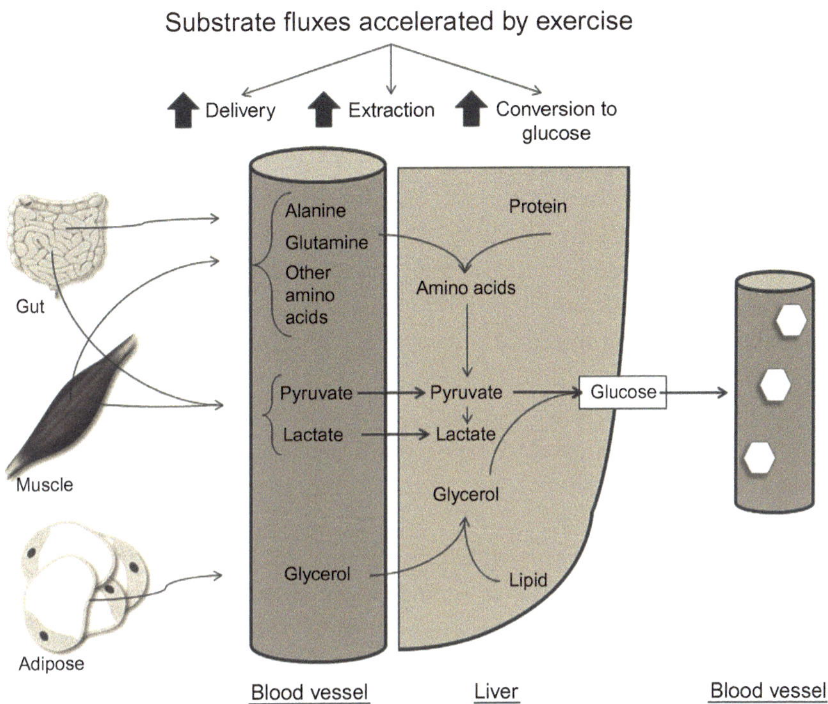

Figure 2 Gluconeogenesis is regulated by gluconeogenic precursor supply to the liver, extraction by the liver, and conversion to glucose within the liver. All these processes are accelerated by physical exercise. (See the color plate.)

hepatic glucose output during exercise.[22] This suggests that the endocrine and hepatic responses to exercise are not due to a feedback signal from accelerated blood glucose removal as once proposed.

Understanding regulation of the liver has proven difficult, particularly during exercise, since it is inaccessible in conscious humans. Moreover, movement makes magnetic resonance spectroscopy of the liver during exercise untenable. The dog has proven to be a useful experimental model for gaining insights into liver function because the portal vein, which perfuses the liver, and hepatic vein, which drains it, can be accessed using implanted catheters. The small size of rodent models is not conducive to these deep abdominal catheterizations or the blood sample volume associated with multiple sampling ports. Despite evidence showing the exquisite sensitivity of the liver to glucagon in humans,[23] a vital role of glucagon for the exercise-induced increase in hepatic glucose output has been slow to gain acceptance. This is because the increment in arterial glucagon with exercise is delayed and dampened with respect to the increase in glucose released from the liver.[2] In fact, depending on the duration and intensity of exercise, arterial glucagon may not increase at all. Glucagon released from the pancreas first perfuses the liver delaying its entry into the peripheral circulation. Since the liver extracts glucagon, the increase in arterial glucagon is attenuated.[24,25] Portal vein glucagon is increased during exercise to a much greater extent than in arterial and hepatic vein blood. The portal vein to arterial glucagon gradient is increased by approximately 10-fold in response to exercise.[24,25] The placement of the liver between the pancreas and general circulation is efficient since it allows for increased glucagon in the blood perfusing the liver without the high glucagon secretion rates needed to rapidly fill the general circulation.

Studies conducted in exercising dogs and humans defined the specific roles of insulin and glucagon in control of hepatic glucose output.[26–33] Studies have shown that the rise in glucagon[32,33] and the fall in insulin[19,33,34] are major determinants of glucose production during moderate exercise. The rise in glucagon is required for the full increment in hepatic glycogenolysis and gluconeogenesis,[32,33] while the fall in insulin[33,34] is necessary for hepatic glycogenolysis. Although changes in glucagon and insulin are individually very important, the interaction of these hormones is an essential component of the stimulus.[2] An increase in glucagon in the physiological context of exercise is considerably more potent than the effects of an experimental increase in glucagon of the same magnitude. A twofold increase in glucagon causes a peak increase in hepatic glucose output of $\sim$1 mg kg^{-1} min^{-1} in the sedentary dog,[35] while the same increase during exercise causes a peak increase of $\sim$5 mg kg^{-1} min^{-1}.[32] Glucagon action is fully manifested during exercise for three reasons (Fig. 3). First, the increased glucose utilization of working muscle prevents the hyperglycemia that accompanies an experimental increase in glucagon. Second, as mentioned earlier, prolonged exercise creates a physiological environment that supports gluconeogenesis.[12] This includes mobilization of gluconeogenic substrates from muscle, adipose, and intestine. Finally, exercise causes a fall in insulin that sensitizes the liver to glucagon.[2]

1.3 Evidence Lacking for Adrenergic Stimulation of Hepatic Glucose Output

The robust increase in arterial catecholamine concentrations that occurs in response to exercise led to a long-standing assumption that norepinephrine and epinephrine stimulate the observed increase in hepatic glucose output.[36] It is now clear that changes in arterial catecholamine concentrations do not

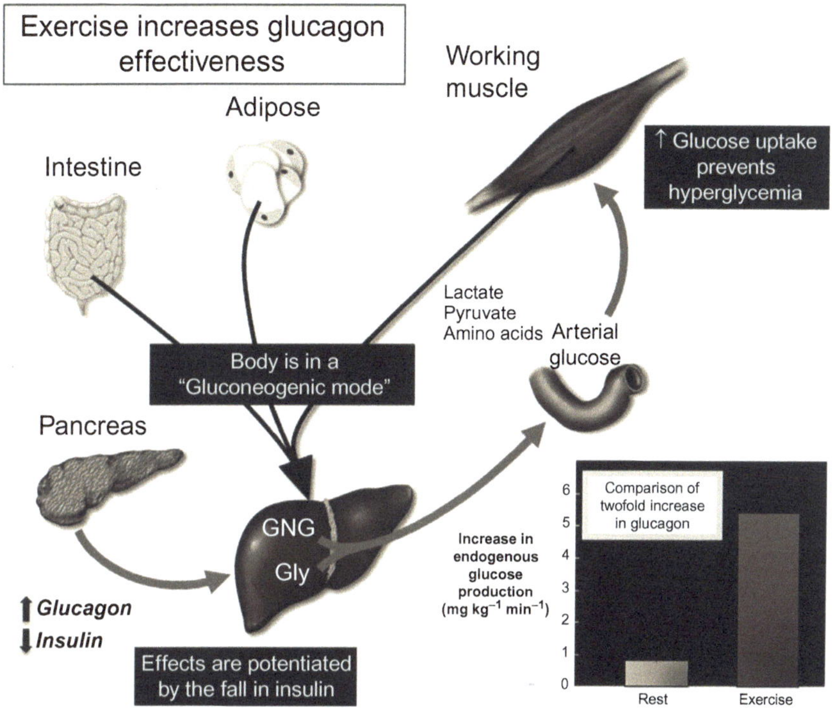

Figure 3 The liver is a metabolic hub where pathways for amino acid, fat, and glucose metabolism are integrated. The integration of these pathways serves to provide energy in the form of glucose, recycle carbon-based metabolites, and prevent nitrogen toxicity. The demand on these pathways is accentuated during exercise and driven by glucagon secreted from pancreatic alpha-cells. (See the color plate.)

translate to changes in concentrations at the hepatocyte. Hepatic norepinephrine spillover (reflecting sympathetic drive) is not increased during moderate exercise, and the portal vein epinephrine concentration at the liver is markedly attenuated as the gastrointestinal tract extracts 50% of it.[37] In line with this, a broad range of experimental approaches have failed to show an appreciable effect of either hepatic sympathetic nerves or circulating epinephrine in stimulation of hepatic glucose output during moderate exercise. There are instances such as high-intensity exercise or exercise in specific populations (e.g., poorly controlled diabetics) where the adrenergic response is unusually high, while the pancreatic hormone response is not.[3] For this reason, the regulation of hepatic glucose output under conditions such as these has been postulated to be different.[38] It has been hypothesized that during high-intensity exercise, control of glucose production shifts from the pancreatic hormones to the catecholamines.[38,39] This is based on two observations. First, circulating blood norepinephrine and epinephrine can increase by 10- to 20-fold,[39] whereas the increase in the glucagon to insulin ratio in *peripheral* blood is considerably less and in some cases undetectable.[31] Second, when high-intensity exercise is performed during a pancreatic clamp (portal insulin and glucagon fixed at basal), hepatic glucose output increases normally even though the increase in *arterial* glucagon is blunted and the fall in insulin is absent.[40]

Despite the evidence cited above, the role of the catecholamines in control of hepatic glucose output with high-intensity exercise (>80% maximum O_2 uptake) remains unclear. Studies that have assessed the role for catecholamines during high-intensity exercise using whole-body pharmacological adrenergic receptor blockade have uniformly been without an effect on hepatic glucose production.[31,39,41] Such studies are difficult to interpret due to the lack of specificity of these pharmacological agents. Intraportal propranolol and phentolamine infusion has been used in a dog model to create selective hepatic adrenergic blockade.[42,43] With this technique, hepatic adrenergic blockade can be achieved without extrahepatic effects. Hepatic adrenergic blockade did not impair the increase in hepatic glucose output or affect blood glucose homeostasis during high-intensity exercise, resulting in a threefold increase in hepatic norepinephrine spillover.[42] Similar results are seen in dogs treated with the β-cell toxin, alloxan.[43] Alloxan-diabetic dogs in poor metabolic control have sevenfold higher rates of hepatic norepinephrine spillover than nondiabetic dogs during moderate exercise. Even with the greater hepatic norepinephrine spillover in diabetic dogs, selective hepatic adrenergic receptor blockade did not affect the exercise-induced increase in hepatic glucose output.[43] Thus, hepatic glucose output in the dog model is not reliant on hepatic adrenergic receptor stimulation even during heavy exercise or in diabetes, which is associated with excessive sympathetic drive.

There are factors involved in the stimulation of glucose production during exercise that are as yet undefined. One could postulate that myokines such as IL-6,[44] retinol-binding protein 4,[45] apelin,[45] and myonectin[46] play a role in regulation of hepatic glucose production. More studies are needed to determine the role of these proteins on liver metabolism.

1.4 Recharging Liver Glycogen Stores after Exercise

Despite the important role of the liver in fuel homeostasis and the added metabolic demands placed on it by exercise, very little is known about how this organ is regulated during the recovery period following exercise. Prior exercise increases the capacity of the liver to consume glucose in response to a simulated meal.[47] Studies following 150 min of exercise in the chronically catheterized dog exhibited a twofold increase in hepatic glucose uptake in response to a twofold increase in glucose load compared to sedentary dogs. These data show that the well-established increase in whole-body glucose tolerance in the postexercise state is due, in part, to an increased ability of liver to take up glucose. Indirect assessments in the anesthetized rabbit support findings in the dog, showing that liver deposition of a fluorescent glucose analog via either an oral bolus or a continuous 120 min intraportal infusion was greater after hind limb contraction.[48] These data were also supported by studies using magnetic resonance spectroscopy that showed that ingestion of ^{13}C-glucose immediately after completion of prolonged moderate exercise in human subjects increased liver glycogen resynthesis by ~0.7 mg kg^{-1} min^{-1} over a period of 4 h of postexercise recovery.[49]

The liver is more insulin sensitive after exercise and this could explain the improved ability of the liver to take up glucose. During a hyperinsulinemic euglycemic clamp, net hepatic glucose output was suppressed to a greater extent following prolonged exercise, compared to their sedentary controls.[50] These results provide a basis for the hypothesis that hepatic insulin sensitivity could be the cause of increased net hepatic glucose uptake during a hepatic glucose load after exercise. This hypothesis was tested in studies in which net hepatic glucose uptake was measured in sedentary and exercised dogs during an increased portal venous glucose supply (simulated meal) with either basal or elevated insulin.[51] The increase in net hepatic glucose uptake (Fig. 4) and fractional glucose extraction with hyperinsulinemia was approximately 50% greater in exercised compared to sedentary dogs. These findings were consistent with subsequent findings in mice that show that acute exercise induced a rapid and pronounced transcriptional effect in the liver and regulated hepatic insulin receptor substrate (IRS) proteins leading to improved cellular insulin signaling.[52] Results from these studies provide a further physiological basis for recommending exercise for patients with insulin resistance.

As discussed earlier, the exercise-induced increase in glucagon and/or decrease in insulin are the major stimuli for the accelerated mobilization of glucose from the liver. The hypothesis that exercise-induced changes in pancreatic hormones could increase the ability of the liver to consume glucose following exercise by depleting hepatic glycogen stores was tested.

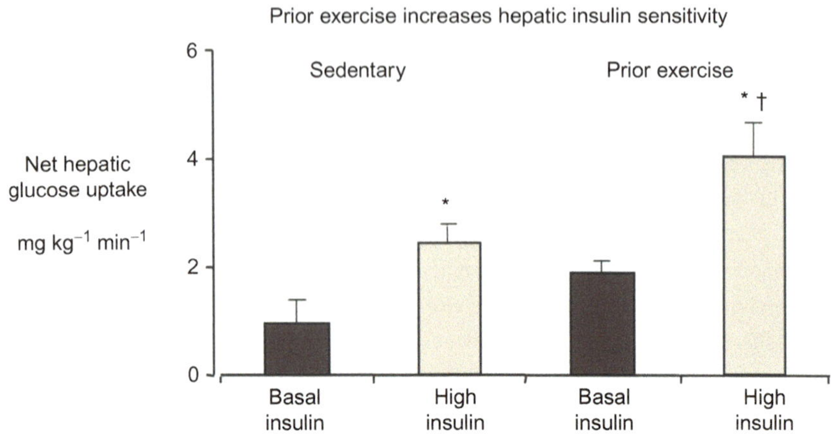

Figure 4 Net hepatic glucose uptake in dogs following 150 min of exercise or an equivalent sedentary period. Glucose was infused into a portal vein to increase the liver glucose supply by twofold, and arterial glucose was clamped at 180 mg dl^{-1}. Insulin was infused at basal rates (0.2 mU kg^{-1} min^{-1}) or at rates designed to simulate a meal (1.2 mU kg^{-1} min^{-1}). $^{*}p \leq 0.05$ versus basal insulin. $^{+}p \leq 0.05$ versus sedentary high insulin. Data are mean $\pm$ SE. *Modified from Pencek* et al.[51] (See the color plate.)

Adaptation of the liver following exercise was tested in dogs in which somatostatin was used to suppress glucagon and insulin. Glucagon and Insulin were replaced at either basal rates or rates that simulated the exercise response.[53] Preventing the response of glucagon and insulin to exercise prevents hepatic glucose output and glycogen breakdown. Simulation of the glucagon and insulin responses to exercise resulted in a threefold increase in hepatic glucose output and marked glycogen breakdown. Despite the differences in glycogen mobilization in the two protocols, hepatic glucose uptake was increased equally in response to a glucose load and hyperinsulinemia, exceeding rates in sedentary dogs. However, when pancreatic hormone responses were simulated and hepatic glucose output was accelerated during exercise, a greater fraction of the glucose consumed by the liver was directed to glycogen.[53] Thus, the increase in insulin-stimulated hepatic glucose uptake observed after exercise is attributable to a factor other than those adaptations resulting from the exercise-induced pancreatic hormone responses. The pancreatic hormone response is a critical determinant of the fate of glucose consumed by the liver, with a greater fraction directed to liver glycogen. This unique finding is consistent with studies that show a greater fraction of the glucose taken up by the liver after prolonged exercise in dogs is metabolized primarily nonoxidatively.[54] A similar result is seen after a prolonged, glycogen-depleting fast.[55]

Exercise also leads to a number of other endocrine changes. It is possible, for example, that the exercise-stimulated glucocorticoid response may prime the liver to take up more glucose since high doses of this hormone can stimulate hepatic glycogen deposition.[56] The growing list of myokines and adipokines, such as adiponectin and irisin, may also have implications for the effects of prior exercise on liver insulin sensitivity and glycogen repletion. Adiponectin released from adipocytes[57] has insulin-sensitizing effects at the liver.[58] Irisin released from muscle has also been associated with improved liver insulin action.[59] Exercise has persistent effects on processes and enzymes involved in liver glucose metabolism that are sustained well after the cessation of exercise.[60] It is possible that some facet of the exercise response activates enzymes involved in hepatic glucose uptake and metabolism.

Clearly, there are effects of prior exercise directly at the liver that facilitate hepatic glucose uptake and glycogen storage. It is important to consider the integrated response of the body when considering the physiological response to ingested glucose. In this regard, replenishment of hepatic glycogen stores is facilitated by intestinal adaptations that cause an increase in absorption of glucose into the portal circulation. Hepatic glucose uptake after exercise is facilitated by increased absorption of ingested glucose.[54,61,62] Through the use of the isotopic glucose analogs 3-*O*-[^{3}H]methylglucose (absorbed via transporter-mediated and passive processes) and L-[^{14}C]glucose (absorbed passively), it was determined that the increase in gut glucose absorption seen following exercise was primarily due to an increase in passive absorption across the intestinal cell wall.[62]

2. THE LIVER RECYCLES CARBONS AND DISPOSES OF EXCESS METABOLITES

The liver plays a vital role in recycling potential energy and thereby conserving it with maximal efficiency. The liver conserves energy by recycling carbon-based molecules and turning them into potential energy by the gluconeogenic pathway. Consistent with the role of glucagon in stimulating hepatic gluconeogenesis, glucagon plays a vital role in stimulating the hepatic extraction of metabolites released during exercise.[63–65] Glucagon stimulates both the N[66] and A[67] amino acid transport systems causing accelerated transport of gluconeogenic amino acids, glutamine and alanine into the liver. Lactate and pyruvate produced by glycolytically and glycerol produced by lipolysis are also recycled into potential energy in the liver gluconeogenic pathway. It is fascinating that gluconeogenic precursors are not only channeled from working muscle[68] and adipose tissue but are also released from nonworking muscle as lactate and the gastrointestinal tract as amino acids.[69,70] Recycling energy from metabolites and amino acids into glucose has an energetic cost. Fatty acid mobilization leads to increased fat oxidation, which provides energy for gluconeogenesis and associated pathways. This fatty acid oxidation, like other vital pathways in the liver during exercise, is driven by the fall in insulin[71] and the increase in glucagon.[65]

3. THE LIVER DETOXIFIES BY CONVERTING EXCESS NITROGEN TO UREA

During exercise, amino acid and AMP deamination lead to an increased formation of NH_3 by skeletal muscle.[72] Moreover, increased protein breakdown from the skeletal muscle[73,74] or the gastrointestinal tract[69,70] results in the release of amino acids into the circulation. The amino acids are only a minor direct fuel source for the muscle. The carbon skeleton of a number of amino acids including alanine, glutamine, glutamate, serine, threonine, and valine is delivered to the liver where the carbons are recycled into glucose. A corollary to the scavenging of amino acid carbons is that the associated nitrogen must be converted to a nontoxic form and excreted. Free or amino acid-associated NH_3 formed during exercise, in large part, enters the urea cycle. Urea is secreted from the liver during exercise and is filtered in the kidney, prior to excretion.

As with mobilization of energy and the recycling of carbons, the incorporation of nitrogen is largely governed by glucagon. Increased glucagon during exercise stimulates the liver uptake of amino acids during exercise.[32,63] In the absence of the rise in glucagon, amino acid concentrations increase. It has been shown in the exercising dog model that the increase in glucagon is required for the transfer of isotopically labeled glutamine nitrogen into urea nitrogen.[63] Thus, there is compelling evidence that glucagon plays an essential role in preventing toxic nitrogen accumulation during times of amino acid formation and delivery to the liver. The role of the exercise-induced increase in glucagon in the integration of metabolic pathways involved in carbohydrate, fat, and amino acid metabolism is illustrated schematically in Fig. 5.

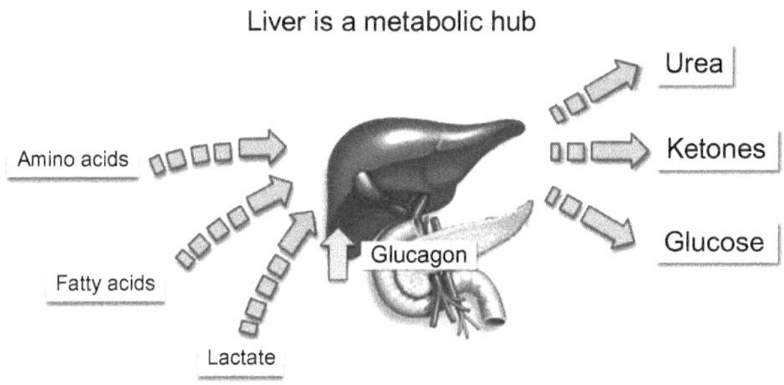

Figure 5 The liver is exquisitely sensitive to the actions of glucagon, exceeding glucagon sensitivity in the sedentary state (inset). Glucagon action is fully manifested during exercise because (i) the increased glucose utilization of working muscle prevents the hyperglycemia that accompanies an experimental increase in glucagon; (ii) prolonged exercise creates a physiological environment that supports gluconeogenesis as gluconeogenic substrates are mobilized from muscle, adipose, and intestine; and (iii) exercise causes a fall in insulin that sensitizes the liver to glucagon. (See the color plate.)

4. HEPATIC ADAPTATIONS TO REGULAR PHYSICAL ACTIVITY

It has become increasingly evident that just as muscle adapts to habitual physical activity, so does the liver. These adaptations that influence macronutrient metabolism are summarized in Fig. 6. The hepatic adaptations may be important in increasing exercise capacity. More significantly, the hepatic adaptations reduce fatty liver, which is a major risk factor for the constellation of conditions that comprise metabolic syndrome. Given the role of liver as an energy balance maintenance system, it is no surprise that it undergoes a dynamic remodeling in response to a metabolically demanding process such as exercise. Physical activity is currently maintained as a first line treatment for obesity and any number of the comorbidities that accompany this disease state. While many studies have examined exercise training, relatively few of these examine specific effects of exercise on liver-related endpoints.

Among the literature focused on liver outcomes in relation to exercise, a significant portion consists of epidemiological associations of exercise or physical activity with general health outcomes or superficial markers of liver function.[75] Typically, these studies conclude that exercise correlates with decreased hepatic fat content[76] and decreased prevalence of nonalcoholic fatty liver disease (NAFLD).[77,78] These studies are valuable for assessment of large medical populations with broad diagnostic criteria but are inherently limited in mechanistic insights.

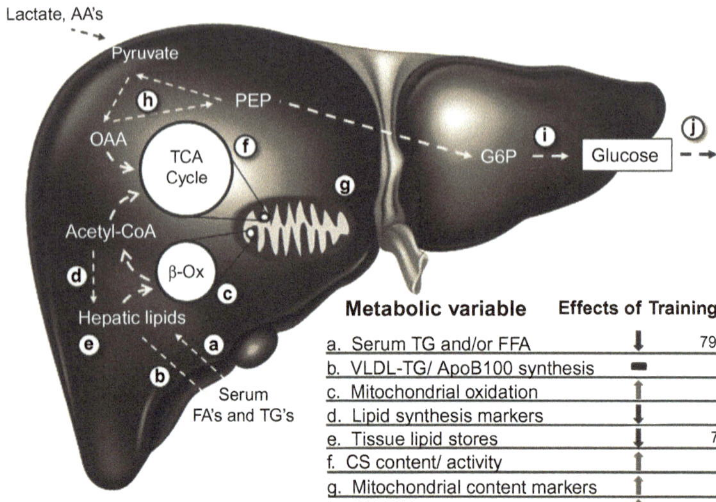

Metabolic variable	Effects of Training	Refs.
a. Serum TG and/or FFA	↓	79–83,90,91,93–97
b. VLDL-TG/ ApoB100 synthesis	–	93
c. Mitochondrial oxidation	↑	79–82
d. Lipid synthesis markers	↓	79–83
e. Tissue lipid stores	↓	79–82,90–93,96
f. CS content/ activity	↑	79,81,82
g. Mitochondrial content markers	↑	79–82,85,86
h. PEPCK	↑	86
i. G6Pase	–	86
j. Suppression of EndoRa by insulin	↑	98

Figure 6 Exercise training elicits complex adaptations in liver metabolic processes.[79–83,85,86,90,93–97] Decreased delivery of substrates (a), constant rates of VLDL-TG synthesis (b), increased mitochondrial oxidation of lipid (c), and decreased lipid anabolic processes (d) may contribute to a net decrease in hepatic lipid stores (e). Increased mitochondrial function (f) and/or content (g) may serve to elevate lipid oxidation rates. An increase in PEPCK levels (h), but not G6Pase (i), may indicate elevated gluconeogenic potential elicited through training. An improved ability of the liver to respond to hormonal stimuli, such as insulin (j), may underlie some of the metabolic adaptations seen. Nuclear transcription and protein translation alterations may underlie adaptive changes of this organ. (See the color plate.)

Animal models have served an invaluable role regarding this general lack of mechanisms consequent to exercise in liver. The Otsuka Long-Evans Tokushima Fatty (OLETF) rat, a rat model of hyperphagic induced obesity and type 2 diabetes, has been carefully characterized in relation to the response to exercise training. In this model, exercise has been used as an intervention and prevention technique for parameters related to NAFLD. Outside of improvements in general measures of glucose tolerance and calculated insulin sensitivity, both study designs demonstrate that exercise seems to modify the liver cellular machinery related to lipid metabolism and gluconeogenesis. One of the underlying causes of decreased liver lipid content and improvements in steatotic phenotypes may be decreased lipid anabolism coupled with increased lipid catabolism.[79–82] This is demonstrated by decreases in the lipogenic proteins fatty acid synthase (Fas) and acetyl-CoA carboxylase (Acc) with relative increases in deactivation of Acc by phosphorylation as well. This decrease in Fas protein persists in a model of swim training in C57Bl6/J mice fed a standard chow or a high-fat diet.[83] An increase in complete oxidation of palmitate to CO_2 in the OLETF model has been interpreted as improved coupling of mitochondrial oxidation to the TCA cycle. Interestingly, rats bred for an intrinsically higher aerobic capacity also show increases in fatty acid oxidation metrics within the liver in absence of training.[84] Whether the intrinsic aerobic capacity is afferent or efferent to hepatic lipid oxidative phenotypes is currently unknown. Other markers such as Cpt-1 activity, β-Had activity, citrate synthase activity, Tr4 protein, and cytochrome *c* oxidase component levels are increased suggesting an increase in mitochondrial function and/or content,[79,80,85–87] which may be dependent on presence of *Pgc-1α*.[86] Alterations in mitochondrial FA oxidation have typically been accompanied by a decrease in extramitochondrial palmitate oxidation in the OLETF rat model.[79,81,87] This decrease in extramitochondrial palmitate oxidation is hypothesized to represent a decrease in peroxisomal contributions to oxidation. This concomitant increase in mitochondrial oxidation and decreased peroxisomal oxidation may be representative of a more "metabolically healthy" lipid oxidation phenotype resulting in decreased oxidative stress,[88] a hypothesized contributor to liver injury in NAFLD,[89] being a potential result.

Among the caveats of evaluating the molecular underpinnings of exercise in liver function is an inability to dissociate effects of a lifestyle intervention from the pleiotropic benefits of weight loss. This proves difficult to resolve in animals without implementation of complicated pair feeding strategies. Despite this challenge, several human studies have shown that exercise, independent of weight loss effects, decreases hepatic lipid content.[90–93] This seems to be independent of lipid delivery to the liver as concentrations of TG and FFA only decrease when weight loss is evident.[79–83,90,91,94–96] Recently, it was confirmed that altered hepatic lipid content is independent of any effects on VLDL-TG or VLDL-ApoB100 secretion rates.[93] Increased utilization of lipid substrates at rest or during exercise may be the underlying cause of these decreases.[90,97] One study specifically delineated the effects of exercise with and without weight loss on liver insulin sensitivity using a modified clamp technique in human subjects.[98] This study concludes that exercise, in absence of weight loss, improves the ability of insulin to suppress glucose production of liver; however, combining exercise with ~6% weight loss in this study led to even further improvements in this metric. This improved liver metabolic profile may be due to an altered localization of adipose tissue from visceral to subcutaneous depots resulting in a lower constitutive provision of FFAs to the portal circulation.[91,92] This hypothesis is far from resolved as metabolic improvements have been reported in the absence of decreased visceral adipose tissue area as well.[90]

As described previously, energy demand from the muscle is met with increased release of glucose from the muscle in a glucagon-dependent manner. Training increases the liver's sensitivity to glucagon.[99–102] This effect is associated with upregulation of glucagon receptors in the liver.[103] The processes contributing to glucagon signaling and gluconeogenesis consume ATP resulting in a reduced AMP/ATP ratio in the liver.[7,104] Decreased energy charge of the liver activates AMPK resulting in a host of molecular cascades leading to a net decrease in lipogenic processes with a complementary increase in lipid oxidation.[105,106] Chronic exercise stimulus does not seem to rely on transient energy charge effects, but rather on a bulk shift in lipid metabolic machinery and oxidative processes. This is seen in the consistent increase in several parameters relating to mitochondria previously noted.

5. SUMMARY

The accelerated metabolic demands of the working muscle cannot be met without a robust response from the liver. If not for the hepatic response, sustained exercise would be impossible. If not for the increased hepatic glucose output with exercise, hypoglycemia would result due to inadequate breakdown of glycogen stores and recycling of metabolites through the gluconeogenic pathway. The energetic demands on the liver are largely met by increased oxidation of fatty acids mobilized from adipose tissue. Adaptations immediately following exercise facilitate the replenishment of glycogen stores. Pancreatic glucagon and insulin responses orchestrate the hepatic response during and immediately following exercise. Like skeletal muscle, liver adapts to repeated demands of exercise by increasing its capacity to produce energy by oxidizing fat. The ability of regular physical activity to increase fat oxidation is protective and can reverse fatty liver disease. Engaging in regular physical exercise has broad ranging positive health implications including those that improve the metabolic health of the liver.

ACKNOWLEDGMENTS

Grants from the NIH are gratefully acknowledged (R01DK054902 and R37 DK050277). A.S.W. received support from the Molecular Endocrinology Training Program (T32 DK07563). The Vanderbilt Diabetes Research and Training Center (P30 DK20593) and Vanderbilt Mouse Metabolic Phenotypiing Centers (U24 DK059637) are acknowledged.

REFERENCES

1. Wasserman DH, Kang L, Ayala JE, Fueger PT, Lee-Young RS. The physiological regulation of glucose flux into muscle in vivo. *J Exp Biol.* 2011;214(pt 2):254–262. http://dx.doi.org/10.1242/jeb.048041. PubMed PMID: 21177945; PubMed Central PMCID: PMC3008632.
2. Wasserman DH. Four grams of glucose. *Am J Physiol Endocrinol Metab.* 2009;296(1): E11–E21. PubMed PMID: 18840763.
3. Sigal RJ, Kenny GP, Wasserman DH, Castaneda-Sceppa C. Physical activity/exercise and type 2 diabetes. *Diabetes Care.* 2004;27(10):2518–2539. PubMed PMID: 15451933.
4. Felig P, Cherif A, Minageref A, Wahren J. Hypoglycemia during prolonged exercise in normal men. *N Engl J Med.* 1982;306:895–900.
5. Wasserman DH, Zinman B. American diabetes association technical review on exercise in individuals with insulin-dependent diabetes mellitus. *Diabetes Care.* 1994;17: 924–937.
6. Wasserman DH. Regulation of glucose fluxes during exercise in the postabsorptive state. *Annu Rev Physiol.* 1995;57:191–218. PubMed PMID: 7778865.
7. Berglund ED, Lee-Young RS, Lustig DG, et al. Hepatic energy state is regulated by glucagon receptor signaling in mice. *J Clin Invest.* 2009;119(8):2412–2422. PubMed PMID: 19662685.
8. Camacho RC, Donahue EP, James FD, Berglund ED, Wasserman DH. Energy state of the liver during short-term and exhaustive exercise in C57BL/6J mice. *Am J Physiol Endocrinol Metab.* 2006;290(3):E405–E408. PubMed PMID: 16219665.
9. Camacho RC, Pencek RR, Lacy DB, James FD, Donahue EP, Wasserman DH. Portal venous 5-aminoimidazole-4-carboxamide-1-beta-D-ribofuranoside infusion overcomes hyperinsulinemic suppression of endogenous glucose output. *Diabetes.* 2005;54(2):373–382. PubMed PMID: 15677495.
10. Pencek RR, Shearer J, Camacho RC, et al. 5-Aminoimidazole-4-carboxamide-1-beta-D-ribofuranoside causes acute hepatic insulin resistance in vivo. *Diabetes.* 2005;54:355–360.
11. Hardie DG, Hawley SA. AMP-activated protein kinase: the energy charge hypothesis revisited. *Bioessays.* 2001;23(12):1112–1119. http://dx.doi.org/10.1002/bies.10009. PubMed PMID: 11746230.
12. Wasserman DH, Lacy DB, Bracy D, Williams PE. Metabolic regulation in peripheral tissues and transition to increased gluconeogenic mode during prolonged exercise. *Am J Physiol.* 1992;263:E345–E354.
13. Foretz M, Viollet B. Regulation of hepatic metabolism by AMPK. *J Hepatol.* 2011;54(4):827–829. http://dx.doi.org/10.1016/j.jhep.2010.09.014. PubMed PMID: 21163246.
14. Koyama Y, Coker RH, Denny JC, et al. Role of carotid bodies in the neuroendocrine response to exercise. *Am J Physiol.* 2001;281:E742–E748.
15. Coker RH, Koyama Y, Lacy DB, Williams PE, Rheaume N, Wasserman DH. Pancreatic innervation is not essential for exercise-induced changes in glucagon and insulin or glucose kinetics. *Am J Physiol.* 1999;277:E1122–E1129.

16. Pedersen BK, Febbraio MA. Muscles, exercise and obesity: skeletal muscle as a secretory organ. *Nat Rev Endocrinol.* 2012;8(8):457–465. http://dx.doi.org/10.1038/nrendo.2012.49. PubMed PMID: 22473333.
17. Barnes TM, Otero YF, Elliott AD, et al. Interleukin-6 amplifies glucagon secretion: coordinated control via the brain and pancreas. *Am J Physiol Endocrinol Metab.* 2014;307(10):E896–E905. http://dx.doi.org/10.1152/ajpendo.00343.2014. PubMed PMID: 25205821; PubMed Central PMCID: PMC4233256.
18. Wasserman DH, Lickley HLA, Vranic M. Interactions between glucagon and other counterregulatory hormones during normoglycemic and hypoglycemic exercise in dogs. *J Clin Invest.* 1984;74:1404–1413.
19. Wasserman DH, Lacy DB, Colburn CA, Bracy DP, Cherrington AD. Efficiency of compensation for the absence of the fall in insulin during exercise. *Am J Physiol.* 1991;261:E587–E597.
20. Berger CM, Sharis PJ, Bracy DP, Lacy DB, Wasserman DH. Sensitivity of exercise-induced increase in hepatic glucose production to glucose supply and demand. *Am J Physiol.* 1994;267:E411–E421.
21. Jenkins AB, Chisholm DJ, Ho KY, Kraegen EW. Exercise induced hepatic glucose output is precisely sensitive to the rate of systemic glucose supply. *Metabolism.* 1985;34:431–434.
22. Fueger PT, Li CY, Ayala JE, et al. Glucose kinetics and exercise tolerance in mice lacking the GLUT4 glucose transporter. *J Physiol.* 2007;582(pt 2):801–812. PubMed PMID: 17495042.
23. Lins PE, Wajngot A, Adamson U, Vranic M, Efendic S. Minimal increases in glucagon levels enhance glucose production in man with partial hypoinsulinemia. *Diabetes.* 1983;32(7):633–636. PubMed PMID: 6134650.
24. Coker RH, Lacy DB, Krishna MG, Wasserman DH. Splanchnic glucagon kinetics in exercising alloxan-diabetic dogs. *J Appl Physiol.* 1999;86(5):1626–1631. PubMed PMID: 10233127.
25. Wasserman DH, Lacy DB, Bracy DP. Relationship between arterial and portal vein immunoreactive glucagon during exercise. *J Appl Physiol.* 1993;75:724–729.
26. Tuttle KR, Marker JC, Dalsky GP, et al. Glucagon, not insulin, may play a secondary role in defense against hypoglycemia. *Am J Physiol.* 1988;254:E713–E719.
27. Wolfe RR, Nadel ER, Shaw JH, Stephenson LA, Wolfe MH. Role of changes in insulin and glucagon in glucose homeostasis in exercise. *J Clin Invest.* 1986;77(3):900–907. PubMed PMID: 2869053.
28. Wasserman DH, Lickley HL, Vranic M. Interactions between glucagon and other counterregulatory hormones during normoglycemic and hypoglycemic exercise in dogs. *J Clin Invest.* 1984;74(4):1404–1413. PubMed PMID: 6148356.
29. Wasserman DH, Lickley HL, Vranic M. Important role of glucagon during exercise in diabetic dogs. *J Appl Physiol.* 1985;59(4):1272–1281. PubMed PMID: 2865245.
30. Hirsch IB, Marker JC, Smith LJ, et al. Insulin and glucagon in the prevention of hypoglycemia during exercise in humans. *Am J Physiol.* 1991;260:E695–E704.
31. Kjaer M, Engfred K, Fernandez A, Secher N, Galbo H. Regulation of hepatic glucose production during exercise in humans: role of sympathoadrenergic activity. *Am J Physiol.* 1993;265:E275–E283.
32. Wasserman DH, Spalding JS, Lacy DB, Colburn CA, Goldstein RE, Cherrington AD. Glucagon is a primary controller of the increments in hepatic glycogenolysis and gluconeogenesis during exercise. *Am J Physiol.* 1989;257:E108–E117.
33. Lavoie C, Ducros F, Bourque J, Langelier H, Chiasson JL. Glucose metabolism during exercise in man: the role of insulin and glucagon in the regulation of hepatic glucose production and gluconeogenesis. *Can J Physiol Pharmacol.* 1997;75(1):26–35. PubMed PMID: 9101062.

34. Wasserman DH, Williams PE, Lacy DB, Goldstein RE, Cherrington AD. Exercise-induced fall in insulin and hepatic carbohydrate metabolism during muscular work. *Am J Physiol.* 1989;256(4 pt 1):E500–E509. PubMed PMID: 2650562.
35. Stevenson RW, Steiner KE, Davis MA, et al. Similar dose responsiveness of hepatic glycogenolysis and gluconeogenesis to glucagon in vivo. *Diabetes.* 1987;36: 382–389.
36. Christensen NJ, Galbo H. Sympathetic nerve activity during exercise. *Annu Rev Physiol.* 1983;45:139–153.
37. Coker RH, Krishna MG, Lacy DB, Zinker B, Wasserman DH. Sympathetic drive to liver and nonhepatic splanchnic tissue during prolonged exercise is increased in diabetes. *Metabolism.* 1997;46:1327–1332.
38. Marliss EB, Simantirakis E, Purdon C, et al. Glucoregulatory and hormonal responses to repeated bouts of intense exercise in normal male subjects. *J Appl Physiol.* 1991;71:924–933.
39. Marliss EB, Purdon C, Halter JB, Sigal RJ, Vranic M. Glucoregulation during and after intense exercise in control and diabetic subjects. In: Devlin J, Horton E, Vranic M, eds. *Diabetes Mellitus and Exercise.* London: Smith-Gordon and Company Limited; 1992:173–188.
40. Sigal RJ, Fisher SF, Halter JB, Vranic M, Marliss EB. The roles of catecholamines in glucoregulation in intense exercise as defined by the islet cell clamp technique. *Diabetes.* 1995;45:148–156.
41. Sigal RJ, Fisher S, Marliss EB. Catecholamines are the primary regulators of glucose production (Ra) during intense exercise: definition using the islet-cell clamp (IC). *Diabetes.* 1994;43(suppl 1):159A.
42. Coker RH, Krishna MG, Lacy DB, Bracy DP, Wasserman DH. Role of hepatic alpha- and beta-adrenergic receptor stimulation on hepatic glucose production during heavy exercise. *Am J Physiol.* 1997;273(5 pt 1):E831–E838. PubMed PMID: 9374667.
43. Coker RH, Lacy DB, Williams PE, Wasserman DH. Hepatic alpha- and beta-adrenergic receptors are not essential for the increase in Ra during exercise in diabetes. *Am J Physiol.* 2000;278:E444–E451.
44. Febbraio MA, Hiscock N, Sacchetti M, Fischer CP, Pedersen BK. Interleukin-6 is a novel factor mediating glucose homeostasis during skeletal muscle contraction. *Diabetes.* 2004;53:1643–1648.
45. Besse-Patin A, Montastier E, Vinel C, et al. Effect of endurance training on skeletal muscle myokine expression in obese men: identification of apelin as a novel myokine. *Int J Obes.* 2014;38(5):707–713. http://dx.doi.org/10.1038/ijo.2013.158. PubMed PMID: 23979219.
46. Seldin MM, Lei X, Tan SY, Stanson KP, Wei Z, Wong GW. Skeletal muscle-derived myonectin activates the mammalian target of rapamycin (mTOR) pathway to suppress autophagy in liver. *J Biol Chem.* 2013;288(50):36073–36082. http://dx.doi.org/10.1074/jbc.M113.500736. PubMed PMID: 24187137; PubMed Central PMCID: PMC3861655.
47. Galassetti P, Coker RH, Lacy DB, Cherrington AD, Wasserman DH. Prior exercise increases net hepatic glucose uptake during a glucose load. *Am J Physiol.* 1999;276: E1022–E1029.
48. Matsuhisha M, Nishizawa I, Ikeda M, et al. Prior muscular contraction enhances disposal of a glucose analog in the liver and muscle. *Metabolism.* 1998;47:44–49.
49. Casey A, Mann R, Banister K, et al. Effect of carbohydrate ingestion on glycogen resynthesis in human liver and skeletal muscle, measured by 13C MRS. *Am J Physiol.* 2000;278:E65–E75.
50. Koyama Y, Galassetti P, Coker RH, et al. Prior exercise and the response to insulin-induced hypoglycemia in the dog. *Am J Physiol.* 2002;282:E1128–E1138.

51. Pencek RR, James FD, Lacy DB, et al. Interaction of insulin and prior exercise in control of hepatic metabolism of a glucose load. *Diabetes*. 2003;52:1897–1903.
52. Hoene M, Lehmann R, Hennige AM, et al. Acute regulation of metabolic genes and insulin receptor substrates in the liver of mice by one single bout of treadmill exercise. *J Physiol*. 2009;587(pt 1):241–252. http://dx.doi.org/10.1113/jphysiol.2008.160275. PubMed PMID: 19001047; PubMed Central PMCID: PMC2670037.
53. Pencek RR, James FD, Lacy DB, Camacho RC, Fueger PT, Wasserman DH. Exercise-induced changes in insulin and glucagon are not required for enhanced glucose uptake by the liver following exercise but influence the fate of glucose. *Diabetes*. 2004;53:3041–3047.
54. Hamilton KS, Gibbons FK, Bracy DP, Lacy DB, Cherrington AD, Wasserman DH. Effect of prior exercise on the partitioning of an intestinal glucose load between splanchnic bed and skeletal muscle. *J Clin Invest*. 1996;98:125–135.
55. Galassetti P, Reed E, Messina A, Lacy DB, Wasserman DH. Effect of prior fast duration on the appearance and disposal of an intraduodenal glucose load. *Am J Physiol*. 1999;276:E543–E552.
56. Long CNH, Katzen B, Fry EG. Adrenal cortex and carbohydrate metabolism. *Endocrinology*. 1940;26:309–351.
57. Bouassida A, Chamari K, Zaouali M, Feki Y, Zbidi A, Tabka Z. Review on leptin and adiponectin responses and adaptations to acute and chronic exercise. *Br J Sports Med*. 2010;44(9):620–630. http://dx.doi.org/10.1136/bjsm.2008.046151. PubMed PMID: 18927166.
58. Tao C, Sifuentes A, Holland WL. Regulation of glucose and lipid homeostasis by adiponectin: effects on hepatocytes, pancreatic beta cells and adipocytes. *Best Pract Res Clin Endocrinol Metab*. 2014;28(1):43–58. http://dx.doi.org/10.1016/j.beem.2013.11.003. PubMed PMID: 24417945.
59. Arias-Loste MT, Ranchal I, Romero-Gomez M, Crespo J. Irisin, a link among fatty liver disease, physical inactivity and insulin resistance. *Int J Mol Sci*. 2014;15(12):23163–23178. http://dx.doi.org/10.3390/ijms151223163. PubMed PMID: 25514415; PubMed Central PMCID: PMC4284758.
60. Dohm GL, Kasperek GJ, Barakat HA. Time course of changes in gluconeogenic enzyme activities during exercise and recovery. *Am J Physiol*. 1985;249:E6–E11.
61. Maehlum S, Felig P, Wahren J. Splanchnic glucose and muscle glycogen metabolism after glucose feeding during post-exercise recovery. *Am J Physiol*. 1978;235:E255–E260.
62. Pencek RR, Koyama Y, Lacy DB, et al. Prior exercise enhances passive absorption of intraduodenal glucose. *J Appl Physiol*. 2003;95:1132–1138.
63. Krishna MG, Coker RH, Lacy DB, Zinker BA, Halseth AE, Wasserman DH. Glucagon response to exercise is critical for accelerated hepatic glutamine metabolism and nitrogen disposal. *Am J Physiol*. 2000;279:E638–E645.
64. Wasserman DH, Williams PE, Lacy DB, Green DR, Cherrington AD. Importance of intrahepatic mechanisms to gluconeogenesis from alanine during prolonged exercise and recovery. *Am J Physiol*. 1988;254:E518–E525.
65. Wasserman DH, Spalding JS, Bracy DP, Lacy DB, Cherrington AD. Exercise-induced rise in glucagon and the increase in ketogenesis during prolonged muscular work. *Diabetes*. 1989;38:799–807.
66. Nissim I, Brosnan M, Yudkoff M, Nissim I, Brosnan JT. Studies of hepatic glutamine metabolism in the perfused rat liver with ^{15}N-labeled glutamine. *J Biol Chem*. 1999;274:28958–28965.
67. Cariappa R, Kilberg MS. Plasma membrane domain localization and transcytosis of the glucagon-induced hepatic system A carrier. *Am J Physiol*. 1992;263(6 pt 1):E1021–E1028. PubMed PMID: 1335691.

68. Ahlborg G, Felig P. Lactate and glucose exchange across the forearm, legs, and splanchnic bed during after prolonged leg exercise. *J Clin Invest.* 1982;69:45–54.
69. Williams BD, Wolfe RR, Bracy DP, Wasserman DH. Gut contributes essential amino acids during exercise. *Am J Physiol.* 1996;270:E85–E90.
70. Wasserman DH, Geer RJ, Williams PE, Lacy DB, Abumrad NN. Interaction of gut and liver in nitrogen metabolism during exercise. *Metabolism.* 1991;40:307–314.
71. Wasserman DH, Lacy DB, Goldstein RE, Williams PE, Cherrington AD. Exercise-induced fall in insulin and the increase in fat metabolism during prolonged exercise. *Diabetes.* 1989;38:484–490.
72. Eriksson LS, Broberg S, Bjorkman O, Wahren J. Ammonia metabolism during exercise in man. *Clin Physiol.* 1985;5:325–336.
73. Felig P, Wahren J. Amino acid metabolism in exercising man. *J Clin Invest.* 1971;50:2703–2711.
74. van Loon LJ. Is there a need for protein ingestion during exercise? *Sports Med.* 2014; 44(suppl 1):S105–S111. http://dx.doi.org/10.1007/s40279-014-0156-z. PubMed PMID: 24791922; PubMed Central PMCID: PMC4008809.
75. Caldwell S, Lazo M. Is exercise an effective treatment for NASH? Knowns and unknowns. *Ann Hepatol.* 2009;8(suppl 1):S60–S66. PubMed PMID: 19381126.
76. Perseghin G, Lattuada G, De Cobelli F, et al. Habitual physical activity is associated with intrahepatic fat content in humans. *Diabetes Care.* 2007;30(3):683–688. http://dx.doi.org/10.2337/dc06-2032. PubMed PMID: 17327341.
77. Church TS, Kuk JL, Ross R, Priest EL, Biltoft E, Blair SN. Association of cardiorespiratory fitness, body mass index, and waist circumference to nonalcoholic fatty liver disease. *Gastroenterology.* 2006;130(7):2023–2030. http://dx.doi.org/10.1053/j.gastro.2006.03.019. PubMed PMID: 16762625.
78. Zelber-Sagi S, Nitzan-Kaluski D, Goldsmith R, et al. Role of leisure-time physical activity in nonalcoholic fatty liver disease: a population-based study. *Hepatology.* 2008;48(6):1791–1798. http://dx.doi.org/10.1002/hep.22525. PubMed PMID: 18972405.
79. Borengasser SJ, Rector RS, Uptergrove GM, et al. Exercise and omega-3 polyunsaturated fatty acid supplementation for the treatment of hepatic steatosis in hyperphagic OLETF rats. *J Nutr Metab.* 2012;2012:268680. http://dx.doi.org/10.1155/2012/268680. PubMed PMID: 21918718; PubMed Central PMCID: PMC3171760.
80. Rector RS, Thyfault JP, Morris RT, et al. Daily exercise increases hepatic fatty acid oxidation and prevents steatosis in Otsuka Long-Evans Tokushima Fatty rats. *Am J Physiol Gastrointest Liver Physiol.* 2008;294(3):G619–G626. http://dx.doi.org/10.1152/ajpgi.00428.2007. PubMed PMID: 18174272.
81. Rector RS, Uptergrove GM, Morris EM, et al. Daily exercise vs. caloric restriction for prevention of nonalcoholic fatty liver disease in the OLETF rat model. *Am J Physiol Gastrointest Liver Physiol.* 2011;300(5):G874–G883. http://dx.doi.org/10.1152/ajpgi.00510.2010. PubMed PMID: 21350190; PubMed Central PMCID: PMC3094141.
82. Linden MA, Fletcher JA, Morris EM, et al. Combining metformin and aerobic exercise training in the treatment of type 2 diabetes and NAFLD in OLETF rats. *Am J Physiol Endocrinol Metab.* 2014;306(3):E300–E310. http://dx.doi.org/10.1152/ajpendo.00427.2013. PubMed PMID: 24326426; PubMed Central PMCID: PMC3920010.
83. Schultz A, Mendonca LS, Aguila MB, Mandarim-de-Lacerda CA. Swimming training beneficial effects in a mice model of nonalcoholic fatty liver disease. *Exp Toxicol Pathol.* 2012;64(4):273–282. http://dx.doi.org/10.1016/j.etp.2010.08.019. PubMed PMID: 20869214.
84. Morris EM, Jackman MR, Johnson GC, et al. Intrinsic aerobic capacity impacts susceptibility to acute high-fat diet-induced hepatic steatosis. *Am J Physiol Endocrinol Metab*

2014;307(4):E355–E364. http://dx.doi.org/10.1152/ajpendo.00093.2014. PubMed PMID: 24961240; PubMed Central PMCID: PMC4137118.

85. Lezi E, Lu J, Burns JM, Swerdlow RH. Effect of exercise on mouse liver and brain bioenergetic infrastructures. *Exp Physiol.* 2013;98(1):207–219. http://dx.doi.org/10.1113/expphysiol.2012.066688. PubMed PMID: 22613742; PubMed Central PMCID: PMC3540163.
86. Haase TN, Ringholm S, Leick L, et al. Role of PGC-1alpha in exercise and fasting-induced adaptations in mouse liver. *Am J Physiol Regul Integr Comp Physiol.* 2011;301(5):R1501–R1509. http://dx.doi.org/10.1152/ajpregu.00775.2010. PubMed PMID: 21832205.
87. Linden MA, Fletcher JA, Morris EM, et al. Treating NAFLD in OLETF rats with vigorous-intensity interval exercise training. *Med Sci Sports Exerc.* 2014;http://dx.doi.org/10.1249/MSS.0000000000000430. PubMed PMID: 24983336; PubMed Central PMCID: PMC4280360.
88. Reddy JK, Hashimoto T. Peroxisomal beta-oxidation and peroxisome proliferator-activated receptor alpha: an adaptive metabolic system. *Annu Rev Nutr.* 2001;21:193–230. http://dx.doi.org/10.1146/annurev.nutr.21.1.193. PubMed PMID: 11375435.
89. Day CP, James OF. Steatohepatitis: a tale of two "hits"? *Gastroenterology.* 1998;114(4):842–845. PubMed PMID: 9547102.
90. Hallsworth K, Fattakhova G, Hollingsworth KG, et al. Resistance exercise reduces liver fat and its mediators in non-alcoholic fatty liver disease independent of weight loss. *Gut.* 2011;60(9):1278–1283. http://dx.doi.org/10.1136/gut.2011.242073. PubMed PMID: 21708823; PubMed Central PMCID: PMC3152868.
91. Johnson NA, Sachinwalla T, Walton DW, et al. Aerobic exercise training reduces hepatic and visceral lipids in obese individuals without weight loss. *Hepatology.* 2009;50(4):1105–1112. http://dx.doi.org/10.1002/hep.23129. PubMed PMID: 19637289.
92. van der Heijden GJ, Wang ZJ, Chu ZD, et al. A 12-week aerobic exercise program reduces hepatic fat accumulation and insulin resistance in obese, Hispanic adolescents. *Obesity.* 2010;18(2):384–390. http://dx.doi.org/10.1038/oby.2009.274. PubMed PMID: 19696755.
93. Sullivan S, Kirk EP, Mittendorfer B, Patterson BW, Klein S. Randomized trial of exercise effect on intrahepatic triglyceride content and lipid kinetics in nonalcoholic fatty liver disease. *Hepatology.* 2012;55(6):1738–1745. http://dx.doi.org/10.1002/hep.25548. PubMed PMID: 22213436; PubMed Central PMCID: PMC3337888.
94. Jenkins NT, Padilla J, Arce-Esquivel AA, et al. Effects of endurance exercise training, metformin, and their combination on adipose tissue leptin and IL-10 secretion in OLETF rats. *J Appl Physiol.* 2012;113(12):1873–1883. http://dx.doi.org/10.1152/japplphysiol.00936.2012. PubMed PMID: 23019312; PubMed Central PMCID: PMC3544496.
95. Marques CM, Motta VF, Torres TS, Aguila MB, Mandarim-de-Lacerda CA. Beneficial effects of exercise training (treadmill) on insulin resistance and nonalcoholic fatty liver disease in high-fat fed C57BL/6 mice. *Braz J Med Biol Res.* 2010;43(5):467–475. PubMed PMID: 20490434.
96. Tamura Y, Tanaka Y, Sato F, et al. Effects of diet and exercise on muscle and liver intracellular lipid contents and insulin sensitivity in type 2 diabetic patients. *J Clin Endocrinol Metab.* 2005;90(6):3191–3196. http://dx.doi.org/10.1210/jc.2004-1959. PubMed PMID: 15769987.
97. Fealy CE, Haus JM, Solomon TP, et al. Short-term exercise reduces markers of hepatocyte apoptosis in nonalcoholic fatty liver disease. *J Appl Physiol.* 2012;113(1):1–6.

http://dx.doi.org/10.1152/japplphysiol.00127.2012. PubMed PMID: 22582214; PubMed Central PMCID: PMC3404833.

98. Coker RH, Williams RH, Yeo SE, et al. The impact of exercise training compared to caloric restriction on hepatic and peripheral insulin resistance in obesity. *J Clin Endocrinol Metab*. 2009;94(11):4258–4266. http://dx.doi.org/10.1210/jc.2008-2033. PubMed PMID: 19808853; PubMed Central PMCID: PMC2775654.
99. Drouin R, Lavoie C, Bourque J, Ducros F, Poisson D, Chiasson JL. Increased hepatic glucose production response to glucagon in trained subjects. *Am J Physiol*. 1998;274(1 pt 1):E23–E28. PubMed PMID: 9458743.
100. Drouin R, Robert G, Milot M, Massicotte D, Peronnet F, Lavoie C. Swim training increases glucose output from liver perfused in situ with glucagon in fed and fasted rats. *Metabolism*. 2004;53(8):1027–1031. PubMed PMID: 15281013.
101. Podolin DA, Wills BK, Wood IO, Lopez M, Mazzeo RS, Roth DA. Attenuation of age-related declines in glucagon-mediated signal transduction in rat liver by exercise training. *Am J Physiol Endocrinol Metab*. 2001;281(3):E516–E523. PubMed PMID: 11500307.
102. Bonjorn VM, Latour MG, Belanger P, Lavoie JM. Influence of prior exercise and liver glycogen content on the sensitivity of the liver to glucagon. *J Appl Physiol*. 2002;92(1):188–194. PubMed PMID: 11744659.
103. Legare A, Drouin R, Milot M, et al. Increased density of glucagon receptors in liver from endurance-trained rats. *Am J Physiol Endocrinol Metab*. 2001;280(1):E193–E196. PubMed PMID: 11120674.
104. Miller RA, Chu Q, Xie J, Foretz M, Viollet B, Birnbaum MJ. Biguanides suppress hepatic glucagon signalling by decreasing production of cyclic AMP. *Nature*. 2013;494(7436):256–260. http://dx.doi.org/10.1038/nature11808. PubMed PMID: 23292513; PubMed Central PMCID: PMC3573218.
105. Ruderman NB, Park H, Kaushik VK, et al. AMPK as a metabolic switch in rat muscle, liver and adipose tissue after exercise. *Acta Physiol Scand*. 2003;178(4):435–442. http://dx.doi.org/10.1046/j.1365-201X.2003.01164.x. PubMed PMID: 12864749.
106. Park H, Kaushik VK, Constant S, et al. Coordinate regulation of malonyl-CoA decarboxylase, sn-glycerol-3-phosphate acyltransferase, and acetyl-CoA carboxylase by AMP-activated protein kinase in rat tissues in response to exercise. *J Biol Chem*. 2002;277(36):32571–32577. http://dx.doi.org/10.1074/jbc.M201692200. PubMed PMID: 12065578.

CHAPTER TEN

Molecular Mechanisms for Exercise Training-Induced Changes in Vascular Structure and Function: Skeletal Muscle, Cardiac Muscle, and the Brain

T. Dylan Olver*, Brian S. Ferguson*, M. Harold Laughlin*,†,‡,1
*Department of Biomedical Sciences, University of Missouri, Columbia, Missouri, USA
†Department of Medical Pharmacology and Physiology, University of Missouri, Columbia, Missouri, USA
‡Dalton Cardiovascular Research Center, University of Missouri, Columbia, Missouri, USA
[1]Corresponding author: e-mail address: laughlinm@missouri.edu

Contents

Abstract

Compared with resting conditions, during incremental exercise, cardiac output in humans is elevated from ~5 to 25 L min^{-1}. In conjunction with this increase, the proportion of cardiac output directed toward skeletal muscle increases from ~20% to 85%, while blood flow to cardiac muscle increases 500% and blood flow to specific brain structures increases nearly 200%. Based on existing evidence, researchers believe that

Progress in Molecular Biology and Translational Science, Volume 135
ISSN 1877-1173
http://dx.doi.org/10.1016/bs.pmbts.2015.07.017

blood flow in these tissues is matched to the increases in metabolic rate during exercise. This phenomenon, the matching of blood flow to metabolic requirement, is often referred to as functional hyperemia. This chapter summarizes mechanical and metabolic factors that regulate functional hyperemia as well as other exercise-induced signals, which are also potent stimuli for chronic adaptations in vascular biology. Repeated exposure to exercise-induced increases in shear stress and the induction of angiogenic factors alter vascular cell gene expression and mediate changes in vascular volume and blood flow control. The magnitude and regulation of this coordinated response appear to be tissue specific and coupled to other factors such as hypertrophy and hyperplasia. The cumulative effects of these adaptations contribute to increased exercise capacity, reduced relative challenge of a given submaximal exercise bout and ameliorated vascular outcomes in patient populations with pathological conditions. In the subsequent discussion, this chapter explores exercise as a regulator of vascular biology and summarizes the molecular mechanisms responsible for exercise training-induced changes in vascular structure and function in skeletal and cardiac muscle as well as the brain.

1. INTRODUCTION

During the transition from rest to dynamic exercise and as exercise intensity increases, the absolute and relative distribution of blood flow changes throughout the entire body. For example, during incremental exercise, as oxygen demands increase, blood flow may increase up to 20- and 5-fold in skeletal and cardiac muscle, respectively. Further, in organs such as the brain, although bulk blood flow may be unchanged, regional flow in specific structures may increase nearly 2-fold.[1,2] Exercise-induced increases in blood flow may increase fluid shear stress (frictional forces along the luminal wall of blood vessels) as well as mediate the local induction and systemic release of angiogenic growth factors. Collectively, chronic exposure to these phenomena, as in exercise training, can alter gene expression, increase arterial diameter, and initiate the formation of new capillaries from preexisting capillaries. Alterations in vascular cell gene expression and the formation of new capillaries may increase blood flow capacity and alter blood flow control.[1] Using available evidence from human and animal research, this chapter will summarize and discuss the aforementioned processes and provide a framework of exercise training-induced adaptations in vascular structure and function in active skeletal muscle, cardiac muscle, and the brain. For detailed discussion of exercise training-induced vascular adaptations, the reader is referred to the following excellent review articles.[1,3,4]

2. THE SKELETAL MUSCLE VASCULATURE AND EXERCISE

Skeletal muscle has an exceptional capacity to respond to changes in energetic requirements. These changes in metabolism observed in the transition from rest to activity are met with increased blood flow for convective oxygen delivery as well as increased diffusion capacity for oxygen extraction ability at the muscle. Whole-body oxygen consumption ($\dot{V}O_2$) is described by the following equation:

$$\dot{V}O_2 = \dot{Q}\ (a - \bar{v}O_2d)$$

Pulmonary $\dot{V}O_2$ increases ~15-fold from rest to maximal activity in average subjects. This is accomplished by ~5-fold increase in cardiac output ($\dot{Q}$) and 3-fold increase in the oxygen extraction by the muscle ($a - \bar{v}O_2d$). Extending this to skeletal muscle where ($\dot{Q}$) represents the blood flow to the active muscle, whole-body exercise can increase $\dot{V}O_2$ approximately 75-fold; corresponding to a 30-fold increase in blood delivery and 2.5-fold increase in oxygen extraction in exercising skeletal muscle. At the single muscle group, maximal exercise generates ~240-fold increase in $\dot{V}O_2$ with ~80-fold increase in blood flow and 3-fold increase in oxygen extraction in the active muscle group.[5,6] The changes in blood flow capacity and capillary exchange capacity induced by exercise training are of paramount importance in the adaptive response to exercise.

To augment blood flow within skeletal muscle in support of an increased metabolic rate, changes in capillarity, i.e., angiogenesis, occur in response to exercise training allowing increased blood delivery as well as an increased transit time for the red blood cells in microcirculation. Exercise training has been shown to be a powerful stimulus for angiogenesis when comparing the skeletal muscle of trained vs. untrained as well as comparisons of skeletal muscle before and after training. Tissue samples from the *vastus lateralis* of moderately trained men ($\dot{V}O_2 = 51.3$ ml 100 g^{-1} min^{-1}) indicate a capillary density of 585 ± 40 per square millimeter (mm^2), while capillary density in those of endurance-trained men $\dot{V}O_2 = 72.0.3$ ml 100 g^{-1} min^{-1}) was 821 ± 28 capillaries mm^{-2}.[7] Examining the same muscle following 8 weeks of cycle training, an increase in $\dot{V}O_{2max}$ of 16% from 49.0 to 56.6 ml 100 g^{-1} min^{-1} was matched by a concomitant 20% increase in capillary density from 329 to 395 capillaries mm^{-2}.[8] Differences in the absolute number of capillaries are likely due to differing methods of obtaining and fixing tissue samples.

2.1 Skeletal Muscle Vascular Remodeling (see Fig. 1)

Regular exposure to bouts of exercise can serve as potent stimuli to induce adaptation in the cardiovascular as well as skeletal muscle systems. Structural vascular adaptations and adaptations to control of vascular resistance in the vascular network mediate improved blood flow capacity, oxygen extraction,

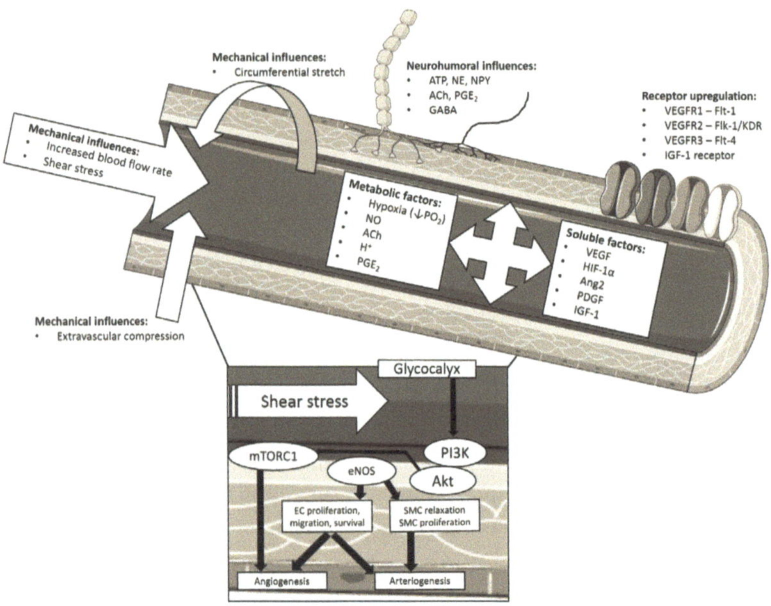

Figure 1 Mechanical and metabolic factors involved in exercise training-induced vascular adaptations. Schematic representation of vascular adaptation in skeletal muscle, coronary, and cerebral vascular tissues. Stimuli include mechanical influences, interluminal metabolites, soluble release factors, and neurohumoral factors. Specific receptor upregulation associated with vascular remodeling and a generalized signaling mechanism are included (inset). Abbreviations: ATP, adenosine triphosphate; NE, norepinephrine; NPY, neuropeptide Y; Ach, acetylcholine, PGE$_2$, prostaglandin E$_2$; GABA, gamma aminobutyric acid; VEGFR1/Flt-1, vascular endothelial growth factor receptor 1; VEGFR2/Flk-1, vascular endothelial growth factor receptor 2; VEGFR3/Flt-4, vascular endothelial growth factor receptor 3; IGF-1 receptor, insulin-like growth factor receptor; NO, nitric oxide; H$^+$, hydrogen ion; VEGF, vascular endothelial growth factor; Ang2, angiopoietin; PDGF, platelet-derived growth factor; IGF-1, insulin-like growth factor; PI3K, phosphatidylinositol-4,5-bisphosphate 3-kinase; Akt, protein kinase B; eNOS, endothelial nitric oxide synthase; mTORC1, mammalian target of rapamycin complex 1; EC, endothelial cell; SMC, smooth muscle cell. (See the color plate.)

and oxygen diffusion capacity. Structural adaptations include arteriogenesis, angiogenesis, and vascular remodeling mechanisms.[9,10] Three types of stimuli are generally thought to contribute to exercise-induced structural adaptation: (1) increased blood flow generating increased fluid shear stress, (2) enhanced local metabolism in the exercising muscle, and (3) mechanical stretch of the tissue causing local and ultimately circulating release of growth factors.

The large increases in metabolism of skeletal muscle during the transition from rest to exercise are only possible because of increased blood flow. The increased blood flow induces pressure, and flow stresses in the vasculature, which increases luminal shear stress. Briefly, shear stress causes the deformation of glycocalyx receptors residing on the luminal side of endothelial cells. This activates calcium ion channels and phospholipase activity in endothelial cells leading to calcium signaling, prostaglandin release, and subsequently cyclic adenosine monophosphate (cAMP)-mediated smooth muscle relaxation. Likewise, luminal shear stress activates endothelial cell phosphatidylinositol-4,5-bisphosphate 3-kinase (PI3K), which phosphorylates and activates protein kinase B (Akt). The induction of Akt mediates endothelial nitric oxide synthase (eNOS) phosphorylation and subsequent nitric oxide (NO) release, which also relaxes smooth muscle. Exercise-induced shear stress mediates acute increases in arterial diameter as well as normal vascular remodeling throughout the affected vasculature.[11]

Training enhances blood flow capacity specific to the trained muscle groups and can affect bulk blood flow delivery at the conduit arteries, local perfusion at the resistance arterioles, and metabolite exchange at the capillaries. Walther *et al.*[12] have reported greater vascular conductance in the trained limbs of swimmers and cyclists compared to sedentary controls. Flow-mediated dilation, vasodilatory capacity, and vascular reactivity were enhanced in trained muscles, and interestingly, cycle training enhanced vascular reactivity of the untrained upper limb conduit and resistance vessels. Greater blood flow capacity in trained athletes has also been observed in a training-specific and activity-dependent manner. The subclavian arteries in the dominant arms of elite tennis players, and athletes with paraplegia, were larger than their contralateral arm or the arms of control subjects. These findings were similarly observed in the brachial arteries of the dominant arms of elite racquetball players versus nondominant arms or healthy inactive controls.[13] The resting arterial diameter and brachial dilatory capacity were greater in the dominant arm relative to nondominant or either arm of the inactive, while the wall thickness of carotid, brachial, and femoral arteries was lower in athletes versus controls. In a similar manner femoral arteries of elite cyclists exhibited increased diameter, while reduced diameters were observed in paraplegia, relative to control.[14] These adaptations may confer specific benefits to exercise capacity in highly trained individuals. The regional differences in vascular adaptation to specific exercise training are an important consideration in evaluating vascular health. As an index of systemic endothelial health, and a potential predictor of cardiovascular disease, brachial artery flow-mediated dilation (FMD) has been utilized as a noninvasive index of whole-body endothelial function. However, in comparing the results of FMD studies in brachial artery versus superficial femoral or popliteal arteries of healthy subjects, it was found that the vasodilatory function of the upper limbs was not predictive of that in the lower limbs.[15]

Arteriogenesis specifically indicates the enlargement of existing arterial vessels either by an increase in vessel diameter, increase in wall thickness, or both. Arterial vessel enlargement is in response to persistent elevations in luminal pressure.[11] Remodeling of arteries largely occurs through adaptation of vascular smooth muscle but includes changes to endothelial cells and fibroblasts to achieve a greater blood flow capacity.[16] Pulsatile blood flow imposes continuous mechanical strain on conduit arteries with increases in both radial wall stress and vessel diameter leading to increases in arterial wall mass.[17,18] Increased blood flow may increase shear stress applied to the endothelial surface which induces similar increases in arterial wall mass.[16] Chronic flow-induced changes in vessel diameter aim to reduce wall shear stress back to baseline values while maintaining distribution control.[19–22] An observed increase in total blood flow capacity to the rat hindlimb, following femoral artery ligation, and in response to treadmill training, was inhibited by administration of the nitric oxide synthase inhibitor L-NG-nitroarginine methyl ester (L-NAME). Coincident to the increases in blood flow, muscle capillarity was unaltered in control and L-NAME groups indicating a nitric oxide dependence for arteriogenesis and increases in total blood flow, while in this setting changes in capillarity occurred through non-nitric oxide-mediated mechanisms.[23]

2.2 Skeletal Muscle Angiogenesis

Exercise training provides stimulus that favors angiogenesis through endothelial cell proliferation, cell migration, intussusception, and vessel sprouting that results in new capillaries.[24] However, although individual bouts of exercise increase proangiogenic stimuli, the resultant capillary development does not manifest without repeated exposure. This seemingly delayed response to putative angiogenic stimuli may ensure that the process of vessel formation, with a high metabolic cost, is not propagated without the correct temporal association of stimuli for a duration sufficient to produce new vessels. The complexity of angiogenesis requires integration of multiple signaling pathways to achieve successful *de novo* vessel formation.

The extensive interplay among mechanical, chemical, and transcriptional mediators is coordinated through the common proangiogenic mediator vascular endothelial growth factor (VEGF). VEGF is a 35- to 450-kDa peptide growth factor with four primary isoforms having been identified in human tissue (VEGF-A—D).[25] Alternate splicing of the human VEGF gene produces four variants of VEGF-A ($VEGF_{121}$, $VEGF_{165}$, $VEGF_{189}$, and $VEGF_{206}$) determined by their respective amino acid product lengths.[26,27] The VEGF-A isoform (commonly referred to as VEGF) predominates and is the most important in postnatal skeletal muscle angiogenesis.[28] VEGF effects are carried out through tyrosine kinase receptor activation, the most prominent being VEGFR-1 (Flt-1) and VEGFR-2 (KDR/FLK-1).[29] Receptor binding induces autophosphorylation of receptor tyrosine residues and subsequent activation of multiple signaling pathways including PI3K, phospholipase C-gamma (PLC-γ), and protein kinase C (PKC).[30] Previous investigation has shown an immediate increase in VEGF mRNA following a single session of treadmill running in rats[31] and during hypoxic or normoxic voluntary leg extension in humans.[32,33] Increases in VEGF with increased muscle activity are congruent with reports from chronically stimulated muscle studies. In rat muscle, 7 days of stimulation for 8 h per day increased both VEGF protein levels and vessel density.[34]

There is a dynamic interplay of angiogenic and angiostatic modulators that regulate adaptation to exercise training. A well-described stimulus of angiogenesis is hypoxia with the upregulation of VEGF mediating the response.[35,36] Cell culture investigation of endothelial cells with reduced PO_2 observed cell proliferation, migration, and tube formation correlated with an upregulation of VEGF, an effect which was reversed by elevating PO_2.[37] Support of a metabolic hypothesis and limited oxygen delivery as modulators of angiogenesis have come from studies employing varied inspired oxygen fractions during treadmill exercise in rats. Reducing the inspired fraction during exercise enhanced the upregulation of VEGF following exercise.[31] Further, hypoxia-induced increases in capillarity subsequent to femoral artery ligation with exercise have also been observed.[38] Hypoxia-inducible factor (HIF-1α) has been a widely investigated intermediary linking the sensing of lowered PO_2 and the angiogenic response. Following hypoxic conditions, HIF-1α mRNA as well as protein levels are elevated. This increased expression has a broad range of effects including stimulating erythropoiesis, increased glycolysis, additional genes stimulating capillary growth[39] and increased transcription of the VEGF gene.[37] Elevations in HIF-1α mRNA and protein have been observed following endurance-type exercise in humans with restricted convective oxygen delivery.[32] However, employing a similar exercise stimulus, Richardson *et al.*[33] found increased VEGF mRNA levels in normoxic and hypoxic conditions that was independent of reduced cellular PO_2. Strong evidence definitively linking HIF-1α to increased VEGF production and subsequent angiogenesis remains to be found. At present, muscle PO_2 during exercise decreases sufficiently to activate HIF-1α; however, this appears to be a contributor to the coordinated angiogenic response rather than the primary regulator.

2.3 Local and Systemic Growth Factors

Supplementary elements important to vascular adaptation are upregulated during exercise. Importantly, mRNA for the VEGF receptor tyrosine kinases VEGFR1 (Flt-1), VEGFR2 (Flk-1/KDR), and VEGF3 (Flt-4) is upregulated following muscular activity[23] which has been shown to translate into increased protein receptor expression. Receptor expression plays an important role in the angiogenic response as highlighted by Milkiewicz *et al.*,[40] chronic skeletal muscle ischemia increased Flk-1 expression which correlated with increased angiogenesis in the presence of high VEGF protein. The negative effects of sustained ischemia were reversed with muscle stimulation through further activation of Flk-1 and downregulation of Flt-1. The upregulation of receptors implicit to angiogenesis potentially enhances tissue responsiveness to VEGF and concentrates vascular growth to those regions.

There are additional factors related to angiogenesis in maturation, tumor growth, arthritis, and atherosclerosis that are subject to the precise balance of angiogenic and angiostatic factors, but may not be requisite for exercise-induced angiogenesis. Among these, placental growth factor (PIGF) is related to VEGF and also signals through receptor tyrosine kinases. However, gene ablation of PIGF with subsequent exercise training revealed no role for PIGF in an exercise-induced angiogenic response.[41] Similarly, angiopoietin 1 (Ang 1) and angiopoietin 2 (Ang 2) signal through the common receptor tyrosine kinase Tie-2 to modulate vessel adaptation.[23] Ang 1 functions to stabilize the existing vasculature,[42] while Ang 2 functions as a competitive antagonist to destabilize it.[43] Following treadmill training, the ratio of Ang 2 to Ang 1 is elevated; however, for capillarity increases to be observed, increased VEGF is requisite.[44]

Acute bouts of exercise increase the release of several factors which prevent the immediate initiation of angiogenesis, i.e., angiostatic factors. The first of these to be characterized was angiostatin which limits angiogenesis in tumor growth.[45] In exercise, soluble VEGFR1 (sFlt-1) is increased by treadmill running in proportion to intensity and is associated with decreases in free VEGF and thus reduced angiogenesis.[46] Platelet factor 4 (PF-4) is a strong inhibitor of angiogenesis that is increased following acute bouts of cycling exercise,[47] repeated maximal cycling bouts,[48] and following 9 months of endurance running.[49] A role for thrombospondin-1 (THBS1) as a modulator of capillary growth in skeletal muscle has been reported for mice deficient in THBS1. These mice exhibit increased capillarity in both cardiac and skeletal muscle as well as an increased exercise capacity.[50] Following an acute bout of exercise, THBS1 mRNA levels are elevated 3- to 4-fold and return to baseline following continued training suggesting an acute inhibition to angiogenesis that is relieved with continued exercise stimulus.[51] The concomitant increase of Forkhead Box "O" transcription factors FoxO1 and FoxO3a has been observed as important determinants in opposing the acute angiogenic factors released following an acute exercise stimulus. These factors promote an angiostatic environment and are decreased following 10 days of repeated exercise further suggesting a role for rapid increases in both angiogenic and angiostatic factors following acute exercise with the latter decreasing in abundance following repeated exercise setting the stage for vascular proliferation.[52] Thus, it is important to assess all of these angiogenic and angiostatic factors stimulated by bouts of exercise to understand the molecular mechanisms involved in structural vascular adaptation. It is clear that complete understanding requires an appreciation of the balance of pro- and antiangiogenic factors in the tissue.

2.4 Functional Adaptations in the Skeletal Muscle Vasculature

Vascular structure (number and size of the tubes) sets the minimal vascular resistance and maximal capillary diffusion capacity. Resting blood flow and capillary diffusion capacity are then established by regional vascular resistance determined by contraction of vascular smooth muscle in the resistance arteries. There are functional vascular adaptations of the processes regulating resistance to blood flow induced by exercise training in conjunction with structural vascular alterations. The increased blood flow capacity attained through exercise training appear to be partially the result of peripheral adaptations in endothelial and vascular smooth muscle phenotype which control microvascular resistance. Changes in control of blood flow can also be centrally mediated, i.e., neurohumoral, or the result of changes in the reactivity of arteries and arterioles through the vascular smooth muscle and endothelium.[1,53] The importance of exercise training-induced adaptation of control of vascular resistance varies through muscle tissue with muscle fiber-type and exercise-specific recruitment patterns facilitating the extent of adaptation.[9]

2.5 **Summary** (see Fig. 2)

Chronic training adaptations, including those in arteriolar density, contribute to the observed increases in blood flow capacity in response to exercise. However, this response occurs despite a nonuniform increase in arteriolar

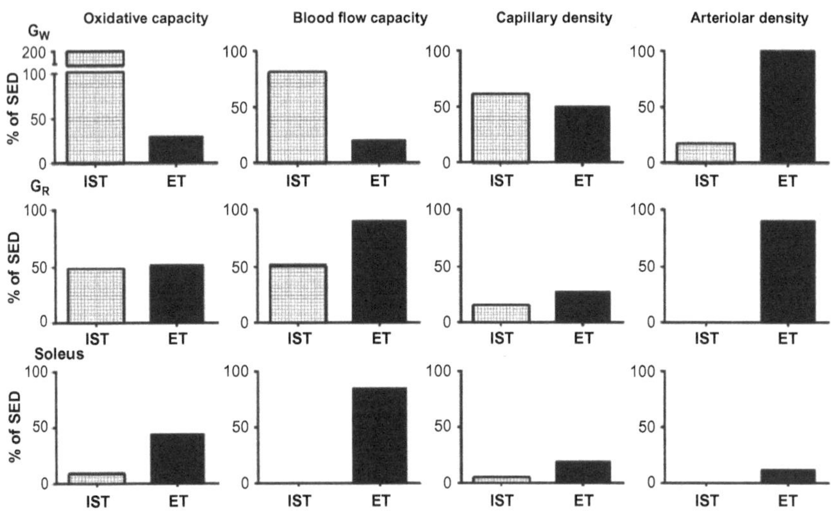

Figure 2 The effects of interval sprint (IST) vs. endurance (ET) exercise training (~10–12 weeks) on oxidative capacity (cytochrome *c* oxidase), blood flow capacity, capillary density, and arteriolar density, in white gastrocnemius (G_W; top panel), red gastrocnemius (G_R; middle panel), and soleus (bottom panel) rat skeletal muscle. Data are presented as % increase above sedentary values. IST increased oxidative and blood flow capacity in G_W. In addition, IST increased, albeit to a lesser extent, capillary density and arteriolar density in G_W as well as oxidative and blood flow capacity in G_R. Conversely, ET increased oxidative and blood flow capacity mostly in G_R and soleus muscle. ET also increased arteriolar density in G_W and G_R muscle. *This graph was adapted from Ref. 1.*

density whereby fast oxidative glycolytic and fast glycolytic muscle (i.e., gastrocnemius) increase arteriolar density,[54] while no significant changes are observed in slow twitch muscle (i.e., soleus).[55] The high density of microvasculature in skeletal muscle coupled with skeletal muscle's great capacity to increase metabolic rate necessitates strict control over vessel formation. Increased capillarity can be observed following endurance training in humans over weeks to months, while in animal models angiogenesis can be induced over the course of days.[11] The potential to increase capillarity in skeletal muscle is further dependent on the initial capillarity such that angiogenic response in glycolytic muscle exceeds that of oxidative. This is potentially related to the proangiogenic stimuli, receptor responsiveness, or a ceiling effect by which oxidative skeletal muscle is near maximally supported at baseline. Contrasting reports of increased capillarity in proportion to changes in oxidative capacity[56–59] are investigations in which chronic stimulation has produced increased capillarity in the absence of significant changes in oxidative capacity.[11,60] In basal conditions, fast-twitch oxidative red gastrocnemius muscle has ~3-fold greater capillary density than fast-twitch glycolytic white.[60] In response to treadmill training following femoral artery ligation (a model of peripheral vascular disease), the fast-twitch red muscle of the gastrocnemius, which receives ~4-fold greater flow in these conditions,[61] had modest increases in VEGF mRNA, while the fast-twitch white portion showed ~6-fold increases in mRNA.[23] Subsequent to these increases in mRNA was a significant increase in the capillarity of fast-twitch white muscle at 10–14 days,[23] while similar training had previously not produced significant changes in capillarity of the fast-twitch red oxidative gastrocnemius.[62] Collectively these results support the assertion that VEGF is integral in the alteration as well as maintenance of vascular density and that this control is correlated with oxidative demand of the muscle. In response to exercise, an exquisite interplay of mechanical and chemical factors coordinates structural and functional adaptation in the peripheral vasculature. Many of the same stimuli provide the impetus for adaptation in the heart and coronary vasculature; however, the precise control and adaptation differ from skeletal muscle.

3. THE HEART AND CORONARY VASCULATURE

Under normal conditions, the heart beats ~100,000 times and pumps ~7200 L of blood per day. This amount of work accounts for ~10–20% of whole-body oxygen consumption. To sustain this workload, the heart relies on continuous blood flow (~250 mL min^{-1} or 0.8 mL min^{-1} g^{-1}) and the oxidative phosphorylation of fatty acids.[3,63] At rest, the arterial oxygen extraction of the heart is more than 2-fold greater than skeletal muscle (~75 vs. 35%).[64,65] To accomplish this level of oxygen extraction, the heart is equipped with a capillary density of ~3500 capillaries per mm^2.[66] As a result of an elevated oxygen extraction at rest, myocardial blood flow must increase in order to meet an increase in oxygen demand, for example, during exercise.[3,64]

The left and right coronary arteries originate at the base of the aorta and feedback to the heart. The left coronary artery branches into the left anterior descending and circumflex arteries. These arteries lie superficially along the heart and are considered epicardial coronary vessels. The left anterior descending and circumflex arteries supply the lateral and anterior walls of the left ventricle and an anterior portion of the septum. The right coronary artery supplies the right ventricle, the posterior wall of the left ventricle, and a posterior portion of the septum. At various segments along the epicardial coronary vessels are smaller arteriole branches that provide blood flow to the microvascular network within the heart. The arterioles of the microvascular network are the primary cite of vascular resistance as well as coronary blood flow control. These arterioles branch into smaller capillaries, which are positioned closely to nearby cardiomyocytes, to optimize nutrient delivery and the removal of metabolic waste products from the cells. Blood flow in the heart is complex and effected by many elements including extravascular arterial compression, perfusion time (i.e., exercise training-induced bradycardia will reduce the duration of systole, thereby decreasing systolic compression of coronary vessels and decreasing resistance to flow), perfusion pressure, autoregulation, neural and humoral factors, the vascular endothelium, and metabolism.[3] During exercise, a combination of these factors interacts and myocardial blood flow increases linearly with oxygen demand as well as heart rate and can increase three to five times above resting values.[64,67–69]

3.1 Exercise Training and the Heart

Exercise training results in myocardial hypertrophy. In fact, relative left ventricle mass is ~20–100% greater in trained vs. untrained inviduals.[70,71] Exercise training-induced increases in left ventricle mass result in simultaneous increases in stroke volume and reductions in heart rate at rest and during submaximal exercise. Accordingly, at a given level of submaximal exercise, heart rate and, therefore, myocardial oxygen consumption and coronary blood flow will be reduced in trained individuals.[3,67]

Although exercise training appears to lower myocardial oxygen demand, it also results in vascular adaptations that increase myocardial oxygen supply. This occurs primarily through an increase in coronary blood flow capacity. As in skeletal muscle, exercise training-induced vascular adaptations can be divided into structural and functional changes that account for the increased coronary blood flow capacity; here we will focus on the structural adaptations.

3.2 Structural Adaptations in the Coronary Vasculature (see Fig. 3)

Evidence from rodents,[72,73] dogs,[74] monkeys,[75] and humans[70,71,76,77] suggests that exercise training increases the diameter of epicardial coronary vessels in proportion to increases in left ventricular mass. Likewise, in exercise-trained swine, it was demonstrated that in resistance vessels

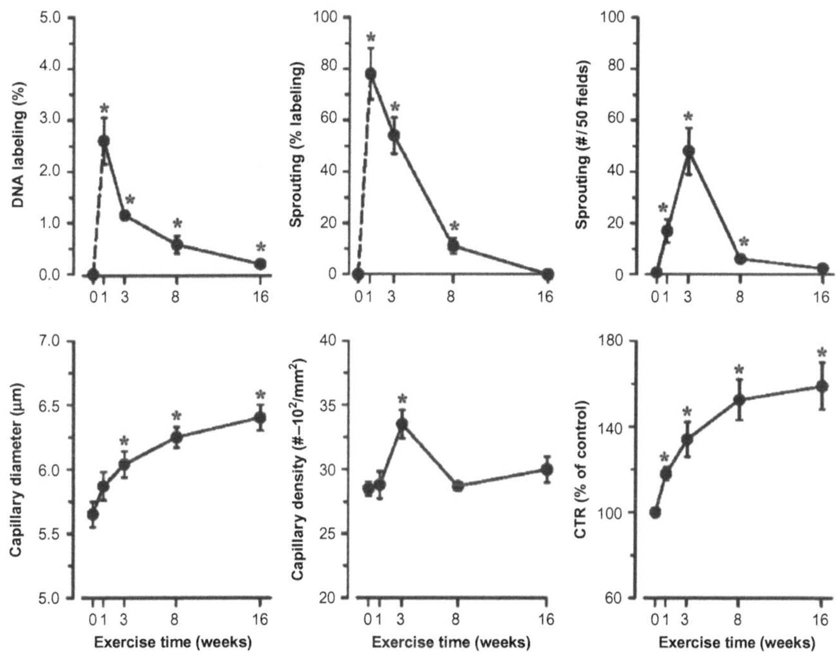

Figure 3 The effect of exercise training in swine at 1, 3, 8, and 16 weeks on DNA labeling (top left), sprouting of new capillaries (top middle), % labeling of sprouts (top right), capillary diameter (bottom left), capillary density (bottom middle), and coronary transport reserve (CTR; bottom right). Data are presented as mean ± SE. *Significantly different from sedentary swine at week 0 ($P < 0.05$). Increases in capillary density are normalized by 16 weeks of training. However, capillary diameter is elevated at 3 weeks and remains elevated over the 16-week training program. *Graph is adapted from Ref. 1 and data are adapted from Ref. 78.*

between the diameters of 40–120 μm; increases in vascular volume were accounted for by increases in arteriole diameter.[78] Exercise training-induced changes in epicardial coronary and arteriole diameter are believed to be the result of structural changes and not changes in vascular tone alone.[77] Increases in coronary blood flow during exercise will theoretically increase fluid shear stress in the coronary vasculature. Arteries may respond to persistent increases in flow, by increasing their diameter chronically to maintain basal levels of shear stress.[79,80] This assumes basal shear stress is a highly regulated variable. However, it is unlikely shear stress alone regulates exercise training-induced changes in arterial diameter, as exercise training also mediates changes in endothelial cell function and vasomotor control in the coronary vasculature.[3]

In addition to increasing epicardial coronary and arteriole diameters, exercise training may also increase myocardial capillary density. In the aforementioned study regarding coronary arteriole diameter in swine, it was also observed that exercise training increased the number of resistance vessels with diameters between 20 and 40 μm.[78] Evidence from guinea pigs[81,82] and dogs[83] as well as from wild (vs. domesticated) rabbits[84] and rats[85] suggests exercise training or physical activity increases capillary numerical density (i.e., capillaries per mm^2). The effect of exercise training on capillary density in the rat heart appears to be related to age, as the rate of angiogenesis exceeds myocardial hypertrophy in young but not old rats.[86,87] However, if capillary density is quantified as the capillary–myocyte ratio, exercise training does not increase capillary density in guinea pigs,[88] dogs,[66,74] or swine.[89] Apart from methodological differences between studies, inconsistencies between the effects of exercise training on capillary numerical density vs. capillary–myocyte ratio may be explained partially by the time of measurement. Consider that exercise training stimulates myocardial hypertrophy; therefore, angiogenesis must occur to maintain capillary density and the rate of angiogenesis must exceed the rate of hypertrophy to increase capillary density. In swine, exercise training increased coronary endothelial cell division and capillary sprouting at 1, 3, and 8 weeks but were not different from sedentary swine at 16 weeks. Also, markers of angiogenesis exceeded myocyte hypertrophy at 3, but not 8, weeks of exercise training, at which point capillary density was similar to sedentary control values. In spite of this finding, total vascular volume increased by 37% in the exercise-trained group.[78] Although left ventricle hypertrophy and capillary growth appear to be coupled in large mammals,[89] it is possible they occur at different rates. Thus, the manner in which capillary density is quantified and when, relative to the angiogenic and hypertrophic processes, a measurement is taken, may impact the results dramatically. Nevertheless, it is clear that exercise training induces myocardial angiogenesis, because left ventricle hypertrophy and capillary growth are matched.

The coupling between exercise training-induced capillary and cardiomyocyte growth may be explained partially by the activation and diverse roles of Akt. Shear stress-induced Akt activation enhances Ang 2 and VEGF activity; both of which are angiogenic factors and can also be induced by hypoxia (or HIF-1α) or cardiomyocyte stretch through a mammalian target of rapamycin complex 1 (mTORC1)-dependent mechanisms. In the presence of VEGF, Ang 2 is involved in the initiation of angiogenic sprouting and both VEGF and Ang 2 are active in vessels undergoing remodeling. Interestingly, Akt signaling is also involved in cardiac hypertrophy; therefore, the effect of Akt on angiogenesis may be one mechanism that couples myocyte and capillary growth.[90,91] Exercise training stimulates angiogenesis through increases in fluid shear stress/Akt signaling, as well as local induction of HIF-1α-mediated factors.[24] Further, because exercise training also stimulates myocardial hypertrophy; instances where the rate of myocyte growth exceeds capillary growth, theoretically, may also induce a mild hypoxic state leading to subsequent HIF-1α-mediated angiogenesis.

3.3 Functional Adaptations in the Coronary Vasculature

Exercise training alters vasomotor control in epicardial coronary and myocardial arterioles. Exercise training decreases sensitivity to α adrenergic stimulation in canine and swine epicardial coronary artery smooth muscle.[92,93] In contrast, in canine myocardial arterioles, both α and β adrenergic receptor sensitivity may be augmented. These findings are complimented by human data that suggest that α and β adrenergic tone is maintained, despite reduced concentrations of circulating catecholamines in trained vs. untrained individuals.[94–98] Likewise, in myocardial arterioles, myogenic tone may also be increased on account of altered calcium-dependent PKC signaling and increased voltage-gated calcium currents through the L-type calcium channels in vascular smooth muscle cells.[99,100] The effects of exercise training on the coronary endothelium are also artery dependent. Specifically, exercise training mediates an early-phase response that improves endothelium-dependent, NO-mediated vasorelaxation. However, this effect may dissipate over the course of a training program in dogs and swine.[92,101] In swine, exercise training increases NO availability/signaling throughout the coronary microvasculature, directly through enhanced eNOS expression, and indirectly through enhanced superoxide dismutase-1 (SOD-1) expression (i.e., decreased NO quenching) and decreased endothelin-1 (ET-1) signaling.[102,103] Although exercise training increases endothelium-dependent vasorelaxation throughout the coronary microvasculature,[99,102] the effect of exercise training on eNOS expression is not uniform throughout the coronary vasculature.[99] Further, the effect of exercise training on eNOS expression in the epicardial arteries and larger arterioles may be transient, as exercise-induced increases in shear stress and eNOS expression are normalized as these vessels undergo structural vascular remodeling and their diameters increase.[78,99,104]

3.4 Summary

Exercise training-induced changes in coronary vascular structure and function can occur within 1 week, but may take longer. Early adaptations are most evident in the epicardial coronary arteries, but these adaptations may subside over the course of long-term training. Conversely, adaptations in the coronary microcirculation may take longer to manifest.[78,105] The effect of different exercise modes and protocols on structural and functional adaptations in the coronary vasculature remains to be elucidated. Athletes that participate in isotonic exercise (i.e., running) as well as isometric exercise (i.e., wrestling) exhibit increased in left ventricle mass, but only the former display increased end diastolic volumes.[70] Assuming that cardiomyocyte hypertrophy and vascularization are linked under normal physiological conditions,[89] an increased left ventricle mass will reflect increased vascular volume as well. With regard to aerobic exercise, coronary vascular adaptations may be intensity-dependent because coronary blood flow increases linearly with oxygen demand and heart rate.

4. THE BRAIN AND CEREBRAL VASCULATURE

The brain is active under all living conditions. It only represents ~2% of total body mass, but accounts for ~20% of whole-body oxygen consumption at rest.[106,107] Brain activity requires energy and relies primarily on the oxidative metabolism of glucose to obtain the requisite amount of adenosine triphosphate (ATP).[108] It typically consumes six molecules of oxygen per molecule of glucose and requires a continuous supply of blood flow.[109,110] In fact, if blood flow to the brain is stopped suddenly, the oxygen content present in the brain, at any given time, is only sufficient to sustain cerebral metabolism for ~1 s by itself.[111] Therefore, in order to sustain or alter neural activity in the brain, cerebral oxygen consumption and cerebral blood flow (CBF) responses must be coordinated precisely.[112] This concept (i.e., matching blood flow to metabolic requirement) is referred to as neurovascular coupling as well as functional hyperemia.[113,114]

The cerebral vasculature is comprised of two sets of large arteries: the internal carotid (responsible for ~75% of CBF) and vertebral arteries (responsible for ~25% of CBF). These two sets of arteries converge to form the circle of Willis; collectively, they are considered extracerebral and are innervated by peripheral sympathetic, parasympathetic, and sensory nerves.[114,115] Alternatively, the branches of the circle of Willis are considered intracerebral, as they give rise to pial arteries, parenchymal arterioles and capillaries. Although the pial arteries receive extrinsic innervation from the peripheral nervous system (similar to the internal carotid and vertebral arteries), the parenchymal arterioles are intrinsically innervated from within the brain neuropil. The close proximity between parenchymal arterioles and neurons as well as perivascular astrocytes facilitates the tight coupling of neural activity and CBF control. This occurs because neurons and perivascular astrocytes release vasoactive signals that bind to receptors on the vascular smooth muscle or the endothelium and stimulate vasodilation or constriction in parenchymal arterioles as well as capillaries.[113,114,116] The surface area of the brain microvasculature is ~100 cm^2 g^{-1}.[117] It is estimated that the capillary and endothelial volumes constitute ~1% and 0.1% of total tissue volume, respectively, and intercapillary distances are ~40 μm.[118] This vascular architecture ensures a close proximity between the microvasculature and neural tissue and permits the rapid transport of nutrients across the blood–brain barrier to sustain neural activity.

4.1 Acute Exercise and the Cerebral Vasculature

For over a century, scientists have been studying the relationship between muscular activity and brain blood flow.[119] During this time, it was discovered that global brain blood flow remains relatively constant at ~50–60 mL 100 g^{-1} min^{-1} under numerous circumstances,[109] providing that $PaCO_2$ does not change and mean arterial pressure remains in the autoregulatory range, from 60 to 150 mmHg.[120] At rest, reductions in $PaCO_2$ or hypocapnia stimulate pial artery vasoconstriction[121,122] and lower brain blood flow values.[116,123] Indeed, during exercise, where $PaCO_2$ values tend to decrease (from ~40 mmHg at rest to ~25 mmHg at maximal exercise), total brain blood flow may remain unchanged or decrease slightly.[124–127] However, evidence from miniature swine,[125,128] dogs,[129] ponies,[124] and humans[126,130,131] suggests regions of the cerebral vasculature vasodilate, possibly in an exercise intensity-dependent manner. Regional changes in brain blood flow appear to be coupled with neural activity[112] and can increase up to ~70% above resting values.[2] Furthermore, chronic exercise training mediates changes in brain vascular structure and function.[4] To understand the aforementioned phenomena, we will discuss brain blood flow control immediately prior to and during exercise. Thereafter, we will address the mechanisms of exercise training-induced adaptations in brain vascular structure and function.

Immediately prior to and at the onset of exercise, heart rate and mean arterial pressure increase rapidly and simultaneously. Such cardiovascular adjustments are initiated by the central nervous system through a process described originally as cortical irradiation.[132] Presently "cortical irradiation" is referred to as "central command" and refers to descending signals from higher brain centers that occur in the absence of physical work or sensory feedback. In essence, central command reflects neural activity during the preparatory phase before and at the onset of exercise.[133,134] The initial activation of central command does not stimulate changes in global brain blood flow. However, it does alter blood flow distribution within the brain, contributing to changes in regional brain blood flow.[133–136] Brain regions that have been implicated in the central command network and may also receive additional blood flow prior to, and at the onset of exercise, include (but are not limited to) the sensorimotor cortex, supplementary cortex, cingulate cortex, insular cortex, medial prefrontal cortex, cerebellum, and thalamic regions. Therefore, it is believed that (i) the aforementioned brain regions activate to elicit the rapid cardiovascular adjustments prior to, and at the onset of exercise; (ii) said neural activity initiates the release of local metabolites (H+, K+, NO, acetylcholine, adenosine, CO_2, lactate, gamma aminobutyric acid, norepinephrine, neuropeptide Y, 5HT, prostaglandins, etc.[113,114]) from neurons and perivascular astrocytes; and (iii) said metabolites are vasoactive, interact with surrounding microvessels and serve to alter blood flow distribution within the brain to facilitate subsequent oxygen and glucose delivery to active brain regions. These vasoactive molecules are able to vasodilate or vasoconstrict and likely in a coordinated and region-specific manner. During exercise, regional vascular resistance (VR = mean arterial pressure divided by regional blood flow) may be unchanged (the rise in mean arterial pressure and rise in flow are equal, in order to sustain a modest increase in neural activity, observed in the Pons), may increase (mean arterial pressure increases and flow remains unchanged, occurs where neural activity does not change; observed in the temporal cortex), or decrease (the rise in mean arterial pressure is smaller than the rise in flow, indicative of vasodilation in order to meet a large increase in neural activity, observed in the cerebellar vermis) in different brain structures.[125] Whether exercise training impacts the release or sensitivity to vasoactive molecules released from perivascular astrocytes and extracerebral neurons is unknown.

Although it is generally accepted that neural activity and, therefore, cerebral metabolism increase during the transition from rest to exercise, whether global brain blood flow increases or decreases remains controversial. However, it is now generally accepted that exercise-induced increases in brain blood flow are regionally specific and related directly to neural activity.[1,126] Therefore, instances where global brain blood flow increased during exercise may reflect exercise bouts that were sufficiently intense to elicit elevated and sustained neural activity, while insufficiently intense to elicit a dramatic rise in $PaCO_2$ and sympathetic nervous system (SNS).[137]

Consistent with the notion that global brain–blood flow remains unaltered, total oxygen uptake in the brain remains relatively constant during steady-state and incremental exercise.[130,138] However, this does not preclude the possibility of regional changes in neural activity, cerebral metabolism, and corresponding changes in CBF responses. This concept is supported by observations that cerebral glucose metabolism[139] and regional flow responses[125,130] are not uniform throughout the brain during exercise. There is compelling evidence, in humans and animals, that during exercise, blood flow increases in extracerebral and intracerebral arteries,[130,140] supplying the frontal, parietal, temporal, and occipital lobes as well as the cerebellum.[137] Most notably, blood flow increases in the sensorimotor, supplementary motor, primary motor, and insular cortices as well as the cerebellar vermis.[2,141,142] However, in the transition from moderate to strenuous exercise, additional neural recruitment may not be widespread and subsequent decreases in $PaCO_2$ (as a result of the increased exercise intensity), increases in SNS activity, and alterations in metabolism may restrain further increases in brain blood flow.[126,140] In the vertebral, posterior cerebral, and cerebellar arteries, blood flow increases during moderate and further during strenuous exercise. However, blood flow in the internal carotid and middle cerebral arteries increase during moderate, but decrease during strenuous exercise. Reductions in regional brain blood flow during exercise coincide with decreased $PaCO_2$ and elevated SNS output.[126,130,131,138,140,143] Whether brain blood flow limits exercise capacity is currently being studied, but remains to be determined.[144]

4.2 Cerebral Vascular Adaptations to Exercise Training

The effects of exercise on the cerebral vasculature are not limited to acute changes in regional brain blood flow. Exercise induces local changes in shear stress, systemic and local changes in angiogenic and neurogenic factors, and may stimulate neurogenesis directly. Therefore, it is assumed that each of the aforementioned mechanisms may contribute to exercise training-induced alterations in cerebral vascular structure and function.

During exercise, regional increases in brain blood flow will theoretically increase the frictional forces along the inner wall of affected blood vessels; this is referred to as fluid shear stress. Although shear stress may result in vasodilation or constriction in cerebral arteries acutely[145]; increases in shear stress mediate chronic changes in arterial diameter[146] and alter endothelial cell phenotype through changes in gene expression.[147] It has been proposed that exercise-induced shear stress contributes to the maintenance of endothelial cell Akt and Akt-dependent eNOS expression in the cerebral vasculature. The expression and activation of eNOS are believed to be critical in maintaining cerebral vascular function and may be involved in the angio- and arteriogenic processes.[147,148]

In addition to fluid shear stress, exercise mediates events systemically and locally that, if repeated (i.e., exercise training), may contribute to angiogenesis and neurogenesis in the brain. For example, acute exercise increases circulating concentrations of insulin-like growth factor (IGF-1), VEGF,[149,150] both of which can bind to receptors in the endothelium and are implicated in the growth of new blood vessels from preexisting blood vessels as well as new neurons in specific regions of the brain (namely the dentate gyrus portion of the hippocampus).[149–151] Likewise, in animal models, exercise training increases mRNA and protein expression of the aforementioned signaling molecules as well as Ang 1, Ang 2, and fibroblast growth factor (FBF) in various regions of the cerebral vasculature.[149,152,153] Based on studies investigating exercise training-induced cerebral vascular angiogenesis in rodent models of health and disease, there is sufficient evidence to support the proposed framework: (i) exercise increases circulating concentrations of angiogenic growth factors (IGF-1, VEGF, FGF, endothelial progenerator cells, etc.) or augments their transcription locally through the induction of HIF-1α, (ii) said growth factors bind to specific receptors on endothelial cells of nearby preexisting blood vessels and initiate signaling cascades that induce the transcription of specialized proteins involved in endothelial cell proliferation and migration, (iii) said growth factors are regulated tightly and act interdependently, as inhibition of calveolin-1[154] or NOS[155] or IGF-1[149] or VEGF[156] attenuates the cerebral vascular remodeling associated with exercise training, and (iv) exercise-induced cerebral vascular remodeling reflects both an increase in the number, diameter, and density of capillaries.[157,158] Evidence from rodent work suggests that exercise-induced angiogenesis occurs in various regions of the brain, including (but not limited to) the cortex,[149] in particular, the primary motor cortex,[159] the cerebellar cortex,[158] and the prefrontal cortex[160] as well as the cerebellum,[157] the striatum,[153] and the hippocampus.[151,161]

Collectively, fluid shear stress and the effects of acute exercise on circulating triggers of angiogenesis, as well as their local induction through the activation of HIF-1α, represent a direct link between exercise and cerebral vascular angiogenesis. However, exercise also stimulates synaptogenesis and neurogenesis, both of which alter cerebral metabolic requirements chronically. The development of new capillaries is coupled to the formation of new neurons, particularly in the dentate gyrus.[149–151] Therefore, the effects of exercise training on neural plasticity and neurogenesis may be associated with simultaneous or subsequent angiogenesis to support changes in neural metabolism. The coupling between neurogenesis and angiogenesis may serve a functional role in maintaining or improving (relative to sedentary or diseased states) the neurovascular unit in newly developed nervous tissue.[162–164]

5. SUMMARY (SEE FIG. 4)

In rodents, neurogenesis and angiogenesis are evident in as few as 3 days after the onset of aerobic exercise.[161] It is important to note that, if continued learning opportunities do not coincide with the aerobic activity, such adaptations may recede.[165] Whether the same is true for humans remains unknown. Bridging the gap between rodent and human investigations is difficult because of differences in species as well as the methodological approaches used to study exercise-induced angiogenesis. In humans, tissue samples are difficult or impossible to obtain; therefore, advanced imaging techniques are used to examine angiogenesis in the brain. However, using such techniques, it is difficult to discern between changes in capillary, neural, astrocyte, or neuropil structures. Further, most studies in humans involve aged or patient populations, rendering it difficult to

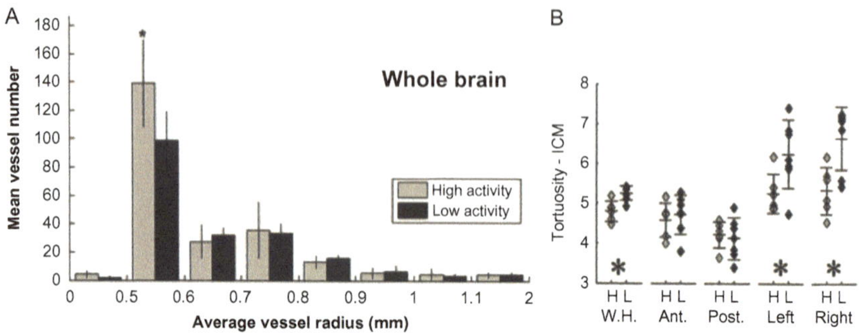

Figure 4 (A) Comparison of high activity (light gray) vs. low activity (dark) groups by the number and size of blood vessels in the whole brain (mean ± SD). (B) Comparing high activity (light gray) vs. low activity (dark) by the tortuosity of the cerebral vasculature (calculated using the inflection count metric (ICM) method; mean ± SD). H indicates high; L indicates low; W.H. indicates whole head; Ant. indicates anterior and Post. indicates posterior. *Significantly different between groups ($P < 0.05$). Older individuals belonging to the high activity group had a greater number of cerebral blood vessels with a radius of ~0.5 mm and less tortuous blood vessels in the left and right cerebral vasculatures. *Adapted from Ref. 171.*

describe normal healthy responses to exercise training. Despite the aforementioned limitations, animal and human studies demonstrate a degree of congruency.[151,166] Recent human investigations suggest that there is a correlation between increased aerobic fitness and greater hippocampal volumes in preadolescents[167] as well as older adults.[168] This appears to be a causative relationship as 3, 6, and 12 months of aerobic exercise training increases hippocampal blood volume progressively in otherwise healthy normal aged and older adults.[168–170] These observations may extend beyond the hippocampus, as aerobic fitness in older adults is also associated with less tortuous blood vessels and a greater number of small caliber blood vessels (radius of ~0.5 mm), throughout the whole brain.[171] Although it is too early to recommend specific exercise prescriptions to optimize cerebral vascular and brain health, preliminary evidence highlights that total exercise volume (i.e., total amount of time spent being physically active) may be more important than exercise intensity for older adults.[172–174] Also, because of the tight coupling between neural activity and acute and chronic cerebral vascular adaptations, exercise that requires global, diverse, and integrated neural output (i.e., exercise that involves acquiring complex and coordinated motor programs, employs a high degree of strategy, consists of repetitive as well as novel stimuli and engages numerous senses, such as taste, touch, scent, hearing, and sight) may elicit more widespread angiogenesis.

REFERENCES

1. Laughlin MH, Davis MJ, Secher NH, et al. Peripheral circulation. *Compr Physiol.* 2012;2(1):321–447.
2. Hiura M, Nariai T, Ishii K, et al. Changes in cerebral blood flow during steady-state cycling exercise: a study using oxygen-15-labeled water with PET. *J Cereb Blood Flow Metab.* 2014;34(3):389–396.
3. Duncker DJ, Bache RJ. Regulation of coronary blood flow during exercise. *Physiol Rev.* 2008;88(3):1009–1086.
4. Schmidt W, Endres M, Dimeo F, Jungehulsing GJ. Train the vessel, gain the brain: physical activity and vessel function and the impact on stroke prevention and outcome in cerebrovascular disease. *Cerebrovasc Dis.* 2013;35(4):303–312.
5. Andersen P, Saltin B. Maximal perfusion of skeletal muscle in man. *J Physiol.* 1985;366:233–249.
6. Richardson RS, Poole DC, Knight DR, et al. High muscle blood flow in man: is maximal O2 extraction compromised? *J Appl Physiol.* 1993;75(4):1911–1916.
7. Brodal P, Ingjer F, Hermansen L. Capillary untrained supply of skeletal muscle fibers and endurance-trained men in. *Am J Physiol Heart Circ Physiol.* 1977;232(6): H705–H712.
8. Andersen P, Henriksson J. Capillary supply of the quadriceps femoris muscle of man: adaptive response to exercise. *J Physiol.* 1977;270:677–690.
9. Laughlin MH, Dunker DJ, Bache RJ. *Control of Blood Flow to Cardiac and Skeletal Muscle during Exercise.* New York, NY: American Physiological Society and Oxford Press; 1996.
10. Laughlin MH, Oltman CL, Bowles DK. Exercise training-induced adaptations in the coronary circulation. *Med Sci Sports Exerc.* 1998;30(3):352–360.
11. Egginton S, Hudlick O. Selective long-term electrical stimulation of fast glycolytic fibres increases capillary supply but not oxidative enzyme activity in rat skeletal muscles. *Exp Physiol.* 2000;85(5):567–574.
12. Walther G, Nottin S, Karpoff L, Pérez-Martin A, Dauzat M, Obert P. Flow-mediated dilation and exercise-induced hyperaemia in highly trained athletes: comparison of the upper and lower limb vasculature. *Acta Physiol.* 2008;193:139–150.
13. Rowley NJ, Dawson EA, Birk GK, et al. Exercise and arterial adaptation in humans: uncoupling localized and systemic effects. *J Appl Physiol.* 2011;110(5):1190–1195.
14. Huonker M, Schmid A, Schmidt-Trucksass A, Grathwohl D, Keul J. Size and blood flow of central and peripheral arteries in highly trained able-bodied and disabled athletes. *J Appl Physiol.* 2003;95(2):685–691.
15. Thijssen DHJ, Rowley N, Padilla J, et al. Relationship between upper and lower limb conduit artery vasodilator function in humans. *J Appl Physiol.* 2011;111(1):244–250.
16. Van Royen N, Piek JJ, Buschmann I, Hoefer I, Voskuil M, Schaper W. Stimulation of arteriogenesis; a new concept for the treatment of arterial occlusive disease. *Cardiovasc Res.* 2001;49:543–553.
17. Koffi I, Safar ME, Labat C, Lacolley P, Benetos A, Mourad JJ. Arterial structural changes with verapamil in spontaneously hypertensive rats. *Am J Hypertens.* 1999;12:732–738.
18. Saleh FH, Jurjus AR. A comparative study of morphological changes in spontaneously hypertensive rats and normotensive Wistar Kyoto rats treated with an angiotensin-converting enzyme inhibitor or a calcium-channel blocker. *J Pathol.* 2006;193:415–420.
19. Furchgott RF. Role of endothelium in responses of vascular smooth muscle. *Circ Res.* 1983;53(5):557–573.
20. Kamiya A, Togawa T. Adaptive regulation of wall shear stress to flow change in the canine carotid artery. *Am J Physiol Heart Circ Physiol.* 1980;239:H14–H21.

21. Zarins CK, Zatina MA, Giddens DP, Ku DN, Glagov S. Shear stress regulation of artery lumen diameter in experimental atherogenesis. *J Vasc Surg*. 1987;5(3):413–420.
22. Glagov S, Zarins CK, Giddens DP, Ku DN. Hemodynamics and atherosclerosis: insights and perspectives gained from studies of human arteries. *Arch Pathol Lab Med*. 1988;112:1018–1031.
23. Lloyd PG, Prior BM, Yang HT, Terjung RL. Angiogenic growth factor expression in rat skeletal muscle in response to exercise training. *Am J Physiol Heart Circ Physiol*. 2003;284:H1668–H1678.
24. Prior BM, Yang HT, Terjung RL. What makes vessels grow with exercise training? *J Appl Physiol*. 2004;97(3):1119–1128.
25. Ferrara N. Role of vascular endothelial growth factor in regulation of physiological angiogenesis. *Am J Physiol Cell Physiol*. 2001;280:C1358–C1366.
26. Ferrara N, Gerber H-P, LeCouter J. The biology of VEGF and its receptors. *Nat Med*. 2003;9(6):669–676.
27. Park JE, Keller G, Ferrara N. The vascular endothelial growth factor (VEGF) isoforms: differential deposition into the subepithelial extracellular matrix and bioactivity of extracellular matrix-bound VEGF. *Mol Biol Cell*. 1993;4:1317–1326.
28. Olfert IM, Howlett RA, Tang K, et al. Muscle-specific VEGF deficiency greatly reduces exercise endurance in mice. *J Physiol*. 2009;587(Pt 8):1755–1767.
29. Neufeld G, Cohen T, Gengrinovitch S, Poltorak Z. Vascular endothelial growth factor (VEGF) and its receptors. *FASEB J*. 1999;13:9–22.
30. Hoier B, Hellsten Y. Exercise-induced capillary growth in human skeletal muscle and the dynamics of VEGF. *Microcirculation*. 2014;21(4):301–314.
31. Breen EC, Johnson EC, Wagner H, Tseng HM, Sung LA, Wagner PD. Angiogenic growth factor mRNA responses in muscle to a single bout of exercise. *J Appl Physiol*. 1996;81(1):355–361.
32. Gustafsson T, Puntschart A, Kaijser L, Jansson E, Sundberg CJ. Exercise-induced expression of angiogenesis-related transcription and growth factors in human skeletal muscle. *Am J Physiol Heart Circ Physiol*. 1999;276(45):H679–H685.
33. Richardson RS, Wagner H, Mudaliar SRD, Henry R, Noyszewski EA, Wagner PD. Human VEGF gene expression in skeletal muscle: effect of acute normoxic and hypoxic exercise. *Am J Physiol Heart Circ Physiol*. 1999;277(46):H2247–H2252.
34. Amaral SL, Linderman JR, Morse MM, Greene AS. Angiogenesis induced by electrical stimulation is mediated by angiotensin II and VEGF. *Microcirculation*. 2001;8:57–67.
35. Tomanek RJ, Schatteman GC. Angiogenesis: new insights and therapeutic potential. *Anat Rec*. 2000;261:126–135.
36. Pugh CW, Ratcliffe PJ. Regulation of angiogenesis by hypoxia: role of the HIF system. *Nat Med*. 2003;9(6):677–684.
37. Shweiki D, Itin A, Soffer D, Keshet E. Vascular endothelial growth factor induced by hypoxia may mediate hypoxia-initiated angiogenesis. *Nature*. 1992;359(29):843–845.
38. Roberts KC, Nixon C, Unthank JL, Lash JM. Femoral artery ligation stimulates capillary growth and limits training-induced increases in oxidative capacity in rats. *Microcirculation*. 1997;4(2):253–260.
39. Semenza GL. Vasculogenesis, angiogenesis, and arteriogenesis: mechanisms of blood vessel formation and remodeling. *J Cell Biochem*. 2007;102:840–847.
40. Milkiewicz M, Hudlicka O, Verhaeg J, Egginton S, Brown MD. Differential expression of Flk-1 and Flt-1 in rat skeletal muscle in response to chronic ischaemia: favourable effect of muscle activity. *Clin Sci*. 2003;105:473–482.
41. Gigante B, Tarsitano M, Cimini V, De Falco S, Persico MG. Placenta growth factor is not required for exercise-induced angiogenesis. *Angiogenesis*. 2004;7(3):277–284.
42. Suri C, Jones PF, Patan S, et al. Requisite role of angiopoietin-1, a ligand for the TIE2 receptor, during embryonic angiogenesis. *Cell*. 1996;87(7):1171–1180.

43. Maisonpierre PC, Suir C, Jones PF, et al. Angiopoietin-2, a natural antagonist for Tie2 that disrupts in vivo angiogenesis. *Science*. 1997;277(5322):55–60.
44. Holash J, Maisonpierre PC, Compton D, et al. Vessel cooption, regression, and growth in tumors mediated by angiopoietins and VEGF. *Science*. 1999;284:1994–1998.
45. O'Reilly MS, Holmgren L, Shing Y, et al. Angiostatin: a novel angiogenesis inhibitor that mediates the suppression of metastases by a Lewis lung carcinoma. *Cell*. 1994;79:315–328.
46. Bailey AP, Shparago M, Gu J-W. Exercise increases soluble vascular endothelial growth factor receptor-1 (sFlt-1) in circulation of healthy volunteers. *Med Sci Monit*. 2006;12(2):CR45–CR50.
47. Wang JS, Jen CJ, Kung HC, Lin LJ, Hsiue TR, Chen HI. Different effects of strenuous exercise and moderate exercise on platelet function in men. *Circulation*. 1994;90(6):2877–2885.
48. Drygas WK. Changes in blood platelet function, coagulation, and fibrinolytic activity in response to moderate, exhaustive, and prolonged exercise. *Int J Sports Med*. 1988;9(1):67–72.
49. Ponjee GA, Janssen GM, van Wersch JW. Prolonged endurance exercise and blood coagulation: a 9 month prospective study. *Blood Coagul Fibrinolysis*. 1993;4(1):21–25.
50. Malek MH, Olfert IM. Global deletion of thrombospondin-1 increases cardiac and skeletal muscle capillarity and exercise capacity in mice. *Exp Physiol*. 2009;94(6):749–760.
51. Olfert IM, Breen EC, Gavin TP, Wagner PD. Temporal thrombospondin-1 mRNA response in skeletal muscle exposed to acute and chronic exercise. *Growth Factors*. 2006;24(4):253–259.
52. Slopack D, Roudier E, Liu STK, Nwadozi E, Birot O, Haas TL. Forkhead BoxO transcription factors restrain exercise-induced angiogenesis. *J Physiol*. 2014;592(Pt 18):4069–4082.
53. Laughlin MH, Roseguini B. Mechanisms for exercise training-induced increases in skeletal muscle blood flow capacity: differences with interval sprint training versus aerobic endurance training. *J Physiol Pharmacol*. 2008;57(suppl 7):71–88.
54. Laughlin MH, Ripperger J. Vascular transport capacity of hindlimb muscles of exercise-trained rats. *J Appl Physiol*. 1987;62:438–443.
55. Laughlin MH, Cook JD, Tremble R, Ingram D, Colleran PN, Turk JR. Exercise training produces nonuniform increases in arteriolar density of rat soleus and gastrocnemius muscle. *Microcirculation*. 2006;13(3):175–186.
56. Laughlin MH. Skeletal muscle blood flow capacity: role of muscle pump in exercise hyperemia. *Am J Physiol Heart Circ Physiol*. 1987;253(5 Pt 2):H993–H1004.
57. Laughlin MH, Korthuis RJ, Sexton WL, Armstrong RB. Regional muscle blood flow capacity and exercise hyperemia in high-intensity trained rats. *J Appl Physiol*. 1988;64(6):2420–2427.
58. Sexton WL, Korthuis RJ, Laughlin MH. High-intensity exercise training increases vascular transport capacity of rat hindquarters. *Am J Physiol Heart Circ Physiol*. 1988;254(2 Pt 2):H274–H278.
59. Sexton WL, Laughlin MH. Influence of endurance exercise training on distribution of vascular adaptations in rat skeletal muscle. *Am J Physiol Heart Circ Physiol*. 1994;266(2 Pt 2):H483–HH490.
60. Gute D, Fraga C, Laughlin MH, Amann JF. Regional changes in capillary supply in skeletal muscle of high-intensity endurance-trained rats. *J Appl Physiol (1985)*. 1996;81(2):619–626.
61. Lloyd PG, Yang HT, Terjung RL. Arteriogenesis and angiogenesis in rat ischemic hindlimb: role of nitric oxide. *Am J Physiol Heart Circ Physiol*. 2001;281: H2528–H2538.

62. Yang HT, Ogilvie RW, Terjung RL. Exercise training enhances basic fibroblast growth factor-induced collateral blood flow. *Am J Physiol Heart Circ Physiol.* 1998;274(43):H2053–H2061.
63. Opie LH, Lopaschuk GD. Fuels: aerobic and anaerobic metabolism. In: Weinberg RW, Bersin J, Aversa F, eds. *Heart Physiology: From Cell to Circulation.* Philadelphia, PA: Lippincott Williams & Wilkins; 2004:306–354.
64. Feigl EO. Coronary physiology. *Physiol Rev.* 1983;63(1):1–205.
65. Laughlin MH, Korthuis RJ, Dunker DJ, Bache RJ. Control of blood flow to cardiac and skeletal muscle during exercise. In: Rowell LB, Shepherd TJ, eds. *Handbook of Physiology, Section 12.* Vol. 16. Bethesda, MD: Oxford Univ Press; 1996:705–769.
66. Laughlin HM, Tomanek RJ. Myocardial capillarity and maximal capacity in exercise-trained dogs. *J Appl Physiol.* 1987;63(4):1481–1486.
67. Heiss HW, Topfer M, Barmeyer J, Wink K, Huber G, Keul J. Studies on the regulation of myocardial blood flow in man II. effects of acute arterial hypoxia. *Clin Cardiol.* 1978;42(1):35–42.
68. Holmberg S, Serzysko W, Varnauskas E. Coronary circulation during heavy exercise in control subjects and patients with coronary heart disease. *Acta Med Scand.* 1971;190(6):465–480.
69. Kitamura K, Charles RJ, Fredarick LG, Taylor HL, Yang W. Hemodynamic correlates of myocardial oxygen consumption during upright exercise. *J Appl Physiol.* 1972;32(4):516–522.
70. Zandrino F, Molinari G, Smeraldi A, Odaglia G, Masperone MA, Sardanelli F. Magnetic resonance imaging of athlete's heart: myocardial mass, left ventricular function, and cross-sectional area of the coronary arteries. *Eur Radiol.* 2000;10:319–325.
71. Kozakova M, Galetta F, Gregorini L, et al. Coronary vasodilator capacity and epicardial vessel remodeling in physiological and hypertensive hypertrophy. *Hypertension.* 2000;36(3):343–349.
72. Tepperman J, Pearlman D. Effects of exercise and anemia on coronary arteries of small animals as revealed by the corrosion-cast technique. *Circ Res.* 1961;9(3):576–584. http://dx.doi.org/10.1161/01.RES.9.3.576.
73. Stevenson JAF, Feleki V, Rechnitzer P, Beaton JR. Effect of exercise on coronary tree size in the rat. *Circ Res.* 1964;15(3):265–269.
74. Wyatt HL, Mitchell J. Influences of physical conditioning and deconditioning on coronary vasculature of dogs. *J Appl Physiol Respir Environ Exerc Physiol.* 1978;45(4): 619–625.
75. Kramsch DM, Aspen AJ, Abramowitz BM, Kreimendahi T, Hood W. Reduction of coronary atherosclerosis by moderate conditioning exercise in monkeys on an atherogenic diet. *N Engl J Med.* 1981;305(25):1483–1489.
76. Windecker S, Allemann Y, Billinger M, et al. Effect of endurance training on coronary artery size and function in healthy men: an invasive followup study. *Am J Physiol Heart Circ Physiol.* 2002;282:2216–2223.
77. Kozakova M, Paterni M, Bartolomucci F, et al. Epicardial coronary artery size in hypertensive and physiologic left ventricular hypertrophy. *Am J Hypertens.* 2007;20(3): 279–284.
78. White FC, Bloor CM, McKirnan MD, Carroll SM. Exercise training in swine promotes growth of arteriolar bed and capillary angiogenesis in heart. *J Appl Physiol.* 1998;85(3):1160–1168.
79. Laughlin MH, McAllister RM. Exercise training-induced coronary vascular adaptations. *J Appl Physiol.* 1992;73:2209–2225.
80. Cai W, Schaper W. Mechanisms of arteriogenesis. *Acta Biochim Biophys Sin (Shanghai).* 2008;40(8):681–692.

81. Petren T, Sjostrand T, Sylven B. Der influss des trainings auf die haufigkeit der cappilaren in herz- und skeltenmuskulature. *Arbeitsphysiologie.* 1930;9:367–386.
82. Petren T, Sylven B. Weitere untersuchungen ubder den einfluss des trainings auf die kappillarisierung der hermuskulator. *Morphol Jahrb.* 1937;80:439–444.
83. Thorner W. Trainingsversuche an hunden. III. Histologische beobachtungen und herz- und skeletmuskeln. *Arbeitsphysiologie.* 1935;8:359–370.
84. Wachtlova M, Rakusan K, Poupa O. The coronary terminal vascular bed in the heart of the hare (Lepus europus) and the rabbit (Oryctolagus domesticus). *Physiol Bohemoslov.* 1965;14:328–331.
85. Wachtlova M, Rakusan K, Roth Z, Poupa O. The terminal vascular bed in the myocardium in the wild rat (Rattus norvegicus) and the laboratory rat (Rattus norvegicus lab). *Physiol Bohemoslov.* 1967;16:548–554.
86. Tomanek J. Effects of age and exercise on the extent of the myocardial capillary bed. *Anat Rec.* 1970;167:55–62.
87. Bloor CM, Leon A. Interaction of age and exercise on the heart and its blood supply. *Lab Invest.* 1970;22:160–165.
88. Hakkila J. Studies on the myocardial capillary concentration in cardiac hypertrophy due to training; an experimental study with guinea pigs. *Ann Med Exp Biol Fenn.* 1955;33:1–82.
89. Breisch EA, White FC, Nimmo E, Mckirnan MD, Bloor CM. Exercise-induced cardiac hypertrophy: a correlation of blood flow and microvasculature. *J Appl Physiol.* 1986;60(4):1259–1267.
90. Matsunaga T, Warltier DC, Tessmer J, Weihrauch D, Simons M, Chilian WM. Expression of VEGF and angiopoietins-1 and -2 during ischemia-induced coronary angiogenesis. *Am J Physiol Heart Circ Physiol.* 2003;285:352–358.
91. Shiojima I, Walsh K. Regulation of cardiac growth and coronary angiogenesis by the Akt/PKB signaling pathway. *Genes Dev.* 2006;20(24):3347–3365.
92. Rogers PJ, Miller TD, Bauer A, et al. Exercise training and responsiveness of isolated coronary arteries. *J Appl Physiol.* 1991;71(6):2346–2351.
93. Oltman CL, Parker LJ, Adams HR, Harold MH. Effects of exercise training on vasomotor of porcine coronary arteries. *Am J Physiol Heart Circ Physiol.* 1992;263: H372–H382.
94. Barbier J, Reland S, Ville N, Rannou-Bekono F, Wong S, Carré F. The effects of exercise training on myocardial adrenergic and muscarinic receptors. *Clin Auton Res.* 2006;16(1):61–65.
95. Blomqvist CG. Cardiovascular adaptations to physical training. *Annu Rev Plant Physiol Plant Mol Biol.* 1983;45:169–189.
96. DiCarlo SE, Blair RW, Bishop VS, Stone HL. Daily exercise enhances coronary resistance vessel sensitivity to pharmacological activation. *J Appl Physiol.* 1989;66(1):421–428.
97. Lowell H. Coronary vascular response to adrenergic stimulation in exercise-conditioned dogs. *J Appl Physiol Respir Environ Exerc Physiol.* 1984;52(2):315–320.
98. Scheel KW, Ingram LA, Wilson JL. Effects of exercise on the coronary and collateral vasculature of beagles with and without coronary occlusion. *Circ Res.* 1981;48(4):523–530.
99. Muller JM, Myers PR, Laughlin MH. Vasodilator responses of coronary resistance arteries of exercise-trained pigs. *Circulation.* 1994;89(5):2308–2314.
100. Bowles DK, Hu Q, Laughlin MH, Sturek M. Exercise training increases L-type calcium current density in coronary smooth muscle. *Med Sci Sport Exerc.* 1998;30(suppl):39.
101. Oltman L, Parker L, Harold M, Parker JL. Endothelium-dependent vasodilation coronary arteries from exercise-trained of proximal pigs. *J Appl Physiol.* 1995;79(1):33–40.

102. Laughlin MH, Pollock JS, Amann JF, Hollis CR, Woodman CR, Price ME. Training induces nonuniform increases in eNOS content along the coronary arterial tree. *J Appl Physiol.* 2014;90:501–510.
103. Jones AW, Rubin LJ, Magliola L. Endothelin-1 sensitivity of porcine coronary arteries is reduced by exercise training and is gender dependent. *J Appl Physiol.* 2014;87:1172–1177.
104. Laughlin MH. Endothelium-mediated control of coronary vascular tone after chronic exercise training. *Med Sci Sports Exerc.* 1995;27(8):1135–1144.
105. Jasperse JL, Laughlin MH. Endothelial function and exercise training: evidence from studies using animal models. *Med Sci Sports Exerc.* 2009;38(3):445–454.
106. Brown AM, Ransom BR. Astrocyte glycogen and brain energy metabolism. *Glia.* 2007;55:1263–1271.
107. Kandel E, Schwartz J, Jessell T. Principles of Neural Science. 4th ed. New York, USA: McGraw Hill; 2000.
108. Howarth C, Gleeson P, Attwell D. Updated energy budgets for neural computation in the neocortex and cerebellum. *J Cereb Blood Flow Metab.* 2012;32(7):1222–1232.
109. Poulsen PH, Smith DF, Ostergaard L, et al. In vivo estimation of cerebral blood flow, oxygen consumption and glucose metabolism in the pig by (15O) water injection, (15O) oxygen inhalation and (18F) dual injections of fluorodeoxyglucose. *J Neurosci Methods.* 1997;77:199–209.
110. Frackowiak RSJ, Herold S, Petty RKH, Morgan-Hughes JA. The cerebral metabolism of glucose and oxygen measured with positron tomography in patients with mitochondrial disease. *Brain.* 1988;111(5):1009–1024.
111. Buxton RB. Interpreting oxygenation-based neuroimaging signals: the importance and the challenge of understanding brain oxygen metabolism. *Front Neuroenerget.* 2010;2:1–16.
112. Raichle ME, Grubb Jr RL, Gado MH, Eichling JO, Ter-Pogossian MM. Correlation between regional cerebral blood flow and oxidative metabolism: in vivo studies in man. *Arch Neurol.* 1976;33:523–526.
113. Hamel E. Perivascular nerves and the regulation of cerebrovascular tone. *J Appl Physiol.* 2006;100(3):1059–1064.
114. Filosa JA, Blanco VM. Neurovascular coupling in the mammalian brain. *Exp Physiol.* 2007;92(4):641–646.
115. Bleys RLAW, Cowen T. Innervation of cerebral blood vessels: morphology, plasticity, age-related, and Alzheimer's disease-related neurodegeneration. *Microsc Res Tech.* 2001;53(2):106–118.
116. Ito H, Kanno I, Ibaraki M, Hatazawa J, Miura S. Changes in human cerebral blood flow and cerebral blood volume during hypercapnia and hypocapnia measured by positron emission tomography. *J Cereb Blood Flow Metab.* 2003;23(6):665–670.
117. Pardrige WM. Brain-type glucose transporter (GLUT-1) is selectively localized to the blood–brain barrier. *J Biol Chem.* 1990;265(29):18035–18040.
118. Duvernoy H, Delon S, Vannson JL. The vascularization of the human cerebellar cortex. *Brain Res Bull.* 1983;11(4):419–480.
119. Roy CS, Sherrington CS. On the regulation of the blood supply of the brain. *J Physiol.* 1890;11(1–2):85–108.
120. Lassen NA. Cerebral blood flow and oxygen consumption in man. *Physiol Rev.* 1959;39(2):183–238.
121. Mirro R, Lowery-Smith L, Armstead WM, Shibata M, Zuckerman SL, Leffler CW. Cerebral vasoconstriction in response to hypocapnia is maintained after ischemia/reperfusion injury in newborn pigs. *Stroke.* 1992;23(11):1613–1616.
122. Coverdale N, Gati JS, Opalevych O, Perrotta A, Shoemaker J. Cerebral blood flow velocity underestimates cerebral blood flow during modest hypercapnia and hypocapnia. *J Appl Physiol.* 2014;117(10):1090–1096.

123. Willie CK, Macleod DB, Shaw AD, et al. Regional brain blood flow in man during acute changes in arterial blood gases. *J Physiol.* 2012;590(14):3261–3275.
124. Manohar M. Regional distribution of brain blood flow during maximal exertion in splenectomized ponies. *Respir Physiol.* 1987;68:77–84.
125. Delp MD, Armstrong RB, Godfrey DA, Laughlin MH, Ross CD, Wilkerson MK. Exercise increases blood flow to locomotor, vestibular, cardiorespiratory and visual regions of the brain in miniature swine. *J Physiol.* 2001;533(3):849–859.
126. Ogoh S, Ainslie PN. Cerebral blood flow during exercise: mechanisms of regulation. *J Appl Physiol.* 2009;107(5):1370–1380. http://dx.doi.org/10.1152/japplphysiol.00573.2009.
127. Trangmar SJ, Chiesa ST, Stock CG, Kalsi KK, Secher NH, González-Alonso J. Dehydration affects cerebral blood flow but not its metabolic rate for oxygen during maximal exercise in trained humans. *J Physiol.* 2014;592(Pt 14):3143–3160.
128. Foreman L, Sanders M, Blair CM. Total and regional cerebral blood flow during moderate and severe exercise in miniature swine. *J Appl Physiol.* 1976;40(2):191–196.
129. Gross PM, Marcus L, Heistad D. Regional distribution of cerebral blood flow during exercise in dogs. *J Appl Physiol Respir Environ Exerc Physiol.* 1980;48(2):213–217.
130. Linkis P, Jorgensen LG, Madsen P, Lassen NA, Secher NH. Dynamic exercise enhances regional cerebral artery mean flow velocity. *J Appl Physiol.* 1995;78(1):12–16.
131. Sato K, Sadamoto T. Different blood flow responses to dynamic exercise between internal carotid and vertebral arteries in women. *J Appl Physiol.* 2010;109(3):864–869.
132. Krogh A, Lindhard L. The regulation of respiration and circulation during the initial stages of muscular work. *J Physiol.* 1913;47(1–2):112–136.
133. Nowak M, Holm S, Biering-Sørensen F, Secher NH, Friberg L. "Central command" and insular activation during attempted foot lifting in paraplegic humans. *Hum Brain Mapp.* 2005;25(2):259–265.
134. Williamson JW. The relevance of central command for the neural cardiovascular control of exercise. *Exp Physiol.* 2010;95(11):1043–1048. http://dx.doi.org/10.1113/expphysiol.2009.051870.
135. Nowak M, Olsen KS, Law I, Jolm S, Paulson OB, Secher NH. Command-related distribution of regional cerebral blood flow during attempted handgrip. *J Appl Physiol.* 1999;86:819–824.
136. Thornton JM, Guz A, Murphy K, et al. Identification of higher brain centres that may encode the cardiorespiratory response to exercise in humans. *J Physiol.* 2014;533.3(2001):823–836.
137. Herholz K, Buskies W, Rist M, Pawlik G, Hollman W, Heiss WD. Regional cerebral blood flow in man at rest and during exercise. *J Neurol.* 1987;234:9–13.
138. Madsen PL, Sperling BK, Warming T, et al. Middle cerebral artery blood velocity and cerebral blood flow and oxygen, uptake during dynamic exercise. *J Appl Physiol.* 1993;74(1):245–250.
139. Vissing J, Andersen M, Diemer NH. Exercise-induced changes in local cerebral glucose utilization in the rat. *J Cereb Blood Flow Metab.* 1996;16(4):729–736.
140. Willie CK, Cowan EC, Ainslie PN, et al. Neurovascular coupling and distribution of cerebral blood flow during exercise. *J Neurosci Methods.* 2011;198(2):270–273.
141. Fink GR, Adams L, Watson JDG, et al. Hyperpnoea during and immediately after exercise in man: evidence of motor cortical involvement. *J Physiol.* 1995;489(3):663–675.
142. Williamson JW, Nobrega ACL, Mccoll R, et al. Activation of the insular cortex during dynamic exercise in humans. *J Physiol.* 1997;503(2):277–283.
143. Hellstrom G, Fischer-Colbrie NG, Wahlgren NG, Jogestrand T. Carotid artery blood flow and middle cerebral artery blood flow velocity during physical exercise. *J Appl Physiol.* 1996;81:413–418.

144. Secher NH, Seifert T, Van Lieshout JJ. Cerebral blood flow and metabolism during exercise: implications for fatigue. *J Appl Physiol.* 2007;104:306–314.
145. Shimoda LA, Norins NA, Jeutter DC, Madden JA. Flow-induced responses in piglet isolated cerebral arteries. *Pediatr Res.* 1996;39(4 Pt1):574–583.
146. Schierling W, Troidl K, Mueller C, et al. Increased intravascular flow rate triggers cerebral arteriogenesis. *J Cereb Blood Flow Metab.* 2009;29(4):726–737.
147. Bolduc V, Thorin-Trescases N, Thorin E. Endothelium-dependent control of cerebrovascular functions through age: exercise for healthy cerebrovascular aging. *Am J Physiol Heart Circ Physiol.* 2013;305(5):H620–H633.
148. Zhang P, Yu H, Zhou N, et al. Early exercise improves cerebral blood flow through increased angiogenesis in experimental stroke rat model. *J Neuroeng Rehabil.* 2013;10(43):1–10.
149. Lopez-Lopez C, LeRoith D, Torres-Aleman I. Insulin-like growth factor I is required for vessel remodeling in the adult brain. *Proc Natl Acad Sci USA.* 2004;101(26):9833–9838.
150. Ding Q, Vaynman S, Akhavan M, Ying Z, Gomez-Pinilla F. Insulin-like growth factor I interfaces with brain-derived neurotrophic factor-mediated synaptic plasticity to modulate aspects of exercise-induced cognitive function. *Neuroscience.* 2006;140(3):823–833.
151. Pereira AC, Huddleston DE, Brickman AM, et al. An in vivo correlate of exercise-induced neurogenesis in the adult dentate gyrus. *Proc Natl Acad Sci USA.* 2007;104(13):5638–5643.
152. Pinilla GF. Spatial learning and physical activity contribute to the induction of fibroblast growth factor: neural substrates for increased cognition associated with exercise. *Neuroscience.* 1998;85(1):53–61.
153. Ding Y, Li J, Zhou Y, Rafols J, Clark J. Cerebral angiogenesis and expression of angiogenic factors in aging rats after exercise. *Curr Neurovasc Res.* 2006;3(1):15–23.
154. Gao Y, Zhao Y, Pan J, et al. Treadmill exercise promotes angiogenesis in the ischemic penumbra of rat brains through caveolin-1/VEGF signaling pathways. *Brain Res.* 2014;1585(109):83–90.
155. Gertz K, Priller J, Kronenberg G, et al. Physical activity improves long-term stroke outcome via endothelial nitric oxide synthase-dependent augmentation of neovascularization and cerebral blood flow. *Circ Res.* 2006;99(10):1132–1140.
156. Fabel K, Tam B, Kaufer D, et al. VEGF is necessary for exercise-induced adult hippocampal neurogenesis. *Eur J Neurosci.* 2003;18:2803–2812.
157. Isaacs KR, Anderson BJ, Alcantara AA, Black JE, Greenough WT. Exercise and the brain: angiogenesis in the adult rat cerebellum after vigorous physical activity and motor skill learning. *J Cereb Blood Flow Metab.* 1992;12(1):110–119.
158. Black JE, Isaacs KR, Anderson BJ, Alcantara AA, Greenough WT. Learning causes synaptogenesis, whereas motor activity causes angiogenesis, in cerebellar cortex of adult rats. *Proc Natl Acad Sci USA.* 1990;87:5568–5572.
159. Swain RA, Harris AB, Wiener EC, et al. Prolonged exercise induces angiogenesis and increases cerebral blood volume in primary motor cortex of the rat. *Neuroscience.* 2003;117(4):1037–1046.
160. Ekstrand J, Hellsten J, Tingström A. Environmental enrichment, exercise and corticosterone affect endothelial cell proliferation in adult rat hippocampus and prefrontal cortex. *Neurosci Lett.* 2008;442(3):203–207.
161. Van der Borght K, Kóbor-Nyakas DE, Klauke K, et al. Physical exercise leads to rapid adaptations in hippocampal vasculature: temporal dynamics and relationship to cell proliferation and neurogenesis. *Hippocampus.* 2009;19(10):928–936.
162. Palmer TD, Willhoite AR, Gage FH. Vascular niche for adult hippocampal neurogenesis. *J Comp Neurol.* 2000;494:479–494.

163. Teng H, Zhang ZG, Wang L, et al. Coupling of angiogenesis and neurogenesis in cultured and endothelial cells and neural progenitor cells after stroke. *J Cereb Blood Flow Metab*. 2009;28(4):764–771.
164. Lange-Asschenfeldt C, Kojda G. Alzheimer's disease, cerebrovascular dysfunction and the benefits of exercise: from vessels to neurons. *Exp Gerontol*. 2008;43(6):499–504.
165. Van Praag H, Shubert T, Zhao C, Gage FH. Exercise enhances learning and hippocampal neurogenesis in aged mice. *J Neurosci*. 2005;25(38):8680–8685.
166. Thomas AG, Dennis A, Bandettini PA, Johansen-Berg H. The effects of aerobic activity on brain structure. *Front Psychol*. 2012;3:86.
167. Chaddock L, Erickson KI, Prakash RS, et al. A neuroimaging investigation of the association between aerobic fitness, hippocampal volume, and memory performance in preadolescent children. *Brain Res*. 2010;1358:172–183.
168. Erickson KI, Prakash RS, Voss MW, et al. Aerobic fitness is associated with hippocampal volume in elderly humans. *Hippocampus*. 2009;19(10):1030–1039.
169. Erickson KI, Voss MW, Prakash RS, et al. Exercise training increases size of hippocampus and improves memory. *Proc Natl Acad Sci USA*. 2011;108(7):3017–3022.
170. Pajonk FG, Wobrock T, Gruber O, et al. Hippocampal plasticity in response to exercise in schizophrenia. *Arch Gen Psychiatry*. 2010;67(2):133–143.
171. Bullitt E, Rahman FN, Smith JK, et al. The effect of exercise on the cerebral vasculature of healthy aged subjects as visualized by MR angiography. *AJNR Am J Neuroradiol*. 2009;30(10):1857–1863.
172. Liu-Ambrose T, Nagamatsu LS, Voss MW, Khan KM, Handy TC. Resistance training and functional plasticity of the aging brain: a 12-month randomized controlled trial. *Neurobiol Aging*. 2012;33(8):1690–1698.
173. Erickson KI, Raji CA, Lopez OL, et al. Physical activity predicts gray matter volume in late adulthood: the Cardiovascular Health Study. *Neurology*. 2010;75(16):1415–1422.
174. Ruscheweyh R, Willemer C, Krüger K, et al. Physical activity and memory functions: an interventional study. *Neurobiol Aging*. 2011;32(7):1304–1319.

CHAPTER ELEVEN

Exercise and Regulation of Bone and Collagen Tissue Biology

Michael Kjaer*^,1, Niklas Rye Jørgensen†, Katja Heinemeier‡, S. Peter Magnusson§

*Institute of Sports Medicine, Department of Orthopedic Surgery, Bispebjerg Hospital and Centre for Healthy Aging, Faculty of Health and Medical Sciences, University of Copenhagen, Copenhagen, Denmark
†Department of Clinical Chemistry, Glostrup Hospital and University of Southern Denmark, Glostrup, Denmark
‡Institute of Biomedical Sciences, Faculty of Health and Medical Sciences and Institute of Sports Medicine, Bispebjerg Hospital, Copenhagen, Denmark
§Musculoskeletal Rehabilitation Research Unit, Department of Physiotherapy and Institute of Sports Medicine, Bispebjerg Hospital, Faculty of Health and Medical Sciences, University of Copenhagen, Copenhagen, Denmark
[1]Corresponding author: e-mail address: michaelkjaer@sund.ku.dk

Contents

Abstract

The musculoskeletal system and its connective tissue include the intramuscular connective tissue, the myotendinous junction, the tendon, the joints with their cartilage and ligaments, and the bone; they all together play a crucial role in maintaining the architecture of the skeletal muscle, ensuring force transmission, storing energy, protecting joint surface and stability, and ensuring the transfer of muscular forces into resulting limb movement. The musculoskeletal connective tissue structure is relatively stable,

Progress in Molecular Biology and Translational Science, Volume 135
ISSN 1877-1173
http://dx.doi.org/10.1016/bs.pmbts.2015.07.008

but mechanical loading and subsequent mechanotransduction and molecular anabolic signaling can result in some adaptation of the connective tissue, its size, its strength, and its mechanical properties, whereby it can improve its capacity by 5–20% with regular physical activity. For several of the mechanically loaded connective tissues, only limited information regarding molecular and cellular signaling pathways and their adaptation to exercise is available. In contrast to tissue responses with exercise, lack of mechanical tissue loading through inactivity or immobilization of the human body will result in a dramatic loss of connective tissue content, structure, and tolerable load within weeks, to a degree (30–40%) that mimics that of contractile skeletal musculature. This illustrates the importance of regular mechanical load in order to preserve the stabilizing role of the connective tissue for the overall function of the musculoskeletal system in both daily activity and exercise.

1. INTRODUCTION

Contractile activity of skeletal muscle cells is a prerequisite for all human physical activity but, at the same time, it is fully dependent on collagen tissue located in the musculoskeletal system in order to result in *de facto* limb movement and locomotor activity. This ranges from the importance of the intramuscular connective tissue (IMCT) in both maintaining the architecture of the skeletal muscle in force transmission between muscle fibers and in passive stiffness of musculature[1] to the role of tendon structures in force transmission and energy storage,[2,3] the importance of ligaments and cartilage in the protection of joint surfaces and joint stability, and the role of bone in ensuring the transfer of muscular forces into resulting limb movement as well as in maintaining the overall bodily structure and position (Fig. 1). Similar to the well-described interaction between the nervous system activity and initiation of muscle contraction, the connective tissues of the musculoskeletal system interact intimately with the contracting muscle cell in a coordinated manner to ensure force transmission resulting in movement and exercise activity.

Bone and the other collagen tissues (IMCT, tendon, ligament, cartilage) share both compositional and adaptive characteristics, and at the same time they differ markedly from—but have some interplay with—changes in contractile skeletal muscle cells (Table 1). First of all, the different collagen tissues contain a large amount of collagen, the most frequent molecule in the human body. Whereas bone, tendon, and ligaments are dominated by collagen type I (and III), cartilage is dominated by collagen type II. Each of the connective tissue types in the musculoskeletal system displays its own unique

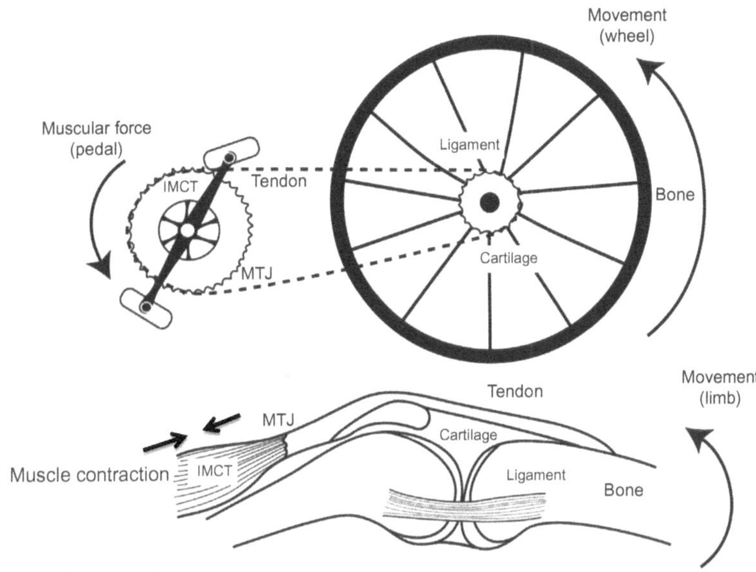

MTJ—myotendinous junction
IMCT—intramuscular connective tissue

Figure 1 A schematic representation of the connective tissue in the musculoskeletal system. The analogy to the transformation of force applied to bicycle pedals, through the chain and the resulting wheel turn, is similar to the muscle cell contraction, the force transmission through the intramuscular connective tissue (IMCT), the myotendinous junction (MTJ), the tendon, over the ligament-stabilized joint with its cartilage, and to the bone tissue, resulting in limb movement. (See the color plate.)

compositional characteristics, whether it be calcification (like in bone), hydrophilic water binding (like in cartilage), aligned fibril structure (like in tendon and ligament), or multidirectional arrangement of fibrils (like in the IMCT). Second, the turnover and thus the renewal of tissue is somewhat slower in the different collagen-rich connective tissues in the musculoskeletal system compared to that of contractile proteins in the skeletal muscle cell itself. However, this does not imply that the connective tissue is not renewed at all, but rather indicates that, in a physiological situation, it renews somewhat slower, and maybe only renews and turns over specific parts of the connective tissue, thus leaving some of the more basic structures of the connective tissue untouched. Third, acute changes in connective tissue as a response to acute bouts of exercise may be more moderate than in skeletal muscle cells. This does not rule out any acute upregulation of molecular signaling pathways by physical activity, but implies a more pronounced stability in its structure where acute bouts of physiological exercise do not

Table 1 Connective Tissues of the Musculoskeletal System and Their Adaptation to Exercise

Tissue	Structure	Major Molecular Components	Regulatory Factors	Tolerable Tissue Load	Improvement with Physical Training	Decline with Inactivity
Bone	Mineralized collagen fiber arranged in specialized structure. Trabecular and cortical bone	Collagen I (90%), (Coll III, V) Hydroxyapatite (calcium, phosphate) BMP	Tensile compress Estrogen PTH PGE	Cortical bone: 130 (tensile) – 200 MPa (compressive) Trabecular bone: <50 MPa	3–5 (−10)% (yrs) (mineralize, microarchitecture strength)	30% (w-m) (mineralize, microarchitecture strength)
Cartilage	Curved collagen fiber network and hydrophilic PGs	Coll II Aggrecan Biglycan COMP	Compress IL-4, IL-10	5–15 MPa (compressive)	5–10% (yrs) (thickness)	30% (w-m) (thickness)
Ligament	Collagen fibrils aligned unidirectionally	Coll I, III (V, XII) Elastin, PGs, decorin, fibronectin	Tensile IGF-I/GH	60–100 MPa (tensile)	20% (yrs) (strength)	30% (w-m) (strength)
Tendon	Collagen fibrils aligned in fascicles	Collagen I (80%), Coll III, V, VI, XII, PGs, fibronectin, elastin	Tensile IGF-I/GH Estrogen	100 MPa (tensile)	10–20% (yrs) (strength, thickness, stiffness (mths))	30% (w-m) (strength, stiffness)

Myotendinous junction (MTJ)	Ridge-like protrusions of tendon into muscle	Collagen I, III, VI, XXII Integrin a7b1	Tensile	50–100 MPa (tensile)	30%? (mths) (surface area)	30% (w-m) (surface area)
Intramuscular connective tissue (IMCT)	Wavy collagen fibrils around muscle cells	Collagen I, III, IV, VI, Coll XII + XIV + elastin (peri), fibronectin + decorin + laminin (endo) Tenascin-C (only exercise)	Tensile IGF-I/GH Estrogen	<100 MPa (tensile)	20%? (yrs) (strength)	30% (w-m) (strength)

An overview of the types and structures of connective tissue in the musculoskeletal system and their adaptation to chronic exercise or inactivity. Loading types are divided into tensile load or compressive load. Improvements in tissue strength/thickness/stiffness/surface area occur as a result of overtraining lasting several months (mths) to many years (yrs), whereas decline in the same parameters occur within some weeks to a few months (w-m). BMP, bone morphogenetic protein; Coll, collagen; COMP, cartilage oligomeric protein; IGF-I/GH, insulin-like growth factor I and growth hormone axis; MPa, mega Pascal; PGs, proteoglycans; w-m, weeks to months; yrs, years; "?" indicates best estimate of improvement with physical training, as this has not directly been measured.

constantly result in large net synthesis or loss of specific proteins. The reason for this might be that several of the connective tissue structures possess a high strength and safety margin toward breakage compared to the maximally developed muscle force. Fourth, the degree of adaptation to chronic exercise (physical training) is more moderate in the connective tissues compared to that seen in both morphology and metabolic capacity in the skeletal muscle cell. Although some differences exist between the different connective tissues subjected to mechanical loading, in regard to structural size or strength of the tissue, only a rise of 3–20% is seen even in response to long-term training over several years[4] (Table 1). This could imply that the strength and structure of connective tissue is already set at a high level and maintained so with even daily activity. Interestingly, the decrease in tissue magnitude (e.g., joint cartilage thickness), mechanical characteristics (e.g., stiffness of tendons), and strength (e.g., bone and ligaments) is quite pronounced with inactivity (often 20–40% within weeks), and thus paralleling the decline in muscle mass and function seen with body and limb immobilization (Table 1). This illustrates a large vulnerability of connective tissue toward lack of mechanical loading, with a resulting loss and deterioration of the matrix structures.

2. IMCT—TRAINING AND DETRAINING

Although the majority of the space in the human skeletal musculature is occupied by the muscle cell, around 5–10% of the space is occupied by IMCT.[5] The latter is classically divided into the endomysium (around the single muscle fiber), the perimysium (around bundles of muscle fibers, fascicles), and the epimysium/fascia (around the entire muscle). Whereas preparation of skeletal muscle in both animals and humans has provided elegant pictures of the structural outline of the IMCT,[6] much less is known regarding the function and cell–matrix interaction of this tissue, mostly due to limitations in isolating, analyzing, and mechanically testing the tissue in a way that represents its role *in vivo*. The force transmission from contracting muscle fibers to the adjacent tendon takes place both through the fibers themselves, as well as by lateral force transmission to adjacent parallel connective tissue network.[7–9] As it has been demonstrated that individual skeletal muscle fibers do not extend throughout the entire length of a given muscle, the muscle depends upon the IMCT in order to function optimally.[9] Primarily, this lateral force transmission depends upon the immediate contact between individual muscle fibers through the endomysium, but also the perimysium has been shown to play an important role.[10] Interestingly, the lateral force transmission has been shown to be diminished with aging, maybe contributing to the decline in muscular force in older age groups.[11] Further, in diseased states such as cerebral palsy, it has been demonstrated that the increased passive stiffness of the skeletal muscle of these patients is primarily due to increased stiffness of the IMCT, rather than to any change in passive stiffness of the skeletal muscle cell itself.[12] When discussing the passive stiffness of skeletal musculature and potential molecules important for this (e.g., titin), it is important to remember that when subjecting the whole skeletal muscle to stretch, the by far largest contributor to passive muscle stiffness at extreme muscle lengths is the IMCT.[5] The composition of the IMCT is dominated by collagen type I and III fibrillar structures, supplemented by a marked amount of collagen type IV being present in the endomysium due to its prominent role in the basal membrane. In the endomysium, the collagen fibrils are grouped in a multidirectional pattern to allow for flexibility in several ways when the muscle fiber is contracting or undergoes morphological changes with training (hypertrophy) or inactivity (atrophy). In the perimysium, "perimysial cables"[13] have been identified that represent highly organized longitudinal structures composed of 25–100 collagen fibrils in bundles running over distances of up to 150 μm within the matrix of the skeletal muscle.[5] Interestingly, the difference between individuals (animals) seems more pronounced for perimysium than for the endomysium, and thus, the perimysium may be the crucial factor for coordination of overall muscle force output for the individual.[14]

The muscle fiber itself contains many myonuclei, but it is important to remember that out of all nuclei of skeletal musculature, almost 50% are non-myonuclei, representing different tissue cells like endothelial cells, nerve cells, myogenic stem cells, and connective tissue cells like fibroblasts. The activity of the fibroblast in the IMCT in relation to exercise has been investigated in both animal and human models. With acute exercise, the protein synthesis of both myofibrillar protein and of collagen (in IMCT) rises approximately twofold and remains elevated for 24–48 h.[15] The upregulation of muscle contractile protein anabolic pathways can be shown to be both intensity and contraction type dependent, whereas the upregulation of matrix protein stimulating pathways to exercise seems more independent of intensity.[16] In humans, experiments within the same individuals who trained at one intensity in one leg and a different intensity in the other leg, the myofibrillar protein synthesis was markedly more elevated with intense exercise than with moderate intensity. In contrast, the exercise-stimulated rise in collagen protein synthesis for the IMCT was identical for both legs independent of training intensity, and this was supported by data on expression of matrix protein anabolic factors.[16] Key factors for stimulation of collagen synthesis in the IMCT are growth hormone/insulin-like growth factor 1 (GH/IGF-1), and in human studies with GH administration, a local upregulation of IGF-1 expression in skeletal muscle (and tendon) led to increase in collagen synthesis in skeletal muscle (and tendon).[17] Further, in animal models where GH is either overexpressed or its receptor is inhibited, a close positive relation between IGF-1 and collagen is observed in the IMCT.[18] This occurred despite no effect of GH administration upon myofibrillar protein synthesis, supporting the view that growth factors may differentiate between contractile muscle cell proteins and IMCT proteins. One hormonal factor that may influence both myofibrillar protein and collagen synthesis is female sex hormones, and in postmenopausal women, a stimulating effect of estradiol upon the exercise-induced rise in protein synthesis has been shown both on myofibrillar as well as on collagen protein in skeletal muscle.[19] The exact molecular and cellular signaling pathways and their adaptation of the IMCT in regard to mechanical loading are relatively unknown.

When heavy, unaccustomed exercise is carried out, delayed onset muscle soreness is developed and several days with muscle pain and reduced function are observed. In humans, stimulation of collagen synthesis in IMCT tissue occurs after heavy bouts of exercise,[20] and only when unfamiliar, electrically induced unphysiological exercise is performed so we find signs of damage in both the IMCT (disruption) and the muscle cell itself (Z-band disruption).[21] Interestingly, when resuming an unaccustomed activity up to several months later, the acute pain and soreness will be markedly reduced (repeated bout effect). In humans, the remodeling of the IMCT plays an important role in muscle regeneration after intense "muscle-damaging" exercise.[22] Whereas immediately after exercise, signs of inflammation (macrophages), injury (heat-shock protein, creatine kinase, Z-band), deadhesion of tissue (tenascin-C), and satellite cell activation are observed. Subsequently, up to 30 days after exercise, an elevation in numbers of macrophages, satellite cells, and factors regulating skeletal muscle matrix (IGF-I, TGF-beta, collagen type I, III, IV, and XII) is observed,[22,23] which seems to signify an attempt to improve connective tissue structure. It is very likely that remodeling of the IMCT is the key factor that explains the "muscle memory" of the intense exercise carried out up to months earlier.

In more severe muscle damage, the importance of not only muscle stem cells/satellite cells but also of fibroblasts has been demonstrated in animal models.[23a] Further, an important role of collagen (type VI) for muscle regeneration and growth has been demonstrated.[24] This supports the view that not only in physiological adaptations to exercise does the IMCT and its cells play an important role, but that fibroblasts of the IMCT are crucial for optimal muscle regeneration after injury. Further, in pathological conditions, the formation of intramuscular fibrosis (like muscle laceration, diabetes, cerebral palsy) also illustrates that a normal homeostasis and balance between the muscle cell and its surrounding connective tissue are important for optimal muscle function.

3. MYOTENDINOUS JUNCTION—TRAINING AND DETRAINING

The connection between muscle and tendon tissue is crucial for force transmission from muscle to tendon, and both in animal and human studies, it has been shown to be a primary site for tissue failure when subjected to strain.[25] Animal studies have demonstrated myotendinous junction (MTJ) to be tendon finger-like processes that merge into skeletal muscle,[26,27] and more lately human studies have confirmed tendon made ridge-like (rather than finger-like) protrusions that interdigitate with groove-like evaginations in the muscle cell.[28] The MTJ is known to consist of actin microfilaments, actin-binding proteins, proteins that link actin bundles to the sarcolemma, transmembrane proteins that link cytoskeletal elements to extracellular components, and proteins that link external lamina to the collagen fibril-rich matrix. Further, a high content of collagen dominated by collagen type I and III is present, and interestingly, the MTJ is the only location in the muscle tendon unit where collagen type XXII is present.[28,29] The exact role of collagen type XXII in relation to mechanical loading remains unclear, but it has been suggested that it is related to integrin alpha 7 expression, and thus to be mechanosensitive. Physical training is known—at least in animals—to cause an increase in the "folding" and thus extent of the area of the MTJ, which results in a larger contact surface area between muscle and tendon, and thus results in a higher tolerable mechanical load.[30] This is interesting, as it suggests multiple ways for the MTJ to improve its strength after physical training. First, the tolerable load may be improved due to contact area expansion between muscle and tendon. Second, a more shear-like rather than simple tensile stress due to the folding may also improve the tolerable tissue load further. Third, the ridge-like tendon form into the muscle cell may contribute to a "grab or trap effect" when muscle contractions shorten but thicken the muscle cell and thus "anchoring" the tendon parts into the muscle.[28] To what extent physical training influence the MTJ in humans has not been documented, and molecular and cellular adaptations remain undescribed.

4. TENDON AND LIGAMENT

Tendon tissue has an essential role in transmitting contractile forces to bone to generate movement, and is therefore uniquely designed to withstand considerable loads (up to ~8 times body weight) during human locomotion.[31,32] Tendon is organized in a strict hierarchical manner.[33] Collagen molecules are organized in a precise pattern to yield the characteristic 67-nm D-periodization that forms collagen fibrils. The collagen molecule is ~300 nm in length and 1.5 nm in diameter,[34] and aggregated molecules of the fibril are stabilized by covalent intramolecular cross-links.[35] The cross-links bind the collagen molecules to one another and thereby confer integrity on the fibril. Groups of fibrils then form fibers known as fascicle bundles, which finally comprise the tendon proper. There are at least 28 different collagen proteins, but tendon and ligament are predominantly made up of type I collagen.[36] The fibrillar collagen is embedded in a hydrophilic extracellular matrix (ECM) consisting of proteoglycans, glycoproteins, and glycosaminoglycans (GAGs), which are involved in the development, organization, and growth control of tendon.[4]

4.1 Force Transmission Within the Tendon

Although the tendon might be functionally regarded as a single force-transmitting structure, it remains unknown if force is transmitted homogeneously throughout the tendon, and therefore if the stress–strain on tendons is uniform. In fact, whether there is a "weak link" in the force transmission, and how it might adapt to loading conditions, remains an enigma. Fascicles from the anterior and posterior portion of the human patellar tendon display noticeably different mechanical properties,[37] and lateral force transmission between adjacent fascicles is relatively small, and therefore, the fascicles can be considered to be functionally independent structures.[38] The fact that sliding can occur between fascicles might be advantageous as, for example, tendons wrap around bones. The interfascicular space contains fibroblasts, capillaries, nerves, and small-diameter fibrils,[38] and it remains unknown if the structures in this space would be adversely affected by disproportionately large shear and/or possible focal adhesions.

The collagen fibril is considered the fundamental force-transmitting unit of the tendon and ligament,[39] although the actual length of fibrils in mature tissues remains an unresolved issue. This lack of fundamental knowledge precludes a detailed understanding of force transmission. In fact, suggestions that fibrils are continuous and discontinuous exist with currently no unequivocal proof of either proposal.[40] Discontinuous fibrils would require force to be transferred between adjacent fibrils, and functionally continuous fibrils would mean that the fibrils assume most of the tensile load. The proposal that fibrils are discontinuous implies that the proteoglycan-associated GAGs found on the surface of the collagen fibrils may be an important transmitter of load. However, it has been shown that the mechanical properties of isolated fascicles from human patellar tendon are unaffected after removal of GAGs.[41] These results suggest that GAGs cannot be considered important mediators of tensile force transmission and that force transmission must either take place through other matrix components or the fibrils must be mechanically continuous.

Fibril morphology has previously been associated with mechanical properties of collagen tissues, where a positive relationship was suggested between the "mass average diameter" of fibrils and ultimate tensile strength of the tissue.[39] Large-diameter fibrils have been thought to possess augmented tensile strength due to greater potential for intrafibrillar cross-links, and smaller fibrils are weaker but more creep resistant due to a higher surface area per unit mass, which allows for increased electrostatic interactions between fibrils and GAGs.[39] More recent data on human patellar tendon, however, suggest that material properties are not closely related to fibril morphology, *per se*.[42]

The tropocollagen molecule comprises three polypeptides arranged as a triple-helical structure stabilized by hydrogen bonds.[43] The collagen molecules are organized in a precise pattern and an important contributor to the mechanical properties of the tendon is the intermolecular cross-links.[35] During loading, the triple helix of the tropocollagen molecule is believed to uncoil, elongate, the gap between longitudinally adjoining molecules of the fibril might increase, or there might be a relative slippage between laterally adjacent molecules.[44–46] Individual collagen molecules have a fracture modulus that far exceeds that of the tendon fibril,[43] and are therefore unlikely the "weak link." However, it is unknown to what extent the proposed gliding mechanisms at the level of the molecule will affect the associated cross-links.[47]

Tendons can store and release energy during locomotion, while ligaments serve to provide joint stability. Although ligaments and tendons have these distinct functions, it appears that they are largely similar in structural and biochemical composition.[48,49] Minor dissimilarities exist in that ligaments have increased concentration of the immature cross-link dihydroxylysinonorleucine, a higher concentration of pyridinoline (36), and a larger number of small fibrils.[48,49]

4.2 Training of Tendon and Ligament

Knowledge of whether and how tendons adapt to increased loading has until recently been somewhat limited. The average tensile stress (force/area) exerted on the tendon will depend on its cross-sectional area (CSA). Therefore, for a given force a larger CSA will yield a reduction in stress and possibly confer a reduction in injury risk. Human tendons, including the patellar and Achilles tendons, typically have a fracture stress of ~100 MPa. However, most tendons are only subjected to stresses of up to 30 MPa,[50] which gives tendons a reasonable safety margin, although the Achilles tendon might experience stresses of up to ~70 MPa.[31]

In humans, cross-sectional data suggest that habitual long-distance running (>5 years) is associated with a markedly greater CSA (22%) of the Achilles tendon compared to that of nonrunners.[51] However, a total training stimulus of ~9 months of running in previously untrained subjects did not result in tendon hypertrophy of the Achilles tendon.[52] This lack of response may relate to the relative unresponsiveness of tendon tissue. On the other hand, it was recently shown that life-long endurance exercise was associated with a 30% greater CSA in elderly men, which lowers the stress on the tendon and potentially reduces the risk of injury.[53] The previously mentioned lack of response to 9 months of endurance training may also relate the absolute load: It has been shown that athletes that subject the Achilles tendon to intermittent high loads have a greater tendon cross-sectional tendon,[54] and it has also been shown that resistance training can result in tendon hypertrophy.[55] A greater tendon cross-section will yield a lower average stress on the tendon when loaded. Curiously, it appears that the tendon hypertrophy is largely regional such that the increase occurs primarily in the ends of the tendon.[54]

Tendon collagen fibrils are the basic force-transducing units of the tendon, and the morphology of the fibrils is commonly believed to largely influence the mechanical properties and function of the tendon. However, currently the picture is not coherent: in animal models, it has been shown that an increase in physical activity may decrease, increase, or leave the fibrils size unchanged.[56–58] In humans, it has been shown that fibril morphology is unaltered by life-long endurance training,[53] while it is influenced by heavy resistance training in patients with tendinopathy.[59]

The intermolecular cross-links may also be an important contributor to the properties of tendon tissue. The cross-links can be divided into two broad groups, the enzymatic and the nonenzymatic. Enzymatic cross-links are formed as a result of the lysyl oxidase (LOX) enzyme acting on the type I collagen molecule.[60] In tendons, the enzymatic cross-link composition changes with maturation, and it has been shown that it is modulated with heavy resistance training.[61] Nonenzymatic cross-links are formed when reducing sugars bind to amino acids and are referred to as advanced glycation end-product (AGE). AGE accumulation is dependent on collagen turnover rates in the specific tissue and therefore naturally collects during the aging process to a higher extent in tissue with low collagen turnover.[62] In tendon tissue, it has been shown that endurance training and resistance training may reduce AGE accumulation.[53,61]

4.3 Detraining of Tendon and Ligament

It is well known that the immobilization-associated drop in maximal muscle strength is greater than that of muscle mass and that both contribute to reduced muscle function.[63,64] Interestingly, recent studies report that unloading affects tendon properties to a greater extent than does muscle loss.[65] Some animal immobilization studies show that tendon stiffness is reduced without a change in tendon size,[66,67] although some have shown opposite results.[68] Human data in young individuals also show that mechanical properties of the tendon decrease without tendon atrophy.[65,69,70] Short-term immobilization has been reported to reduce tendon modulus by up to 30% in 23 days[65] and up to 58% after 90 days.[71] With respect to the fibril morphology, it has been shown that there is a decrease in the number and average diameter of collagen fibrils with immobilization.[72,73] In addition, the interfibrillar spacing has been reported to increase with immobilization.[74] The reduced fibril density may explain the loss of tendon material properties, while tendon CSA remains unchanged due to increased interstitial water and/or other ECM components.[65,75] It was recently shown in humans that a 2-week immobilization period resulted in reduced tendon stiffness, although the fibril morphology was unaffected. Such rapid changes in mechanical properties may be related to changes in enzymatically derived cross-links.[76] It has been proposed that tension on the tendon is a key factor in maintaining homeostasis,[77] and this rapid change in mechanical properties in response to inactivity underscores this notion.

4.4 Collagen Synthesis and Turnover—Regulation in Tendon

As noted above, tendon mechanical properties and CSA have been shown to increase in response to training in humans and animals. This indicates that mechanical stimuli can lead to adaptive responses of the tendon cells, resulting in changes in the ECM. However, the mechanisms responsible for these adjustments are still debated, and a relatively large discrepancy exists between results from animal and human studies.

One hypothesis is that mechanical loading of tendon tissue during exercise or training initiates a signaling cascade that stimulates the cells located in the tissue to increase their production of matrix proteins, ultimately leading to tendon hypertrophy. This phenomenon—termed "mechanotransduction"—is well established and described *in vitro*.[78] Cell culture studies on tendon and ligament fibroblasts show that fibroblasts respond to mechanical stretch by increasing their production and secretion of certain growth factors that in turn act on the fibroblasts to induce expression and synthesis of collagen.[78] Growth factors involved in this signaling cascade include transforming growth factor b1 (TGFb1) and connective tissue growth factor (CTGF).[79,80] In addition, more indirect evidence suggests that IGF-I could act as a link between mechanical load and collagen synthesis in tendon tissue.[81,82]

Several experiments on small animals suggest that this mode of mechanotransduction is involved in tendon adaptation to loading. At least intense tendon loading in rats, by electrically induced muscle training or synergist ablation, leads to substantial increases in mRNA expression of collagen-inducing growth factors, IGF-I and TGFb1, in parallel with increased mRNA expression of collagen type I and III in the tendon tissue.[3,83,84] In addition, repeated bouts of treadmill running have been shown to elevate levels of IGF-I protein in rat tendons.[82] Thus, it seems likely that the tendon cells respond to loading by increasing growth factor production, and that the action of these growth factors leads to induction of collagen expression. However, a causal link between the increased expression of these growth factors and the increase in collagen expression has not been proven, and the exact molecular signaling pathways involved in transforming mechanical signals into biochemical signals in tendons are still largely unknown.[85]

In support of a comparable loading-induced tendon collagen synthesis in adult humans, microdialysis studies show increased levels of markers for collagen synthesis in the peritendinous tissue surrounding the Achilles tendon tissue, in response to both acute exercise and long-term training.[86,87] However, these microdialysis data likely reflect the collagen synthesis at the very periphery of the tendon, or even outside the tendon, rather than that of the actual tendon tissue. A more direct way of measuring collagen synthesis in human tendon tissue is to trace incorporation of labeled amino acids in tendon tissue. With this approach, an increased rate of collagen synthesis was observed in patellar tendons of young men in response to acute kicking exercise.[15] However, several other studies using the same technique, and the same exercise model, could not confirm this loading-induced collagen synthesis in adult human tendon.[88–91]

With regard to gene expression of growth factors and collagen, results in adult humans are not consistent with the dramatic responses seen in rodents in response to tendon loading (discussed above). Three studies have investigated gene expression in human patellar tendon tissue in response to acute exercise. Two of these investigations found decreased or unchanged growth factor and collagen mRNA expression in tendon biopsies from the midportion of the tendon,[92,93] while one study found modest increases in collagen and CTGF mRNA expression in tissue from the proximal part of the patellar tendon in response to exercise.[88] In other words, the response of adult human tendon tissue to acute loading does not seem to mimic that of rodent tendon tissue. This suggests that adult human tendon tissue is far less responsive than that of small animals, and such differences may relate to the fact that rats and mice are still in a growth phase at the point when they are typically used in experiments (typically at 10–12 weeks of age for rats),[3,83,84] and consequently, their tendons may have more potential for adaptation than adult human tendons.

Due to the diverging results in tendon research, it is currently debated to what degree acute exercise influences rates of collagen production in the adult human tendon and at what basal rate tendon tissue is actually metabolized. A recent study on tendon tissue turnover has taken advantage of the so-called carbon-14 (^{14}C) bomb-pulse method. With this method, ^{14}C, left over from nuclear bomb tests performed in the 1950–1960s, can be traced in tissues with slow turnover. Large amounts of ^{14}C were retained in healthy Achilles tendon core from adults born during the peak of the ^{14}C bomb pulse, and this gives clear evidence of an extremely slow rate of Achilles tendon tissue turnover in adult life.[94] In comparison, no remnants of ^{14}C from the bomb pulse were found in muscle tissue from the same persons, indicating rapid turnover of this tissue.[94] The evidence of slow tendon turnover is supported by two earlier studies, in which accumulation of pentosidine (nonenzymatic cross-link) and D-aspartate was measured to estimate long-term tissue turnover in adult human biceps tendon[95] and mature horse tendon,[96] respectively.

The apparently slow rate of tendon tissue renewal fits poorly with data from studies investigating acute tendon collagen synthesis rates with use of labeled amino acids. These studies indicate relatively high basal rates of tendon collagen synthesis, almost resembling the synthesis rate of myofibrillar protein in skeletal muscle.[15] However, the measure of acute incorporation of labeled amino acids suffers from the complication that all new synthesis of collagen is detected, even though this newly synthesized collagen may often be rapidly degraded and thus never incorporated into the tissue matrix.[97] Hypothetically, a potentially large production of excess collagen, which is broken down relatively quickly and never incorporated into the more permanent tissue structures (i.e., collagen fibrils), will contribute to the measured synthesis rate. If this hypothesis holds true, the findings of high basal levels of tendon protein synthesis (measured with amino acid tracers) could be reconciled with a relatively permanent tendon matrix in adult human and horse tendon.[94–96]

It does, however, remain difficult to reconcile a very slow tissue turnover with the fact that human tendons can hypertrophy in response to long-term loading (e.g., Ref. 55), as the hypertrophic response indicates some degree of synthetic activity. One possible explanation is that loading-induced tendon growth takes place at the very periphery of the tendon. This could be reconciled both with the ^{14}C bomb-pulse data, showing very low turnover rates in the tendon core, and with the fact that microdialysis experiments consistently indicate a loading-induced collagen synthesis in peritendinous tissue.[86,94] This hypothesis is supported by recent data on 6-month-old mice, which showed that overload-induced plantaris tendon hypertrophy was based on growth and cell proliferation only in the most superficial layers of tendon tissue, while the "original" core tendon remained relatively constant.[98] A greater potential for growth at the tendon periphery is further supported by an early study that showed greater levels of IGF-I protein expression in cells located in the rat Achilles tendon periphery compared to those located in the deeper part of the tendon.[82] Also, more recent studies suggest greater potential for growth and cell proliferation in the superficial parts of the tendon in rodents.[99,100] In other words, it may be speculated that a new layer of collagenous matrix is added, comparable to a tree ring, when the tendon grows in response to loading.

An alternative explanation for the diverging results in tendon research, with regard to the adaptability of tendon tissue to loading and the overall metabolic activity, could be that large differences exist between different types of tendons. Data from horses show that high-load tendons have slower turnover than tendons subjected to more moderate loads.[96] Though perhaps counterintuitive, it may be speculated that high-load tendons simply cannot "afford" to have a constant remodeling going on as this may reduce strength. Therefore, the high-load Achilles tendon may well have slower turnover than tendons that are loaded less, such as the patellar tendon. This could explain why patellar tendon hypertrophy is seen relatively consistently in response to long-term loading (e.g., Ref. 55), while data on Achilles tendon hypertrophy in relation to loading are far less consistent.[52] In addition, there may well be regional differences within individual tendons. As an example, training-induced hypertrophy of the patellar tendon is in most cases seen exclusively at the proximal and distal parts, and not the mid-tendon (e.g., Ref. 55). It remains to be investigated if the apparent differences in responsiveness between different tendons, and between regions within tendons, are connected to different levels of tissue turnover.

Finally, a potential explanation for exercise-induced hypertrophy could be that it is merely a consequence of increased water content and not an actual accrual of collagen matrix. This leaves the training-induced increase in tendon stiffness to be explained, and potentially, an increase in stiffness could happen in the absence of collagen accrual if cross-linking of the existing matrix was augmented by loading. In favor of this hypothesis, one study in rats showed a substantial increase in mRNA expression of the collagen cross-linking enzyme LOX in response to 4 days of strength training.[3]

5. CARTILAGE—TRAINING AND DETRAINING

The joint protecting function of cartilage is ensured by its fibrous network with fibrils primarily consisting of type II collagen (more than 90%). In addition, noncollagen molecules especially proteoglycans (PGs) that contain hydrophilic GAGs ensure high water content. This provides the basis for a tissue that is resistant to compression. The turnover of the main component—collagen type II—in cartilage is extremely low, and matrix metalloproteinase (MMP)-initiated degradation of cartilage is counteracted by the presence of small leucine-rich proteins like biglycan, fibromodulin, decorin, fibronectin, cartilage oligomeric protein (COMP), and matrilines.[101] With mechanical loading of cartilage, the interaction between the chondrocyte and fibroblast growth factor 2 results in a synthesis of matrix proteins through the mitogen-activated protein kinase (MAPK) pathways.[102] Further, mechanical stimulation of the cartilage induces a release from chondrocytes of interleukin-4 (IL-4) that can influence aggrecan and MMP-13 and act as an anti-inflammatory, combined with the release of IL-10 that can counteract IL-1 and tumor necrosis factor alpha (TNF-α) to amplify the anti-inflammatory response. IL-10 has been demonstrated to be released from articular cartilage after acute exercise in patients with knee osteoarthritis[103] and could potentially be associated with the symptom diminishing effect of exercise in knee and hip osteoarthritis.

Chronic moderate mechanical loading of joints will lead to thickening of the cartilage in animals, with a higher content of proteoglycans, but unchanged levels of collagen.[104,105] In humans, regular exercise in children and adolescents resulted in increased amount of joint cartilage,[106] and in adult elite endurance athletes, the area of the weight-bearing joint surface was increased compared to sedentary controls, whereas no difference in cartilage thickness was observed.[106] Acute exercise resulted in a short-term deformation of the cartilage—decreasing the thickness of loaded area but compensatory thickening of other areas so the total volume was maintained—and this acute change occurred in a similar fashion whether individuals were sprint trained, strength trained, or untrained.[106,107] This indicates that training does not change thickness or deformation capacity of cartilage, but if it changes the tolerable mechanical load is unknown.

In running dogs, overloading of the cartilage could be achieved by extreme distances, resulting in a decline of GAG content and destruction of the collagen structure and organization.[105] This leads to loss of collagen and loss of chondrocytes. It is known that ADAMTS 4/5 (a disintegrin and metalloproteinase with thrombospondin motifs) and MMP-13 are involved in cartilage damage and development of osteoarthritis, and interestingly, MMP-13 is overexpressed in mechanically overloaded tendons. Thus, this opens the possibility that development of osteoarthritis is connected to mechanical overloading of the joint in some predisposed individuals.

Inactivity leads to a reduction in cartilage thickness as demonstrated in paraplegic and long-term immobilized individuals.[106] Animal studies support the view that this reduction is associated with reduced content of proteoglycans, and that the number of cross-links was decreased with softening of the cartilage.[105] Remobilization resulted in some reestablishment of cross-links and proteoglycan accumulation, but not in all joint areas a full recovery was observed.

6. BONE

Bone is ideally designed to withstand large mechanical forces, and it thereby supports the primary purposes of the skeleton: (1) to provide a framework for the attachment of muscles, tendons, and other tissues thereby supporting motion and postural position of the body, (2) to protect vital organs such as the heart, brain, bone marrow, and reproductive organs, and (3) to support metabolic function where it serves as storage for essential minerals including calcium and phosphorus, as well as acting as an endocrine organ secreting hormones regulating systemic metabolic functions.[108] Bone is a composite material. The ECM consists of mineral, collagen (primarily type I, but also type III and V), water, and noncollagenous proteins. The collagen forms a scaffold of bone matrix. Collagen fibers are tightly connected through intra- and intermolecular cross-links. Into the collagen scaffold, hydroxyapatite crystals are attached and form the mineral phase of bone. Hydroxyapatite is a naturally occurring calcium-containing mineral that is enriched with magnesium, carbonate acid, phosphate, and other trace elements. The collagen composite provides the elastic properties of bone while the hydroxyapatite provides the support for the collagen matrix giving bone its strength and mechanical resistance, thus making bone a strong yet flexible structure.[108] Furthermore, collagen fibers are arranged in layers with the fibers in the individual layers arranged at different angles, thus resembling plywood providing even more strength to the tissue.

Already in the nineteenth century, the anatomist Wolff observed that bone could adapt to the mechanical stimuli in the environment. Almost a century later, Frost formulated the "Mechanostat" theory in which he proposed that there is a set point of mechanical strains around which bone mass and structure remains constant. Moreover, if bone strains increase above the set point, bone formation occurs resulting in stiffer bone. In contrast, if bone experiences strains lower than the set point, bone loss will occur.[109] However, it was later shown that the set point can be altered or modified by systemic factors such as hormones (parathyroid hormone (PTH), sex hormones, etc.).[110]

6.1 Mechanosensing in Bone

Due to the unique properties of bone, strenuous physical activity only results in a small degree of deformation of the skeletal tissue of up to 0.3% (3000 microstrain). This level of deformation is common across a number of species.[111] It has been somewhat controversial what the actual physical stimulus to bone cells is, and most *in vitro* studies employ fluid flow for the investigation of the response of bone cells to mechanical stimuli. However, it is much more likely that bone cells are exposed to a variety of mechanical stimuli, altogether creating a highly diverse physical environment with the biological response being an integral of the different stimuli. Also, the nature of the stimulus may very well be different between the different types of bone cells. Osteoblasts reside in the bone marrow and on the endost and are exposed to pressure in the medullary cavity. In contrast, osteocytes are located deep within the bone tissue where they form a three-dimensional network with processes that are interconnected through the canalicular network surrounded by thin layers of pericellular fluid separating the processes from walls of the canaliculi. Thus, osteocytes experience dynamic fluid flow pressure along the processes. Also, shear forces may affect the cells together with dynamic electric fields as the interstitial fluid passes charged bone crystals.[112] Finally, osteoclasts and osteoclast precursors may be exposed to mechanical stimuli as they reside in the bone marrow, where dynamic pressure can stimulate them.

A number of cellular components in bone cells can act as mechanosensors. First, deformation of the membrane due to shear stress and pressure is transmitted to the cytoskeleton and further to the proteins forming the cell–matrix adhesion.[113] Through various pathways, the deformation of the cytoskeleton induces responses in the nucleus and ultimately protein expression. In osteoblasts, stretch or strain of the cell membrane can activate a number of ion channels.[114] Primary cilia are expressed on the surface of some osteocytes[115] and may sense fluid flow thereby activating ion fluxes primarily through PCI and TRPV4 with subsequent modulation of the Wnt signaling pathway. Thus, the various mechanisms involved in sensing mechanical signals in bone presumably promotes an integrated signal based on the many types of stimuli affecting the bone cells involved.

6.2 Mechanotransduction in Bone

Osteocytes are ideally located to not only sense but also communicate responses to mechanical stimuli throughout the osteocyte syncytium. Also, they can amplify the response and provide a directional effect taking into account which specific local area of the bone that is being loaded. Thus, bending of bone results in enhanced bone structure on the compressed side of the bone, while bone mass is reduced on the contralateral side of the bone. Multiple pathways are involved in translation of the mechanical stimulus to biological responses in bone, including calcium signaling, MAPK, Wnt, and RhoA/ROCK pathways, with the Wnt signaling pathway being one of the most extensively described. Wnt ligands bind to Frizzled and LRP-5/6 receptors and induce accumulation of β-catenin in the osteoblast cytoplasm and furthermore translocation into the nucleus, where β-catenin acts as a transcriptional coactivator of osteoblastic transcription factors such as RUNX2. Thereby, it promotes commitment of mesenchymal stem cells to preosteoblasts, osteoblast proliferation and differentiation, and subsequently formation of mineralized matrix. This may explain why exercise drives the mesenchymal stem cells into the osteoblastic lineage of cells,[116] while immobilization/disuse increases adipose tissue in the bone marrow compartment.[117] Moreover, Wnt signaling downregulates osteoclast formation and activity by modifying the OPG and RANKL expression in osteoblasts and osteocytes.[118] Sclerostin, a product of osteocytes, is constitutively released and inhibits the Wnt signaling pathway, thereby inhibiting bone formation. In contrast, loading reduces sclerostin release by osteocytes, thus reversing the inhibition of the Wnt pathway with subsequent increase in bone formation and reduction of bone resorption. Sclerostin inhibitors are currently being tested in clinical trials for the treatment of osteoporosis.

Other important intracellular pathways activated upon mechanical stimulation of bone cells include the prostaglandin E2 (PGE2) pathway. Perception of strain induces release of PGE2 which subsequently binds to its receptors EP2 and EP4 and, together with integrin, inhibits GSK-3β and Akt activation, resulting in accumulation of β-catenin.[119]

Osteoblast also has the ability to communicate directly with one another upon mechanical stimulation through intercellular calcium transients that involve either release of adenosine triphosphate and activation of P2 purinergic receptors on neighboring cells or passage of different molecules through gap junctions connecting the cells.[120] Also, osteoblasts can propagate mechanically induced signals to osteoclasts via purinergic signaling,[121] though it is controversial whether osteoblasts and osteoclasts are present at the remodeling site at the same time. Interestingly, osteoclasts themselves actually decrease resorbing activity when exposed to direct mechanical stimulation such as pressure.[122] Also, the PTH type 1 receptor[123] has been shown to be involved in mediating the effect of loading on bone cells.

Recently, studies have demonstrated that microRNAs (miRNAs) influence the response of osteoblasts to mechanical loading. miRNAs regulate Runx2 expression with both positive and negative effects on skeletal morphogenesis and osteoblast formation.[124] miRNA-135 and miRNA-26a can inhibit osteoblast formation through the BMP-2/Smad signaling pathway,[125] while miRNA-29a induces Wnt signaling during osteoblast differentiation.[126] Thus, miRNAs clearly regulate osteoblast differentiation and activity and may be important players in regulation of skeletal responses to mechanical loading.

6.3 Training of Bone

Studies have consistently shown that loads resulting in high strains induce bone formation in the loaded areas, while areas with lower peak strains have lower levels of bone formation or even increased bone resorption.[127] While bone and bone cells are clearly responsive to mechanical stimulation, the mode and nature of the stimulus are critical to the anabolic response. A number of animal studies have demonstrated that only dynamic stimuli can create an osteogenic response, whereas static loading does not induce the formation of new bone.[128] It has also been shown that changes in remodeling are dependent on the strain amplitude/magnitude,[129] as well as the number and interval between the loading cycles.[130] The anabolic response can also be desensitized and periods of rest between the load cycles can therefore increase the osteogenic response.[131] However, if the magnitude of the load is too big or the number of loading cycles is too high, this may result in fatigue of the bone tissue resulting in tissue damage, also called microdamage. Thus, instead of inducing bone formation, microdamage will result in a series of cellular events including osteocyte apoptosis and induction of bone resorption locally.[132]

Interestingly, periods of rest between loading cycles increase bone formation.[133] This indicates that there is some kind of desensitization mechanism in the mechanotransductory apparatus. This fits well with the underlying mechanisms, where several of the receptors and pathways involved in mechanotransduction are known to desensitize, including the P2 purinergic receptors.[120]

A number of studies have used various exercise interventions to investigate which exercise was the most optimal to stimulate bone growth or to preserve bone mass. Exercises can load bones mechanically, either when bones are weight bearing (aerobics, running, walking, etc.) or when engaged in resistance training (weight lifting and other resistance exercises). Also, studies have employed more artificial ways of exercising bone such as vibration plates. A recent Cochrane review evaluated the evidence for the effect of exercise on preventing and treating osteoporosis in postmenopausal women.[134] The study concluded that exercise provides a relatively small, yet statistically significant effect on bone density in postmenopausal women compared with control groups. The most effective type of exercise intervention on femoral bone mass appeared to be nonweight-bearing exercise with high force such as progressive resistance strength training for the lower limbs. For increasing the vertebral bone mass, the most effective intervention was exercises that combined different types of dynamic loading. The effect size was relatively small as the increase in bone mass was only around 3% compared with control groups. Moreover, the risk of fracture across all exercise groups was not significantly different in the exercise groups when compared to the control groups.[134] This fits well with previous *in vivo* studies that have shown that the number of loading cycles required to maximally stimulate bone formation is surprisingly small. This was demonstrated in an avian ulnar loading model where only four loading cycles per day were sufficient to prevent the bone loss associated with immobilization. In contrast, increasing the number of loading cycles from 36 to 1800 per day did not give any additional anabolic response to the loading,[130] demonstrating that the loading necessary to maintain the bone tissue is relatively light.

Whole-body vibration has been mentioned as a way to mechanically stimulate bone anabolic effects in patients that are otherwise immobilized. It potentially activates mechanotransduction and osteogenesis, but studies have shown varying results. This may be due to differences in the stimulation protocols, such as number of loading cycles, amplitude, and frequency. Overall, training duration ranged from 6 to 18 months and frequency from once to twice per week. Loading time was from 4 to 20 min per day. Increases in bone mass varied from 0% to 4% over 6–8 months of training.[135] However, even though the effect on bone mass was minimal, balance was improved and the number of falls decreased significantly over 18 months[136] indicating that benefits other than increases in bone mass can be obtained by vibration plate stimulation of the musculoskeletal system. Vibration plate training may therefore be beneficial in patients that are exercise intolerant or with limited mobility. However, the optimal treatment protocol has not yet been established in terms of frequency, amplitude, treatment duration, mode of vibration, etc.

Other factors (systemic) also seem to influence the local effects of exercise. Age and sex clearly influence the anabolic response. Adolescents have a clearly better bone response to exercise than adults. Also, the effect is higher in premenopausal compared to postmenopausal women.[137] This may be because the number of viable osteocytes decreases in the aging skeleton, thus making the skeleton less sensitive to mechanical stimuli and thereby less prone to respond to the mechanotransductory signals. Moreover, older men have better effect than older women,[138] indicating a regulatory effect of sex hormones.

6.4 Detraining of Bone

The effect of detraining, and even unloading, on bone mass is far greater than that of loading. Osteocyte health and survival is clearly dependent on a constant use of the skeleton. If bone is immobilized, osteocyte apoptosis increases resulting in the loss of osteocytes in the immobilized bone.[139] It is well established that the bone mass gained will be lost upon detraining, so training should be continued to maintain what has been gained.[140] From different conditions of unloading, it has been seen that unloading and disuse lead to substantial increases in bone resorption and subsequently bone loss—this is true for astronauts not exposed to gravity, patients with spinal cord injury, and after exercise detraining. Also, elderly people with conditions followed by partial or total immobilization all lose bone at a high rate. In a recent study in patients with acute spinal cord injury, the mean loss in bone mineral density over the first 4 months was around 8.3%, with large individual differences.[141] However, some studies have shown that total immobilization can result in loss of bone mass of up to 30–40%. The structural changes in bone during unloading are many. Again, in a study of patients with spinal cord injury, cortical thickness was 20–40% lower than in healthy controls.[142] Trabecular density was also found to be reduced substantially in patients with spinal cord injury with annual losses of 14%[143] though with huge interindividual variations (from 2% to 80%). During spaceflight, astronauts experienced mean losses in trabecular density of 4% at the lumbar spine and 12% at the proximal femur, though again with considerable individual variability.[144]

In summary, the consequence of physical activity on the adult skeleton is preservation, not acquisition. Thus, it seems that exercise in adults only results in minor increases in bone mass and quality, but detraining/disuse results in rapid and substantial losses of bone mass. Therefore, in order to optimize bone mass, exercise should be started when the skeleton is still growing to optimize the effect, and exercise and loading of bone should be continued throughout life.

7. CONCLUSION

The collagen tissues of the musculoskeletal system all the way from IMCT, over the MTJ, through the tendon, passing the joints with its cartilage and ligaments and ending with the bone, play a crucial role in maintaining the architecture of the skeletal muscle, ensuring force transmission, storing energy, protecting joint surface and stability, and ensuring the transfer of muscular forces into resulting limb movement. The overall structure of the musculoskeletal connective tissue is relatively stable, but mechanical loading and subsequent mechanotransduction and molecular anabolic signaling can result in some adaptation of the connective tissue, its size, its strength, and its mechanical properties, whereby it can improve its capacity by 5–20% with regular physical activity. For several of the mechanically loaded connective tissues, the exact molecular and cellular signaling pathways remain to be explored in detail. In contrast to the adaptive responses in connective tissue to regular exercise, lack of mechanical tissue loading through inactivity or immobilization of the human body will result in a dramatic loss of connective tissue content, structure, and tolerable load within weeks, to a degree (30–40%) that mimics that observed in contractile skeletal musculature. This illustrates the importance of regular mechanical load in order to preserve the stabilizing role of the connective tissue for the overall function of the musculoskeletal system in both daily activity and exercise.

REFERENCES

1. Meyer GA, Lieber RL. Elucidation of extracellular matrix mechanics from muscle fibers and fiber bundles. *J Biomech*. 2011;44(4):771–773.
2. Magnusson SP, Narici MV, Maganaris CN, Kjaer M. Human tendon behaviour and adaptation, in vivo. *J Physiol*. 2008;586(1):71–81.
3. Heinemeier KM, Olesen JL, Haddad F, et al. Expression of collagen and related growth factors in rat tendon and skeletal muscle in response to specific contraction types. *J Physiol*. 2007;582(pt 3):1303–1316.
4. Kjaer M. Role of extracellular matrix in adaptation of tendon and skeletal muscle to mechanical loading. *Physiol Rev*. 2004;84(2):649–698.
5. Lieber RL, Ward SR. Cellular mechanisms of tissue fibrosis. 4. Structural and functional consequences of skeletal muscle fibrosis. *Am J Physiol Cell Physiol*. 2013;305: C241–C252.
6. Trotter JA, Purslow PP. Functional morphology of the endomysium in series fibered muscles. *J Morphol*. 1992;212(2):109–122.

7. Patel TJ, Lieber RL. Force transmission in skeletal muscle: from actomyosin to external tendons. *Exerc Sport Sci Rev*. 1997;25:321–363.
8. Huijing PA. Muscle as a collagen fiber reinforced composite: a review of force transmission in muscle and whole limb. *J Biomech*. 1999;32(4):329–345.
9. Street SF. Lateral transmission of tension in frog myofibers: a myofibrillar network and transverse cytoskeletal connections are possible transmitters. *J Cell Physiol*. 1983;114(3):346–364.
10. Passerieux E, Rossignol R, Letellier T, Delage JP. Physical continuity of the perimysium from myofibers to tendons: involvement in lateral force transmission in skeletal muscle. *J Struct Biol*. 2007;159(1):19–28.
11. Zhang C, Gao Y. Effects of aging on the lateral transmission of force in rat skeletal muscle. *J Biomech*. 2014;47(5):944–948.
12. Smith LR, Lee KS, Ward SR, Chambers HG, Lieber RL. Hamstring contractures in children with spastic cerebral palsy result from a stiffer extracellular matrix and increased in vivo sarcomere length. *J Physiol*. 2011;589(pt 10):2625–2639.
13. Borg TK, Caulfield JB. Morphology of connective tissue in skeletal muscle. *Tissue Cell*. 1980;12(1):197–207.
14. Purslow PP. The structure and functional significance of variations in the connective tissue within muscle. *Comp Biochem Physiol A Mol Integr Physiol*. 2002;133(4):947–966.
15. Miller BF, Olesen JL, Hansen M, et al. Coordinated collagen and muscle protein synthesis in human patella tendon and quadriceps muscle after exercise. *J Physiol*. 2005;567(pt 3):1021–1033.
16. Holm L, van Hall G, Rose AJ, et al. Contraction intensity and feeding affect collagen and myofibrillar protein synthesis rates differently in human skeletal muscle. *Am J Physiol Endocrinol Metab*. 2010;298(2):E257–E269.
17. Doessing S, Heinemeier KM, Holm L, et al. Growth hormone stimulates the collagen synthesis in human tendon and skeletal muscle without affecting myofibrillar protein synthesis. *J Physiol*. 2010;588(pt 2):341–351.
18. Nielsen RH, Clausen NM, Schjerling P, et al. Chronic alterations in growth hormone/insulin-like growth factor-I signaling lead to changes in mouse tendon structure. *Matrix Biol*. 2014;34:96–104.
19. Hansen M, Skovgaard D, Reitelseder S, Holm L, Langbjerg H, Kjaer M. Effects of estrogen replacement and lower androgen status on skeletal muscle collagen and myofibrillar protein synthesis in postmenopausal women. *J Gerontol A Biol Sci Med Sci*. 2012;67(10):1005–1013.
20. Crameri RM, Langberg H, Teisner B, et al. Enhanced procollagen processing in skeletal muscle after a single bout of eccentric loading in humans. *Matrix Biol*. 2004;23(4):259–264.
21. Crameri RM, Aagaard P, Qvortrup K, Langberg H, Olesen J, Kjaer M. Myofibre damage in human skeletal muscle: effects of electrical stimulation versus voluntary contraction. *J Physiol*. 2007;583(pt 1):365–380.
22. Mackey AL, Brandstetter S, Schjerling P, et al. Sequenced response of extracellular matrix deadhesion and fibrotic regulators after muscle damage is involved in protection against future injury in human skeletal muscle. *FASEB J*. 2011;25(6):1943–1959.
23. Saclier M, Yacoub-Youssef H, Mackey AL, et al. Differentially activated macrophages orchestrate myogenic precursor cell fate during human skeletal muscle regeneration. *Stem Cells*. 2013;31(2):384–396.
23a. Murphy MM, Lawson JA, Mathew SJ, Hutcheson DA, Kardon G. Satellite cells, connective tissue fibroblasts and their interactions are crucial for muscle regeneration. *Development*. 2011;138:3625–3637.
24. Urciuolo A, Quarta M, Morbidoni V, et al. Collagen VI regulates satellite cell self-renewal and muscle regeneration. *Nat Commun*. 2013;4:1964.

25. Tidball JG, Salem G, Zernicke R. Site and mechanical conditions for failure of skeletal muscle in experimental strain injuries. *J Appl Physiol.* 1993;74(3):1280–1286.
26. Roffino S, Carnino A, Chopard A, Mutin M, Marini JF. Structural remodeling of unweighted soleus myotendinous junction in monkey. *C R Biol.* 2006;329(3):172–179.
27. Ciena AP, Luques IU, Dias FJ, Yokomizo de Almeida SR, Iyomasa MM, Watanabe IS. Ultrastructure of the myotendinous junction of the medial pterygoid muscle of adult and aged Wistar rats. *Micron.* 2010;41(8):1011–1014.
28. Knudsen AB, Larsen M, Mackey AL, et al. The human myotendinous junction: an ultrastructural and 3D analysis study. *Scand J Med Sci Sports.* 2015;25:e116–e123.
29. Trotter JA. Structure-function considerations of muscle-tendon junctions. *Comp Biochem Physiol A Mol Integr Physiol.* 2002;133(4):1127–1133.
30. Kojima H, Sakuma E, Mabuchi Y, et al. Ultrastructural changes at the myotendinous junction induced by exercise. *J Orthop Sci.* 2008;13(3):233–239.
31. Finni T, Komi PV, Lepola V. In vivo human triceps surae and quadriceps femoris muscle function in a squat jump and counter movement jump. *Eur J Appl Physiol.* 2000;83(4–5):416–426.
32. Magnusson SP, Aagaard P, Dyhre-Poulsen P, Kjaer M. Load-displacement properties of the human triceps surae aponeurosis in vivo. *J Physiol.* 2001;531(pt 1):277–288.
33. Kastelic J, Galeski A, Baer E. The multicomposite structure of tendon. *Connect Tissue Res.* 1978;6(1):11–23.
34. Kadler KE, Holmes DF, Trotter JA, Chapman JA. Collagen fibril formation. *Biochem J.* 1996;316(pt 1):1–11.
35. Bailey AJ. Molecular mechanisms of ageing in connective tissues. *Mech Ageing Dev.* 2001;122(7):735–755.
36. Riley GP. Gene expression and matrix turnover in overused and damaged tendons. *Scand J Med Sci Sports.* 2005;15(4):241–251.
37. Haraldsson BT, Aagaard P, Krogsgaard M, Alkjaer T, Kjaer M, Magnusson SP. Region-specific mechanical properties of the human patella tendon. *J Appl Physiol.* 2005;98(3):1006–1012.
38. Haraldsson BT, Aagaard P, Qvortrup K, et al. Lateral force transmission between human tendon fascicles. *Matrix Biol.* 2008;27(2):86–95.
39. Parry DA, Barnes GR, Craig AS. A comparison of the size distribution of collagen fibrils in connective tissues as a function of age and a possible relation between fibril size distribution and mechanical properties. *Proc R Soc Lond B Biol Sci.* 1978;203(1152):305–321.
40. Provenzano PP, Vanderby Jr R. Collagen fibril morphology and organization: implications for force transmission in ligament and tendon. *Matrix Biol.* 2006;25(2):71–84.
41. Svensson RB, Hassenkam T, Hansen P, Kjaer M, Magnusson SP. Tensile force transmission in human patellar tendon fascicles is not mediated by glycosaminoglycans. *Connect Tissue Res.* 2011;52:415–421.
42. Hansen P, Haraldsson BT, Aagaard P, et al. Lower strength of the human posterior patellar tendon seems unrelated to mature collagen cross-linking and fibril morphology. *J Appl Physiol.* 2010;108(1):47–52.
43. Buehler MJ. Nanomechanics of collagen fibrils under varying cross-link densities: atomistic and continuum studies. *J Mech Behav Biomed Mater.* 2008;1(1):59–67.
44. Fratzl P, Misof K, Zizak I, Rapp G, Amenitsch H, Bernstorff S. Fibrillar structure and mechanical properties of collagen. *J Struct Biol.* 1998;122(1–2):119–122.
45. Mosler E, Folkhard W, Knorzer E, Nemetschek-Gansler H, Nemetschek T, Koch MH. Stress-induced molecular rearrangement in tendon collagen. *J Mol Biol.* 1985;182(4):589–596.
46. Sasaki N, Odajima S. Elongation mechanism of collagen fibrils and force-strain relations of tendon at each level of structural hierarchy. *J Biomech.* 1996;29(9):1131–1136.

47. Svensson RB, Mulder H, Kovanen V, Magnusson SP. Fracture mechanics of collagen fibrils: influence of natural cross-links. *Biophys J.* 2013;104(11):2476–2484.
48. Rumian AP, Wallace AL, Birch HL. Tendons and ligaments are anatomically distinct but overlap in molecular and morphological features—a comparative study in an ovine model. *J Orthop Res.* 2007;25(4):458–464.
49. Suzuki D, Takahashi M, Abe M, Nagano A. Biochemical study of collagen and its crosslinks in the anterior cruciate ligament and the tissues used as a graft for reconstruction of the anterior cruciate ligament. *Connect Tissue Res.* 2008;49(1):42–47.
50. Ker RF, Alexander RM, Bennett MB. Why are mammalian tendons so thick? *J Zoo Lond.* 1988;216:309–324.
51. Rosager S, Aagaard P, Dyhre-Poulsen P, Neergaard K, Kjaer M, Magnusson SP. Load-displacement properties of the human triceps surae aponeurosis and tendon in runners and non-runners. *Scand J Med Sci Sports.* 2002;12(2):90–98.
52. Hansen P, Aagaard P, Kjaer M, Larsson B, Magnusson SP. Effect of habitual running on human Achilles tendon load-deformation properties and cross-sectional area. *J Appl Physiol.* 2003;95(6):2375–2380.
53. Couppe C, Svensson RB, Grosset JF, et al. Life-long endurance running is associated with reduced glycation and mechanical stress in connective tissue. *Age (Dordr).* 2014;36(4):9665.
54. Couppe C, Kongsgaard M, Aagaard P, et al. Habitual loading results in tendon hypertrophy and increased stiffness of the human patellar tendon. *J Appl Physiol.* 2008;105(3):805–810.
55. Kongsgaard M, Reitelseder S, Pedersen TG, et al. Region specific patellar tendon hypertrophy in humans following resistance training. *Acta Physiol (Oxf).* 2007;191(2):111–121.
56. Michna H. Morphometric analysis of loading-induced changes in collagen-fibril populations in young tendons. *Cell Tissue Res.* 1984;236(2):465–470.
57. Patterson-Kane JC, Parry DA, Birch HL, Goodship AE, Firth EC. An age-related study of morphology and cross-link composition of collagen fibrils in the digital flexor tendons of young thoroughbred horses. *Connect Tissue Res.* 1997;36(3):253–260.
58. Patterson-Kane JC, Firth EC, Parry DA, Wilson AM, Goodship AE. Effects of training on collagen fibril populations in the suspensory ligament and deep digital flexor tendon of young thoroughbreds. *Am J Vet Res.* 1998;59(1):64–68.
59. Kongsgaard M, Qvortrup K, Larsen J, et al. Fibril morphology and tendon mechanical properties in patellar tendinopathy: effects of heavy slow resistance training. *Am J Sports Med.* 2010;38(4):749–756.
60. Bailey AJ, Paul RG, Knott L. Mechanisms of maturation and ageing of collagen. *Mech Ageing Dev.* 1998;106(1–2):1–56.
61. Kongsgaard M, Kovanen V, Aagaard P, et al. Corticosteroid injections, eccentric decline squat training and heavy slow resistance training in patellar tendinopathy. *Scand J Med Sci Sports.* 2009;19(6):790–802.
62. Verzijl N, DeGroot J, Oldehinkel E, et al. Age-related accumulation of Maillard reaction products in human articular cartilage collagen. *Biochem J.* 2000;350(pt 2):381–387.
63. Narici MV, Maganaris CN. Plasticity of the muscle-tendon complex with disuse and aging. *Exerc Sport Sci Rev.* 2007;35(3):126–134.
64. Suetta C, Hvid LG, Justesen L, et al. Effects of ageing on human skeletal muscle after immobilisation and re-training. *J Appl Physiol.* 2009;107:1172–1180.
65. de Boer MD, Maganaris CN, Seynnes OR, Rennie MJ, Narici MV. Time course of muscular, neural and tendinous adaptations to 23 day unilateral lower-limb suspension in young men. *J Physiol (Lond).* 2007;583(pt 3):1079–1091.
66. Matsumoto F, Trudel G, Uhthoff HK, Backman DS. Mechanical effects of immobilization on the Achilles' tendon. *Arch Phys Med Rehabil.* 2003;84(5):662–667.

67. Rumian AP, Draper ER, Wallace AL, Goodship AE. The influence of the mechanical environment on remodelling of the patellar tendon. *J Bone Joint Surg Br.* 2009;91(4):557–564.
68. Eliasson P, Fahlgren A, Pasternak B, Aspenberg P. Unloaded rat Achilles tendons continue to grow, but lose viscoelasticity. *J Appl Physiol.* 2007;103(2):459–463.
69. Kinugasa R, Hodgson JA, Edgerton VR, Shin DD, Sinha S. Reduction in tendon elasticity from unloading is unrelated to its hypertrophy. *J Appl Physiol.* 2010;109(3):870–877.
70. Shin D, Finni T, Ahn S, et al. Effect of chronic unloading and rehabilitation on human Achilles tendon properties: a velocity-encoded phase-contrast MRI study. *J Appl Physiol.* 2008;105(4):1179–1186.
71. Reeves ND, Maganaris CN, Ferretti G, Narici MV. Influence of 90-day simulated microgravity on human tendon mechanical properties and the effect of resistive countermeasures. *J Appl Physiol.* 2005;98(6):2278–2286.
72. Majima T, Yasuda K, Tsuchida T, et al. Stress shielding of patellar tendon: effect on small-diameter collagen fibrils in a rabbit model. *J Orthop Sci.* 2003;8(6):836–841.
73. Nakagawa Y, Hayashi K, Yamamoto N, Nagashima K. Age-related changes in biomechanical properties of the Achilles tendon in rabbits. *Eur J Appl Physiol Occup Physiol.* 1996;73(1–2):7–10.
74. Nakagawa Y, Totsuka M, Sato T, Fukuda Y, Hirota K. Effect of disuse on the ultrastructure of the achilles tendon in rats. *Eur J Appl Physiol Occup Physiol.* 1989;59(3):239–242.
75. Tsuchida T, Yasuda K, Kaneda K, et al. Effects of in situ freezing and stress-shielding on the ultrastructure of rabbit patellar tendons. *J Orthop Res.* 1997;15(6):904–910.
76. Boesen AP, Dideriksen K, Couppe C, et al. Tendon and skeletal muscle matrix gene expression and functional responses to immobilisation and rehabilitation in young males: effect of growth hormone administration. *J Physiol.* 2013;591(pt 23):6039–6052.
77. Majima T, Yasuda K, Fujii T, Yamamoto N, Hayashi K, Kaneda K. Biomechanical effects of stress shielding of the rabbit patellar tendon depend on the degree of stress reduction. *J Orthop Res.* 1996;14(3):377–383.
78. Chiquet M, Gelman L, Lutz R, Maier S. From mechanotransduction to extracellular matrix gene expression in fibroblasts. *Biochim Biophys Acta.* 2009;1793(5):911–920.
79. Yang G, Crawford RC, Wang JH. Proliferation and collagen production of human patellar tendon fibroblasts in response to cyclic uniaxial stretching in serum-free conditions. *J Biomech.* 2004;37(10):1543–1550.
80. Schild C, Trueb B. Mechanical stress is required for high-level expression of connective tissue growth factor. *Exp Cell Res.* 2002;274(1):83–91.
81. Abrahamsson SO, Lohmander S. Differential effects of insulin-like growth factor-I on matrix and DNA synthesis in various regions and types of rabbit tendons. *J Orthop Res.* 1996;14(3):370–376.
82. Hansson HA, Engstrom AM, Holm S, Rosenqvist AL. Somatomedin C immunoreactivity in the Achilles tendon varies in a dynamic manner with the mechanical load. *Acta Physiol Scand.* 1988;134(2):199–208.
83. Heinemeier KM, Olesen JL, Schjerling P, et al. Short-term strength training and the expression of myostatin and IGF-I isoforms in rat muscle and tendon: differential effects of specific contraction types. *J Appl Physiol.* 2007;102(2):573–581.
84. Olesen JL, Heinemeier KM, Haddad F, et al. Expression of insulin-like growth factor I, insulin-like growth factor binding proteins, and collagen mRNA in mechanically loaded plantaris tendon. *J Appl Physiol.* 2006;101(1):183–188.
85. Gumucio JP, Sugg KB, Mendias CL. TGF-beta superfamily signaling in muscle and tendon adaptation to resistance exercise. *Exerc Sport Sci Rev.* 2015;43:93–99.

86. Langberg H, Skovgaard D, Petersen LJ, Bulow J, Kjaer M. Type I collagen synthesis and degradation in peritendinous tissue after exercise determined by microdialysis in humans. *J Physiol.* 1999;521(pt 1):299–306.
87. Langberg H, Rosendal L, Kjaer M. Training-induced changes in peritendinous type I collagen turnover determined by microdialysis in humans. *J Physiol.* 2001;534(pt 1):297–302.
88. Dideriksen K, Sindby AK, Krogsgaard M, Schjerling P, Holm L, Langberg H. Effect of acute exercise on patella tendon protein synthesis and gene expression. *SpringerPlus.* 2013;2(1):109.
89. Hansen M, Kongsgaard M, Holm L, et al. Effect of estrogen on tendon collagen synthesis, tendon structural characteristics, and biomechanical properties in postmenopausal women. *J Appl Physiol.* 2009;106(4):1385–1393.
90. Hansen M, Miller BF, Holm L, et al. Effect of administration of oral contraceptives in vivo on collagen synthesis in tendon and muscle connective tissue in young women. *J Appl Physiol.* 2009;106(4):1435–1443.
91. Petersen SG, Miller BF, Hansen M, Trappe TA, Kjaer M, Holm L. Exercise and NSAID: effect on muscle protein synthesis in knee osteoarthritis patients? *Med Sci Sports Exerc.* 2010;43:425–431.
92. Heinemeier KM, Bjerrum SS, Schjerling P, Kjaer M. Expression of extracellular matrix components and related growth factors in human tendon and muscle after acute exercise. *Scand J Med Sci Sports.* 2013;23(3):e150–e161.
93. Sullivan BE, Carroll CC, Jemiolo B, et al. Effect of acute resistance exercise and sex on human patellar tendon structural and regulatory mRNA expression. *J Appl Physiol.* 2009;106(2):468–475.
94. Heinemeier KM, Schjerling P, Heinemeier J, Magnusson SP, Kjaer M. Lack of tissue renewal in human adult Achilles tendon is revealed by nuclear bomb (14)C. *FASEB J.* 2013;27(5):2074–2079.
95. Bank RA, TeKoppele JM, Oostingh G, Hazleman BL, Riley GP. Lysylhydroxylation and non-reducible crosslinking of human supraspinatus tendon collagen: changes with age and in chronic rotator cuff tendinitis. *Ann Rheum Dis.* 1999;58(1):35–41.
96. Thorpe CT, Streeter I, Pinchbeck GL, Goodship AE, Clegg PD, Birch HL. Aspartic acid racemization and collagen degradation markers reveal an accumulation of damage in tendon collagen that is enhanced with aging. *J Biol Chem.* 2010;285(21):15674–15681.
97. McAnulty RJ, Laurent GJ. Collagen synthesis and degradation in vivo. Evidence for rapid rates of collagen turnover with extensive degradation of newly synthesized collagen in tissues of the adult rat. *Coll Relat Res.* 1987;7(2):93–104.
98. Gumucio JP, Phan AC, Ruehlmann DG, Noah AC, Mendias CL. Synergist ablation induces rapid tendon growth through the synthesis of a neotendon matrix. *J Appl Physiol.* 2014;117:1287–1291. jap 00720 02014.
99. Tan Q, Lui PP, Lee YW. In vivo identity of tendon stem cells and the roles of stem cells in tendon healing. *Stem Cells Dev.* 2013;22(23):3128–3140.
100. Mendias CL, Gumucio JP, Bakhurin KI, Lynch EB, Brooks SV. Physiological loading of tendons induces scleraxis expression in epitenon fibroblasts. *J Orthop Res.* 2012;30(4):606–612.
101. Heinegard D. Proteoglycans and more—from molecules to biology. *Int J Exp Pathol.* 2009;90(6):575–586.
102. Vincent TL, McLean CJ, Full LE, Peston D, Saklatvala J. FGF-2 is bound to perlecan in the pericellular matrix of articular cartilage, where it acts as a chondrocyte mechanotransducer. *Osteoarthritis Cartilage.* 2007;15(7):752–763.
103. Helmark IC, Mikkelsen UR, Borglum J, et al. Exercise increases interleukin-10 levels both intraarticularly and peri-synovially in patients with knee osteoarthritis: a randomized controlled trial. *Arthritis Res Ther.* 2010;12(4):R126.

104. Saadat E, Lan H, Majumdar S, Rempel DM, King KB. Long-term cyclical in vivo loading increases cartilage proteoglycan content in a spatially specific manner: an infrared microspectroscopic imaging and polarized light microscopy study. *Arthritis Res Ther.* 2006;8(5):R147.
105. Arokoski JP, Jurvelin JS, Vaatainen U, Helminen HJ. Normal and pathological adaptations of articular cartilage to joint loading. *Scand J Med Sci Sports.* 2000;10(4):186–198.
106. Eckstein F, Hudelmaier M, Putz R. The effects of exercise on human articular cartilage. *J Anat.* 2006;208(4):491–512.
107. Eckstein F, Lemberger B, Gratzke C, et al. In vivo cartilage deformation after different types of activity and its dependence on physical training status. *Ann Rheum Dis.* 2005;64(2):291–295.
108. Rosen CJ. *Primer on the Metabolic Bone Diseases and Disorders of Mineral Metabolism.* New York: Wiley-Blackwell; 2013.
109. Frost HM. Bone's mechanostat: a 2003 update. *Anat Rec A Discov Mol Cell Evol Biol.* 2003;275(2):1081–1101.
110. Price JS, Sugiyama T, Galea GL, Meakin LB, Sunters A, Lanyon LE. Role of endocrine and paracrine factors in the adaptation of bone to mechanical loading. *Curr Osteoporos Rep.* 2011;9(2):76–82.
111. Rubin CT, Lanyon LE. Dynamic strain similarity in vertebrates; an alternative to allometric limb bone scaling. *J Theor Biol.* 1984;107(2):321–327.
112. Pollack SR, Salzstein R, Pienkowski D. The electric double layer in bone and its influence on stress-generated potentials. *Calcif Tissue Int.* 1984;36(suppl 1):S77–S81.
113. Katsumi A, Orr AW, Tzima E, Schwartz MA. Integrins in mechanotransduction. *J Biol Chem.* 2004;279(13):12001–12004.
114. Duncan RL, Hruska KA, Misler S. Parathyroid hormone activation of stretch-activated cation channels in osteosarcoma cells (UMR-106.01). *FEBS Lett.* 1992;307(2):219–223.
115. Coughlin TR, Voisin M, Schaffler MB, Niebur GL, McNamara LM. Primary cilia exist in a small fraction of cells in trabecular bone and marrow. *Calcif Tissue Int.* 2015;96(1):65–72.
116. David V, Martin A, Lafage-Proust MH, et al. Mechanical loading down-regulates peroxisome proliferator-activated receptor gamma in bone marrow stromal cells and favors osteoblastogenesis at the expense of adipogenesis. *Endocrinology.* 2007;148(5):2553–2562.
117. Zayzafoon M, Gathings WE, McDonald JM. Modeled microgravity inhibits osteogenic differentiation of human mesenchymal stem cells and increases adipogenesis. *Endocrinology.* 2004;145(5):2421–2432.
118. Xiong J, Onal M, Jilka RL, Weinstein RS, Manolagas SC, O'Brien CA. Matrix-embedded cells control osteoclast formation. *Nat Med.* 2011;17(10):1235–1241.
119. Qin YX, Hu M. Mechanotransduction in musculoskeletal tissue regeneration: effects of fluid flow, loading, and cellular-molecular pathways. *Biomed Res Int.* 2014;2014:863421.
120. Jorgensen NR, Henriksen Z, Brot C, et al. Human osteoblastic cells propagate intercellular calcium signals by two different mechanisms. *J Bone Miner Res.* 2000;15(6):1024–1032.
121. Jorgensen NR, Henriksen Z, Sorensen OH, Eriksen EF, Civitelli R, Steinberg TH. Intercellular calcium signaling occurs between human osteoblasts and osteoclasts and requires activation of osteoclast P2X7 receptors. *J Biol Chem.* 2002;277(9):7574–7580.
122. Wiltink A, Nijweide PJ, Scheenen WJ, Ypey DL, Van Duijn B. Cell membrane stretch in osteoclasts triggers a self-reinforcing Ca2+ entry pathway. *Pflugers Arch.* 1995;429(5):663–671.

123. Maycas M, Ardura JA, de Castro LF, Bravo B, Gortazar AR, Esbrit P. Role of the parathyroid hormone type 1 receptor (PTH1R) as a mechanosensor in osteocyte survival. *J Bone Miner Res*. 2014;30:1231–1244.
124. Lian JB, Stein GS, van Wijnen AJ, et al. MicroRNA control of bone formation and homeostasis. *Nat Rev Endocrinol*. 2012;8(4):212–227.
125. Luzi E, Marini F, Sala SC, Tognarini I, Galli G, Brandi ML. Osteogenic differentiation of human adipose tissue-derived stem cells is modulated by the miR-26a targeting of the SMAD1 transcription factor. *J Bone Miner Res*. 2008;23(2):287–295.
126. Kapinas K, Kessler CB, Delany AM. miR-29 suppression of osteonectin in osteoblasts: regulation during differentiation and by canonical Wnt signaling. *J Cell Biochem*. 2009;108(1):216–224.
127. Lambers FM, Schulte FA, Kuhn G, Webster DJ, Muller R. Mouse tail vertebrae adapt to cyclic mechanical loading by increasing bone formation rate and decreasing bone resorption rate as shown by time-lapsed in vivo imaging of dynamic bone morphometry. *Bone*. 2011;49(6):1340–1350.
128. Burr DB, Robling AG, Turner CH. Effects of biomechanical stress on bones in animals. *Bone*. 2002;30(5):781–786.
129. Rubin CT, Lanyon LE. Regulation of bone mass by mechanical strain magnitude. *Calcif Tissue Int*. 1985;37(4):411–417.
130. Rubin CT, Lanyon LE. Regulation of bone formation by applied dynamic loads. *J Bone Joint Surg Am*. 1984;66(3):397–402.
131. Srinivasan S, Weimer DA, Agans SC, Bain SD, Gross TS. Low-magnitude mechanical loading becomes osteogenic when rest is inserted between each load cycle. *J Bone Miner Res*. 2002;17(9):1613–1620.
132. Kennedy OD, Laudier DM, Majeska RJ, Sun HB, Schaffler MB. Osteocyte apoptosis is required for production of osteoclastogenic signals following bone fatigue in vivo. *Bone*. 2014;64:132–137.
133. Robling AG, Burr DB, Turner CH. Partitioning a daily mechanical stimulus into discrete loading bouts improves the osteogenic response to loading. *J Bone Miner Res*. 2000;15(8):1596–1602.
134. Howe TE, Shea B, Dawson LJ, et al. Exercise for preventing and treating osteoporosis in postmenopausal women. *Cochrane Database Syst Rev*. 2011;7:CD000333.
135. Gusi N, Raimundo A, Leal A. Low-frequency vibratory exercise reduces the risk of bone fracture more than walking: a randomized controlled trial. *BMC Musculoskelet Disord*. 2006;7:92.
136. von Stengel S, Kemmler W, Engelke K, Kalender WA. Effects of whole body vibration on bone mineral density and falls: results of the randomized controlled ELVIS study with postmenopausal women. *Osteoporos Int*. 2011;22(1):317–325.
137. Nikander R, Sievanen H, Heinonen A, Daly RM, Uusi-Rasi K, Kannus P. Targeted exercise against osteoporosis: a systematic review and meta-analysis for optimising bone strength throughout life. *BMC Med*. 2010;8:47.
138. Maddalozzo GF, Snow CM. High intensity resistance training: effects on bone in older men and women. *Calcif Tissue Int*. 2000;66(6):399–404.
139. Basso N, Heersche JN. Effects of hind limb unloading and reloading on nitric oxide synthase expression and apoptosis of osteocytes and chondrocytes. *Bone*. 2006;39(4):807–814.
140. Dalsky GP. The role of exercise in the prevention of osteoporosis. *Compr Ther*. 1989;15(9):30–37.
141. Edwards WB, Schnitzer TJ, Troy KL. The mechanical consequence of actual bone loss and simulated bone recovery in acute spinal cord injury. *Bone*. 2014;60:141–147.

142. Lee TQ, Shapiro TA, Bell DM. Biomechanical properties of human tibias in long-term spinal cord injury. *J Rehabil Res Dev.* 1997;34(3):295–302.
143. Frey-Rindova P, de Bruin ED, Stussi E, Dambacher MA, Dietz V. Bone mineral density in upper and lower extremities during 12 months after spinal cord injury measured by peripheral quantitative computed tomography. *Spinal Cord.* 2000;38(1):26–32.
144. Lang T, LeBlanc A, Evans H, Lu Y, Genant H, Yu A. Cortical and trabecular bone mineral loss from the spine and hip in long-duration spaceflight. *J Bone Miner Res.* 2004;19(6):1006–1012.

CHAPTER TWELVE

Exercise and the Regulation of Endocrine Hormones

Anthony C. Hackney[*,†,‡,1], **Amy R. Lane**[‡]
[*]Department of Exercise and Sport Science, University of North Carolina, Chapel Hill, North Carolina, USA
[†]Department of Nutrition, Gillings School of Public Health, University of North Carolina, Chapel Hill, North Carolina, USA
[‡]Curriculum in Human Movement Science, Department of Allied Health Sciences, University of North Carolina, Chapel Hill, North Carolina, USA
[1]Corresponding author e-mail address: ach@email.unc.edu

Contents

Abstract

The endocrine system has profound regulatory effects within the human body and thus the ability to control and maintain appropriate function within many physiological systems (i.e., homeostasis). The hormones associated with the endocrine system utilize autocrine, paracrine, or endocrine actions on the cells of their target tissues within these physiologic systems to adjust homeostasis. The introduction of exercise as a stressor to disrupt homeostasis can greatly amplify and impact the actions of these hormones. To that end, the endocrine response to an acute exercise session occurs in a progression of phases with the magnitude of the response being relative to the exercise work intensity or volume. Various physiologic mechanisms are considered responsible for these responses, although not all are completely understood or elucidated. Chronic exercise training does not eliminate the acute exercise response but may attenuate the overall effect of the responsiveness as the body adapts in a positive fashion to the training

Progress in Molecular Biology and Translational Science, Volume 135
ISSN 1877-1173
http://dx.doi.org/10.1016/bs.pmbts.2015.07.001

stimulus. Regrettably, an excessive intensity and/or volume of training may lead to maladaptation and is associated with inappropriate endocrine hormonal responses. The mechanisms leading to a deleterious maladaptive state are not well understood and require additional research for elucidation.

1. INTRODUCTION

Humans have a variety of hormone-like substances that exhibit endocrine, paracrine, and autocrine actions on the tissues of the body. Physiologically these substances are involved in a multitude of regulatory roles for processes such as growth and development, reproduction, metabolism, hydration, cardiovascular adjustments, immune responses, and stress reactivity. Research studies indicate that physical activity and exercise can have powerful effects on the production of these hormone-like substances and thus the processes these substances regulate.[1–4] In most situations, the effects of exercise on the endocrine aspects of the body are of a positive nature and aid in improved functional aspects of tissues and organs resulting in improved health and/or performance. Unfortunately, however, when exercise is executed incorrectly or poorly developed, exercise training programs are implemented counter-productive endocrine responses and negative physiological consequences can occur.

Individuals involved with exercise for the intent of improving health and well-being may typically carry out a training "exercise prescription" comprised of 3–5 days a week of activity, for 20–60 min in duration at a low-moderate to moderately vigorous intensity.[5] This 1.5–5 h commitment per week is challenging for many ordinary people to achieve. For athletes who participate in sports, this volume of activity might represent what they typically do in a single day of their exercise training program. Such large dosages of exercise in athletes, when administered chronically bring about large and substantial physiological adjustment and adaptation, which many experts feel is necessary to enhance sporting performance to the highest levels.[6,7] Regardless of whether one is an ordinary person or an athlete involved in exercise, there are profound changes in the endocrine system that allow the body to accommodate and adjust to the physiological needs of exercise.

To this end, this chapter will address the responses and adjustments—adaptation of the endocrine system with respect to exercise and exercise training. There are many forms and types of exercise activity and to a degree the type of hormonal response is influenced by the energy systems demands and predominating physiological systems necessary to perform the activity. For this latter reason and as a means to organize the dialog on this topic, the endocrine responses will be presented in a dichotomous approach dealing with activities of cardiovascular–aerobic type (i.e., endurance based) versus activities of a muscular power–anaerobic type (resistance–strength based).

In assessing and quantifying exercise training, it is critical to understand the concept of exercise dosages. Exercise dosage consists of "how much" exercise an individual is being exposed. The components which are manipulated to modulate the dosage are the duration (volume) of an exercise session, the intensity of effort within a respective exercise session, and the frequency of exercise sessions per day-week (if dealing with a training program). Intensity is the most complex of these components. Characteristically, the perception of how difficult intense exercise is has been linked to cardiovascular responses, usually expressed as a percentage of maximal heart rate (HR_{max}) or maximal oxygen uptake ($\dot{V}O_{2max}$).[8,9] For endurance-based exercise activities, there is a high reliance on the cardiovascular system and aerobic energy metabolism to determine performance capacity. This means for such activities the expressing of intensity relative to $\dot{V}O_{2max}$ or HR_{max} is appropriate. On the other hand, resistance-based exercise activities are highly explosive in muscular force–power output and performance is far less dependent upon cardiovascular functioning and much more so on anaerobic energy production pathways.[5,9] In the latter such exercise activities, since resistance exercise training is a predominant component, the expression of exercise intensity is based upon percentage of maximal performance relative to a resistance task (i.e., the lifting of a specific amount of weight). The reference criterion here is the 1 repetition maximum (1RM). This represents the maximal weight–force that can be generated for that task and signifies an intensity capacity of 100%.[10] Table 1 provides a summary and explanation of exercise activities intensity issues.

2. ACUTE EXERCISE HORMONE RESPONSES

2.1 What Happens During Exercise

The acute responses of the major physiological systems to a single bout of exercise can be substantial and are usually proportional to the intensity of the exercise, although it is important to recognize that the relationship of this proportional response(s) is not always perfectly linear.[2] The hormonal

Table 1 Endurance and Resistance-Based Exercises Classified by Different Levels of Intensity

Category Term	Effort Perception	Relative Intensity	Energy Pathway Predominating	Representative Duration (min)	Other Terminology
Endurance exercise activities					
Light exercise	Easy	<35% VO_{2max}	Aerobic	>30	Short-term, submaximal
Moderate exercise	Modest difficulty	<70>35% VO_{2max}	Aerobic	30–180	Submaximal, prolonged
Heavy exercise	Difficult	<100>70% VO_{2max}	Aerobic–anaerobic	≤120	Submaximal, prolonged, high intensity
Maximal exercise	Strenuous	100% VO_{2max}	Aerobic–anaerobic	<15	Maximal or max, high intensity
Supramaximal exercise	Extremely strenuous	>100% VO_{2max}	Anaerobic	<1	All-out, power
Resistance exercise activities					
Submaximal exercise	Modest difficulty	<70>35% 1RM	Aerobic–anaerobic	<1	Submaximal
Maximal exercise	Extremely strenuous	~100% 1RM	Anaerobic	<0.1	All-out, power
Supramaximal exercise	Extremely strenuous	>100% 1RM	Anaerobic	≤0.1	Negatives

VO_{2max}—maximal oxygen uptake; 1RM—one repetition maximum.

changes to exercise occur for various physiological reasons: (a) to induce cardiovascular adjustments, (b) to activate energy production pathways and mobilize energy substrates, (c) to facilitate maintenance of adequate hydration, and (d) to some extent as part of stress reactivity. Table 2 illustrates the general changes for a variety of hormones in response to endurance versus resistance exercise activities.

Many of these hormonal responses are not independent of one another but are highly interrelated. To help illustrate this point, an explanatory model is presented, i.e., the "Hormonal Exercise Response Model" (HERM). This model explains that the hormonal responses to exercise are a series of three interactive phases.[2,11,12,15]

The first phase of this model deals with the hormonal response immediately at the onset of exercise, with these responses taking just seconds to occur. These responses revolve around the increased sympathetic nervous system activation that occurs with the onset of body motion. This increased sympathetic nervous system activity can also be a result of anticipation to the stress of the ensuing exercise—which is most certainly the case in sport competition scenarios. This increased sympathetic nervous system activity results in catecholamine (norepinephrine) release at target tissues directly, as well as elevations in circulating catecholamine from so-called sympathetic "spillover" effects.[11,15,16] This effect is further amplified by the sympathetic connection to the adrenal medullary gland which in turn adds to the circulating catecholamine (epinephrine) response.[15,17] Concurrent with these sympathetic-adrenal actions, pancreatic insulin secretion begins to be inhibited, while glucagon secretion becomes stimulated.[15] This entire process involves a feed-forward mechanism of the central nervous system to drive these initial responses, although the events are also modified by peripheral afferent neural input from sensory receptors, in particular, those of skeletal muscle.[13,17]

The intermediate or secondary phase takes slightly longer to develop but is still typically very fast beginning usually in much less than a minute from the onset of exercise. In this stage, the hypothalamus begins the process of releasing its hormones such as thyrotropin-releasing factor, corticotropin-releasing factor (CRF), and growth hormone-releasing factor in an attempt to provoke changes at the anterior pituitary gland to stimulate the release of specific hormones. As the pituitary begins to respond to the hypothalamic stimulus, its "trophic hormones" begin to be added to the circulation, and these hormones in turn begin to act upon their specific peripheral target glands to stimulate additional hormonal release. One of the most rapidly

Table 2 Physiological Responses to an Acute Exercise Session Based upon Whether the Activity Is Predominately Endurance or Resistance Exercise

Hormone	Exercise–Activity Predominating Training Component		Mechanism of Cellular Actions (Direct Gene Activation or 2nd Messenger)
	Endurance	**Resistance**	
Adrenocorticotrophic hormone (ACTH)	↑	↑	2nd messenger
Aldosterone	↑	↑	Direct gene activation
Angiotensin	↑	↑	2nd messenger
Antidiuretic hormone (ADH)	↑	↑↓	2nd messenger
Cortisol	↑	↑	Direct gene activation
Dehydroepiandrosterone (DHEA)	↑	↑	Direct gene activation
β-Endorphin	↑	↑	2nd messenger
Epinephrine (adrenaline)	↑↑	↑↑	2nd messenger
Estrogens	↑	↑	Direct gene activation
Follicle-stimulating hormone (FSH)	↑↓, nc	↑↓, nc	2nd messenger
Glucagon	↑	↑	2nd messenger
Growth hormone (GH)	↑	↑	2nd messenger
Insulin	↓	↑↓, nc	2nd messenger
Insulin-like growth factor-1 (IGF-1)	↑, nc	↑, nc	2nd messenger
Leptin	↑↓, nc	↑↓, nc	2nd messenger
Luteinizing hormone (LH)	↑↓, nc	↑↓, nc	2nd Messenger
Norepinephrine (noradrenaline)	↑↑	↑↑	2nd messenger
Progesterone	↑	↑	Direct gene activation
Prolactin (PRL)	↑	↑	2nd messenger
Testosterone	↑	↑	Direct gene activation
Thyroxine (T4)	↑	↑	Direct gene activation
Triiodothyronine (T3)	↑	↑	Direct gene activation
Vitamin D	↑	?	Direct gene activation

The hormone change denoted here are relative to before versus immediately after the activity.[11–14]
↑, increase; ↑↑, large increase; ↓, decrease; ↑↓, possible increase or decrease; nc, no change; ?, unresolved.

acting players in this cascade of events is the hypothalamic–pituitary–adrenocortical interaction where CRF brings about adrenocorticotrophic hormone release and that in turn brings about cortisol release.[12,13,18]

If the duration of an exercise session continues, there is a transition beyond the intermediate phase into a third phase of response which is a more prolonged state of events. In this third phase, the responses of the sympathetic-adrenal axis are being augmented by other hormones from the anterior and posterior pituitary (e.g., antidiuretic hormone, growth hormone, prolactin) and the peripheral endocrine glands subordinated to pituitary regulation (testosterone, thyroxine, triiodothyronine, insulin-like growth factor-1).[2,12,15] As fluids shift from the vascular space and total body water stores are compromised due to sweating for heat dissipation, the renin–angiotensin–aldosterone system (RAAS) is activated (inducing vasoconstrictive actions and water resorption).[8,11] Additionally, during this phase, the skeletal muscle begins to release selected cytokines (e.g., interleukin-6 [IL-6]), hormonal-like agents, into the circulation which affect other hormones to be released (IL-6 $\Rightarrow$ cortisol) which can have actions to signal energy substrate mobilization.[19]

Phases 1 and 2 of the model propose that neural factors are the primary stimuli regulating the hormonal responses to exercise; however, in the third phase of the model, there is an ever-increasing influence of the humoral and hormonal factors that regulate the overall responses due to the changes in the "internal milieu."[15] This shifting of primary regulatory factors allows an increasing reliance upon feedback rather than feed-forward control mechanism to determine the magnitude of the hormonal response. The influence of humoral and hormonal stimuli in modulating the hormonal levels is magnified as the exercise duration is extended and energy substrate availability issues causes shifts in energy fuel usage (i.e., decreased carbohydrate $\Rightarrow$ increased lipid), or hydration issues (i.e., hemoconcentration and/or dehydration) begin compromising the thermoregulatory ability and leading to greater heat storage within the body affecting hormonal responses (e.g., increased heat storage $\Rightarrow$ increased norepinephrine, epinephrine).[8,13,17] Interestingly, the core temperature changes with exercise (and via ambient environment amplification) results in an exceedingly greater hormonal response for a number of key endocrine agents.[20] This HERM model, while not perfectly inclusive, does provide an organized framework for the endocrine responses to exercise activities and illustrate the highly interactive and complexity of these responses.[2,11,12,15]

3. ACUTE EXERCISE

3.1 Why It Happens—Mechanism of Responses to Exercise

The prior discussion of the HERM centers on hormonal changes occurring to bring about physiological consequences that allow one to adjust to and mediate the homeostatic disruptions of an exercise session. These responses are not all brought about by glandular-tissue production. In fact, the basic mechanisms by which hormone levels (i.e., concentrations) in the blood change during exercise are associated with alterations in vascular fluid content, rate of hormonal metabolic clearance (MCR), and glandular-tissue secretion responses.

Vascular fluid content alterations are a result of plasma fluid moving into or out of the vascular spaces. During exercise, it is common for plasma volume "shifts" to occur, specifically a loss of fluid, resulting in a hemoconcentration (increased levels) of any substance trapped within the blood vessel. This would apply to large molecular weight hormones. Typically, in a 30–90 min exercise session a 10–15% hemoconcentration effect can be seen. Obviously, this is going to be exacerbated if the ambient environment condition creates a greater thermal load and/or an individual does not consume fluids during exercise. Conversely, during the recovery from exercise (when ample fluids are consumed), there tends to be a rebound effect and fluid moves into the vascular space to cause a hemodilution effect on substances such as hormones.[14,21]

The rate of MCR of a hormone is in reference to its removal from the blood. Hormones are "taken up" by receptor mechanisms at their target tissues which can implement physiological actions at the tissue (see discussion below). Additionally though, some clearance is through the degradation—deactivation removal process at tissues such as the hepatocytes, or via renal filtration.[14]

Glandular secretion involves the endocrine gland producing the hormone in response to a stimuli and the gland secreting more of the hormone in the blood. This can typically result in increases in blood levels; but it is possible for the rate of secretory increases and the rate of MCR removal to match one another. In this case, the level of the hormone in the blood does not seem to increase, but actually the turnover has and the target tissues are being presented with greater amounts of the hormone.[22]

Many exercise scientists believed that if hormonal levels increase in the blood, then the physiological actions of that hormone are going to be activated. This is, however, a gross over simplification of physiological events. Hormonal concentration in the blood is a key fundamental determinant of activation of physiological actions, but not the only. In addition, for a hormone to initiate and activate a physiological process there must be (a) adequate numbers of target tissue cells expressing functional receptors for the hormone, (b) adequate numbers of functional receptors on the cells, (c) a high affinity level of the receptors on the cells for the hormone, and (d) sufficient postreceptor amplification mechanisms within the cells.[22,23] This last point is highly critical and can be influenced by the behavior and actions of the athlete. As a simple illustration: increase testosterone ⇒ increase anabolism; only if adequate amino acids have been consumed in the diet and are available for cellular mechanisms (biochemical pathways, enzymes) to use in the protein synthesis process. The amplification mechanisms are essential for a positive adaptation to occur in skeletal muscle, arguably the most critical tissue—organ affecting exercise capacity.[14,24]

4. CHRONIC EXERCISE

4.1 Adaptive Responses to Exercise Training

The general responses of the body's various physiological systems to an exercise session after performing a progressive exercise training program (chronic exposure) are similar to those as before such training. In other words, an acute bout of exercise after chronic training is still a stimulus to the physiology of the body. However, at all exercise intensities, less than the maximal level, such responses are typically attenuated to some extent. The greater the training adaption incurred due to the training program, the greater the attenuation of the response. The exception to this occurrence is maximal or supramaximal exercise. In such situations, the training adaptations result in a greater level of workload performed at maximal or supramaximal efforts—for example, the athlete can run further or run faster over a fixed distance or lift a greater maximal amount weight, etc. The greater absolute maximal workload achievement in turn produces a greater physiological stimulus and thus comparable (or greater) maximal responses in the measures/parameters of the physiological systems; in other words, typically not an attenuation of responses. A noted allowance to this generalization is HR_{max}, as it is common for this parameter to be slightly lower or unchanged after a well-executed training program even when there is substantial and further cardiovascular adaptation.[8,9,13]

With respect to the endocrine system, typically after an adaptation and physical improvement to a training program, an exercise session (endurance- or resistance-based activities) still provokes a hormonal response. But, just as noted above, the responses tend to be attenuated. These mitigated responses come about by a greater sensitivity of target tissue to the hormonal stimulus and because the level of neural, humoral, and hormonal stimuli disturbances in the blood that influence the various endocrine glands is greatly diminished.[13] In relation to the former point of sensitivity, in response to an exercise training program, many target tissues increase the expression of functional hormone receptors, receptor affinity for hormones becomes increased, and postreceptor amplification mechanisms in the cells of target tissues are typically increased. Essentially, all these changes result in a target tissue needing less amount of a hormone to bring about a physiological outcome/change.

5. CHRONIC EXERCISE AND PERFORMANCE

5.1 Maladaptations Responses to Exercise Training

If exercise training programs are excessive in nature (excessive intensity, excessive volume, or both), or an athlete has too many additional life stresses compounding their situation during appropriate training, it is possible for inappropriate hormone responses (maladaptations) to occur. Such responses are usually associated with overreaching or overtraining scenarios. A detailed explanation of the overreaching–overtraining concepts is beyond the scope of this discussion, but the reader is directed to the excellent position statement published by the European Congress of Sport Science (see Ref. 25) for clarification on the training continuum and these concepts. In brief, however, if an individual is subjected to gradual increases of "training overload" followed by a period of time for rest and recovery, they will adapt in a positive fashion and promote adaptations leading to an improved performance capacity. If the training overload stimulus is too much, or inadequate rest occurs, then the individual may not be able to adapt and his/her performance declines. The person may progress from "normal" training to experiencing "overreaching" and then in due course "overtraining." If not corrected, this progression may ultimately develop into and display the clinical characteristics of the overtraining syndrome (OTS) (see Table 3), which is a serious medical condition.[25–28] Review of Table 3 indicates that there are several hormonal manifestations associated with development of the OTS. It should be noted that recently it has been proposed to use the

Table 3 The Major Symptoms of Overtraining Separated into Four Major Categories[2,4,25]

Physiological—performance	
Decreased performance	Decreased body fat
Decreased muscular strength	Increased $\dot{V}O_2$ at submaximal loads
Increased muscle soreness	Changes in heart rate (rest, exercise)
Prolonged recovery periods	Loss of appetite
Chronic fatigue	Gastrointestinal disturbances
Psychological	
Feelings of depression	General apathy
Difficulty concentrating	Emotional instability
Fear of competition	Excitation—irritability
Restlessness—loss of sleep	Anorexic behavior
Immunological dysfunction	
Increased susceptibility to infection	Increased severity of minor infections
Decreased functional activity of neutrophils	Decreased total lymphocyte counts
Reduced response to mitogens	Decreased production of immunoglobulin
Biochemical alterations	
Decreased hemoglobin	Negative nitrogen balance
Increased urea levels	Decreased free testosterone levels
Decreased ratio of free testosterone to cortisol ratio of more than 30%	Elevated cortisol levels

term "unexplained underperformance syndrome" rather than the OTS in order to account for the role life stresses can play as just training load alone may not be the total causative factor in the syndrome development.[29–32]

The research in exercisers (i.e., athletes) who are overreaching–overtraining suggests that the endocrine system's hormonal responses seem to occur in two phases: an initial hyperactivity phase, followed by a latter hypoactivity phase (see Table 4).[27,28,33,34] In the hyperactivity phase, elevations in the circulating levels of hormones such as adrenocorticotrophic hormone, cortisol, prolactin, and catecholamines have been reported at rest and/or in response to an acute exercise session,[27,33–36] although these hormonal findings are not completely universal.[25,36] This hyperactivity phase may be reflective of the "overreaching" status in the training continuum. Interestingly, it should be noted that in some situations during "overreaching," if the athlete is given short-term rest and recovery

Table 4 Pathophysiologic Findings in Hyperactivity Versus Hypoactivity Phases of the Overtraining Syndrome (OTS)

Parasympathetic, Hypoactivity	Sympathetic, Hyperactivity
Decrease in physical performance	Decrease in physical performance
Easily fatigued	Easily fatigued
Depression	Hyperexcitability
Lack of motivation to exercise train	Lack of motivation to exercise train (?)
Normal sleep—excessive sleep	Disturbed sleep—insomnia
Normal constant weight	Weight loss
Low resting HR	Increased resting HR and blood pressure
Hypoglycemia during exercise	Slow recovery of HR and BP after exercise
Loss of competitive desire	Loss of competitive desire
Gonadal-reproductive function normal (?)	Amenorrhea in women Hypogonadism in men
Increased incidence of infections	Increased incidence of infections
Decreased maximal lactate response to exercise	Decreased maximal lactate response to exercise

BP, blood pressure; HR, heart rate; ?, too little, or conflicting, data exist to make conclusive statement.[25–32]

(~2 weeks), they may actually compensate with greater than normal adaptations and enhancements of performance.[25] The hypoactivity phase seems to more closely correspond to the "overtraining" status and/or with the occurrence of the OTS. For this phase, certain hormones (adrenocorticotrophic hormone, catecholamines, cortisol, growth hormone, follicle-stimulating hormone, luteinizing hormone, testosterone, thyroids) are found to be suppressed.[35–38] The development of the hypoactivity phase appears to be the more serious outcome as it may require months for the individual to rest and recover in order to regain normal endocrine function and performance capacity.[35] Space limitations do not allow an extensive discussion of this topic or the contrasting aspects of overtraining characteristics and symptomology in endurance versus resistance-based activities (see Refs. 25 and 28 for discussions).

Currently, it is unclear as to what is the physiological mechanism that induces the OTS. The most prevailing thoughts suggest that clinically the syndrome may reflect some degree of the exhaustion stage of Selye's General Adaptation Syndrome.[39]

Several prevailing theories exist as to why the syndrome develops. Currently, the most compelling theory is that put forth by Smith[31,32] who proposes that with overtraining there is an excessive musculo-skeletal loading (compounded by inadequate rest and recovery) resulting in tissue damage; hence, local and systemic inflammation develops and becomes excessive. Proinflammatory cytokines (hormone-like substances with endocrine actions), part of the immune system, such as IL-6, tumor necrosis factor-α, and IL-1β in turn act upon multiple levels of the hypothalamic–pituitary–adrenal axis, most notably the hypothalamic paraventricular nucleus where CRF production occurs.[31,32] The neuropeptide CRF influences adrenocorticotrophic hormone secretion (and the whole hypothalamic–pituitary–adrenal axis) and can also affect mood, sexual, and additional immune functions,[33,35] either directly acting on brain sensitive regions or indirectly via the sympathetic nervous system.[26,27,35] Furthermore, the cytokine changes noted may facilitate a suppression of the cell-mediated aspects of the adaptive immune system. This creates an increased risk for infection and the so-called "sick response or sickness behavior" which is associated with many of the characteristics found in overtrained athletes.[30,31,40] As these systems interact, the negative events of overtraining intensify, and the athlete spirals downward from being healthy and physically fit to syndrome development and psychophysiologically compromised.[25,41–43]

6. CELLULAR AND MOLECULAR ASPECTS OF EXERCISE ENDOCRINOLOGY

6.1 Background Basics

As noted in Section 1, exercise training is an important contributor to a better quality of life (i.e., health and fitness) and improved sporting–physical performance of individuals through its impact on a variety of physiological systems. The role of the endocrine system on these adaptive responses, however, is not just at the systemic organismal physiologic level. A multitude of cellular and molecular events also occur and are regulated by the endocrine system. The specific means of action for these cellular–molecular events is dependent upon the major chemical classification and activation mechanisms of hormones (e.g., steroidal—direct gene activation vs. peptide—secondary messenger activation; see Table 2).

Steroid hormones (and thyroidal hormones) belong to a family of agents that signal through the nuclear receptors that act as transcriptional factors.[44] Structurally, all of these hormones are small lipophilic molecules that can pass through cell membranes freely and even reach the chromatin directly. These hormones interact with their associated nuclear receptors, which form a superfamily comprising two broad classes, Type I and Type II receptors.[44] Type I nuclear receptors reside in the cytosol through interactions with heat-shock proteins. When hormones bind to Type I receptors, the receptors are activated, released from the heat-shock complex, and translocate to the cell nucleus, where they bind to DNA sequences known as hormone-responsive elements and initiate transcription. Type II receptors reside in the nucleus even in the absence of a ligand and bind to *cis*-elements as heterodimers. The activated receptor interacts with the hormone response element, and the transcription process is initiated as with Type I receptors. Furthermore, evidence also exists for the presence of cell membrane-bound receptor mechanisms for some steroidal hormones.[45] The circulating levels of the steroid hormones are dependent upon the individual rate of biosynthesis, the amount of their binding proteins in the circulation (which affects how much free biologically active hormone is available), and the level of degradation or activation (via receptors) through the MCR process.[46]

The other major class of hormones, peptides, is similar to other proteins in that they are encoded by one to two genes for different subunits.[47] These hormones are frequently released as "prohormones" into the circulation and transformed into active hormonal forms through steps of intracellular processing and posttranscriptional modifications. The actions of these hormones are mediated through interactions with cell membrane receptors, which when activated begins a cascade of reactions in the target cells usually involving phosphorylations—typically referred to as the secondary messenger process (e.g., involving cAMP, cGMP, calcium, or other mediating agents). The action of peptide hormones is normally faster than that of steroid hormones at the target cells because the immediate effects of these hormones are mediated by enzymatic processes in the cytoplasm without involving gene transcription and the synthesis of new proteins.[48] Furthermore, the titer and biological activity of these hormones are not moderated by specific binding proteins as seen with most steroid hormones.

6.2 Epigenetics and Exercise

Growing evidence points to exercise induced epigenetic events or alterations as being a critical mechanism for effecting changes and responses in physiological systems.[49] In simplistic terms, epigenetics alterations are changes that occur in the DNA or the chromatin structures of a cell that influence the transcription of genes independent of their primary sequence (i.e., controlling how genes are regulated without changing the actual gene nucleotide sequence). Common epigenetic changes induced by exercise are (a) the expression of different types of microRNAs (miRNAs), (b) histone modifications (such as methylation and acetylation), and (c) DNA methylation–demethylation.[50] In brief, miRNAs are a group of small noncoding RNA molecules that generally function to mitigate or silence protein translation. Histone modifications are posttranslational alterations on the lysine-rich tail region of histones (especially H3 and H4 histones). The DNA methylation (demethylation) is a reversible process which is catalyzed by a family of DNA methyltransferases (DNMTs) enzymes (see Fig. 1).[51] A more detailed explanation of the specific steps involved with these epigenetic processes is beyond the scope of this chapter due to space limitations, but the reader is directed to Ref. 51.

Relative to exercise, studies point to the important role DNA methylation and histone modification have in regulating key enzymes associated with steroidal hormone biosynthesis and modification, or actual expression of receptors (e.g., STAR—steroidogenic acute regulatory protein, a transport protein that is the rate-limiting step in the production of steroid hormones).[52] Additionally, genes encoding peptide hormones or their receptors are potential epigenetic targets. Evidence points to some, but not all, peptide

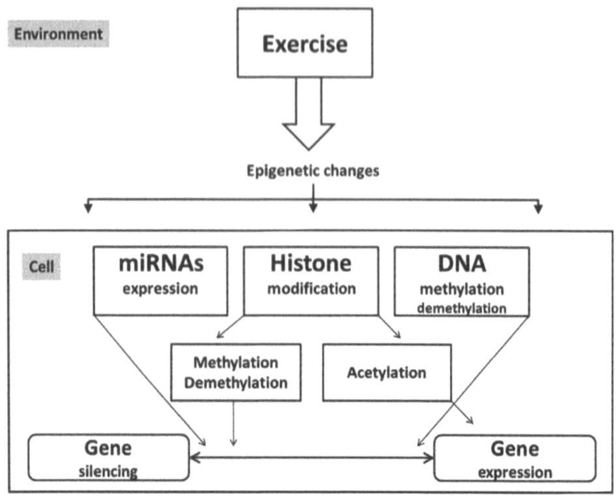

Figure 1 Proposed mechanism by which exercise can influence and develop epigenetic changes.

hormone genes being influenced by DNA methylation and/or miRNAs. Interestingly, epigenetic regulations of genes encoding peptide hormones or their receptors are largely related to developmental stage- and tissue-specific function or the development of metabolic or neural disorders. There does not seem to be an entirely analogous situation for steroid-type hormones.[53]

Finally, it is well established that endocrine glands and their target tissues/organs, because they function to maintain homeostasis in the body, must be highly responsive to environmental changes. The hormones released by these glands have an important role in relating and relaying environmental conditions as they signal the type, severity, and duration of different environmental cues and stresses which can disrupt homeostasis. Hormones, therefore, act as epigenetic signals in developmental programming and can allow transmission of nongenomic factors important in developing the optimal phenotype for functioning and survival.[54] This suggests that hormones are influenced by epigenetic environmental factors (e.g., exercise, or the lack of it) and also serve as a means to induce epigenetic alterations; hence "it is a two way street."

7. CONCLUSIONS

In the endocrinological study of exercise, technological advancements and increasing rigorous experimental approaches have allowed for new discoveries in the existence of new hormones, hormonal interactions, and the hormonal-like effects of many substances released from both glandular and nonglandular tissues. For these reasons, the last 20 years have been an extremely exciting time in this field of endeavor. There is still a tremendous amount of uncertainty and questions in need of addressing. In particular, the scientists who work in exercise aspects of endocrinology need to focus on not just "what" happens with exercise, but "why" it happens, and what are the "consequences" of it happening. Researchers have well defined what the acute and chronic effects of exercise (and exercise training) are to either increase or decrease the presence of many known hormones; but, the specific regulatory events to induce such actions and outcomes from the hormonal change induced are not completely understood and are an issue of continued debate in the science community.[21,55–57] The next-generation exercise endocrinologists must develop studies to address such points of "why," "how," and the "consequences of" (especially in light of the new epigenetic aspects of exercise adaptations[58]) if we are to gain a true understanding of how the integral and interacting aspect of acute and chronic exercise exposure completely affects the endocrine system and the processes under the control of the hormones released by the system.

REFERENCES

1. Bunt JC. Hormonal alterations due to exercise. *Sports Med.* 1986;3(33):1–345.
2. Hackney AC. Exercise as a stressor to the neuroendocrine system. *Medicina.* 2006;42(10):788–797.
3. Galbo H. The hormonal response to exercise. *Diabetes Metab Rev.* 1986;1(4):385–408.
4. Urhausen A, Gabriel H, Kindermann W. Blood hormones as markers of training stress and overtraining. *Sports Med.* 1995;20:251–276.
5. Garber CE, Blissmer B, Deschenes MR, et al. American College of Sports Medicine (ACSM) position stand: quantity and quality of exercise for developing and maintaining cardiorespiratory, musculoskeletal, and neuromotor fitness in apparently healthy adults: guidance for prescribing exercise. *Med Sci Sports Exerc.* 2011;43(7):334–1359.
6. Joyner MJ. Physiologic limiting factors and distance running: influence of gender and age on record performances. *Exerc Sport Sci Rev.* 1993;21:103–133.
7. Joyner MJ, Coyle EF. Endurance exercise performance: the physiology of champions. *J Physiol.* 2008;586(1):35–44.
8. Rowell LB. Cardiovascular adjustments to thermal stress. In: Lindsey M, Hester R, eds. *Handbook of Physiology, The Cardiovascular System, Peripheral Circulation and Organ Blood Flow (Supplement 8).* New York, NY: Wiley-Blackwell; 1983:967–1023.
9. Saltin B, Astrand PO. Maximal oxygen uptake in athletes. *J Appl Physiol.* 1967;23:353–358.
10. Kraemer W, Fleck S. *Optimizing Strength Training: Designing Nonlinear Periodization Workouts.* Champaign, IL: Human Kinetics Publishing; 2007.
11. Galbo H. *Hormonal and Metabolic Adaptation to Exercise.* Stuttgart, Germany: Thieme Verlag; 1983.

12. Hackney AC. Stress and the neuroendocrine system: the role of exercise as a stressor and modifier of stress. *Expert Rev Endocrinol Metab*. 2006;1(6):783–792.
13. Viru A, Viru M. *Biochemical Monitoring of Sport Training*. Champaign, IL: Human Kinetics; 2001.
14. Viru A. *Hormonal Ensemble in Exercise—Hormones in Muscular Activity*, vol. 1. Boca Raton, FL: CRC Press; 1985.
15. Christensen NJ, Galbo H. Sympathetic nervous activity during exercise. *Annu Rev Plant Physiol Plant Mol Biol*. 1983;45:139–145.
16. Kjaer M, Secher NH, Galbo H. Physical stress and catecholamine release. *Baillieres Clin Endocrinol Metab*. 1987;1(2):279–298.
17. Viru A. Plasma hormones and physical exercise. *Int J Sports Med*. 1992;13(3):201–209.
18. Davies CT, Few JD. Effects of exercise on adrenocortical function. *J Appl Physiol*. 1973;35(6):887–891.
19. Steenberg A, Fischer CP, Keller C, Møller K, Pedersen BK. IL-6 enhances plasma IL-1ra, IL-10, and cortisol in humans. *Am J Physiol Endocrinol Metab*. 2003;285(2): E433–E437.
20. Radomski MW, Cross M, Buguet A. Exercise-induced hyperthermia and hormonal responses to exercise. *Can J Physiol Pharmacol*. 1998;76:547–552.
21. Hackney AC, Viru A. Research methodology: issues with endocrinological measurements in exercise science and sport medicine. *J Athl Train*. 2008;43(6):631–639.
22. Goodman HM. Endocrinology concepts for medical students. *Adv Physiol Educ*. 2001;25:213–224.
23. Widmaier EP. Metabolic feedback in mammalian endocrine systems. *Horm Metab Res*. 1992;24:147–153.
24. Brooks G, Fahey TD, White TP, Booth F. *Exercise Physiology: Human Bioenergetics and Its Application*. 4th ed. New York, NY: McGraw Hill Company; 2005.
25. Meeusen R, Duclos M, Gleeson M, Rietjens G, Steinacker J, Urhausen A. Prevention, diagnosis and treatment of the Overtraining Syndrome: ECSS position statement 'Task Force'. *Eur J Sport Sci*. 2006;6:1–14.
26. Lehmann M, Foster C, Keul J. Overtraining in endurance athletes: a brief review. *Med Sci Sports Exerc*. 1993;25:854–862.
27. Lehmann M, Knizia K, Gastmann U, et al. Influence of 6-week, 6 days per week training on pituitary function in recreational athletes. *Br J Sports Med*. 1993;27:186–192.
28. Lehmann M, Lormes W, Opitz-Gress A, et al. Training and overtraining: an overview and experimental results in endurance sports. *J Sports Med Phys Fitness*. 1997;37:7–17.
29. Budgett R. Fatigue and underperformance in athletes: the overtraining syndrome. *Br J Sports Med*. 1998;32:107–110.
30. Robson-Ansley PJ, Blannin A, Gleeson M. Elevated plasma interleukin-6 levels in trained male triathletes following an acute period of intensive interval training. *Eur J Appl Physiol*. 2007;99:353–360.
31. Robson-Ansley JP, Lakier-Smith L. Causes of extreme fatigue in underperforming athletes—a synthesis of recent hypotheses and reviews. *S Afr J Sports Med*. 2006;18:108–114.
32. Smith LL. Tissue trauma: the underlying cause of the overtraining syndrome? *J Strength Cond Res*. 2004;18:185–193.
33. Raglin J, Barzdukas A. Overtraining in athletes: the challenge of prevention—a consensus statement. *Health Fit J*. 1999;3:27–31.
34. Viru A. The mechanism of training effects: a hypothesis. *Int J Sports Med*. 1985;5(5):219–227.
35. Fry AC, Steinacker JM, Meeusen R. Endocrinology of overtraining. In: Kraemer WJ, Rogol AD, eds. *The Endocrine System in Sports and Exercise*. Oxford, UK: Blackwell Publishing; 2005:584–593.

36. Hackney AC. Hormonal changes at rest in overtrained endurance athletes. *Biol Sport.* 1991;8:49–56.
37. Barron JL, Noakes TD, Levy W, Smith C, Millar RP. Hypothalamic dysfunction in overtrained athletes. *J Clin Endocrinol Metab.* 1985;60:803–806.
38. Hackney AC. Effects of endurance exercise on the reproductive system of men: the "exercise-hypogonadal male condition" *J Endocrinol Invest.* 2008;31:932–938.
39. Selye H. A syndrome produced by diverse nocuous agents. *Nature.* 1936;138:32–33.
40. Robson PJ. Elucidating the 'unexplained underperformance syndrome' in endurance athletes. *Sports Med.* 2003;33:771–781.
41. Hackney AC, Battaglini C. The overtraining syndrome: neuroendocrine imbalances in athletes. *Braz J Biomotricity.* 2007;1(2):34–44.
42. Kreider R, Fry AC, O'Toole M. Overtraining in sport: terms, definitions, and prevalence. In: Kreider R, Fry AC, O'Toole M, eds. *Overtraining in Sport.* Champaign, IL: Human Kinetics; 1998.
43. Hackney AC, Koltun KJ. The immune system and overtraining in athletes: clinical implications. *Acta Clin Croat.* 2012;51(4):633–641.
44. Evans RM. The steroid and thyroid hormone receptor super-family. *Science.* 1988;240:889–895.
45. Maggiolini M, Picard D. The unfolding stories of GPR30, a new membrane-bound estrogen receptor. *J Endocrinol.* 2010;204(2):105–114.
46. Novac N, Heinzel T. Nuclear receptors: overview and classification. *Curr Drug Targets Inflamm Allergy.* 2004;3:335–346.
47. Pearson RK, Anderson B, Dixon JE. Molecular biology of the peptide hormone families. *Endocrinol Metab Clin North Am.* 1993;22:753–774.
48. Jansen RP. Endocrine response in the fallopian tube. *Endocr Rev.* 1984;5:525–551.
49. Barrès R, Yan J, Egan B, et al. Acute exercise remodels promoter methylation in human skeletal muscle. *Cell Metab.* 2012;15(3):405–411.
50. Eccleston A, Dewitt N, Gunter C, Marte B, Nath D. Introduction epigenetics. *Nature.* 2007;447:396–440.
51. Ntanasis-Stathopoulos J, Tzanninis JG, Philippou A, Koutsilieris M. Epigenetic regulation on gene expression induced by physical exercise. *J Musculoskelet Neuronal Interact.* 2013;13(2):133–146.
52. Hiroi H, Christenson LK, Strauss JF. III Regulation of transcription of the steroidogenic acute regulatory protein (StAR) gene: temporal and spatial changes in transcription factor binding and histone modification. *Mol Cell Endocrinol.* 2004;215:119–126.
53. Zang X, Ho SM. Epigenetics meets endocrinology. *J Mol Endocrinol.* 2011;46(1): R11–R32.
54. Fowden AL, Forhead AJ. Hormones as epigenetic signals in developmental programming. *Exp Physiol.* 2009;94(6):607–625.
55. West DW, Phillips SM. Anabolic processes in human skeletal muscle: restoring the identities of growth hormone and testosterone. *Phys Sportsmed.* 2010;38(3):97–104.
56. Vingren JL, Kraemer WJ, Ratamess NA, Anderson JM, Volek JS, Maresh CM. Testosterone physiology in resistance exercise and training: the up-stream regulatory elements. *Sports Med.* 2010;40(12):1037–1053.
57. Fry AC, Kraemer WJ. Resistance exercise overtraining and overreaching: neuroendocrine responses. *Sports Med.* 1997;23(2):106–129.
58. Phillips BE, Williams JP, Gustafsson T, et al. Molecular networks of human muscle adaptation to exercise and age. *PLoS Genet.* 2013;9(3):1–15 [e1003389].

CHAPTER THIRTEEN

Exercise and Regulation of Adipokine and Myokine Production

Sven W. Görgens*, Kristin Eckardt†, Jørgen Jensen‡, Christian A. Drevon†, Jürgen Eckel*,§,1

*Paul-Langerhans-Group for Integrative Physiology, German Diabetes Center (DDZ), Auf'm Hennekamp, Düsseldorf, Germany
†Department of Nutrition, Institute of Basic Medical Sciences, Faculty of Medicine, University of Oslo, Oslo, Norway
‡Department of Physical Performance, Norwegian School of Sport Sciences, Oslo, Norway
§German Center for Diabetes Research (DZD e.V.), Düsseldorf, Germany
[1]Corresponding author: e-mail address: eckel@uni-duesseldorf.de

Contents

Abstract

Skeletal muscle and white adipose tissue are the largest organs in the human body and both tissues act as endocrine organs capable of secreting many bioactive molecules. There has been some confusion about nomenclature and we suggest that the name myokine should be restricted to a protein or molecule secreted from myocytes, whereas the term adipokine should be used to describe proteins and molecules secreted from adipocytes. In fact, many myokines are also produced by adipocytes and we propose to name them adipo-myokines. Many adipo-myokines produced by skeletal muscle or adipose tissue are influenced by exercise. Therefore, it is likely that adipo-myokines may contribute in the mediation of the health benefits of exercise and physical inactivity probably leads to an altered adipo-myokine profile, which could provide a potential mechanism for the association between sedentary behavior and many chronic diseases. Within this review, we evaluate the effects of acute and chronic exercise on myokine, adipokine, and adipo-myokine production. By using the adipo-myokine concept and including both skeletal muscle and adipose tissue, an attempt is made to gain a global view on the beneficial effects of different exercise programs and the underlying pathways.

Progress in Molecular Biology and Translational Science, Volume 135
ISSN 1877-1173
http://dx.doi.org/10.1016/bs.pmbts.2015.07.002

1. INTRODUCTION

Skeletal muscle plays a key role in postural control, locomotion, and other physiological tasks requiring mechanical activity based on muscle fiber contraction. As the largest organ in the body, the energy production and consumption by skeletal muscle are fundamental for metabolic control and homeostasis. More recently, skeletal muscle has gained considerable interest as an endocrine organ and the release of myokines from contracting muscle is assumed to be at least partly responsible for the health-promoting effects of physical activity[1] that protects against major chronic, low-grade inflammatory diseases like type 2 diabetes, insulin resistance, the metabolic syndrome, and many others. Myokines are part of a complex network of interorgan communication and exert both auto-/paracrine and endocrine effects.[1] The most extensively studied myokine is interleukin 6 (IL-6), which is profoundly upregulated during acute exercise and assumed to play a key role in the anti-inflammatory effect of acute exercise.[2] However, it is now evident that skeletal muscle secretes hundreds of myokines, with many of them being regulated by muscle contraction.[3] At present, for the majority of these molecules, the biological function has remained elusive. An additional complexity results from the broad spectrum of physical activity used in lifestyle intervention studies and profound differences in the exercise training programs. The textbox summarizes some of the major acute and chronic exercise training programs that were used in the studies considered for this review.

In this chapter, we have used the term acute exercise to describe structured exercise mostly performed in laboratory settings under controlled intensity and duration. A single bout of acute exercise will normally not cause major adaptations in skeletal muscles or other organs, but acute exercise can increase insulin sensitivity for the following 24–48 h. In this chapter, we divide training as endurance and strength training and do not put much attention on the fact strength as well as endurance training can be varied unlimited by changing intensity, duration, and volume. Acute endurance exercise in laboratory settings is normally performed at intensities corresponding to ~70% of maximal oxygen consumption for duration of 30–60 min. Strength training is normally performed with heavy loads related to personal maximal strength. Intensity of strength training can be related to maximal strength in the specific exercise (1-RM, 1 repetition maximum). Intensity can also be related to the weight that can be lifted maximally, e.g., 8 times (8-RM). Strength training at intensities of 70–80% of 1-RM (~8-RM) is known to improve strength. In studies of acute strength exercise loads of 70–80%, 1-RM is normally used and sets are repeated 3–5 times with 2–3 min rest between in sets. Relating training load to individual strength (or VO_{2max}) normalizes to some degree physiological responses. However, well-trained subjects tolerate high-intensity training better than untrained subjects and myokine responses may vary. Data from young subjects are not always descriptive for old subjects.

In addition to being the paradigm myokine, IL-6 is a pro-inflammatory adipo-cytokine and produced by adipose tissue in substantial amounts. In fact, many myokines are also produced by adipose tissue and we suggested to call them adipo-myokines[4] (see Fig. 1). These molecules are mediators of exercise and mediators of inflammation, most likely exerting as of yet incompletely understood time- and concentration-dependent functions. To gain a more comprehensive view on the molecular pathways involved in physical activity, studies of myokines, adipokines, and adipo-myokines, including those produced by adipose tissue, are required. In fact, exercise training has been reported to reduce central adiposity independent of overall weight changes,[5] and this may represent an additional mechanism of the anti-inflammatory action of acute exercise. Unfortunately, only very limited information is available regarding direct effects of exercise on adipokine and adipo-myokine production.

In this review, by focusing on the adipo-myokine concept and including both skeletal muscle and adipose tissue, an attempt is made to gain a more global view of the beneficial effects of different exercise programs and the

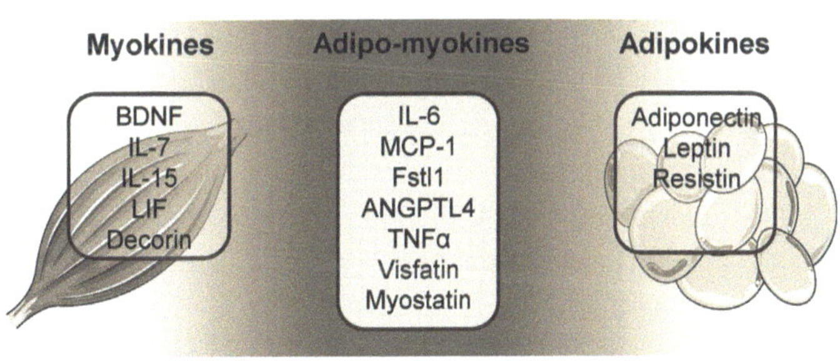

Figure 1 The adipo-myokine concept. A search of original articles in PubMed was performed for the major exercise-regulated myokines and adipokines to identify molecules that were produced and secreted in both tissues. The term adipo-myokines was used for proteins fulfilling both of these criteria. The search terms we used were "skeletal muscle" or "adipose tissue," "myokine" or "adipokine," and "exercise." (See the color plate.)

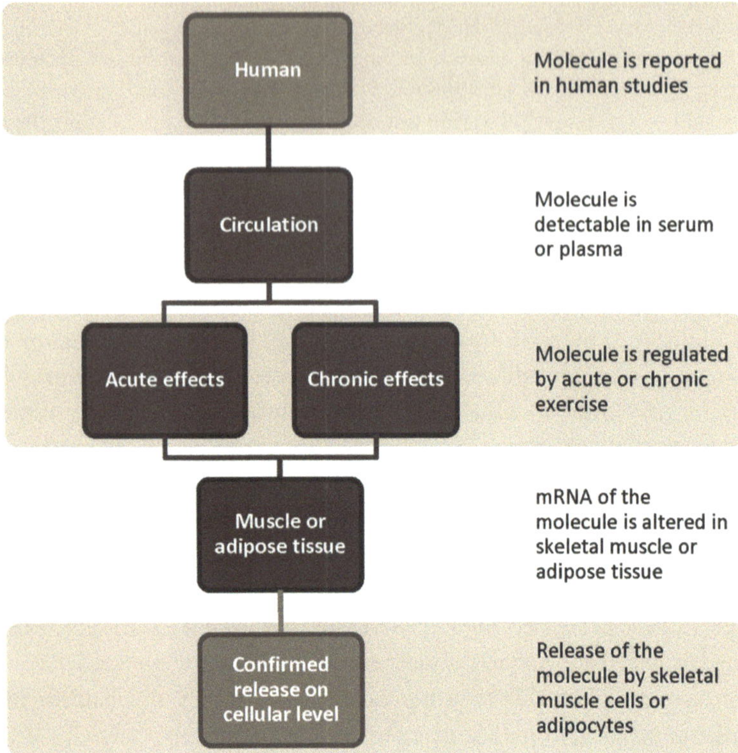

Figure 2 Scheme for the approach to selecting published studies. A search of original articles in PubMed was performed to identify myokines, adipokines, or adipo-myokines, which satisfied all of the following criteria: The molecule must be human and detectable in serum or plasma. Furthermore, the molecule must be regulated by either acute or chronic exercise and expressed in skeletal muscle or adipose tissue. The protein has to be secreted by adipocytes or myocytes. The search terms we used were "skeletal muscle" or "adipose tissue," "myokine" or "adipokine," and "exercise." (See the color plate.)

underlying pathways. The general approach for selecting published studies is outlined in Fig. 2. Thus, we only considered human studies and molecules that are detectable in the circulation. This does not exclude auto-/paracrine effects of adipo-myokines, but emphasizes the cross talk between fat and muscle and the potential mediators of this process.

2. SKELETAL MUSCLE AS THE SOURCE OF MYOKINES AND ADIPO-MYOKINES AFTER ACUTE AND CHRONIC EXERCISE

IL-6 is the prototype of an adipo-myokine meaning that it is expressed and released by both skeletal muscle and adipose tissue.[4] It acts in an auto-/

paracrine fashion within skeletal muscle and endocrine in a hormone-like fashion to mediate metabolic and anti-inflammatory effects.[6] IL-6 has been shown to trigger an anti-inflammatory cascade by promoting the induction of anti-inflammatory factors, such as IL-10 and IL-1ra (IL-1 receptor agonist), and to inhibit the production of the pro-inflammatory cytokines IL-1ß and TNFα.[1] Moreover, IL-6 plays a role in hypertrophic muscle growth.[7] Skeletal muscle releases IL-6 into the circulation in response to acute muscle contractions. The increase of circulating IL-6 in response to acute exercise occurs in an exponential manner and the maximum is reached at the end of the exercise session.[6] The magnitude by which IL-6 levels increase is related to the type of exercise, its duration, intensity, and the amount of muscle mass engaged in the exercise but is not affected by muscle damage.[6,8,9] Moreover, preexercise muscle glycogen stores impact on the amount of IL-6 released during exercise.[10,11] The most pronounced increase of plasma IL-6 has been observed with running, which involves several large muscle groups.[8] Based on data from several epidemiological studies, a negative association between the extent of regular physical activity and the basal plasma IL-6 levels has been described; basal plasma IL-6 concentration is more closely associated with physical inactivity than other cytokines associated with the metabolic syndrome.[8] However, data on the effect of regular exercise on circulating IL-6 concentrations are more controversial. Aerobic training for 10 months with overweight elderly subjects (age $\geq$64) reduced basal plasma levels of IL-6, whereas strength training had no effect.[12] A study with severely obese humans that used a combination of hypocaloric diet and regular physical activity for 15 weeks reported decreased basal plasma levels of IL-6 and IL-6 mRNA expression in skeletal muscle along with a 13% reduction of body weight.[13] However, other studies observed no changes in circulating IL-6 concentrations in response to chronic exercise.[14,15] In summary, IL-6 is released from contracting human skeletal muscle, and exercise-induced IL-6 has metabolic effects in human, and those include effects on insulin-stimulated glucose disposal and fatty acid oxidation.

Another member of the IL-6 superfamily has been classified as a myokine, namely leukemia inhibitory factor (LIF). LIF has multiple biological functions such as induction of satellite cell proliferation, which is essential for muscle hypertrophy and regeneration after muscle damage.[16] Broholm *et al.* reported in young healthy volunteers that LIF mRNA expression in skeletal muscle increased following acute aerobic exercise (4.5-fold)[17] and heavy resistance exercise (9-fold).[16] However, LIF protein level in skeletal muscle tissue was not changed,[17] and LIF could not be detected in plasma in the same study.[16] Due to the paucity of data, future studies should address regulation of plasma LIF levels during acute or chronic exercise to determine whether the increased LIF mRNA expression in the muscle also is reflected in the circulation.

More recently, interleukin 7 (IL-7) was described as a novel exercise-regulated myokine. It belongs to the interleukin superfamily 2 and is required for T-cell and B-cell development, but its relevance in nonimmune cells has not been sufficiently explored. A study by Haugen *et al.* has shown that IL-7 is expressed and secreted by primary human skeletal muscle cells.[18] *In vitro* experiments suggested that IL-7 increased migration of satellite cells without affecting their proliferation.[18] There are only a few studies which have examined the effect of acute or chronic exercise on circulating IL-7 concentrations. Andersson *et al.* observed a robust increase in plasma IL-7 levels following a 90-min soccer games in elite female soccer players.[19] In young healthy men, acute resistance exercise increased plasma IL-7 levels 30 min postexercise, while 12 weeks of chronic resistance training had no effect on circulating IL-7 concentrations.[20] Skeletal muscle mRNA expression of IL-7 was reported to be increased after 11 weeks of strength training in healthy males.[18] It has been concluded that IL-7 is an exercise-regulated myokine and that it may play a role in the regulation of muscle cell development.

Another member of the interleukin superfamily 2 is interleukin 15 (IL-15). The potential effect of exercise on IL-15 expression and secretion by skeletal muscle is still unclear. In marathon runners, Nieman *et al.* found no alteration of IL-15 mRNA expression in skeletal muscle immediately after a 3-h treadmill run.[21] Similarly, another study performed with healthy, physically active men reported no difference in skeletal muscle IL-15 mRNA expression and circulating IL-15 after 3 h of cycling.[22] Furthermore, chronic endurance training for 12 weeks had no effect on plasma IL-15 levels and skeletal muscle mRNA expression. Interestingly, increased IL-15 protein content was found in skeletal muscle in this study.[22] In contrast, Tamura *et al.* have shown that 30-min treadmill running at 70% of maximum heart rate promoted a significant increase in circulating IL-15 levels in untrained healthy young men.[23] Another study reported increased plasma IL-15 levels directly after acute resistance exercise in young, healthy but previously inactive volunteers, whereas no change of plasma IL-15 concentrations was observed after chronic resistance training for 10 weeks.[24] In support of these results, a study by Nielsen *et al.* described increased skeletal muscle mRNA expression of IL-15 after acute heavy resistance exercise in healthy, normally active men. However, no changes in skeletal muscle IL-15 protein content or plasma IL-15 were found.[25] Studies investigating the effect of exercise on IL-15 in overweight and obese humans are still missing. So far, it remains an open question which type of exercise is able to impact on IL-15 and how important are duration and intensity of the training program, as well as preexercise activity level and health status of the participants.

Myostatin is a member of the transforming growth factor beta (TGF-β) superfamily that negatively regulates skeletal muscle size and was the first described myokine.[26]

When subjected to moderate aerobic exercise training for 6 months, overweight or obese men displayed a reduction of skeletal muscle expression and circulating levels of myostatin.[27] Moreover, acute endurance[28] or resistance exercise[29–31] was shown to attenuate myostatin mRNA expression in skeletal muscle. Importantly, myostatin is antagonized by different factors such as follistatin[32] and decorin.[33] Following acute resistance exercise, Dalbo *et al.* have found no effect on skeletal muscle follistatin mRNA expression in lean young and old men,[31] while another study performed with postmenopausal women reported enhanced follistatin mRNA expression.[30] In addition, it was recently shown that decorin mRNA expression is enhanced in skeletal muscle after chronic exercise combining strength and endurance training.[34] A study by Heinemeier *et al.* reported increased decorin mRNA expression in skeletal muscle following an acute endurance exercise,[35] and decorin plasma levels were increased after acute resistance exercise.[34]

Another antagonist of proteins of the TGF-β superfamily is follistatin-like 1 (FSTL1) which is a secreted glycoprotein belonging to the follistatin family of proteins.[36] Studies indicate that FSTL1 may have beneficial effects on ischemia–reperfusion injury in muscle and heart tissue associated with antiapoptosis.[37,38] We have recently shown that FSTL1 is a myokine expressed and secreted by primary human skeletal muscle cells. Moreover, FSTL1 plasma levels were increased after acute endurance exercise performed by young healthy men.[39] In line with these results, Norheim *et al.* reported elevated *FSTL1* mRNA expression in skeletal muscle from healthy male volunteers after chronic strength training.[40] We can conclude that members of TGF-β superfamily and their specific antagonist are strongly regulated by acute and chronic exercise and are playing a role in exercise-related restructuring processes of skeletal muscle.

Brain-derived neurotrophic factor (BDNF) also has been described as an exercise-induced myokine,[41] although the protein and its receptor are most abundantly expressed in the brain.[42] Acute endurance exercise increases plasma BDNF levels,[41,43,44] specifically with high-intensity exercise.[45] In addition, Matthews *et al.* found significantly increased BDNF mRNA and protein abundance in skeletal muscle after acute endurance exercise.[41] Moreover, circulating BDNF concentrations increased in response to chronic endurance training.[46] On the other hand, a study using resistance exercise could not find an effect on BDNF plasma levels after acute or chronic exercise.[47] In humans, it should be noted that 70–80% of plasma BDNF originates from the brain during both rest and after exercise, suggesting the brain as the major source of this factor.[48] It might be that muscle-derived BDNF acts primarily within skeletal muscle tissue, e.g., inducing lipid oxidation via AMPK activation,[41] whereas brain-derived BDNF may act more systemically and is potentially involved in the apparent beneficial effects of exercise with regard to Alzheimer's disease, depression or impaired cognitive function.[49]

Angiopoietin-like protein 4 (ANGPTL4) is detected in skeletal muscle as well as in adipose tissue and is therefore classified as adipo-myokine which is also regulated by exercise.[50] Importantly, the induction of plasma ANGPTL4 after exercise may depend on the increase of free fatty acids. In healthy, untrained male volunteers, circulating ANGPTL4 was increased after acute endurance exercise.[50] Catoire *et al.* observed significantly increased plasma ANGPTL4 levels in healthy men after one-legged cycling exercise.[51] Interestingly, ANGPTL4 mRNA is more highly induced in the nonexercising leg than in the exercising leg.[51] However, plasma ANGPTL4 levels were not changed after 2 weeks of intense endurance exercise training, or by 12 weeks of endurance training.[51] In line with these results, Norheim *et al.* reported that acute endurance exercise significantly increased plasma ANGPTL4 levels and skeletal muscle ANGPTL4 mRNA. However, a chronic exercise intervention combining strength and endurance training for 12 weeks had no effect on basal plasma ANGPTL4 concentrations and skeletal muscle ANGPTL4 mRNA content.[52] In conclusion, ANGPTL4 is highly induced in muscle in response to exercise. However, adipose tissue and the liver may contribute more than muscle to the exercise-induced increase in circulating ANGPTL4.[52]

The chemokine MCP-1 is another adipo-myokine which is regulated by exercise in muscle. Among other functions, it plays an important role in the recruitment of monocytes and T lymphocytes into tissues.[53] Acute resistance exercise strongly increased MCP-1 mRNA expression in skeletal muscle of young male volunteers 2 h postexercise.[54] These results are supported by another study that observed a significant increase in MCP-1 protein level in skeletal muscle after acute resistance training in young and elderly healthy subjects.[55] Furthermore, cycling at 70% VO_{2max} for 40 min enhanced MCP-1 mRNA expression in skeletal muscle of lean, obese, and type 2 diabetic subjects.[56] Kraemer *et al.* observed in young lean males increased plasma MCP-1 levels after an acute bout of resistance exercise that was normalized 30 min postexercise.[20] Another study found increased MCP-1 plasma levels after acute treadmill running at different intensities with significantly higher plasma levels of MCP-1 following high-intensity compared to moderate-intensity trials.[57] These results suggest that the exercise-induced production of MCP-1 is more influenced by the intensity of exercise than by exercise-induced muscle damage, but it is not clear what role MCP-1 plays under exercising conditions. With regard to chronic resistance training, no effect on MCP-1 protein expression in the muscle of young and elderly healthy subjects was observed after 12 weeks.[55] Moreover, plasma MCP-1 levels were not changed after 12 weeks of low-intensity resistance training in a study with sedentary, lean, elderly women.[58] However, other studies observed a decrease in MCP-1 plasma levels after chronic strength training which could be related to the observed reduction of body fat mass.[13,59]

The pro-inflammatory molecule TNFα is an early mediator of local inflammatory responses as well as initiator of the systemic acute phase response. It is produced by adipose tissue, and circulating TNFα levels are positively correlated with body fat mass.[1] However, TNFα mRNA is also detectable in skeletal muscle but no difference in its expression was found between overweight and lean subjects[60] or type 2 diabetics and BMI-matched controls,[61] while elevated TNFα mRNA content in skeletal muscle was reported in elderly compared to young humans.[62] Interestingly, Febbraio *et al.* demonstrated that TNFα is not released by skeletal muscle after acute exercise in either healthy subjects or patients with type 2 diabetes.[63] In contrast, chronic resistance or endurance training significantly reduced TNFα mRNA and protein in skeletal muscle but had no effect on circulating TNFα concentrations.[62,64] More importantly, a large body of evidence shows an inverse relationship between plasma TNFα levels and the amount of physical activity even in healthy lean subjects as reviewed by Golbidi *et al.*[65] In contrast to the observation that regular moderate exercise reduces pro-inflammatory cytokines such as TNFα, high-intensive training cause a temporary depression of various aspects of immune function and an increase in systemic inflammation during the postexercise period (~3–24 h). TNFα levels increase only during a very intensive exercise (such as marathon running) in response to muscle damage.[66] It is well established that muscle repair and regeneration following acute muscle injury involve a tissue-remodeling, growth-promoting local inflammation. The initial inflammatory response seems to be required for the positive muscle repair process.[67]

Nicotinamide phosphoribosyl transferase (NAMPT) is an ubiquitously expressed NAD biosynthetic enzyme that occurs in an either intracellular (iNAMPT) or extracellular (eNAMPT/visfatin) form.[68] In mammals, NAMPT is responsible for the first and rate-limiting step in the conversion of nicotinamide to nicotinamide dinucleotide (NAD^+) in the NAD^+ salvage pathway. Costford *et al.* reported a twofold higher NAMPT expression in skeletal muscle of athletes as compared to sedentary obese, nonobese, and type 2 diabetic subjects.[69] Moreover, 3 weeks of endurance training enhanced NAMPT mRNA expression and protein content more than twofold in skeletal muscle of nonobese sedentary individuals. A study by Brandauer *et al.* using one-legged endurance exercise training for 3 weeks confirmed these results and reported a specific upregulation of NAMPT in the exercising leg.[70] Interestingly, acute exercise (3 h of cycling at 60% VO_{2max}) had no effect on skeletal muscle NAMPT expression and circulating eNAMPT/visfatin was found unchanged.[71] However, an acute bout of high-intensity exercise was shown to increase plasma eNAMPT/visfatin immediately after the challenge,[72] suggesting that acute release of eNAMPT/visfatin could be intensity-dependent. Interestingly, Costford *et al.* found that skeletal muscle NAMPT protein was negatively correlated with body fat, while in obesity and type 2 diabetes circulating eNAMPT/visfatin were reported to be elevated compared to controls.[69] In addition, several chronic exercise intervention studies conducted with obese volunteers reported reductions in plasma eNAMPT/visfatin levels.[73–75] Taken together, these observations suggest a potential impact of the adipose tissue on circulating levels of eNAMPT/visfatin and an independent regulation of intracellular NAMPT and circulating eNAMPT/visfatin, which awaits further investigations.

Although adiponectin is a classical adipokine, and therefore will be discussed in more detail in the next section, it is also expressed in skeletal muscle.[76,77] In obesity and insulin resistance, plasma adiponectin levels are lower, whereas skeletal muscle expression of adiponectin receptors AdipoR1 and AdipoR2 are increased.[78] A study by Bluher *et al.* found that chronic endurance training increased both plasma adiponectin levels as well as expression of AdipoR1 and AdipoR2 in skeletal muscle of overweight/obese subjects with impaired glucose metabolism and in lean healthy controls. However, in severely obese subjects, a combination of endurance exercise and diet increased plasma adiponectin levels but had no effect on AdipoR1 and AdipoR2 mRNA expression in skeletal muscle.[13] Interestingly, patients with chronic heart failure displayed elevated adiponectin and lower AdipoR1 mRNA expression in skeletal muscle which was normalized by a combined endurance and resistance exercise training for 4 months.[79] So far, no other studies have investigated the regulation of adiponectin expression in skeletal muscle with exposure to various types of exercise in different groups of patients.

Leptin is a classical adipokine that has also been found to be expressed in skeletal muscle.[80] More importantly, release of leptin by human skeletal muscle has been described by two studies. Wolsk *et al.* have shown that leptin is released by human skeletal muscle *in vivo*[81] and reported a release of ~0.8 ng per min per 100 g tissue from adipose tissue and ~0.5 ng per min per 100 g tissue from skeletal muscle. These data suggest that the contribution of skeletal muscle to whole-body leptin production could be substantial in lean humans because of the greater muscle mass compared to fat mass. In contrast, Lappas *et al.* observed a more than 10-fold higher release of adiponectin from adipose tissue explants as compared to skeletal muscle explants.[82]

Another set of genes that are highly expressed in cultured myotubes encodes proteins related to extracellular matrix (ECM) in the resting state.[40] In particular, proteoglycans and proteins related to their metabolism, like secreted protein, acidic and rich in cysteine, collagen alpha-1, lumican, gelsolin, cathepsin B, D, H, and L1, fibronectin, tissue inhibitor metalloproteinase 1, and ECM protein 1, are expressed also in muscle biopsies and are enhanced during strength training.[40] The enhanced expression of these ECM-related proteins is likely due to restructuring of the muscle tissue in response to increased mechanical demands. Moreover, some glycoproteins like serglycin may regulate storage of numerous proteases, growth factors, and chemokines and could be important in addition to their structural functions in ECM.[83]

In summary, skeletal muscle tissues produce a wide range of different molecules *in vivo* that can be classified as myokines or adipo-myokines.

3. EXERCISE AND THE PRODUCTION OF ADIPOKINES BY ADIPOSE TISSUE

While numerous studies have investigated the effect of different types of exercise training on circulating levels of adipose tissue-derived factors, only a few have addressed the adipose tissue-specific expression of these factors.

Adiponectin is one of the best characterized classical adipokines and has been shown to increase fatty acid oxidation and glucose uptake in skeletal muscle and inhibit gluconeogenesis in liver.[84] In macrophages, adiponectin inhibits expression and secretion of TNFα while increasing the production of anti-inflammatory cytokines such as IL-10.[85] Moreover, it exerts cardio-protective effects and is inversely related to BMI.[86] Studies on the regulation of adiponectin by exercise have shown divergent results, although most studies indicate no effect of exercise. Two reports have investigated adiponectin mRNA expression in adipose tissue following acute exercise and found either increased[87] or decreased[60] expression after cycling in overweight, obese, and lean subjects with no difference between the groups. Also, plasma adiponectin levels remained unchanged after acute exercise at moderate intensity in healthy[88,89] or overweight/obese participants.[89–92] On the other hand, Saunders *et al.* reported increased adiponectin plasma levels following an acute bout of exercise in sedentary obese men.[93] With regard to chronic exercise intervention, most studies using endurance type training with moderate intensity reported no change in plasma levels of adiponectin,[88,94–97] while mRNA expression in adipose tissue was found either increased[94,96] or unmodified.[97] In contrast, two studies that used chronic endurance exercise protocols reported increased plasma adiponectin levels in obese young women[98] and in overweight patients with impaired glucose metabolism[78] after the intervention period. In studies using resistance exercise for chronic intervention, the expression of adiponectin mRNA in adipose tissue as well as circulating levels was found unchanged in obese men[99] and women.[100] Interestingly, Fatouros *et al.* reported in overweight elderly men that resistance training at high intensity for 6 months increased plasma adiponectin levels while a moderate-intensity training had no effect.[101] Similar results were reported for acute resistance exercise in overweight elderly men.[102] These data suggest that the intensity of exercise may be an important factor in the regulation of adiponectin concentration in plasma.

Leptin is another classical adipokine predominantly secreted from adipocytes into the circulation to regulate energy homeostasis.[103] Following acute endurance exercise, leptin mRNA in adipose tissue was found unchanged[104] or decreased in lean and overweight subjects.[60] Moreover, plasma leptin levels analyzed immediately after an exercise challenge were found to be no different compared to preexercise levels,[89,105,106] while several studies have shown a delayed decrease of circulating leptin levels in healthy active men 24 and 48 h postexercise.[100,107–109] On the other hand, acute resistance exercise at different intensities had no effect on circulating leptin levels in overweight elderly men.[102] Also, preexercise training status of the participants has no effect on leptin levels after a single weight training session.[106] Following chronic intervention of $\geq$12 weeks leptin mRNA expression in adipose tissue remained unchanged after endurance training in obese subjects.[94,97] Leptin plasma concentrations are mostly found to be decreased[97,98,105,110] but also body weight and fat mass of the participants were reduced after the intervention periods. However, Polak *et al.* reported declined leptin plasma level which remained significant after adjusting for BMI and fat mass,[97] suggesting independent effects of the training besides reduction of body weight. Moreover, studies using resistance training reported decreased plasma leptin levels without changes in body weight and fat mass and no effect on adipose tissue mRNA expression of leptin.[99,100] So far, the available data suggest that plasma leptin levels are decreased in response to exercise training, while adipose tissue mRNA expression seems to be not affected.

TNFα is a major inflammatory cytokine that is highly expressed in adipose tissue in obese conditions and plays a role in the pathogenesis of insulin resistance. Following acute exercise, TNFα mRNA expression in adipose tissue was found to be unchanged in overweight, obese, and lean subjects.[60,87] Interestingly, Christiansen *et al.* reported higher TNFα mRNA expression in the postexercise period, 2.5 h after completing the exercise bout. With regard to circulating TNFα levels, either no effect of acute exercise was found[60,92,111] or increased plasma TNFα concentrations were reported.[87,89] Several studies on chronic exercise using either endurance[94,97] or resistance training[99] found no effect on adipose tissue mRNA expression of TNFα, whereas one study reported lower TNFα mRNA expression in obese humans.[112] Circulating TNFα levels were mostly reported to be reduced[98,100,113] by chronic exercise. However, not all studies conducted on obese humans have found an effect of long-term exercise on plasma TNFα concentrations.[97,110,114] In summary, chronic exercise seems to be able to reduce circulating TNFα levels without an effect on adipose tissue expression.

As discussed above, IL-6 is the prototype of an adipo-myokine. It is well established that the increase of circulating IL-6 upon exercise is mainly due to its release from skeletal muscle.[6] However, subjects with insulin resistance, obesity, and type 2 diabetes display chronically elevated serum levels of IL-6.[115–117] An important source of circulating IL-6 in obesity is the expanding visceral adipose tissue mass. Expression of IL-6 by macrophages within the adipose tissue is dependent on activation of the NFκB signaling pathway, whereas intramuscular IL-6 expression is regulated by different signaling cascades.[1] To clarify a potential contribution of adipose tissue to increased IL-6 levels after acute endurance exercise, Hojbjerre *et al.* have analyzed microdialysates from abdominal adipose tissue in overweight and lean males.[60] Interestingly, they observed increased IL-6 release from adipose tissue in the postexercise phase but not during exercise in both groups. Moreover, IL-6 mRNA expression in adipose tissue was increased during and postexercise. Similar results were observed by Christiansen.[87] Following chronic endurance training with moderate weight loss, no reduction of IL-6 mRNA expression in adipose tissue of obese subjects was found and basal IL-6 concentrations remained unaffected.[94,97] In the case of chronic resistance exercise, no reduction of adipose tissue IL-6 mRNA expression or basal circulating IL-6 in obese subjects was observed.[99,100] Another study that combined chronic endurance training and a hypocaloric diet reported decreased IL-6 mRNA expression in adipose tissue as well as reduced plasma IL-6 levels.[13] However, it has to be mentioned that this lifestyle intervention also decreased body weight by 13% and that the reduction of IL-6 mRNA expression and circulating levels could be related to the reduced fat mass.

In addition to the above discussed adipokines/adipo-myokines, only a few studies have investigated other adipose tissue-derived factors in response to the exercise. Resistin is an inflammatory biomarker and potential mediator of obesity-associated diseases. It is positively correlated to percent fat mass and waist circumference, and evidence suggests that this adipokine causes endothelial dysfunction by promoting oxidative stress and down-regulating the production of nitric oxide.[106,118] Acute endurance training had no effect on circulating resistin levels up to 48 h postexercise in overweight males.[90] Similarly, in another study with lean and overweight volunteers, resistin mRNA expression in adipose tissue was not affected.[60] A lifestyle intervention combining chronic endurance training and hypocaloric diet with obese patients for 3 months had also no effect on plasma resistin concentrations.[119] Interestingly, a study using acute resistance exercise observed that the exercise effect on plasma resistin levels was dependent on the training status of the subjects. While a reduction was observed in participants that performed regular weight training ($\geq$1 h, 3 times/week, for 6 months before the study), no effect was found in sedentary males or in active runners (running $\geq$15 miles/week for 6 months before the study).[106]

With regard to MCP-1, which has been discussed above in more detail, studies have found no effect of chronic endurance training on adipose tissue mRNA expression despite reductions of circulating MCP-1 levels.[94] Also, chronic resistance training in obese women left MCP-1 mRNA expression in adipose tissue unaltered but circulating levels were not measured in this study.[100]

The expression of NAMPT/Visfatin in human adipose tissue has been shown by several groups.[71,120] So far, only one study has investigated adipose tissue NAMPT expression after exercise and reported an increase after an acute bout of endurance training which was not accompanied by elevated plasma eNAMPT/visfatin concentrations. Reports on the effect of chronic exercise on circulating eNAMPT/visfatin levels are not consistent. Obese nondiabetic subjects were reported to have higher eNAMPT/visfatin levels compared to controls which were reduced by 12 weeks of aerobic[74,75] or aerobic plus resistance exercise[73] accompanied by decreased body weight. However, Jorge *et al.* reported increased plasma eNAMPT/visfatin levels after 12 weeks of exercise (endurance training, resistance training, or a combination of both) in obese type 2 diabetic patients. Importantly, body weight was not changed in the course of this study. Based on these data, it is most likely that exercise-induced reduction of plasma eNAMPT/visfatin is the result of weight loss and body composition changes.

Few reports are available on ANGPTL4 in human adipose tissue with exercise. In a study by Norheim *et al.*, it was reported that 30 min after an acute bout of endurance exercise, ANGPTL4 mRNA expression was significantly higher in overweight dysglycemic subjects compared to lean controls at the beginning of the intervention study.[52] Interestingly, they found that ANGPTL4 mRNA expression was much higher in adipose tissue than muscle. Following 12 weeks of combined endurance and resistance training, adipose tissue mRNA expression of ANGPTL4 as well as basal plasma levels was unchanged in both groups. In contrast, another chronic endurance training study in obese but otherwise healthy participants found significantly reduced mRNA expression in adipose tissue but increased circulating

ANGPTL4 levels.[121] Clearly, more studies are required to understand the regulation of ANGPTL4 in response to various types of exercise in different types of subjects as well as the contribution of the different tissues to circulating ANGPTL4.

4. CONCLUSION

There are hundreds of secretory proteins released from skeletal muscle as well as adipose tissue with an array of biological effects on most organs in the body. It is a striking observation that many of the myokines and adipokines are expressed and secreted from many different tissues. It seems as if there are very few of these adipo-myokines that are exclusively expressed in one organ. However, considering that both adipose tissue and skeletal muscle tissues are often closely associated with the same phenotypes or outcomes, it is likely that the pattern of signal molecules released from these two tissues have marked physiological effects of essential importance for health as well as well-being. The original expectation that we would be able to describe discrete signatures of secretory proteins from muscle and adipose has to be replaced by a much more complicated model.

A single bout of exercise is predominantly characterized by the secretion of myokines and adipo-myokines by the working skeletal muscles, which exert a variety of autocrine and endocrine effects (Fig. 3). Acute induction of myokines like myostatin, IL-7, decorin, and LIF is involved in the regulation of muscle growth and may play a role in exercise-related restructuring of skeletal muscle. Muscle-derived IL-6 has metabolic effects on insulin-stimulated glucose disposal and fatty acid oxidation. Furthermore, acute high levels of circulating IL-6 provide an anti-inflammatory environment after exercise by induction of IL-1ra and IL-10, and inhibition of TNFα production. One the other hand, regular exercise training is associated with reduced levels of adipose tissue-derived pro-inflammatory cytokines, which are linked to low-grade systemic inflammation and low whole-body insulin sensitivity (Fig. 3). Chronic exercise reduces visceral fat mass and has been reported to reduce central adiposity independent of overall weight changes. The anti-inflammatory effects of regular exercise training may be mediated by both, the reduction of body fat mass and the induction of an anti-inflammatory environment associated with each single

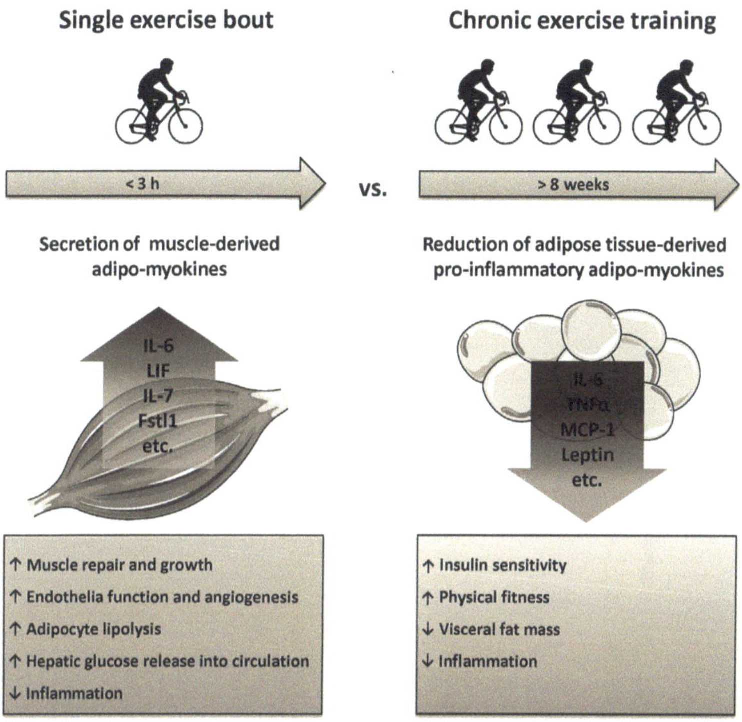

Figure 3 Differential effects of acute and chronic exercise. After acute exercise, a high number of myokines are secreted by skeletal muscle exerting a variety of endocrine effects. Acute induction of myokines like myostatin, IL-7, decorin, and LIF is involved in the regulation of muscle hypertrophy and may play a role in exercise-related restructuring of skeletal muscle. The high level of circulating IL-6 after exercise induces an anti-inflammatory environment by inducing the production of IL-1ra and IL-10, and also inhibits TNFα production. Furthermore, IL-6 has metabolic effects, by affecting insulin-stimulated glucose disposal and fatty acid oxidation. The myokine FSTL1 has protective effects on ischemia–reperfusion injury in muscle and heart tissue. On the other hand, regular exercise training induces a reduction of adipose tissue-derived pro-inflammatory cytokines like IL-6, TNFα, and MCP-1, which are associated with low-grade systemic inflammation and a reduction of whole-body insulin sensitivity. Exercise training has been reported to reduce central adiposity independent of overall weight changes, and this may represent an additional mechanism of the anti-inflammatory action of chronic exercise training. (See the color plate.)

bout of exercise. In summary, regular exercise reduces the risk of chronic metabolic diseases and various mechanisms may contribute to this beneficial effect, including decreased production of adipose tissue-derived proinflammatory cytokines and increased production of anti-inflammatory myokines from contracting muscle.

ACKNOWLEDGMENTS

We apologize to the many authors of original research whose publications we could not cite owing to space restrictions. The secretarial assistance of Birgit Hurow (Paul-Langerhans Group for Integrative Physiology, German Diabetes Center, Düsseldorf, Germany) is gratefully acknowledged. K.E. is supported by the Deutsche Forschungsgemeinschaft (German Research Foundation; EC 440/2-1). Furthermore, the work was supported by the Ministerium für Wissenschaft und Forschung des Landes Nordrhein-Westfalen (Ministry of Science and Research of the State of North Rhine-Westphalia), the Bundesministerium für Gesundheit (Federal Ministry of Health), the German Academic Exchange Service (Project 57068553), and in part by a grant from the Bundesministerium für Bildung und Forschung (German Federal Ministry of Education and Research) to the Deutsches Zentrum für Diabetesforschung (DZD e.V.; German Center for Diabetes Research).

Conflict of Interests: The authors have no conflict of interest in relation to the contents of this chapter.

REFERENCES

1. Eckardt K, Gorgens SW, Raschke S, Eckel J. Myokines in insulin resistance and type 2 diabetes. *Diabetologia*. 2014;57:1087–1099.
2. Pedersen BK, Febbraio MA. Muscles, exercise and obesity: skeletal muscle as a secretory organ. *Nat Rev Endocrinol*. 2012;8:457–465.
3. Raschke S, Eckardt K, Bjorklund HK, Jensen J, Eckel J. Identification and validation of novel contraction-regulated myokines released from primary human skeletal muscle cells. *PLoS One*. 2013;8:e62008.
4. Raschke S, Eckel J. Adipo-myokines: two sides of the same coin—mediators of inflammation and mediators of exercise. *Mediators Inflamm*. 2013;2013:320724.
5. Lee S, Kuk JL, Davidson LE, et al. Exercise without weight loss is an effective strategy for obesity reduction in obese individuals with and without type 2 diabetes. *J Appl Physiol (1985)*. 2005;99:1220–1225.
6. Pedersen BK. Muscle as a secretory organ. *Compr Physiol*. 2013;3:1337–1362.
7. Serrano AL, Baeza-Raja B, Perdiguero E, Jardi M, Munoz-Canoves P. Interleukin-6 is an essential regulator of satellite cell-mediated skeletal muscle hypertrophy. *Cell Metab*. 2008;7:33–44.
8. Fischer CP. Interleukin-6 in acute exercise and training: what is the biological relevance? *Exerc Immunol Rev*. 2006;12:6–33.
9. Willoughby DS, McFarlin B, Bois C. Interleukin-6 expression after repeated bouts of eccentric exercise. *Int J Sports Med*. 2003;24:15–21.
10. Keller C, Steensberg A, Pilegaard H, et al. Transcriptional activation of the IL-6 gene in human contracting skeletal muscle: influence of muscle glycogen content. *FASEB J*. 2001;15:2748–2750.
11. Steensberg A, Febbraio MA, Osada T, et al. Interleukin-6 production in contracting human skeletal muscle is influenced by pre-exercise muscle glycogen content. *J Physiol*. 2001;537:633–639.
12. Kohut ML, McCann DA, Russell DW, et al. Aerobic exercise, but not flexibility/resistance exercise, reduces serum IL-18, CRP, and IL-6 independent of beta-blockers, BMI, and psychosocial factors in older adults. *Brain Behav Immun*. 2006;20:201–209.

13. Bruun JM, Helge JW, Richelsen B, Stallknecht B. Diet and exercise reduce low-grade inflammation and macrophage infiltration in adipose tissue but not in skeletal muscle in severely obese subjects. *Am J Physiol Endocrinol Metab*. 2006;290:E961–E967.
14. Rokling-Andersen MH, Reseland JE, Veierod MB, et al. Effects of long-term exercise and diet intervention on plasma adipokine concentrations. *Am J Clin Nutr*. 2007;86:1293–1301.
15. Bruunsgaard H, Bjerregaard E, Schroll M, Pedersen BK. Muscle strength after resistance training is inversely correlated with baseline levels of soluble tumor necrosis factor receptors in the oldest old. *J Am Geriatr Soc*. 2004;52:237–241.
16. Broholm C, Laye MJ, Brandt C, et al. LIF is a contraction-induced myokine stimulating human myocyte proliferation. *J Appl Physiol (1985)*. 2011;111:251–259.
17. Broholm C, Mortensen OH, Nielsen S, et al. Exercise induces expression of leukaemia inhibitory factor in human skeletal muscle. *J Physiol*. 2008;586:2195–2201.
18. Haugen F, Norheim F, Lian H, et al. IL-7 is expressed and secreted by human skeletal muscle cells. *Am J Physiol Cell Physiol*. 2010;298:C807–C816.
19. Andersson H, Bohn SK, Raastad T, Paulsen G, Blomhoff R, Kadi F. Differences in the inflammatory plasma cytokine response following two elite female soccer games separated by a 72-h recovery. *Scand J Med Sci Sports*. 2010;20:740–747.
20. Kraemer WJ, Hatfield DL, Comstock BA, et al. Influence of HMB supplementation and resistance training on cytokine responses to resistance exercise. *J Am Coll Nutr*. 2014;33:247–255.
21. Nieman DC, Davis JM, Henson DA, et al. Carbohydrate ingestion influences skeletal muscle cytokine mRNA and plasma cytokine levels after a 3-h run. *J Appl Physiol (1985)*. 2003;94:1917–1925.
22. Rinnov A, Yfanti C, Nielsen S, et al. Endurance training enhances skeletal muscle interleukin-15 in human male subjects. *Endocrine*. 2014;45:271–278.
23. Tamura Y, Watanabe K, Kantani T, Hayashi J, Ishida N, Kaneki M. Upregulation of circulating IL-15 by treadmill running in healthy individuals: is IL-15 an endocrine mediator of the beneficial effects of endurance exercise? *Endocr J*. 2011;58:211–215.
24. Riechman SE, Balasekaran G, Roth SM, Ferrell RE. Association of interleukin-15 protein and interleukin-15 receptor genetic variation with resistance exercise training responses. *J Appl Physiol (1985)*. 2004;97:2214–2219.
25. Nielsen AR, Mounier R, Plomgaard P, et al. Expression of interleukin-15 in human skeletal muscle effect of exercise and muscle fibre type composition. *J Physiol*. 2007;584:305–312.
26. McPherron AC, Lawler AM, Lee SJ. Regulation of skeletal muscle mass in mice by a new TGF-beta superfamily member. *Nature*. 1997;387:83–90.
27. Hittel DS, Axelson M, Sarna N, Shearer J, Huffman KM, Kraus WE. Myostatin decreases with aerobic exercise and associates with insulin resistance. *Med Sci Sports Exerc*. 2010;42:2023–2029.
28. Harber MP, Crane JD, Dickinson JM, et al. Protein synthesis and the expression of growth-related genes are altered by running in human vastus lateralis and soleus muscles. *Am J Physiol Regul Integr Comp Physiol*. 2009;296:R708–R714.
29. Kim JS, Cross JM, Bamman MM. Impact of resistance loading on myostatin expression and cell cycle regulation in young and older men and women. *Am J Physiol Endocrinol Metab*. 2005;288:E1110–E1119.
30. Dieli-Conwright CM, Spektor TM, Rice JC, Sattler FR, Schroeder ET. Hormone therapy and maximal eccentric exercise alters myostatin-related gene expression in postmenopausal women. *J Strength Cond Res*. 2012;26:1374–1382.
31. Dalbo VJ, Roberts MD, Sunderland KL, et al. Acute loading and aging effects on myostatin pathway biomarkers in human skeletal muscle after three sequential bouts of resistance exercise. *J Gerontol A Biol Sci Med Sci*. 2011;66:855–865.

32. Amthor H, Nicholas G, McKinnell I, et al. Follistatin complexes myostatin and antagonises myostatin-mediated inhibition of myogenesis. *Dev Biol.* 2004;270:19–30.
33. Miura T, Kishioka Y, Wakamatsu J, et al. Decorin binds myostatin and modulates its activity to muscle cells. *Biochem Biophys Res Commun.* 2006;340:675–680.
34. Kanzleiter T, Rath M, Gorgens SW, et al. The myokine decorin is regulated by contraction and involved in muscle hypertrophy. *Biochem Biophys Res Commun.* 2014;450:1089–1094.
35. Heinemeier KM, Bjerrum SS, Schjerling P, Kjaer M. Expression of extracellular matrix components and related growth factors in human tendon and muscle after acute exercise. *Scand J Med Sci Sports.* 2013;23:e150–e161.
36. Shi Y, Massague J. Mechanisms of TGF-beta signaling from cell membrane to the nucleus. *Cell.* 2003;113:685–700.
37. Ouchi N, Oshima Y, Ohashi K, et al. Follistatin-like 1, a secreted muscle protein, promotes endothelial cell function and revascularization in ischemic tissue through a nitric-oxide synthase-dependent mechanism. *J Biol Chem.* 2008;283:32802–32811.
38. Ogura Y, Ouchi N, Ohashi K, et al. Therapeutic impact of follistatin-like 1 on myocardial ischemic injury in preclinical models. *Circulation.* 2012;126:1728–1738.
39. Gorgens SW, Raschke S, Holven KB, Jensen J, Eckardt K, Eckel J. Regulation of follistatin-like protein 1 expression and secretion in primary human skeletal muscle cells. *Arch Physiol Biochem.* 2013;119:75–80.
40. Norheim F, Raastad T, Thiede B, Rustan AC, Drevon CA, Haugen F. Proteomic identification of secreted proteins from human skeletal muscle cells and expression in response to strength training. *Am J Physiol Endocrinol Metab.* 2011;301:E1013–E1021.
41. Matthews VB, Astrom MB, Chan MH, et al. Brain-derived neurotrophic factor is produced by skeletal muscle cells in response to contraction and enhances fat oxidation via activation of AMP-activated protein kinase. *Diabetologia.* 2009;52:1409–1418.
42. Huang EJ, Reichardt LF. Neurotrophins: roles in neuronal development and function. *Annu Rev Neurosci.* 2001;24:677–736.
43. Ferris LT, Williams JS, Shen CL. The effect of acute exercise on serum brain-derived neurotrophic factor levels and cognitive function. *Med Sci Sports Exerc.* 2007;39:728–734.
44. Gold SM, Schulz KH, Hartmann S, et al. Basal serum levels and reactivity of nerve growth factor and brain-derived neurotrophic factor to standardized acute exercise in multiple sclerosis and controls. *J Neuroimmunol.* 2003;138:99–105.
45. Schmidt-Kassow M, Schadle S, Otterbein S, et al. Kinetics of serum brain-derived neurotrophic factor following low-intensity versus high-intensity exercise in men and women. *Neuroreport.* 2012;23:889–893.
46. Griffin EW, Mullally S, Foley C, Warmington SA, O'Mara SM, Kelly AM. Aerobic exercise improves hippocampal function and increases BDNF in the serum of young adult males. *Physiol Behav.* 2011;104:934–941.
47. Goekint M, De Pauw K, Roelands B, et al. Strength training does not influence serum brain-derived neurotrophic factor. *Eur J Appl Physiol.* 2010;110:285–293.
48. Rasmussen P, Brassard P, Adser H, et al. Evidence for a release of brain-derived neurotrophic factor from the brain during exercise. *Exp Physiol.* 2009;94:1062–1069.
49. Mattson MP. Energy intake and exercise as determinants of brain health and vulnerability to injury and disease. *Cell Metab.* 2012;16:706–722.
50. Kersten S, Lichtenstein L, Steenbergen E, et al. Caloric restriction and exercise increase plasma ANGPTL4 levels in humans via elevated free fatty acids. *Arterioscler Thromb Vasc Biol.* 2009;29:969–974.
51. Catoire M, Alex S, Paraskevopulos N, et al. Fatty acid-inducible ANGPTL4 governs lipid metabolic response to exercise. *Proc Natl Acad Sci USA.* 2014;111:E1043–E1052.

52. Norheim F, Hjorth M, Langleite TM, et al. Regulation of angiopoietin-like protein 4 production during and after exercise. *Physiol Rep.* 2014;2:e12109.
53. Qin S, LaRosa G, Campbell JJ, et al. Expression of monocyte chemoattractant protein-1 and interleukin-8 receptors on subsets of T cells: correlation with trans-endothelial chemotactic potential. *Eur J Immunol.* 1996;26:640–647.
54. Vella L, Caldow MK, Larsen AE, et al. Resistance exercise increases NF-kappaB activity in human skeletal muscle. *Am J Physiol Regul Integr Comp Physiol.* 2012;302: R667–R673.
55. Della Gatta PA, Garnham AP, Peake JM, Cameron-Smith D. Effect of exercise training on skeletal muscle cytokine expression in the elderly. *Brain Behav Immun.* 2014;39:80–86.
56. Tantiwong P, Shanmugasundaram K, Monroy A, et al. NF-kappaB activity in muscle from obese and type 2 diabetic subjects under basal and exercise-stimulated conditions. *Am J Physiol Endocrinol Metab.* 2010;299:E794–E801.
57. Peake JM, Suzuki K, Hordern M, Wilson G, Nosaka K, Coombes JS. Plasma cytokine changes in relation to exercise intensity and muscle damage. *Eur J Appl Physiol.* 2005;95:514–521.
58. Ogawa K, Sanada K, Machida S, Okutsu M, Suzuki K. Resistance exercise training-induced muscle hypertrophy was associated with reduction of inflammatory markers in elderly women. *Mediators Inflamm.* 2010;2010:171023.
59. Glintborg D, Christensen LL, Kvorning T, et al. Strength training and testosterone treatment have opposing effects on migration inhibitor factor levels in ageing men. *Mediators Inflamm.* 2013;2013:539156.
60. Hojbjerre L, Rosenzweig M, Dela F, Bruun JM, Stallknecht B. Acute exercise increases adipose tissue interstitial adiponectin concentration in healthy overweight and lean subjects. *Eur J Endocrinol.* 2007;157:613–623.
61. Plomgaard P, Nielsen AR, Fischer CP, et al. Associations between insulin resistance and TNF-alpha in plasma, skeletal muscle and adipose tissue in humans with and without type 2 diabetes. *Diabetologia.* 2007;50:2562–2571.
62. Greiwe JS, Cheng B, Rubin DC, Yarasheski KE, Semenkovich CF. Resistance exercise decreases skeletal muscle tumor necrosis factor alpha in frail elderly humans. *FASEB J.* 2001;15:475–482.
63. Febbraio MA, Steensberg A, Starkie RL, McConell GK, Kingwell BA. Skeletal muscle interleukin-6 and tumor necrosis factor-alpha release in healthy subjects and patients with type 2 diabetes at rest and during exercise. *Metabolism.* 2003;52:939–944.
64. Gielen S, Adams V, Mobius-Winkler S, et al. Anti-inflammatory effects of exercise training in the skeletal muscle of patients with chronic heart failure. *J Am Coll Cardiol.* 2003;42:861–868.
65. Golbidi S, Laher I. Exercise induced adipokine changes and the metabolic syndrome. *J Diabetes Res.* 2014;2014:726861.
66. Pedersen BK, Akerstrom TC, Nielsen AR, Fischer CP. Role of myokines in exercise and metabolism. *J Appl Physiol.* 2007;103:1093–1098.
67. Paulsen G, Mikkelsen UR, Raastad T, Peake JM. Leucocytes, cytokines and satellite cells: what role do they play in muscle damage and regeneration following eccentric exercise? *Exerc Immunol Rev.* 2012;18:42–97.
68. Revollo JR, Korner A, Mills KF, et al. Nampt/PBEF/Visfatin regulates insulin secretion in beta cells as a systemic NAD biosynthetic enzyme. *Cell Metab.* 2007;6:363–375.
69. Costford SR, Bajpeyi S, Pasarica M, et al. Skeletal muscle NAMPT is induced by exercise in humans. *Am J Physiol Endocrinol Metab.* 2010;298:E117–E126.
70. Brandauer J, Vienberg SG, Andersen MA, et al. AMP-activated protein kinase regulates nicotinamide phosphoribosyl transferase expression in skeletal muscle. *J Physiol.* 2013;591:5207–5220.

71. Frydelund-Larsen L, Akerstrom T, Nielsen S, Keller P, Keller C, Pedersen BK. Visfatin mRNA expression in human subcutaneous adipose tissue is regulated by exercise. *Am J Physiol Endocrinol Metab.* 2007;292:E24–E31.
72. Ghanbari-Niaki A, Saghebjoo M, Soltani R, Kirwan JP. Plasma visfatin is increased after high-intensity exercise. *Ann Nutr Metab.* 2010;57:3–8.
73. Choi KM, Kim JH, Cho GJ, Baik SH, Park HS, Kim SM. Effect of exercise training on plasma visfatin and eotaxin levels. *Eur J Endocrinol.* 2007;157:437–442.
74. Haus JM, Solomon TP, Marchetti CM, et al. Decreased visfatin after exercise training correlates with improved glucose tolerance. *Med Sci Sports Exerc.* 2009;41:1255–1260.
75. Lee KJ, Shin YA, Lee KY, Jun TW, Song W. Aerobic exercise training-induced decrease in plasma visfatin and insulin resistance in obese female adolescents. *Int J Sport Nutr Exerc Metab.* 2010;20:275–281.
76. Delaigle AM, Jonas JC, Bauche IB, Cornu O, Brichard SM. Induction of adiponectin in skeletal muscle by inflammatory cytokines: in vivo and in vitro studies. *Endocrinology.* 2004;145:5589–5597.
77. Van Berendoncks AM, Garnier A, Beckers P, et al. Functional adiponectin resistance at the level of the skeletal muscle in mild to moderate chronic heart failure. *Circ Heart Fail.* 2010;3:185–194.
78. Bluher M, Bullen Jr JW, Lee JH, et al. Circulating adiponectin and expression of adiponectin receptors in human skeletal muscle: associations with metabolic parameters and insulin resistance and regulation by physical training. *J Clin Endocrinol Metab.* 2006;91:2310–2316.
79. Van Berendoncks AM, Garnier A, Beckers P, et al. Exercise training reverses adiponectin resistance in skeletal muscle of patients with chronic heart failure. *Heart.* 2011;97:1403–1409.
80. Wang J, Liu R, Hawkins M, Barzilai N, Rossetti L. A nutrient-sensing pathway regulates leptin gene expression in muscle and fat. *Nature.* 1998;393:684–688.
81. Wolsk E, Mygind H, Grondahl TS, Pedersen BK, van Hall G. Human skeletal muscle releases leptin in vivo. *Cytokine.* 2012;60:667–673.
82. Lappas M, Yee K, Permezel M, Rice GE. Release and regulation of leptin, resistin and adiponectin from human placenta, fetal membranes, and maternal adipose tissue and skeletal muscle from normal and gestational diabetes mellitus-complicated pregnancies. *J Endocrinol.* 2005;186:457–465.
83. Kolset SO, Pejler G. Serglycin: a structural and functional chameleon with wide impact on immune cells. *J Immunol.* 2011;187:4927–4933.
84. Yamauchi T, Iwabu M, Okada-Iwabu M, Kadowaki T. Adiponectin receptors: a review of their structure, function and how they work. *Best Pract Res Clin Endocrinol Metab.* 2014;28:15–23.
85. Moschen AR, Wieser V, Tilg H. Adiponectin: key player in the adipose tissue-liver crosstalk. *Curr Med Chem.* 2012;19:5467–5473.
86. Han SH, Quon MJ, Kim JA, Koh KK. Adiponectin and cardiovascular disease: response to therapeutic interventions. *J Am Coll Cardiol.* 2007;49:531–538.
87. Christiansen T, Bruun JM, Paulsen SK, et al. Acute exercise increases circulating inflammatory markers in overweight and obese compared with lean subjects. *Eur J Appl Physiol.* 2013;113(6):1635–1642.
88. Bobbert T, Wegewitz U, Brechtel L, et al. Adiponectin oligomers in human serum during acute and chronic exercise: relation to lipid metabolism and insulin sensitivity. *Int J Sports Med.* 2007;28:1–8.
89. Ferguson MA, White LJ, McCoy S, Kim HW, Petty T, Wilsey J. Plasma adiponectin response to acute exercise in healthy subjects. *Eur J Appl Physiol.* 2004;91:324–329.
90. Jamurtas AZ, Theocharis V, Koukoulis G, et al. The effects of acute exercise on serum adiponectin and resistin levels and their relation to insulin sensitivity in overweight males. *Eur J Appl Physiol.* 2006;97:122–126.

91. Bouassida A, Lakhdar N, Benaissa N, et al. Adiponectin responses to acute moderate and heavy exercises in overweight middle aged subjects. *J Sports Med Phys Fitness.* 2010;50:330–335.
92. Numao S, Katayama Y, Hayashi Y, Matsuo T, Tanaka K. Influence of acute aerobic exercise on adiponectin oligomer concentrations in middle-aged abdominally obese men. *Metabolism.* 2011;60:186–194.
93. Saunders TJ, Palombella A, McGuire KA, Janiszewski PM, Despres JP, Ross R. Acute exercise increases adiponectin levels in abdominally obese men. *J Nutr Metab.* 2012;2012:148729.
94. Christiansen T, Paulsen SK, Bruun JM, Pedersen SB, Richelsen B. Exercise training versus diet-induced weight-loss on metabolic risk factors and inflammatory markers in obese subjects: a 12-week randomized intervention study. *Am J Physiol Endocrinol Metab.* 2010;298:E824–E831.
95. Hulver MW, Zheng D, Tanner CJ, et al. Adiponectin is not altered with exercise training despite enhanced insulin action. *Am J Physiol Endocrinol Metab.* 2002;283: E861–E865.
96. Moghadasi M, Mohebbi H, Rahmani-Nia F, Hassan-Nia S, Noroozi H. Effects of short-term lifestyle activity modification on adiponectin mRNA expression and plasma concentrations. *Eur J Sport Sci.* 2013;13:378–385.
97. Polak J, Klimcakova E, Moro C, et al. Effect of aerobic training on plasma levels and subcutaneous abdominal adipose tissue gene expression of adiponectin, leptin, interleukin 6, and tumor necrosis factor alpha in obese women. *Metabolism.* 2006;55:1375–1381.
98. Kondo T, Kobayashi I, Murakami M. Effect of exercise on circulating adipokine levels in obese young women. *Endocr J.* 2006;53:189–195.
99. Klimcakova E, Polak J, Moro C, et al. Dynamic strength training improves insulin sensitivity without altering plasma levels and gene expression of adipokines in subcutaneous adipose tissue in obese men. *J Clin Endocrinol Metab.* 2006;91:5107–5112.
100. Phillips MD, Patrizi RM, Cheek DJ, Wooten JS, Barbee JJ, Mitchell JB. Resistance training reduces subclinical inflammation in obese, postmenopausal women. *Med Sci Sports Exerc.* 2012;44:2099–2110.
101. Fatouros IG, Tournis S, Leontsini D, et al. Leptin and adiponectin responses in overweight inactive elderly following resistance training and detraining are intensity related. *J Clin Endocrinol Metab.* 2005;90:5970–5977.
102. Fatouros IG, Chatzinikolaou A, Tournis S, et al. Intensity of resistance exercise determines adipokine and resting energy expenditure responses in overweight elderly individuals. *Diabetes Care.* 2009;32:2161–2167.
103. Munzberg H, Morrison CD. Structure, production and signaling of leptin. *Metabolism.* 2014;64(1):13–23.
104. Keller P, Keller C, Steensberg A, Robinson LE, Pedersen BK. Leptin gene expression and systemic levels in healthy men: effect of exercise, carbohydrate, interleukin-6, and epinephrine. *J Appl Physiol (1985).* 2005;98:1805–1812.
105. Perusse L, Collier G, Gagnon J, et al. Acute and chronic effects of exercise on leptin levels in humans. *J Appl Physiol (1985).* 1997;83:5–10.
106. Varady KA, Bhutani S, Church EC, Phillips SA. Adipokine responses to acute resistance exercise in trained and untrained men. *Med Sci Sports Exerc.* 2010;42:456–462.
107. Essig DA, Alderson NL, Ferguson MA, Bartoli WP, Durstine JL. Delayed effects of exercise on the plasma leptin concentration. *Metabolism.* 2000;49:395–399.
108. Olive JL, Miller GD. Differential effects of maximal- and moderate-intensity runs on plasma leptin in healthy trained subjects. *Nutrition.* 2001;17:365–369.
109. Yang CB, Chuang CC, Kuo CS, Hsu CH, Tsao TH. Effects of an acute bout of exercise on serum soluble leptin receptor (sOB-R) levels. *J Sports Sci.* 2014;32:446–451.

110. O'Leary VB, Marchetti CM, Krishnan RK, Stetzer BP, Gonzalez F, Kirwan JP. Exercise-induced reversal of insulin resistance in obese elderly is associated with reduced visceral fat. *J Appl Physiol (1985)*. 2006;100:1584–1589.
111. Harris RA, Padilla J, Hanlon KP, Rink LD, Wallace JP. The flow-mediated dilation response to acute exercise in overweight active and inactive men. *Obesity (Silver Spring)*. 2008;16:578–584.
112. Baturcam E, Abubaker J, Tiss A, et al. Physical exercise reduces the expression of RANTES and its CCR5 receptor in the adipose tissue of obese humans. *Mediators Inflamm*. 2014;2014:627150.
113. Abubaker J, Tiss A, Abu-Farha M, et al. DNAJB3/HSP-40 cochaperone is downregulated in obese humans and is restored by physical exercise. *PLoS One*. 2013;8: e69217.
114. Nicklas BJ, Ambrosius W, Messier SP, et al. Diet-induced weight loss, exercise, and chronic inflammation in older, obese adults: a randomized controlled clinical trial. *Am J Clin Nutr*. 2004;79:544–551.
115. Dandona P, Aljada A, Bandyopadhyay A. Inflammation: the link between insulin resistance, obesity and diabetes. *Trends Immunol*. 2004;25:4–7.
116. Duncan BB, Schmidt MI, Pankow JS, et al. Low-grade systemic inflammation and the development of type 2 diabetes: the atherosclerosis risk in communities study. *Diabetes*. 2003;52:1799–1805.
117. Kern PA, Ranganathan S, Li C, Wood L, Ranganathan G. Adipose tissue tumor necrosis factor and interleukin-6 expression in human obesity and insulin resistance. *Am J Physiol Endocrinol Metab*. 2001;280:E745–E751.
118. Codoner-Franch P, Alonso-Iglesias E. Resistin: insulin resistance to malignancy. *Clin Chim Acta*. 2014;438C:46–54.
119. de Luis DA, Aller R, Izaola O, Sagrado MG, Conde R. Influence of ALA54THR polymorphism of fatty acid binding protein 2 on lifestyle modification response in obese subjects. *Ann Nutr Metab*. 2006;50:354–360.
120. Berndt J, Kloting N, Kralisch S, et al. Plasma visfatin concentrations and fat depot-specific mRNA expression in humans. *Diabetes*. 2005;54:2911–2916.
121. Cullberg KB, Christiansen T, Paulsen SK, Bruun JM, Pedersen SB, Richelsen B. Effect of weight loss and exercise on angiogenic factors in the circulation and in adipose tissue in obese subjects. *Obesity (Silver Spring)*. 2013;21:454–460.

CHAPTER FOURTEEN

Exercise and the Regulation of Inflammatory Responses

Jacob Allen, Yi Sun, Jeffrey A. Woods[1]
Department of Kinesiology and Community Health, University of Illinois at Urbana-Champaign, Urbana, Illinois, USA
[1]Corresponding author: e-mail address: woods1@illinois.edu

Contents

Abstract

Exercise initiates a cascade of inflammatory events, which ultimately lead to long-term effects on human health. During and after acute exercise in skeletal muscle, interactions between immune cells, cytokines, and other intracellular components, create an inflammatory milieu responsible for the recovery and adaption from an exercise bout. In the systemic circulation, cytokines released from muscle (myokines) mediate metabolic and inflammatory processes. Moderate exercise training results in improvements in systemic inflammation, evident by reductions in acute phase proteins. The anti-inflammatory effects of regular exercise include actions dependent and independent of changes in adipose tissue mass. Future research should encompass approaches, which attempt to integrate other, less-recognized physiological processes with acute and long-term inflammatory changes. This will include investigation into metabolic, endocrine, and immune components of various tissues and organs.

Progress in Molecular Biology and Translational Science, Volume 135
ISSN 1877-1173
http://dx.doi.org/10.1016/bs.pmbts.2015.07.003

ABBREVIATIONS

CRP C-reactive protein
GPx glutathione peroxidase
IL interleukin
IL-10 interleukin-10
IL-1β interleukin-1 beta
ROS reactive oxygen species
SOD superoxide dismutase
TLR toll-like receptor
TNF-α tumor necrosis factor-alpha

1. A BRIEF HISTORY OF INFLAMMATION AND ITS UNDERLYING RELATIONSHIP TO EXERCISE

In the first century medical encyclopedia, *De Medicina*, the Roman physician Conrelius Celsus first introduced four cardinals signs of inflammation, "*Rubor et tumor cum colore et dolore*"; or "redness and swelling with heat and pain." Nearly two millennia later, in the mid-1800s, Rudolf Virchow introduced the fifth cardinal sign of inflammation, "*function laesa*"; or "loss of function," which described the inadequate functionality of cells when exposed to a stressful stimuli.[1] Virchow proposed that these five characteristics are a product of a larger inflammatory process that combats "cellular" stress. These insights by Virchow broke sharply away from the traditional view of "humor imbalance" as the mediator of health and sickness, and thus helped usher in a new investigative era into inflammatory processes.[1] Building on the work of Virchow and others, Eli Metchnikoff, in 1892, proposed that inflammatory responses are not only vital for host defense but also imperative for natural tissue homeostasis.[2] Importantly, Metchnikoff suggested that the innate immune system (i.e., macrophages and neutrophils) might induce a broad range of remodeling processes, which are not necessarily detrimental to a host organism.[1,2] Following this, and just before the turn of the nineteenth century, Robert Koch and Louis Pasteur proposed the germ theory of diseases, a vital step for understanding microbes as major inducers of acute inflammation.[1]

Over a century after Koch and Pasteur, a far more expansive understanding of the underlying inflammatory processes of health and disease has been revealed. Presently, inflammation is recognized as "a perpetual and essential immune response that maintains tissue homeostasis under a variety of noxious conditions."[1] Inflammatory process can be separated into four distinct components: inducers, sensors, mediators, and targets.[1] Together, various combinations of these components invoke inflammation that is vital for host defense from pathogens as well as repair from other internal disturbances. Unfortunately, the necessity of the inflammatory insult to combat such conditions may, paradoxically, lead to negative consequences within an organism. More specifically, chronic low-grade inflammation disrupts tissue homeostasis in ways that drive the progression of chronic conditions, such as diabetes, atherosclerosis, autoimmune diseases, and cancer.[3] With such devastating consequences, researchers and physicians have and will continue to search for ways to combat such chronic inflammatory insults.

Promisingly, a potent and long-lasting anti-inflammatory therapy exists in the form of physical exercise. This is supported by a vast number of epidemiological studies, which show that the long-term exercise training and/or increased physical activity greatly reduce the occurrence of chronic inflammatory diseases.[3–5] Interestingly, however, exercise only exerts its anti-inflammatory effect after an intricate and time-dependent inflammatory cascade that begins with largely proinflammatory actions. Such acute inflammatory actions, if not resolved, may lead to immunosuppression and long-term pathologies in chronic exercisers (i.e., upper respiratory tract infections).[6,7] In light of these variable effects, accounting for intensity, duration, and recovery from exercise is key to understanding the time course and resolution of the inflammatory response.

To help address how exercise regulates inflammation, this chapter will aim to unravel both the molecular mechanisms and systemic inflammatory effects of both acute and long-term exercise training. We will begin by discussing the timeline of acute, inflammatory actions within the primary tissue that responds to exercise skeletal muscle. Next, we will explore the acute, systemic inflammatory effects of acute exercise. Throughout this section, readers will also be exposed to the major inflammatory regulators (i.e., cytokines), as well as the possible sources and mechanisms which regulate systemic inflammation. In Section 2 of this chapter, we will explore effects of long-term exercise training on inflammatory processes. Here, we will also aim to unravel the roles of exercise in regulating inflammation in tissues and organ systems outside of skeletal muscle, including adipose tissue; a well-established inflammatory mediator. Throughout all sections of this chapter, we will include evidence from both humans and animal models, which together will help integrate mechanistic data with clinical applications of exercise training.

2. EXERCISE AND ACUTE INFLAMMATION IN SKELETAL MUSCLE

Muscular contraction is required for locomotion in mammals, thus skeletal muscle seems the most appropriate tissue to begin the investigation into the inflammatory processes regulated by exercise. This section will discuss the roles of metabolic and muscular injury by-products in inducing inflammatory events within skeletal muscle during and after acute exercise. We will also explore possible mechanisms by which inflammatory processes initiate long-term adaptation within skeletal muscles.

A large portion of exercise-induced inflammation within the muscle can be traced to mitochondrial uncoupling and the subsequent induction of reactive oxygen species (ROS). Such accumulation results in numerous downstream affects within the myocyte (muscle cell) and the surrounding tissue.[8] Namely, exercise-induced ROS generation may initiate a series of redox-sensitive intracellular signaling events. For example, NF-kappaB (NFkB) and activator protein-1 (AP-1), both potent transcription factors involved in inflammation, are activated after exhaustive exercise.[9] The relationship between exercise-induced ROS and these transcription factors became apparent when a potent xanthine oxidase inhibitor, allopurinol, was sufficient to attenuate NFkB binding in rat skeletal muscle.[10] Other proteins linked to inflammatory processes after exercise include the family of stress-activated mitogen-activated protein kinases (MAPKs), which together initiate processes important for inflammation, muscular adaptation, and metabolic control within skeletal muscle.[11,12] Interestingly, the MAPK family (p38, erk 1/2, jnk, and erk 5) are differentially activated depending on the exercise modality.[10,13] For example, it appears that concentric exercise induces phosphorylation of $MAPK^{erk\ 1/2}$ but not $MAPK^{p38}$. On the other hand, eccentric exercise can concurrently activate both kinases, leading to separate downstream signaling events.[14]

The differentially activated MAPK pathways highlight the complexity of muscular pathways and the subsequent inflammatory processes that ensues after exercise. These pleiotropic effects are highly dependent on the mode of muscular contractions (eccentric vs. concentric) as well as the intensity, the duration, and the novelty of an exercise task. In addition to the production of ROS, these factors also contribute to the degree of muscle damage after exercise. Current knowledge indicates that muscle damage is initiated when myofibrils are stretched during contraction.[8] An immediate response to muscle injury is the release of a vast array of damage-associated molecular patterns (DAMPs), which are released into the extracellular environment in response to trauma.[15] The result is integrity disruption within the myocyte, which includes disturbances in the sarcoplasmic reticulum, transverse tubules, and myofibrillar proteins.[8,16] Excessive damage to these areas can lead to alterations in calcium homeostasis and subsequent proteolytic or inflammatory events.[8] Damage to proteins and membranes within these areas also result in an increase in inflammatory agents such as prostaglandins, substance P, and inflammatory cytokines, which promote the migration of innate immune effectors (i.e., macrophages and neutrophils) to the damaged area.[17–19]

3. THE RESOLUTION OF INFLAMMATION WITHIN SKELETAL MUSCLE: A COORDINATED INFLAMMATORY RESPONSE

The promotion of ROS and DAMPs by exercise can and has been viewed as a deleterious response. However, there is growing evidence that many of the pro-oxidative and proinflammatory processes that occur after acute exercise may be vital for the long-term adaptive responses to exercise training.[8] The temporal regulation of antioxidant enzymes and anti-inflammatory agents, which initiate metabolic adaptation and tissue repair within skeletal muscle, is evidence of such a phenomenon. The adaptive responses within skeletal muscle are modulated partially by antioxidant enzymes and innate immune cells, mainly neutrophils and macrophages, which produce a pattern of signals involved in satellite cell activation, matrix remodeling, and neovasculature formation.[20]

Despite their negative connotation, ROS are important for the induction of endogenous antioxidant enzymes such as superoxide dismutase (SOD), glutathione peroxidase (GPx), and catalase within skeletal muscle and other tissues. Such antioxidant enzymes serve many protective functions within the myocyte and other tissues.[8] For example, reduced glutathione acts as an electron donor and engages various biological roles, which include the detoxification of electrophilic xenobiotics and the modulation of redox-sensitive pathways.[21] These antioxidant pathways also lead to reduction in proinflammatory cytokines (i.e., tumor necrosis factor-alpha (TNF-α)) produced by skeletal muscle or the associated immune cells.[22]

The exercise-induced proinflammatory response and subsequent immune cell recruitment and polarization may also be required for repair and adaptive processes within skeletal muscle. Evidence of this stems from studies showing that the nonsteroidal anti-inflammatory drugs disrupt both the early translational responses within skeletal muscle[23] and the resolving lipid mediator response.[24] A dynamic, time-dependent polarization of macrophages during the tissue repair process further exemplifies the importance of the initial inflammatory response to exercise. A switch from an initial M1 proinflammatory phenotype 24 h after eccentric exercise, to an M2, anti-inflammatory phenotype 48–72-h after exercise, characterizes this phenomenon.[25] It is hypothesized that the initial recruitment of M1 macrophages is vital for the phagocytosis of necrotic cells as well as the proliferation of myogenic cells.[15,25] The effector molecules (i.e., ROS and Reactive Nitrogen Species (RNS)) and inflammatory cytokines (interleukin-1 beta (IL-1β), TNF-α, and IL-6) produced by M1 macrophages are the most likely candidates that promote necrotizing and inflammatory processes.[25]

The subsequent induction of M2 macrophages is perhaps the least understood of the inflammatory processes after exercise, but nevertheless is vital for muscle repair and regeneration. Namely, muscle recovery is largely dependent on the MAP kinase phosphatase-1 (MKP-1), which promotes p38 MAPK downregulation in macrophages, and subsequently promotes M1 to M2 macrophage differentiation.[26] Additionally, AMP-activated protein kinase (AMPK) alpha1 and cAMP response element-binding protein may also play integral roles in macrophage phenotype switching during muscle regeneration.[27] More recently, a unique population of T-regulatory cells has also been shown to be a key component to muscle repair and regeneration. More specifically, there may exist a "muscle-specific" T-regulatory cell population which is located in close proximity to regenerating satellite cells, exhibits anti-inflammatory and regenerative properties (IL-10 and TGF-β), and produces growth factors (i.e., Amphiregulin) that regulate myocyte regeneration.[28]

4. BEYOND THE MUSCLE: ACUTE EXERCISE AND SYSTEMIC INFLAMMATION

Acute exercise can also induce changes in the inflammatory milieu in areas beyond the skeletal muscle. Massive changes in circulating cytokines (although interestingly not the prototypical proinflammatory cytokines TNF-α and IL-1β) after acute exercise are evidence of this. Notably, some of these cytokines are postulated to be released directly from skeletal muscle, and thus have been termed "myokines."[29] In addition to proposed metabolic effects,[30] one particular myokine, IL-6, has pleiotropic effects on systemic immune function.[31] More specifically, IL-6 initiates the release of anti-inflammatory cytokines such as IL-1 receptor antagonist (IL-1RA) and IL-10.[32] IL-1RA binds to the IL-1 receptor (IL-1R), which inhibits the potent and deleterious effects of proinflammatory cytokines, IL-1β and IL-1α.[33] Meanwhile, IL-10 exerts anti-inflammatory effects on a variety of immune cells including reducing the expression of Major Histocompatibility Protein (MHC) molecules, cell adhesion molecules, and costimulatory molecules (CD80 and CD86) on antigen-presenting cells.[33,34] IL-10 also downregulates proinflammatory cytokines, T cells, and other effector cells, thereby limiting the capacity of these cells to maintain a prolonged inflammatory response.[34]

The anti-inflammatory effects of IL-6 were clearly demonstrated when a recombinant version of the cytokine (infused at levels observed during exercise) initiated a subsequent increase in circulating IL-1RA, IL-10, and cortisol levels.[35] However, the precise mechanisms of how exercise-induced release of IL-6 leads to anti-inflammatory cytokine production are still unclear. One prevailing theory is that exercise-induced IL-6 initiates production of IL-1RA through peripheral blood mononuclear cells and/or resident tissue macrophages.[33] Regardless of the mechanism, the acute and potent induction of anti-inflammatory cytokines after exercise has profound effects on inflammation and immune function.

The inflammatory response following an exercise bout may also initiate proinflammatory and cell death responses within immune cells and effected tissue, but this largely depends on the modality and the intensity of the exercise.[36] Examples of these phenomena have been demonstrated in humans and rodents, and likely involve numerous mechanisms. One example of such a response was shown by an increase in circulating IL-17 and IL-23 after an intensive endurance exercise event.[37] Briefly, IL-17 and IL-23 represent active components of the TH17 axis and have numerous inflammatory effects on various tissues. IL-17 is the major proinflammatory effector of TH17 cells, and IL-23 promotes the differentiation to an IL-17 producing TH17 phenotype.[38] A postulated mechanism by which IL-17 may initiate an inflammatory response is through activation and recruitment of neutrophils through the chemokine IL-8.[37] Perhaps not surprisingly then, IL-17 and IL-23 production after exercise were strongly correlated with neutrophil activation marker myeloperoxidase, and muscle damage marker myoglobin (Mb), indicating a possible role of these cytokines in regulating neutrophil activation and subsequently muscle damage.[37] The source and cause of IL-17 and IL-23 release after exercise is less understood, but may be linked to IL-6 production.

Systemic inflammation after exercise may also be dependent on other unique events that occur after exercise. Notably, researchers have linked intense, prolonged exercise with an increase in circulating lipopolysaccharide (LPS), a major component of the outer membrane of Gram-negative bacteria.[39,40] Termed "endotoxemia," the release of enteric Gram-negative bacteria and/or its associated cell membrane components ultimately leads to a systemic inflammatory response through activation of pattern-recognition receptors (PRRs) on various cell types.[41] The release of enteric bacteria into circulation has been linked to a loss of barrier integrity in the gastrointestinal tract (GI) partially caused by aberrant function or activity of the epithelial cytoskeleton, tight junction proteins (TJPs), and/or Paneth cells.[42–44] The disruption in the epithelial barrier is likely a result of splanchnic hypoperfusion, a conserved evolutionary adaptation to enhance blood perfusion to tissues required for movement and respiration.[42]

The regulation of inflammation and immune trafficking within the GI after exercise may also be traced to local host–microbe interactions. For instance, Uchida *et al.* showed that the administration of the bacterial protein flagellin (FG) after exhaustive treadmill exercise exacerbated systemic inflammation in mice as measured by circulating TNF-α.[45] Moreover, these authors found that the inflammatory response may be exacerbated by the catecholamine, epinephrine, through an upregulation of the bacterial FG receptor, toll-like receptor 5 (TLR-5), on GI epithelial tissue.[45] Together, these data suggest that the GI tract and associated immune tissue may be especially sensitive to GI barrier disruption during and after acute exercise, as intraluminal bacteria exposed to epithelial cells and immune cells can induce an exacerbated inflammatory response. Despite this, data regarding the regulation of inflammation by enteric microbes during and after acute exercise are still lacking.

Despite the increases in circulating endotoxin and inflammatory cytokines that can occur after strenuous exercise, the acute immunosuppressive effects of exercise are still quite significant. Starkie *et al.* first demonstrated such an effect by showing that a single endurance exercise bout significantly reduced endotoxin-induced TNF-α production in humans.[46] A likely mechanism behind this blunted inflammatory response may be a result of alterations in PRRs on innate immune cells. Indeed, Lancaster *et al.* demonstrated a significant decrease in TLR (1, 2, and 4) receptor expression on circulating monocytes 1.5 h after intensive endurance exercise.[47] This finding was later confirmed by others.[48,49] Unfortunately, the mechanisms regulating TLR expression on circulating leukocytes after exercise are still not well understood, but may be related to core body temperature, stress hormones, anti-inflammatory cytokines, and/or heat shock proteins.[50]

As discussed briefly in the previous sections, circulating hormones have significant effects on the inflammatory process after exercise. The most studied of these hormones are catecholamines (epinephrine and norepinephrine) and glucocorticoids (cortisol), likely due their relative ease of manipulation and well-defined actions on the immune system. These hormones are substantially increased after exercise by activation of the sympathetic nervous system and the hypothalamic–pituitary–adrenal axis in dose- and time-dependent fashions. Cortisol (the major glucocorticoid in humans) can induce genomic and nongenomic anti-inflammatory effects, thus reducing the production of major inflammatory cytokines.[51,52] Catecholamines serve a parallel anti-inflammatory action by downregulating the production of inflammatory cytokines from circulating leukocytes.[53] However, as previously discussed, catecholamines may have direct or indirect pro-inflammatory actions within other areas of the body, most notably the GI tract.[45]

5. SUMMARY OF ACUTE EXERCISE AND INFLAMMATION

In summary, an acute exercise bout initiates a complex, time-dependent cascade of inflammatory events, which depends largely on the mode, intensity, duration, and familiarity of the exercise bout (Fig. 1). The inflammatory mediators which regulate this response act upon various tissues, most notably the skeletal muscle. Here, the inflammatory cascade is characterized by an initial proinflammatory response (~1.5–24 h after exercise) which is followed by an anti-inflammatory muscle regenerative response (~24–72 h after exercise). Acute exercise also initiates "inflammatory" processes in the circulation, evident by significant rises in circulating myokines (i.e., IL-6) followed by a subsequent increase in circulating anti-inflammatory cytokines (IL-10 and IL-1RA). Acute exercise also acts on immune cell receptor presentation, evidenced by a significant reduction of TLRs on circulating monocytes 1.5 h after strenuous exercise. It is currently unclear how an acute inflammatory response to exercise initiates long-term adaptive responses. Nevertheless, it is plausible that the acute anti-inflammatory processes of exercise may propagate into some of the well-known beneficial effects of exercise training, which will be discussed in the section 6.

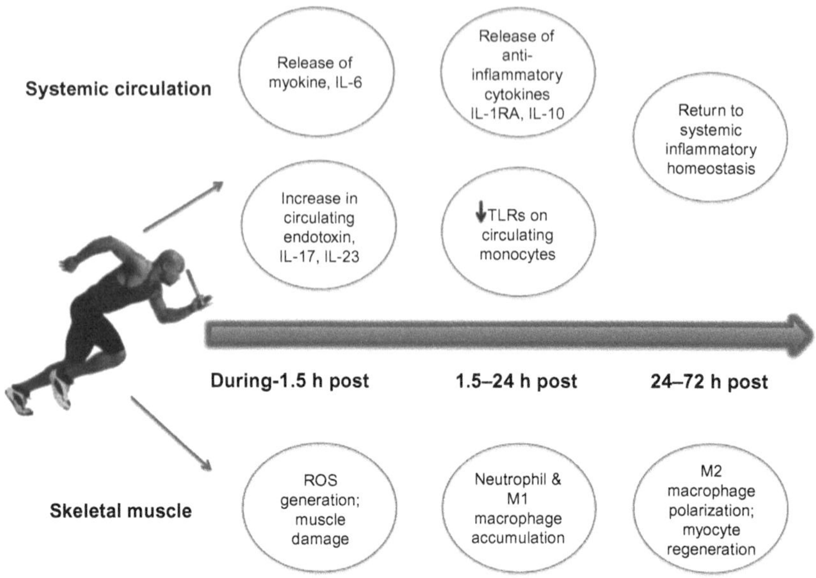

Figure 1 The effects of acute exercise on skeletal muscle and systemic inflammation. *Abbreviations*: IL, interleukin; ROS, reactive oxygen species; TLR, toll-like receptor; ra, receptor antagonist. (See the color plate.)

6. EXERCISE TRAINING AND CHRONIC INFLAMMATION

Studies have demonstrated that chronic inflammation may increase the risk of disability and mortality even in people who do not have clinical disease.[5] For instance, for individuals with infection, C-reactive protein (CRP) levels may be 1000-fold higher than the standard values.[54] CRP is a hepatic acute phase protein that is commonly used as a biomarker of systemic inflammation and risk factor for cardiovascular disease (CVD) if levels are >3 mg/L[55] and, importantly, its levels have been correlated with frailty, morbidity, and mortality.[56,57] Ridker *et al.* have shown that the CRP levels predict CVD effectively in middle-aged and older population.[58,59] In addition, increased CRP is related to diabetes,[60] heart failure,[61] physical disability,[62] and other diseases. Therefore, studying ways of diminishing chronic inflammation is critical for reducing the risk of inflammation-associated diseases.

Regular physical activity has the potential to improve chronic inflammation.[63] A large number of cross-sectional studies have consistently reported an inverse association between self-reported physical activities or objectively measured aerobic capacity with inflammatory blood biomarkers.[64] More definitive evidence comes from intervention studies. Regular moderate exercise training has been shown to act in an anti-inflammatory fashion in a number of states where inflammation is chronic in nature including aging,[65] obesity,[66] metabolic disease,[67] spinal cord injury,[68] and stroke[69] among others. For example, we found that 10 months of cardiovascular exercise training significantly reduce serum CRP in a cohort of community-dwelling older adults.[70] Many,[70–72,61,73] but not all,[74,75] intervention studies have demonstrated reduced inflammatory biomarkers. The evidence seems to indicate that if the exercise intervention is of long enough duration (>3 months) and of sufficient intensity then reductions in inflammatory biomarkers will be realized. Moreover, exercise-induced reductions in inflammation seem to occur at a higher rate in studies that have utilized participants with elevated inflammatory markers to start with (e.g., obese, aged, and diabetic).

7. POTENTIAL MECHANISMS OF THE EFFECT OF EXERCISE TRAINING ON ANTI-INFLAMMATION

While the measurement of blood inflammatory biomarkers in people is informative, they do not shed light on tissue-specific inflammation or its causes. This is important because local tissue inflammation results in an increase in blood inflammatory biomarkers. Animal models assist in localizing inflammatory defects to particular tissues of interest and will aid in the identification of tissue-specific anti-inflammatory mechanisms associated with regular exercise. There are several potential mechanisms (Fig. 2), whereby regularly performed exercise may dampen chronic inflammation and these include exercise-induced reductions in body fat, especially visceral fat, increased production and secretion of anti-inflammatory cytokines from contracting muscle, downregulation of TLRs on monocytes and macrophages, and adaptations in intracellular generation of ROS.[33]

Much of the work focusing on the mechanisms, whereby exercise reduces inflammation has implicated adipose tissue. The current thinking is that physical inactivity and elevated caloric intake result in adipose tissue and adipocyte hypertrophy. As adipocytes grow beyond a critical size, the distance for oxygen diffusion becomes too great and the cells undergo hypoxic stress leading to necrosis or cell death. Necrosis is a potent stimulus for inflammatory responses as the innate immune system attempts to clean up the cellular debris and initiate adaptive and reparative response in the

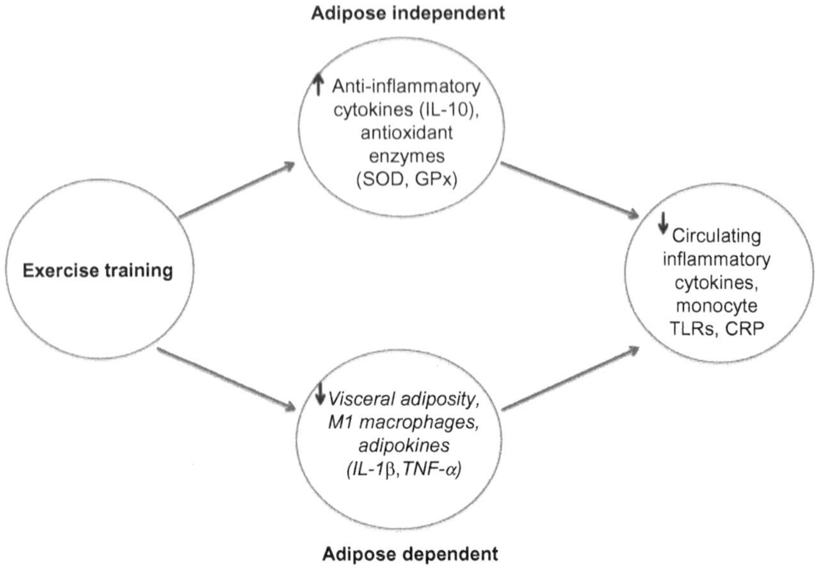

Figure 2 The effects of long-term, moderate intensity exercise training on inflammatory markers and immune mediators. The anti-inflammatory effects may manifest from both adipose-dependent and -independent mechanism. *Abbreviations*: IL, interleukin; SOD, superoxide dismutase; GPx, glutathione peroxidase. (See the color plate.)

inflamed adipose.[76] This effect is particularly damaging in visceral adipose tissue which seems to have a high inflammatory potential relative to subcutaneous fat. Exercise, by creating a caloric imbalance and mobilizing fat from adipose for fuel, reduces the size of the adipocytes thereby reducing hypoxic stress and inflammation. Indeed, we have found that 6 or 12 weeks of treadmill exercise training can significantly reduce both systemic and adipose tissue inflammation in mice fed a high-fat diet, suggesting that exercise-induced anti-inflammatory actions in adipose tissue are responsible for the reduced systemic inflammation in this model.[77]

Exercise and physical activity can also invoke anti-inflammatory actions independent of changes in fat mass. It has been suggested that, in the context of exercise, transient elevations in IL-6 coming from exercising skeletal muscle actually act in an anti-inflammatory fashion by inducing anti-inflammatory cytokines such as IL-1 receptor antagonist (IL-1RA) and soluble IL-10 which act to antagonize the actions of the quintessential proinflammatory cytokines IL-1β and TNF-α.[31] IL-6 also stimulates the release of cortisol which is an anti-inflammatory hormone.[35] Thus, IL-6 seems to be an important molecule that is elevated in response to inflammatory stimuli and contributes to regulation of inflammatory reactions, but its measurement is frequently used as a biomarker for inflammation. There is also one study showing that aerobic exercise training decreases mononuclear cell production of TNF-α and IL-1α (atherogenic cytokines) and increases IL-4, IL-10, and TGF-β1 (atheroprotective cytokines) in individuals with high risks of heart disease.[78]

TLRs are highly evolutionarily conserved transmembrane proteins that play an important role in the detection of microbial molecular patterns and endogenous "danger signals," such as those induced by tissue damage.[79] Activation of TLRs results in the production of proinflammatory cytokines. Acute exercise can result in a reduction in the expression of TLR on blood monocytes[50] which would desensitize these cells to inflammatory stimuli thus contributing an anti-inflammatory effect. However, in animal experiments performed in our lab, exercise training failed to reduce the behavioral and inflammatory effects of a wide range of doses of LPS; a TLR4 ligand comprised of cell wall components of Gram-negative bacteria.[80]

The anti-inflammatory effect of exercise training may also be the result of modulation of intracellular signaling pathways mediated by nitric oxide (NO) and ROS. During exercise, increased production levels of NO and ROS are important in inducing anti-inflammatory defense mechanisms[81] and have long-term effects on muscle gene expression. Gielen *et al.* have reported that after exercise training, there is a significant reduction of NO synthase expression in skeletal muscle.[82] The adaptive responses of redox pathways in response to exercise training protect skeletal muscle from exposure to the increase of ROS following exercise.[81] In addition, TNF-α levels in skeletal muscle will be changed by ROS, which may reduce inflammation. Therefore, exercise-induced adaptation in redox-sensitive pathways may also attenuate inflammation.

8. SUMMARY

Exercise exerts a pleiotropic, time-dependent cascade of inflammatory events, which have numerous roles in health and disease. These include interactions between innate and adaptive immune cells, cytokines, and other intracellular components, which under appropriate conditions, provide an inflammatory milieu optimal for recovery, regeneration, and adaptation from an exercise bout. The inflammatory response to an acute exercise bout, however, does not necessarily provoke one type of "inflammatory environment," as the mode intensity and duration of exercise are vital components of the exercise-induced inflammatory response. Nevertheless, it is evident that exercise training, over time, exerts anti-inflammatory actions through several distinct mechanisms. These include actions dependent and independent of changes in adipose tissue mass. Despite the mounting evidence regarding the anti-inflammatory potential of exercise, a mechanistic understanding into the key mediators and processes behind such response is still to be determined. As such, future research should encompass multicomponent approaches to exercise immunology, which attempt to integrate other, less-recognized physiological processes with acute and long-term inflammatory changes. This will include investigation into the metabolic, endocrine, and immune components of various tissues and organs, including the brain and GI. Ultimately, a more comprehensive understanding into the regulation of inflammation during and after exercise may be established.

REFERENCES

1. Medzhitov R. Inflammation 2010: new adventures of an old flame. *Cell.* 2010;140:771–776.
2. Tauber AI. Metchnikoff and the phagocytosis theory. *Nat Rev Mol Cell Biol.* 2003;4:897–901.
3. Handschin C, Spiegelman BM. The role of exercise and PGC1alpha in inflammation and chronic disease. *Nature.* 2008;454:463–469.
4. Lemanne D, Cassileth B, Gubili J. The role of physical activity in cancer prevention, treatment, recovery, and survivorship. *Oncology.* 2013;27:580–585.
5. Beavers KM, Brinkley TE, Nicklas BJ. Effect of exercise training on chronic inflammation. *Clin Chim Acta.* 2010;411:785–793.
6. Walsh NP, Gleeson M, Shephard RJ, et al. Position statement. Part one: immune function and exercise. *Exerc Immunol Rev.* 2011;17:6–63.
7. Nieman DC, Henson DA. Role of endurance exercise in immune senescence. *Med Sci Sports Exerc.* 1994;26:172–181.
8. Niess AM, Simon P. Response and adaptation of skeletal muscle to exercise—the role of reactive oxygen species. *Front Biosci.* 2007;12:4826–4838.
9. Hollander J, Fiebig R, Gore M, Ookawara T, Ohno H, Ji LL. Superoxide dismutase gene expression is activated by a single bout of exercise in rat skeletal muscle. *Pflugers Arch—Eur J Physiol.* 2001;442:426–434.
10. Gomez-Cabrera MC, Borras C, Pallardo FV, Sastre J, Ji LL, Vina J. Decreasing xanthine oxidase-mediated oxidative stress prevents useful cellular adaptations to exercise in rats. *J Physiol.* 2005;567:113–120.
11. Kramer HF, Goodyear LJ. Exercise, MAPK, and NF-kappaB signaling in skeletal muscle. *J Appl Physiol.* 2007;103:388–395.
12. Kim JS, Saengsirisuwan V, Sloniger JA, Teachey MK, Henriksen EJ. Oxidant stress and skeletal muscle glucose transport: roles of insulin signaling and p38 MAPK. *Free Radic Biol Med.* 2006;41:818–824.
13. Widegren U, Wretman C, Lionikas A, Hedin G, Henriksson J. Influence of exercise intensity on ERK/MAP kinase signalling in human skeletal muscle. *Pflugers Arch—Eur J Physiol.* 2000;441:317–322.

14. Wretman C, Lionikas A, Widegren U, Lannergren J, Westerblad H, Henriksson J. Effects of concentric and eccentric contractions on phosphorylation of MAPK(erk1/2) and MAPK(p38) in isolated rat skeletal muscle. *J Physiol.* 2001;535:155–164.
15. Kharraz Y, Guerra J, Mann CJ, Serrano AL, Munoz-Canoves P. Macrophage plasticity and the role of inflammation in skeletal muscle repair. *Mediat Inflamm.* 2013;2013:491497.
16. Proske U, Morgan DL. Muscle damage from eccentric exercise: mechanism, mechanical signs, adaptation and clinical applications. *J Physiol.* 2001;537:333–345.
17. Proske U, Allen TJ. Damage to skeletal muscle from eccentric exercise. *Exerc Sport Sci Rev.* 2005;33:98–104.
18. Smith LL, Anwar A, Fragen M, Rananto C, Johnson R, Holbert D. Cytokines and cell adhesion molecules associated with high-intensity eccentric exercise. *Eur J Appl Physiol.* 2000;82:61–67.
19. Peake J, Nosaka K, Suzuki K. Characterization of inflammatory responses to eccentric exercise in humans. *Exerc Immunol Rev.* 2005;11:64–85.
20. Rigamonti E, Zordan P, Sciorati C, Rovere-Querini P, Brunelli S. Macrophage plasticity in skeletal muscle repair. *Biomed Res Int.* 2014;2014:560629.
21. Sen CK. Glutathione homeostasis in response to exercise training and nutritional supplements. *Mol Cell Biochem.* 1999;196:31–42.
22. Sen CK, Khanna S, Reznick AZ, Roy S, Packer L. Glutathione regulation of tumor necrosis factor-alpha-induced NF-kappa B activation in skeletal muscle-derived L6 cells. *Biochem Biophys Res Commun.* 1997;237:645–649.
23. Markworth JF, Vella LD, Figueiredo VC, Cameron-Smith D. Ibuprofen treatment blunts early translational signaling responses in human skeletal muscle following resistance exercise. *J Appl Physiol.* 2014;117:20–28.
24. Markworth JF, Vella L, Lingard BS, et al. Human inflammatory and resolving lipid mediator responses to resistance exercise and ibuprofen treatment. *Am J Physiol Regul Integr Comp Physiol.* 2013;305:R1281–R1296.
25. Minari AL, Oyama LM, Dos Santos RV. Downhill exercise-induced changes in gene expression related with macrophage polarization and myogenic cells in the triceps long head of rats. *Inflammation.* 2014;38:209–217.
26. Perdiguero E, Kharraz Y, Serrano AL, Munoz-Canoves P. MKP-1 coordinates ordered macrophage-phenotype transitions essential for stem cell-dependent tissue repair. *Cell Cycle.* 2012;11:877–886.
27. Ruffell D, Mourkioti F, Gambardella A, et al. A CREB-C/EBPbeta cascade induces M2 macrophage-specific gene expression and promotes muscle injury repair. *Proc Natl Acad Sci USA.* 2009;106:17475–17480.
28. Burzyn D, Kuswanto W, Kolodin D, et al. A special population of regulatory T cells potentiates muscle repair. *Cell.* 2013;155:1282–1295.
29. Ostrowski K, Rohde T, Zacho M, Asp S, Pedersen BK. Evidence that interleukin-6 is produced in human skeletal muscle during prolonged running. *J Physiol.* 1998;508(pt 3):949–953.
30. Pal M, Febbraio MA, Whitham M. From cytokine to myokine: the emerging role of interleukin-6 in metabolic regulation. *Immunol Cell Biol.* 2014;92:331–339.
31. Petersen AM, Pedersen BK. The anti-inflammatory effect of exercise. *J Appl Physiol.* 2005;98(4):1154–1162.
32. Petersen AM, Pedersen BK. The role of IL-6 in mediating the anti-inflammatory effects of exercise. *J Physiol Pharmacol.* 2006;57(suppl 10):43–51.
33. Gleeson M, Bishop NC, Stensel DJ, Lindley MR, Mastana SS, Nimmo MA. The anti-inflammatory effects of exercise: mechanisms and implications for the prevention and treatment of disease. *Nat Rev Immunol.* 2011;11(9):607–615.

34. Maynard CL, Weaver CT. Diversity in the contribution of interleukin-10 to T-cell-mediated immune regulation. *Immunol Rev*. 2008;226:219–233.
35. Steensberg A, Fischer CP, Keller C, Moller K, Pedersen BK. IL-6 enhances plasma IL-1ra, IL-10, and cortisol in humans. *Am J Physiol Endocrinol Metab*. 2003;285: E433–E437.
36. Suzuki K, Yamada M, Kurakake S, et al. Circulating cytokines and hormones with immunosuppressive but neutrophil-priming potentials rise after endurance exercise in humans. *Eur J Appl Physiol*. 2000;81:281–287.
37. Sugama K, Suzuki K, Yoshitani K, Shiraishi K, Kometani T. IL-17, neutrophil activation and muscle damage following endurance exercise. *Exerc Immunol Rev*. 2012;18:116–127.
38. Korn T, Bettelli E, Oukka M, Kuchroo VK. IL-17 and Th17 cells. *Annu Rev Immunol*. 2009;27:485–517.
39. Bosenberg AT, Brock-Utne JG, Gaffin SL, Wells MT, Blake GT. Strenuous exercise causes systemic endotoxemia. *J Appl Physiol*. 1988;65:106–108.
40. Brock-Utne JG, Gaffin SL, Wells MT, et al. Endotoxaemia in exhausted runners after a long-distance race. *S Afr Med J*. 1988;73:533–536.
41. Jialal I, Rajamani U. Endotoxemia of metabolic syndrome: a pivotal mediator of meta-inflammation. *Metab Syndr Relat Disord*. 2014;12:454–456.
42. Van Wijck K, Lenaerts K, Grootjans J, et al. Physiology and pathophysiology of splanchnic hypoperfusion and intestinal injury during exercise: strategies for evaluation and prevention. *Am J Physiol Gastrointest Liver Physiol*. 2012;303:G155–G168.
43. Thuijls G, Derikx JP, de Haan JJ, et al. Urine-based detection of intestinal tight junction loss. *J Clin Gastroenterol*. 2010;44:e14–e19.
44. Thuijls G, de Haan JJ, Derikx JP, et al. Intestinal cytoskeleton degradation precedes tight junction loss following hemorrhagic shock. *Shock*. 2009;31:164–169.
45. Uchida M, Oyanagi E, Kawanishi N, et al. Exhaustive exercise increases the TNF-alpha production in response to flagellin via the upregulation of toll-like receptor 5 in the large intestine in mice. *Immunol Lett*. 2014;158:151–158.
46. Starkie R, Ostrowski SR, Jauffred S, Febbraio M, Pedersen BK. Exercise and IL-6 infusion inhibit endotoxin-induced TNF-alpha production in humans. *FASEB J*. 2003;17:884–886.
47. Lancaster GI, Khan Q, Drysdale P, et al. The physiological regulation of toll-like receptor expression and function in humans. *J Physiol*. 2005;563:945–955.
48. Oliveira M, Gleeson M. The influence of prolonged cycling on monocyte toll-like receptor 2 and 4 expression in healthy men. *Eur J Appl Physiol*. 2010;109:251–257.
49. Simpson RJ, McFarlin BK, McSporran C, Spielmann G, o Hartaigh B, Guy K. Toll-like receptor expression on classic and pro-inflammatory blood monocytes after acute exercise in humans. *Brain Behav Immun*. 2009;23:232–239.
50. Gleeson M, McFarlin B, Flynn M. Exercise and toll-like receptors. *Exerc Immunol Rev*. 2006;12:34–53.
51. Prigent H, Maxime V, Annane D. Science review: mechanisms of impaired adrenal function in sepsis and molecular actions of glucocorticoids. *Crit Care*. 2004;8:243–252.
52. Cupps TR, Fauci AS. Corticosteroid-mediated immunoregulation in man. *Immunol Rev*. 1982;65:133–155.
53. Kitamura H, Shiva D, Woods JA, Yano H. Beta-adrenergic receptor blockade attenuates the exercise-induced suppression of TNF-alpha in response to lipopolysaccharide in rats. *Neuroimmunomodulation*. 2007;14:91–96.
54. Pepys MB. C-reactive protein fifty years on. *Lancet*. 1981;1:653–657.
55. Koenig W. High-sensitivity C-reactive protein and atherosclerotic disease: from improved risk prediction to risk-guided therapy. *Int J Cardiol*. 2013;168(6):5126–5134.
56. Lai HY, Chang HT, Lee YL, Hwang SJ. Association between inflammatory markers and frailty in institutionalized older men. *Maturitas*. 2014;79(3):329–333.

57. Zhang W, He J, Zhang F, et al. Prognostic role of C-reactive protein and interleukin-6 in dialysis patients: a systematic review and meta-analysis. *J Nephrol*. 2013;26(2):243–253.
58. Ridker PM, Cushman M, Stampfer MJ, Tracy RP, Hennekens CH. Inflammation, aspirin, and the risk of cardiovascular disease in apparently healthy men. *N Engl J Med*. 1997;336:973–979.
59. Ridker PM, Rifai N, Pfeffer MA, et al. Inflammation, pravastatin, and the risk of coronary events after myocardial infarction in patients with average cholesterol levels. Cholesterol and Recurrent Events (CARE) Investigators. *Circulation*. 1998;98:839–844.
60. Pradhan AD, Manson JE, Rifai N, Buring JE, Ridker PM. C-reactive protein, interleukin 6, and risk of developing type 2 diabetes mellitus. *JAMA*. 2001;286:327–334.
61. Cesari M, Penninx BW, Newman AB, et al. Inflammatory markers and onset of cardiovascular events: results from the Health ABC study. *Circulation*. 2003;108:2317–2322.
62. Penninx BW, Kritchevsky SB, Newman AB, et al. Inflammatory markers and incident mobility limitation in the elderly. *J Am Geriatr Soc*. 2004;52:1105–1113.
63. Nicklas BJ, You T, Pahor M. Behavioural treatments for chronic systemic inflammation: effects of dietary weight loss and exercise training. *CMAJ*. 2005;172:1199–1209.
64. Lavie CJ, Church TS, Milani RV, Earnest CP. Impact of physical activity, cardiorespiratory fitness, and exercise training on markers of inflammation. *J Cardiopulm Rehabil Prev*. 2011;31:137–145.
65. Nicklas BJ, Hsu FC, Brinkley TJ, et al. Exercise training and plasma C-reactive protein and interleukin-6 in elderly people. *J Am Geriatr Soc*. 2008;56(11):2045–2052.
66. Park YM, Myers M, Vieira-Potter VJ. Adipose tissue inflammation and metabolic dysfunction: role of exercise. *Mo Med*. 2014;111(1):65–72.
67. Lancaster GI, Febbraio MA. The immunomodulating role of exercise in metabolic disease. *Trends Immunol*. 2014;35(6):262–269.
68. Neefkes-Zonneveld CR, Bakkum AJ, Bishop NC, van Tulder MW, Janssen TW. The effect of long-term physical activity and acute exercise on markers of systemic inflammation in persons with spinal cord injury: a systematic review. *Arch Phys Med Rehabil*. 2015;96(1):30–42.
69. Austin MW, Ploughman M, Glynn L, Corbett D. Aerobic exercise effects on neuroprotection and brain repair following stroke: a systematic review and perspective. *Neurosci Res*. 2014;87C:8–15.
70. Vieira VJ, Hu L, Valentine RJ, et al. Reduction in trunk fat predicts cardiovascular exercise training-related reductions in C-reactive protein. *Brain Behav Immun*. 2009;23(4):485–491.
71. Nimmo MA, Leggate M, Viana JL, King JA. The effect of physical activity on mediators of inflammation. *Diabetes Obes Metab*. 2013;15(suppl 3):51–60.
72. Pai JK, Pischon T, Ma J, et al. Inflammatory markers and the risk of coronary heart disease in men and women. *N Engl J Med*. 2004;351:2599–2610.
73. Blankenberg S, Luc G, Ducimetiere P, et al. Interleukin-18 and the risk of coronary heart disease in European men: the Prospective Epidemiological Study of Myocardial Infarction (PRIME). *Circulation*. 2003;108:2453–2459.
74. Gray SR, Baker G, Wright A, Fitzsimons CF, Mutrie N, Nimmo MA. The effect of a 12 week walking intervention on markers of insulin resistance and systemic inflammation. *Prev Med*. 2009;48:39–44.
75. Church TS, Earnest CP, Thompson AM, et al. Exercise without weight loss does not reduce CRP: the INFLAME study. *Med Sci Sports Exerc*. 2010;42(4):708–716.
76. Trayhum P. Hypoxia and adipose tissue function and dysfunction in obesity. *Physiol Rev*. 2013;93(1):1–21.
77. Vieira VJ, Valentine RJ, Wilund KR, Antao N, Baynard T, Woods JA. Effects of exercise and low-fat diet on adipose tissue inflammation and metabolic complications in obese mice. *Am J Physiol Endocrinol Metab*. 2009;296(5):E1164–E1171.

78. Smith JK, Dykes R, Douglas JE, Krishnaswamy G, Berk S. Long-term exercise and atherogenic activity of blood mononuclear cells in persons at risk of developing ischemic heart disease. *JAMA*. 1999;281:1722–1727.
79. Mifsud EJ, Tan AC, Jackson DC. TLR agonists as modulators of the innate immune response and their potential as agents against infectious disease. *Front Immunol*. 2014;5:79.
80. Martin SA, Pence BD, Greene RM, et al. Effects of voluntary wheel running on LPS-induced sickness behavior in aged mice. *Brain Behav Immun*. 2013;29:113–123.
81. Scheele C, Nielsen S, Pedersen BK. ROS and myokines promote muscle adaptation to exercise. *Trends Endocrinol Metab*. 2009;20:95–99.
82. Gielen S, Adams V, Mobius-Winkler S, et al. Anti-inflammatory effects of exercise training in the skeletal muscle of patients with chronic heart failure. *J Am Coll Cardiol*. 2003;42:861–868.

CHAPTER FIFTEEN

Exercise and the Regulation of Immune Functions

Richard J. Simpson[1], Hawley Kunz, Nadia Agha, Rachel Graff
Department of Health and Human Performance, Laboratory of Integrated Physiology, University of Houston, Houston, Texas, USA
[1]Corresponding author: e-mail address: rjsimpson@uh.edu

Contents

Abstract

Exercise has a profound effect on the normal functioning of the immune system. It is generally accepted that prolonged periods of intensive exercise training can depress immunity, while regular moderate intensity exercise is beneficial. Single bouts of exercise evoke a striking leukocytosis and a redistribution of effector cells between the blood compartment and the lymphoid and peripheral tissues, a response that is mediated by increased hemodynamics and the release of catecholamines and glucocorticoids following the activation of the sympathetic nervous system and the hypothalamic–pituitary–adrenal axis. Single bouts of prolonged exercise may impair T-cell, NK-cell, and neutrophil function, alter the Type I and Type II cytokine balance, and blunt immune responses to primary and recall antigens *in vivo*. Elite athletes frequently report symptoms associated with upper respiratory tract infections (URTI) during periods of heavy training and competition that may be due to alterations in mucosal immunity, particularly reductions in secretory immunoglobulin A. In contrast, single bouts of moderate intensity exercise are "immuno-enhancing" and have been used to effectively increase vaccine responses in "at-risk" patients. Improvements in immunity due to regular

Progress in Molecular Biology and Translational Science, Volume 135
ISSN 1877-1173
http://dx.doi.org/10.1016/bs.pmbts.2015.08.001

exercise of moderate intensity may be due to reductions in inflammation, maintenance of thymic mass, alterations in the composition of "older" and "younger" immune cells, enhanced immunosurveillance, and/or the amelioration of psychological stress. Indeed, exercise is a powerful behavioral intervention that has the potential to improve immune and health outcomes in the elderly, the obese, and patients living with cancer and chronic viral infections such as HIV.

1. INTRODUCTION

The immune system is a complex network of cells and molecules that function to protect the host from invading microorganisms, prevent disease, and facilitate wound healing. Although the immune system is generally divided into two broad branches, innate (nonspecific, natural) and adaptive (repetitive, specific) immunity, it is important to note that both arms of the immune system often work synergistically in the overall immune response. For instance, innate immune cells help facilitate specific (memory) immune responses through antigen presentation, while adaptive immune cells may release cytokines and other messenger molecules that regulate innate immune cell function.

Interest in the effects of exercise on the immune system stemmed from the early work of David Nieman, who showed that individuals engaged in regular exercise of a moderate intensity reported fewer symptoms associated with upper respiratory tract infections (URTI) compared to sedentary peers, while, conversely, those engaged in frequent high-volume exercise training appeared to be at a greater risk of infection than those who remained sedentary.[1,2] Studies concerned with the effects of exercise on the immune system have focused on the impact of acute bouts of exercise as well as the chronic effects of exercise training. While athletes have been the focus in many of these studies, there is also great interest in how exercise training can improve immune function in the elderly and in diseased patients. In this chapter, we describe the effects of acute and chronic exercise on immune responses and briefly discuss their potential health and clinical implications.

2. THE EFFECTS OF ACUTE EXERCISE ON IMMUNE CELL NUMBER AND COMPOSITION

A single bout of exercise has a profound effect on the total number and composition of circulating leukocytes. It is not uncommon for the total leukocyte count to increase two- to threefold after even brief (order of minutes) dynamic exercise, whereas prolonged endurance exercise (0.5–3 h) may cause the leukocyte count to increase up to fivefold.[3,4] Although increased leukocyte numbers are often indicative of infection and/or inflammation, the exercise-induced leukocytosis is known to be a transient phenomenon, with normal leukocyte and leukocyte subtype counts typically returning to preexercise levels within 6–24 h after exercise cessation. The predominant cells mobilized with exercise are neutrophils and lymphocytes with a smaller contribution being made from monocytes.[3,4] During the early stages of exercise recovery (within 30–60 min of exercise cessation), a rapid reduction in the blood lymphocyte count (lymphocytopenia) occurs concomitantly with a sustained neutrophilia (elevated blood neutrophil count). The exercise-induced lymphocytopenia is believed by some to have potential clinical ramifications by leaving individuals vulnerable to illness during exercise recovery (termed the "open-window" hypothesis). In response to endurance-based exercise in particular, the blood lymphocyte count may reach clinically low levels ($<1.0 \times 10^9$/l) by falling 30–50% below preexercise values and may remain diminished up to 6 h later.[3,4]

While exercise evokes a general leukocytosis, the response across the major leukocyte subtypes is not uniform (for a more detailed account, see Walsh *et al.*[3] and Simpson[4]). Moreover, many discrete neutrophil, monocyte, and lymphocyte subtypes also show preferential levels of cell trafficking in response to a single exercise bout. There are three main characteristics that are shared by all leukocyte subsets that are preferentially redeployed following a single bout of exercise.[4] First, exercise-mobilized cells tend to have increased cytotoxic/effector functions and have a mature/differentiated phenotype. For instance, exercise preferentially mobilizes cytotoxic cells such as NK-cells, CD8+ T-cells, and γδ T-cells,[5] and nonlymphocyte cells that have high effector functions and maturation profiles such as $CD16^{pos}$ monocytes[6] and $CD16^{neg}$ neutrophils.[7] In contrast, those subtypes with early maturation profiles and/or limited cytotoxic/effector functions such as CD4+ T-cells, B cells, and classical ($CD16^{neg}$) monocytes are redeployed in relatively fewer numbers.[3] Even within those cell types, such as CD8+ T-cells, that are considered highly cytotoxic as a whole, there is still a preferential mobilization of those subsets shown to have greater relative cytotoxic functions (i.e., $KLRG1^{pos}/CD28^{neg}/CCR7^{neg}$).[8–10] Second, those leukocyte subtypes that are preferentially redeployed with exercise tend to exhibit phenotypes associated with tissue migration. Exercise-mobilized leukocytes express high levels of integrins and intracellular adhesion molecules[11] and a range of chemokine receptors (i.e., CXCR2, CXCR3, and CXCR5) that have ligands for activated endothelium.[12] Third, the leukocyte subtypes that are preferentially redeployed with exercise have increased expression levels of adrenoreceptors (β_2-ARs) and glucocorticoid receptors, and are therefore highly responsive to catecholamines and cortisol. Indeed, similar cell types redeployed by exercise can also be mobilized with β-agonist (i.e., epinephrine or isoproterenol) infusion,[5,13,14] and the use of non-selective β-blockers (propranolol) abrogates the response.[13] Moreover, certain leukocyte subtypes that preferentially egress the blood compartment during exercise recovery (particularly monocytes and lymphocytes) are very responsive to glucocorticoids,[15] indicating that leukocyte trafficking between the blood and tissues is strongly influenced by both sympathetic nervous system and hypothalamic–pituitary–adrenal (HPA) axis activation.

2.1 Factors Responsible for Exercise-Induced Leukocytosis

The mechanisms responsible for the redeployment of discrete leukocyte subtypes with exercise are multifactorial. Among the main contributors is the shear stress-induced demargination of leukocytes from the vascular, pulmonary, hepatic, and/or splenic reservoirs that accompanies exercise-induced increases in cardiac output and blood flow.[4] Equally as important is the action of catecholamines and glucocorticoids, which bind to adrenoreceptors and glucocorticoid receptors that are expressed by the exercise-responsive leukocytes causing their mobilization and egress from the blood compartment both during and after exercise.

2.1.1 Leukocyte Demargination

Intravascular leukocytes are either in free circulation or adhered (marginated) to the vascular endothelium that lines the entire circulatory system.[16] Acute exercise causes substantial increases in hemodynamics such as increases in cardiac output, vascular vasodilation, and blood flow, which place greater mechanical forces on the endothelium causing leukocytes to demarginate and enter the free-flowing circulation.[4] This is coupled with greater levels of shear stress within the capillary structures that contain marginated leukocytes, which can also sever leukocyte–endothelial cell interactions and drive more leukocytes into the peripheral circulation.[4] Moreover, increased lymphatic flow with exercise is also likely to contribute to this leukocytosis as lymph is emptied into the blood via the thoracic duct. These hemodynamic factors account for the majority of the leukocyte demargination that occurs with exercise, particularly for neutrophils and most monocytes. However, hemodynamics may not be responsible for the vast majority of lymphocytes and some monocytes that are redeployed with exercise, which appear to have a greater reliance on catecholamine-mediated mechanisms.[14]

2.1.2 The Effects of Stress Hormones

Increased sympathetic nervous system activity and the resulting secretion of catecholamines (i.e., the hormone epinephrine and the neurotransmitter norepinephrine) are largely responsible for the mobilization of certain lymphocyte and monocyte subtypes during exercise. Moreover, exercise increases activity of the HPA axis, causing the release of corticotropin-releasing hormone, adrenocorticotrophic hormone, and cortisol, which are known to have a profound effect on leukocyte trafficking.[15,17] Additionally, the release of catecholamines and glucocorticoids appears to affect the kinetics of the leukocyte subtype response to exercise. Increased activation of the sympathetic–adrenal–medullary axis and the resulting release of catecholamines evokes a profound lymphocytosis within minutes of engaging in dynamic exercise, whereas the release of cortisol from the adrenal cortex causes leukocyte (neutrophil) numbers to continue their increase, often reaching peak values within a few hours after exercise cessation.

2.1.3 Origins and Destinations of the Leukocytes Redeployed by Acute Exercise

The demargination of leukocytes from the walls of the blood vessels under the influence of shear stress due to increases in blood pressure and cardiac output is an accepted mechanism for exercise-induced increases in the peripheral leukocyte count.[3] However, the leukocytosis that occurs with exercise is often much greater than what could feasibly be provided by the dislodging of leukocytes from the endothelium. Thus, the liver, lung, and spleen are likely to be major sources of the leukocytes mobilized into the circulation with exercise, as all harbor very large leukocyte reservoirs. Moreover, the lymph nodes, intestines, bone marrow, thymus gland, and even skeletal muscle also contain large numbers of white cells and may also be responsible for deploying certain leukocyte subsets into the circulation during or after exercise. The likely origin and destination of the redeployed leukocyte subtypes will vary from one leukocyte subtype to another. For instance, neutrophils may enter the circulation from the marginated pool or the bone marrow, while lymphocyte demargination may occur from the spleen under the influence of catecholamines.[4] As the cells mobilized with exercise tend to have a more mature/differentiated profile, the primary (i.e., bone marrow, thymus), and even some secondary (i.e., lymph nodes), lymphoid organs are unlikely to contribute greatly to the initial exercise-induced leukocytosis. However, the bone marrow likely contributes to the sustained neutrophilia during the recovery phase after prolonged exercise, while the lymph nodes and thymus may be responsible for the restoration of the blood lymphocyte count following the transient lymphocytopenia.

3. ACUTE EXERCISE AND IMMUNE FUNCTION

3.1 Innate Immune Responses to Acute Exercise

The innate arm of the immune system incorporates both cells and soluble factors. Innate cells, such as neutrophils, are attracted to sites of infection or inflammation via chemotaxis and ingest and destroy microbes by phagocytosis. The microbes are then attacked and digested within intracellular vacuoles by granular lytic enzymes (degranulation) and reactive oxygen species (oxidative/respiratory burst). Acute moderate intensity exercise has been reported to enhance neutrophil chemotaxis but not their ability to adhere to the endothelium,[18] which is the first step in neutrophil migration to sites of injury or infection. Regarding the phagocytic activity of innate immune cells, the vast majority of studies have reported that neutrophil phagocytosis is enhanced immediately after a single exercise bout, while monocyte phagocytosis may increase after prolonged submaximal[19] but not exhaustive exercise.[20] Spontaneous neutrophil degranulation may increase after acute exercise, but neutrophil degranulation in response to bacterial stimulation appears to be impaired.[21] Neutrophil oxidative burst is greatly influenced by the intensity and duration of the exercise bout, as it was reported that cycling exercise at 50% and 80% of $\dot{V}O_{2max}$ enhanced and attenuated neutrophil oxidative burst, respectively.[22] Moreover, during exercise recovery, neutrophil oxidative burst continues to be enhanced after moderate intensity exercise but remains impaired after exhaustive or prolonged exercise.[23,24]

NK-cell cytotoxicity is measured by coculturing human peripheral blood mononuclear cells (PBMCs, which include NK-cells) with a tumor target cell line, with the number of dead target cells being quantified relative to the number of NK-cells in the PBMC fraction. While a single bout of exercise causes an increase in NK-cell cytotoxicity that is quickly followed by a delayed suppression during exercise recovery, it has become apparent that changes in the proportion of NK-cells among the PBMC fraction are largely responsible for the biphasic response in NK-cell function. Indeed, changes in NK-cell function with exercise mirror the changes in NK-cell number, often resulting in the number of dead target cells per NK-cell to remain unchanged.[25] However, more recently, Bigley *et al.*[26] challenged this dogma by examining the effects of exercise on NK-cell cytotoxicity using a wide range of tumor target cells. While supporting the finding that exercise does not affect the per-cell killing of the K562 target cell (the cell line that is used in practically all previous studies), they observed increases in NK-cell function (on a per NK-cell basis) during the recovery phase of exercise against tumor cell lines of lymphoma and multiple myeloma origin.[26] Moreover, the increase in NK-cell function could be explained by proportional shifts of discrete NK-cell subsets that overexpress activating receptors, such as NKG2C, for idiosyncratic ligands expressed by the target cells.[26] Thus, it remains unclear if exercise affects the functional capability of NK-cells at the individual cell level, or whether changes in NK-cell function merely reflect exercise-induced alterations in NK-cell number and subset distribution.

3.2 Adaptive Immune Responses to Acute Exercise

3.2.1 Immune Responses Determined In Vitro *and* Ex Vivo

T-cells proliferate in response to an antigenic challenge to form multiple effector T-cell clones. These expanded cells are then able to recognize the antigen that caused the initial response and destroy any infected cell that displays properties of that antigen on its surface. T-cell function is typically modeled *in vitro* using a mitogen (i.e., PHA) or an antigen to trigger cell replication. Several studies have reported that T-cell proliferation decreases both during and after exercise.[3] This appears to be a consistent effect whenever mitogens are used to activate T-cells regardless of exercise modality, intensity, or duration. Moreover, reductions in the migratory and homing properties of T-cells have been reported during the recovery phase of prolonged exercise. Bishop *et al.*[27] found that CD4+ and CD8+ T-cell migration toward the supernatants released from human rhinovirus-infected bronchial epithelial cells was decreased 1 h after completing a 2-h treadmill run at 60% $\dot{V}O_{2max}$. Although this may indicate that acute exercise compromises T-cell function, it is debatable whether T-cell proliferation is truly impaired during or after exercise, with many researchers arguing that mitogens elicit a very general T-cell response and are not indicative of the antigen-specific properties of memory T-cells. Indeed, when T-cells are stimulated with viral peptides against common viral antigens such as cytomegalovirus (CMV) and Epstein–Barr Virus (EBV), T-cell activation and proliferation is greatly enhanced after 30 min of steady-state exercise,[28] with many of the expanded T-cell clones being specific to the antigen that was used to activate them.[28] Further, although the T-cell proliferation index is mathematically adjusted for the number of T-cells in the PBMC fraction[29] (or, in some cases, isolated T-cells[30] or NK-cell-depleted cells[31] are used instead of mixed PBMCs), as with NK-cells, it is likely that exercise-induced alterations in the proportion of broad (i.e., CD4+ T-cells, CD8+ T-cells, γδ T-cells) and more discrete T-cell subsets (i.e., naïve, central memory, effector memory (EM), and high differentiated or senescent cells) will confound T-cell functional measures after exercise.[8,10] For instance, although mitogen-activated proliferation and migration of blood T-cells appear to fall during exercise recovery, the T-cell subsets that have the largest tissue migrating and effector functions may have already egressed the peripheral blood compartment *in vivo*.

3.2.2 Immune Responses Determined In Vivo

To overcome limitations associated with the *in vitro* experiments that have been used to assess T-cell function after exercise, several studies have investigated the effects of exercise on immune responses to antigen challenge *in vivo*. Trained triathletes were given an intradermal inoculation containing several recall antigens, including tetanus, diphtheria, and streptococcus, following a half-Ironman race.[32] The diameter of the resulting oedema at the inoculation site was measured 48 h later as an indicator of the delayed-type hypersensitivity (DTH) response. It was found that the DTH response was significantly lower in the triathletes compared to both nonexercising triathletes and moderately trained healthy men.[32] This indicates that single bouts of prolonged endurance exercise may inhibit immune responses to recall antigens (i.e., memory T-cell responses). More recently, Harper Smith *et al.*[33] showed that prolonged endurance exercise also inhibits immune responses to the topically applied novel antigen, diphenylcyclopropenone (DPCP). Separate groups of participants received a primary DPCP exposure 20 min after 2 h of running at 60% VO_{2max} or 2 h of seated rest. Participants received a DPCP challenge 4 weeks later on the upper inner arm and oedema (skinfold thickness) and erythema (redness) were measured under blinded conditions 24 and 48 h later. The exercised participants displayed significantly reduced responses to the DPCP challenge at both 24 and 48 h postexposure, evident by smaller changes in skinfold thickness and redness. A subset of the participants, now presensitized to DPCP, completed both the exercise and seated rest conditions and also demonstrated exercise-induced impairments in the DPCP antigen recall response. The primary antigen response was ~50% lower in exercisers compared to controls, whereas the recall response was ~20% lower in the exercise versus control condition, indicating that the immune response to a novel antigen is more susceptible to exercise-induced impairment than the memory recall response.[33]

Some researchers have contended that a single bout of acute exercise is "immuno-enhancing" and have tried to harness this natural stress response to augment vaccine efficacy. Acute exercise involving whole-body dynamic exercise, or localized eccentrically based resistance exercise designed to evoke a local inflammatory response at the site of vaccination, have been used. Vaccine response is measured in two ways: the production and presence of antibodies against the vaccine strain in the serum, and the memory T-cell response (i.e., proliferation, cytokine secretion) to antigen stimulation *in vitro*. In a review of nine studies that examined the effects of acute exercise on vaccine responses, Pascoe *et al.*[34] reported that a single bout of moderate intensity exercise mostly enhances immune responses to vaccination. Among the vaccines used in these studies were the influenza, tetanus toxoid, diphtheria, pneumococcal and meningococcal vaccines, with the subjects ranging from young, healthy adults to community-dwelling elderly. Interestingly, the majority of these studies reported that exercise enhanced the response against those vaccine strains that elicited the poorest response in the control group, indicating that immune responses to those vaccine antigens with low immunogenicity are most likely to be enhanced by exercise. Moreover, the response to certain vaccine strains may also differ between males and females.[34] In summation, it appears that acute exercise is capable of both augmenting and inhibiting adaptive immune responses *in vivo* with the intensity, duration, and modality of exercise likely to serve as governing factors.

3.2.3 Mucosal Immunity

The mucosal immune system is responsible for protecting the mucosal surfaces of the respiratory tract, nasal passages, and the intestines. The effect of acute exercise on mucosal immunity has been widely studied,[35] with the majority of studies focused on changes in the secretion of immunoglobin A (SIgA) as determined in saliva. At present, there is little consensus on the impact of acute exercise on SIgA due to the inconsistencies in the literature.[3] It does appear that a number of factors may influence the response including training status, intensity and duration of the exercise bout, the method of saliva collection, sleep patterns, nutrition, and diurnal variation.[3] Conversely, saliva flow rate has consistently been shown to decrease after exercise of moderate to high intensity, which may be due to a withdrawal of the inhibitory effects of the parasympathetic nervous system.[3]

3.3 Factors Affecting the Immune Response to a Single Exercise Bout

There are many factors that may affect the immune response to acute exercise, including training status, repeated exercise bouts, and extreme environments (for a more detailed discussion of these topics, see the textbook edited by Gleeson *et al.*[36]), but the factors that are particularly prominent are: intensity and duration of the exercise bout, age, nutritional status, and infection history. Here, we discuss how these factors can influence the immune response to acute exercise.

3.3.1 Exercise Intensity and Duration

The intensity and duration of the exercise bout has a profound impact on both innate and adaptive immune responses.[3] Changes in blood lymphocyte numbers during and after exercise are mostly dependent on the intensity of exercise, whereas exercise duration has a stronger influence on the neutrophil and total leukocyte count. The egress of lymphocytes during the recovery phase of exercise that results in lymphocytopenia is also greater following high-intensity exercise.[3,8,37] Exercise intensity largely influences the blood monocyte count, although it is not uncommon for monocyte numbers to continue increasing during recovery from very prolonged exercise. Neutrophil function is also affected by exercise intensity, with moderate workloads found to increase respiratory burst activity, and more severe exercise associated with reductions in respiratory burst.[22–24] Most studies have shown that NK-cell activity, when expressed on a per-cell basis, does not appear to change much after acute exercise,[38] unless the bout is intense and prolonged, in which case it may be impaired for up to several hours after exercise.[25] Changes in T-cell function after acute exercise appear to be proportional to the exercise intensity and duration; Nieman *et al.*[29] observed 50% and 25% reductions in T-cell proliferation after 45 min of treadmill running at 80% and 50% $\dot{V}O_{2max}$, respectively. However, because exercise intensity and duration also affect the composition of T-cell and NK-cell subsets, it is possible that this underpins the observed functional changes. With that being said, a recent study reported impaired *in vivo* immune responses to

DPCP after 2 h of running at 60% VO_{2peak}, but not after 30 min of running at 80% or 60% VO_{2peak}.[39] So while the altered composition of discrete lymphocyte subtypes after exercise may confound many immune cell function measures determined *in vitro*, evidence from the *in vivo* studies indicates that there are genuine effects of exercise intensity and duration on the normal functioning of the immune system.

Antimicrobial proteins that are found in saliva and protect the upper respiratory tract from invading microorganisms usually increase during acute bouts of exercise, although the effects of acute exercise on SIgA responses are largely inconsistent. A recent study by Kunz *et al.*[40] reported that the secretion and/or concentration of SIgA, lactoferrin, lysozyme, LL-37, HNP 1–3, and alpha amylase increased after 30 min of steady-state cycling exercise at intensities corresponding to −5%, +5%, or +15% of the individual blood lactate threshold, although only alpha amylase was sensitive to the intensity of the exercise bout.

3.3.2 Aging

Aging is associated with profound declines in immunity that are described by the canopy term, immunosenescence.[41] It has become apparent that older individuals (>50 years) have an impaired ability to redeploy certain leukocyte subtypes in response to a single exercise bout. Moreover, T-cell proliferation to mitogens following exercise is significantly lower in older compared to younger exercisers, while aging does not appear to affect NK-cell mobilization and function in response to a single exercise bout.[42,43] Apart from NK-cells, there is very little information available on the effects of aging on the innate immune response to acute exercise.[3] Although older adults typically exercise at lower absolute and relative exercise intensities, which could be a confounding factor, aging has been shown to impair the recruitment of certain leukocyte subpopulations to the blood compartment even when similar physiological responses to acute maximal and steady-state exercise are observed between the age groups.[43,44] Moreover, even when there are no differences in the relative and absolute exercise intensities between young and older subjects, older exercisers were still found to mobilize substantially fewer CD8+ T-cells than their younger counterparts.[44] In an analysis of CD8+ T-cell subsets, it was found that this effect was attributable to a lower mobilization of naïve cells, possibly due to the reductions in thymic output that are accompanied with aging.[44] Interestingly, this study also reported that older exercisers infected with CMV (discussed in more detail below) were able to mobilize comparable numbers of CD8+ T-cells as their young counterparts, and substantially greater cell numbers compared to their age-matched, noninfected counterparts.[44] Thus, it appears that infection history is a major confounding factor of apparent age-related differences in the immune response to exercise.

A number of factors could be involved in the differences seen in leukocytosis with age. Reductions in absolute T-cell number combined with age-related declines in cardiac output may result in fewer cells being mobilized into the circulation due to shear stress. Moreover, while the catecholamine response to acute exercise is not affected by age, the sensitivity of β2-ARs appears to be blunted,[45] suggesting that older individuals have a higher threshold for catecholamine-induced immune cell redeployment.

3.3.3 Nutritional Status

The effect of nutritional interventions on the immune response to acute exercise has been studied extensively. For a more thorough overview, see Davison and Simpson[46] or Walsh *et al.*[47] While many studies have examined the impact of macro-and micronutrient intake on the immune responses to exercise, they have yielded mostly equivocal results.[46,47] However, there is consensus that carbohydrate availability has a profound impact on the immune responses to acute exercise. Performing a single bout of exercise in a state of glycogen depletion, or after a few days of a low CHO diet, evokes larger immune perturbations (increased catecholamine/glucocorticoid response, greater leukocyte redistribution, and depressed immune function) compared to exercise on a normal or high CHO diet (or when not glycogen depleted).[46,47] T-cell, NK-cell, and neutrophil functions have all been reported to decline by a greater extent following exercise under glycogen-depleted conditions. It is likely that CHO availability impacts immunity during exercise via indirect mechanisms, as exercising under conditions of low CHO intake or glycogen depletion evokes a larger stress response. In comparison to similar exercise under conditions of normal CHO availability, increased HPA axis activation (i.e., cortisol secretion), catecholamine release, and muscle IL-6 production and release are often observed,[46,47] all of which profoundly impact the composition and function of leukocyte populations in blood. Acute CHO ingestion in the region of 30–60 g/h is usually sufficient to blunt the cortisol response and reductions in immune function during prolonged exercise.[46,47] These include blunting the inhibitory effects of exercise on T-cell proliferation, T-cell migration toward infected cell culture supernatants, and neutrophil phagocytosis/oxidative burst activity.

3.3.4 *Infection History*

The human immune system is largely shaped by the pathogens that have previously infected the host. The idea that infection history may govern immune responses to acute exercise is new, with recent attention being paid to CMV, a β-herpes virus that, by 40 years of age, infects 50–80% of the US population. Latent CMV infection has been shown to have somewhat dichotomous effects on immune cell distribution following a single exercise bout. While total CD8+ T-cells and γδ T-cells are redeployed in relatively greater numbers,[44,48,49] NK-cell mobilization and egress is markedly reduced in people with CMV.[50,51] This effect appears to be due to the impact CMV has on the proportion of discrete CD8+ T-cell and NK-cell subsets. People with CMV have more KLRG1+, EM, and CD45RA expressing effector memory cells (EMRA) T-cell subsets that are known to be highly responsive to exercise,[8,10] but also have greater proportions of NKG2C+ NK-cell subsets, that have been shown to be poorly responsive to exercise and catecholamine stimulation *in vitro*.[51] CMV also enhances NK-cell function at rest, although any enhancements in NK-cell function due to exercise appear to occur only in people without CMV.[51] Conversely, changes in neutrophil and monocyte number and function after exercise are not influenced by CMV serostatus.[52] Although CMV-infected individuals typically harbor other latent herpesvirus infections [i.e., EBV, herpes simplex virus-1 (HSV-1)], the effects of CMV on CD8+ and γδ T-cell redeployment appear to be independent of these coinfections.[48,53]

4. CHRONIC EXERCISE AND IMMUNE FUNCTION

Regular exercise has the potential to exert both positive and deleterious effects on the normal functioning of the immune system. Studies concerned with the effects of chronic exercise on immunity have focused mostly on athletes or the elderly, with the overall goals of ascertaining the extent of immune decline in athletes due to excessive exercise training or identifying the factors responsible for improving immunity in the elderly or the immunocompromised due to regular moderate intensity exercise training.

4.1 The Effects of High Volume Exercise Training

It is generally accepted that excessive high intensity, high-volume exercise training can contribute to transient, and sometimes long-term, depressions in immunity that can increase infection risk. Single bouts of prolonged

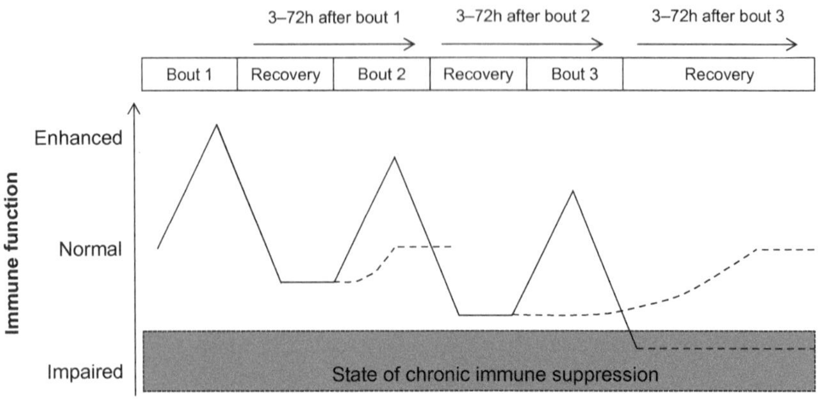

Figure 1 The "open-window" hypotheses adapted from the original model proposed by Pedersen and Ullum.[54] A single bout of exercise is associated with an initial enhancement in immune function that is quickly followed by a transient period of immune depression (the "open window") that can last for 3–72 h after the initial exercise bout (although in most cases immune function is restored to normal levels within 24 h). It is believed that this "open window" leaves the host susceptible to opportunistic infections. If a second bout of exercise is performed during the window (i.e., without adequate recovery), then the exercise-induced enhancement in immunity is blunted and the post-exercise immune depression is more severe and prolonged (i.e., the window is opened wider and for longer), rendering the athlete more susceptible to infection. The response is exacerbated with subsequent exercise bouts that, if completed without adequate recovery, may result in a state of chronic immune suppression (a third bout is used here for illustrative purposes, but in reality, it is likely to be many more). (See the color plate.)

endurance exercise are believed to leave an "open window" for opportunistic infections that may last up to 72 h postexercise[54] (Fig. 1). When repeated bouts of acute strenuous exercise are performed without adequate recovery, this "open window" is prolonged and may cause the transient depressions in immunity after acute exercise to cumulate, resulting in a chronic state of impaired immunity. In addition, the physical and psychological stress of high-intensity training, and travel associated with competition, may increase pathogen exposure and increase the incidence of viral reactivation, placing additional burden on the immune system.[47] Although this "open window" is believed to underpin the increased reporting of URTI symptoms in athletes following prolonged bouts of endurance exercise, it has been debated whether or not these symptoms are attributable to actual infections. Indeed, Spence *et al.*[55] obtained throat and nasopharyngeal swabs from athletes reporting symptoms of URTI and found that pathogens were isolated in fewer than 30% of the cases, suggesting that noninfectious factors, such as local inflammation which occurs as a result of increased respiration, and exposure to irritants, such as environmental air pollutants, may also be involved.[47,3]

4.2 Potential Factors Involved in Exercise-Induced Immune Depression

It has consistently been shown that those engaging in frequent bouts of prolonged strenuous exercise report more symptoms associated with URTI, particularly during the subsequent hours and days following major sporting events and competitions.[3,47] Although the evidence to suggest that these self-reported symptoms are due to actual infections is sketchy, there is evidence that high-volume exercise training can depress certain aspects of immune function that may leave athletes susceptible to opportunistic infections.[56]

4.2.1 Immune Cell Frequency and Function

Declines in the number of circulating immune cells are known to increase infection susceptibility. Total leukocyte and leukocyte subset numbers tend to be similar between athletes and healthy age-matched controls, although endurance athletes (particularly runners) may present with lower resting lymphocyte numbers. Horn *et al.*[57] collected peripheral blood samples from over 2000 elite Australian athletes spanning a 10-year period. They found that total leukocyte, monocyte, and neutrophil counts were lowest among cyclists and triathletes. Around 17% of cyclists and triathletes were neutropenic ($<2.0 \times 10^9$/l) compared to 5% across all other sports. Lymphocytopenia, although transiently observed after prolonged bouts of acute exercise, was only evident in 2% of athlete resting blood samples across all sports.[57] Longitudinal training studies have also reported declines in total NK-cells and CD8+ T-cells after 12 weeks and 6 months of intensive training in swimmers and cyclists, respectively.[58,59]

Neutrophil respiratory burst[23] and NK-cell activity[60] have been reported to decrease due to heavy exercise training. Moreover, athletes undertaking prolonged periods of intensive training may also present with impairments in adaptive immunity, with reports on reduced T-cell proliferation, numbers of circulating Type 1 T-cells, and stimulated B-cell immunoglobulin synthesis.[58,61,62] A common practice among athletes is to "overreach," which is characterized by intensified short-term training periods incorporated into their schedules to enhance future performance. Studies that have tracked athletes during periods of overreaching have observed discernible declines in neutrophil and monocyte oxidative burst activity, T-cell proliferation, CD4+/CD8+ T-cell ratios, antibody synthesis, and NK-cell function.[56]

Although these declines are not indicative of clinical immune deficiency, they may leave athletes more susceptible to opportunistic infections during these heavy training periods.[56] A recent study reported that naïve T-cell number and thymic output were drastically reduced in elite athletes and that their peripheral immune compartments closely resembled those of an older adult population.[63] This could have important implications for the long-term health of the high-performance athlete as low naïve T-cell numbers due to reductions in thymic output could impair adaptive immune responses to novel pathogens. Functional declines in adaptive immunity due to prolonged periods of intensive exercise training appear to be related to alterations in the pro- and anti-inflammatory cytokine balance and elevated plasma stress hormone levels, particularly cortisol.[56]

4.2.2 Latent Viral Reactivation

Few studies have looked at the effects of infection history and immune responses to exercise training. There is evidence that latent herpesviruses, such as EBV, can reactivate during periods of high-intensity/volume exercise training.[64] Gleeson *et al.*[64] reported that EBV serostatus was associated with self-reported symptoms of URTI in a group of elite swimmers following a 30-day period of intensive training. Moreover, EBV DNA was detectable in saliva (indicative of viral shedding) prior to the manifestation of URTI symptoms, causing the authors to speculate that the two could be causally related.[64] In contrast, a recent study found that prior exposure to both CMV and EBV exerted protective effects against URTI incidence over a 4-month winter training period in a large number of student athletes,[65] although this study did not link infection risk with viral shedding. It is possible that having CMV, due to the increased frequency of effector T-cells and highly cytotoxic NK-cells, strengthens immunosurveillance. However, when immune integrity is breached and CMV or EBV is allowed to reactivate from their latent states, it can overburden the immune system and increase the risk for opportunistic infections. Indeed, latent viral reactivation has been frequently documented in astronauts during spaceflight missions and is associated with shifts in the Type 1 T-helper cells (Th1) and Type 2 T-helper cells (Th2) cytokine balance.[66] Additionally, viral reactivation has been linked with acute and chronic stress[67,68] that may be exacerbated by frequent training and competition.[47]

4.2.3 Nutritional Status

The impact of nutrition and supplementation on maintaining immune health during exercise has been extensively studied.[46,47] It is widely accepted that reductions in nutrient availability and poor dietary practices during periods of heavy training are involved in the etiology of exercise-induced immune depression.[46,47] Inadequate fluid intake may lead to dehydration, which can impair the normal functioning of the immune system. Low substrate availability (i.e., muscle glycogen depletion, decreased blood glucose concentration) due to malnutrition may augment the stress response to exercise, particularly cortisol, which can have direct immunosuppressive effects. Moreover, many cells of the immune system have high-energy requirements and reduced substrate availability, such as glucose, can limit their capacity to function effectively. Athletes are therefore advised to consume sufficient fluids during exercise and recovery and maintain a diet that is high in CHO, not only to facilitate athletic performance, but to also inhibit the deleterious effects of cortisol. Indeed, consuming 30–60 g of CHO every hour during or shortly after prolonged exercise has been shown to reduce cortisol release and blunt exercise-induced immune perturbations.[46,47] Finally, athletes with known nutritional deficiencies may benefit from supplementation, although there is little scientific evidence supporting the use of many so-called "immune-enhancing" supplements at preventing alterations in immunity following both acute exercise bouts and chronic exercise training.[46,47]

4.2.4 Mucosal Immune Function

There is consensus that intensified periods of exercise training is associated with a decrease in the concentration and secretion of SIgA, which have often been linked with subsequent URTI symptoms, particularly when saliva samples are collected frequently over several months during the training period. In a study of elite swimmers over a 7-month training season, the preseason and pretraining SIgA concentration was found to predict the number of infectious episodes.[69] Moreover, when saliva samples were collected on a weekly basis from elite yachtsmen over a 12-month period, significant reductions in SIgA were observed in the 3 weeks prior to clinically confirmed respiratory infection.[70] It was further reported that SIgA values less than 40% of the individual's "normal" value were associated with a 50% chance of contracting a respiratory infection within 3 weeks.[70]

4.3 The Effects of Moderate Intensity Exercise Training

In contrast to the excessive exercise behaviors typically practiced by highly competitive athletes, regular exercise training of a moderate intensity is believed to exert beneficial effects on immune function. Immunity is generally well maintained up until age 50; therefore, the effects of regular exercise training on immune function has typically been studied in older adults,[41] although many more studies are examining the effects of moderate exercise training in obese and diseased patients, including those living with cancer and HIV. Moreover, the exercise mode (i.e., aerobic, resistance, or combinations) most likely to exert the greatest influence on immunity is debatable, but is likely dependent on the subject/patient group and any underlying disease. Although it is beyond the scope of this chapter to discuss all of these factors in turn, this section briefly highlights those aspects of immunity that typically change with aging and are likely to be positively impacted by regular moderate intensity exercise training.

4.4 Potential Factors Involved in Exercise-Induced Immune Enhancement

4.4.1 Immune Biomarkers and Function

Regular moderate intensity exercise has been linked with enhanced vaccine responses,[71,72] lower numbers of exhausted/senescent T-cells,[73] increased T-cell proliferation,[2,74] lower levels of circulating inflammatory cytokines,[75] increased neutrophil phagocytic activity,[76] lowered inflammatory response to bacterial challenge,[77] greater NK-cell cytotoxic activity,[2,78] and increased IL-2 production,[79] all of which indicate that regular moderate intensity exercise is capable of improving, or at least maintaining, immunity across the life span.[41,80] Moreover, many studies done in animals have also revealed positive effects of moderate intensity exercise training on immune responses and outcomes to viral infections and cancer.[81,82] Hallmark features of immunosenescence include an inverted CD4+/CD8+ T-cell ratio and an increased frequency and proportion of senescent T-cells.[41] It has been hypothesized that regular exercise may facilitate the selective apoptosis of these "older" senescent T-cells allowing them to be replaced by "younger" T-cells capable of responding to novel antigens.[83] Indeed, high fitness levels have been associated with more naïve and less senescent cells in healthy men.[73] Furthermore, regular exercise may help prolong or reinvigorate thymic activity. In both mice and humans, IL-7 therapy has been shown to increase thymic mass and activity,[84] and exercise, in turn, has been shown to increase plasma levels of IL-7[85] that is most likely released by contracting skeletal muscle.[86] Thus, exercise may play an instrumental role in the acquisition of "new recruits," and reduce the burden on the existing T-cell pool to expand and maintain T-cell diversity.[80]

There are, however, discrepancies between the longitudinal exercise-training studies and those studies that have used a cross-sectional design to compare physically active with sedentary people. While the positive differences in many markers of immune function between exercisers and nonexercisers are striking, exercise-training interventions have oftentimes failed to change immune function measures in previously sedentary individuals.[41] A common problem with many of the longitudinal studies is that immune function has been assessed before and after exercise training in previously sedentary but otherwise healthy people. It is therefore not surprising that very few of these studies show enhancements in immune function in these healthy subjects.[41] Indeed, training studies in cancer survivors,[87] the obese,[88] and those living with HIV[89] have reported positive effects of exercise on the immune system. Moreover, improvements in vaccine response to exercise training have provided the strongest evidence to date that the plasticity of the immune system can be positively affected by exercise in nondiseased people. Compared to nonexercising controls, antibody titers to the H1N1 and H3N2 strains of the influenza A virus were enhanced in older adults immunized with a trivalent influenza vaccine before and after a 10-month aerobic exercise-training intervention.[71] Woods *et al.*[72] also found that community-dwelling elderly subjects randomized to a 10-month cardiovascular exercise-training program had increased seroprotection rates compared to controls who performed flexibility/stretching exercise up to 24 weeks after receiving the influenza vaccine (Fig. 2).

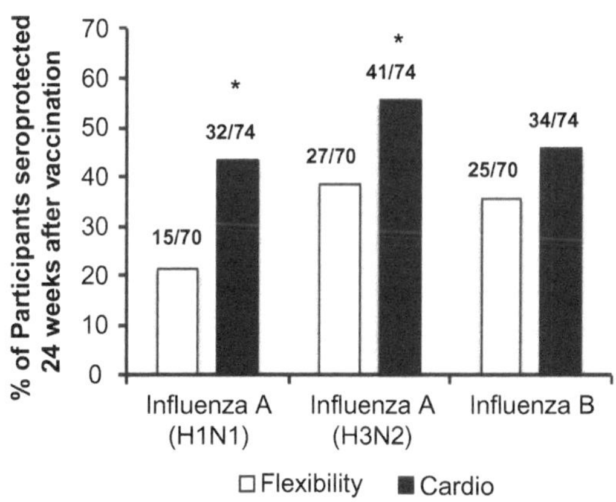

Figure 2 Compared to 10 months of flexibility training, 10 months of cardiovascular training increased the percentage of participants who achieved seroprotection (defined as a hemagglutination inhibition titer >40) at 24 weeks posttrivalent influenza vaccine. Statistically significant difference from the flexibility training group indicated by *. *Data from Woods* et al.[72]

4.4.2 Reducing Inflammation

Systemic low-grade inflammation is involved in the etiology of many chronic disease states and regular exercise may reduce the risk of disease due to its anti-inflammatory effects. Various mechanisms have been proposed to explain means by which exercise reduces the low-grade, systemic inflammation associated with obesity and sedentary behavior.[90] Regular exercise reduces visceral fat mass, the accumulation of which results in elevated production of proinflammatory adipokines.[90] The release of anti-inflammatory cytokines following an acute bout of exercise may also contribute to the reduction in systemic inflammation. IL-6 released from skeletal muscle during exercise results in a subsequent increase in IL-10 and IL-1 receptor antagonist, both of which are considered anti-inflammatory.[91] Further, the hormones released during exercise have anti-inflammatory properties: cortisol acts as an anti-inflammatory mediator[92] and adrenaline downregulates the production of the inflammatory cytokines IL-1β and TNF.[93] Exercise also downregulates surface expression of TLRs on monocytes and macrophages, and in turn mitigates their downstream inflammatory cascades.[94] Exercise may also promote the switching of M_1-type inflammatory macrophages to anti-inflammatory M_2-type and reduce the infiltration of macrophages into adipose tissue, resulting in a reduction in the production of inflammatory cytokines.[95]

4.4.3 Indirect Mechanisms

Regular exercise may also positively affect the immune system via indirect mechanisms, particularly due to its effects on regulating body composition. Reducing fat mass may prevent chronic low-grade inflammation by inhibiting the inflammatory pathways described earlier. Moreover, exercise-induced reductions in the cholesterol content of cell membranes that accompany fat loss may improve T-cell receptor signaling and the translocation of MHC molecules for antigen presentation.[96] The preferential mobilization of cells with effector phenotypes during acute exercise, combined with enhancements in cardiovascular function, may improve immunity by increasing the frequency and efficiency of immune cell trafficking between the circulation, lymphoid, and peripheral tissues.[41] Exercise may enhance immunity by reducing stress, which has been associated with frequent infections, fewer naïve cells, an inverted CD4:CD8 ratio,[97] latent viral reactivation,[67,68] and overall immune decline. Finally, regular exercise may also increase the body's antioxidant defense system,[98] preventing oxidative DNA damage to lymphocytes and other immune cells.

5. SUMMARY

The effects of exercise on the normal functioning of the immune system are profound. Although immune responses to single exercise bouts are transient, it is likely that these effects cumulate over time and form the immunological adaptations to chronic exercise training. With regard to exercise "dose," the generally accepted hypothesis is that prolonged periods of intensive exercise training can depress immunity, while regular moderate intensity exercise is beneficial. Studies in athletes are done with the purpose of identifying immunological underpinnings of infection risk so that they may be counteracted by an appropriate intervention: be it nutritional, pharmacological, or behavioral. In contrast, moderate exercise-training interventions are used to positively modulate the plasticity of the immune system. Exercise is a powerful behavioral intervention that is being used in earnest to improve immune and health outcomes in the elderly and the obese, as well as patients living with cancer and chronic viral infections such as HIV. While studies focused on the immune response to acute exercise are mostly descriptive, the effects of single exercise bouts on immunity may also have clinical implications. For instance, acute exercise increases the numbers of CD34+ hematopoietic stem cells[99] and viral-specific T-cells[28,44] in the circulation, which may increase the yield of these cells for therapeutic use in stem cell transplantation and adoptive immunotherapy, respectively. Moreover, acute exercise has already proven to be a simple and economical method to improve vaccine responses, and the implementation of such methods in "at-risk" populations such as the elderly could have important implications for public health.

REFERENCES

1. Nieman DC, Johanssen LM, Lee JW. Infectious episodes in runners before and after a roadrace. *J Sports Med Phys Fitness*. 1989;29(3):289–296.
2. Nieman DC, Henson DA, Gusewitch G, et al. Physical activity and immune function in elderly women. *Med Sci Sports Exerc*. 1993;25(7):823–831.
3. Walsh NP, Gleeson M, Shephard RJ, et al. Position statement. Part one: immune function and exercise. *Exerc Immunol Rev*. 2011;17:6–63.
4. Simpson RJ. The effects of exercise on blood leukocyte numbers. In: Gleeson M, Bishop NC, Walsh NP, eds. *Exercise Immunology*. Oxford, UK, New York, USA: Routledge; 2013:64–105.
5. Anane LH, Edwards KM, Burns VE, et al. Mobilization of gammadelta T lymphocytes in response to psychological stress, exercise, and beta-agonist infusion. *Brain Behav Immun*. 2009;23(6):823–829.

6. Steppich B, Dayyani F, Gruber R, Lorenz R, Mack M, Ziegler-Heitbrock HW. Selective mobilization of CD14(+)CD16(+) monocytes by exercise. *Am J Physiol Cell Physiol.* 2000;279(3):C578–C586.
7. Peake J, Wilson G, Hordern M, et al. Changes in neutrophil surface receptor expression, degranulation, and respiratory burst activity after moderate- and high-intensity exercise. *J Appl Physiol.* 2004;97(2):612–618.
8. Campbell JP, Riddell NE, Burns VE, et al. Acute exercise mobilises CD8+ T lymphocytes exhibiting an effector-memory phenotype. *Brain Behav Immun.* 2009;23(6):767–775.
9. Simpson RJ, Cosgrove C, Chee MM, et al. Senescent phenotypes and telomere lengths of peripheral blood T-cells mobilized by acute exercise in humans. *Exerc Immunol Rev.* 2010;16:36–51.
10. Simpson RJ, Florida-James GD, Cosgrove C, et al. High-intensity exercise elicits the mobilization of senescent T lymphocytes into the peripheral blood compartment in human subjects. *J Appl Physiol.* 2007;103(1):396–401.
11. Simpson RJ, Florida-James GD, Whyte GP, Guy K. The effects of intensive, moderate and downhill treadmill running on human blood lymphocytes expressing the adhesion/activation molecules CD54 (ICAM-1), CD18 (beta2 integrin) and CD53. *Eur J Appl Physiol.* 2006;97(1):109–121.
12. Bosch JA, Berntson GG, Cacioppo JT, Dhabhar FS, Marucha PT. Acute stress evokes selective mobilization of T cells that differ in chemokine receptor expression: a potential pathway linking immunologic reactivity to cardiovascular disease. *Brain Behav Immun.* 2003;17(4):251–259.
13. Schedlowski M, Hosch W, Oberbeck R, et al. Catecholamines modulate human NK cell circulation and function via spleen-independent beta 2-adrenergic mechanisms. *J Immunol.* 1996;156(1):93–99.
14. Dimitrov S, Lange T, Born J. Selective mobilization of cytotoxic leukocytes by epinephrine. *J Immunol.* 2010;184(1):503–511.
15. Okutsu M, Suzuki K, Ishijima T, Peake J, Higuchi M. The effects of acute exercise-induced cortisol on CCR2 expression on human monocytes. *Brain Behav Immun.* 2008;22(7):1066–1071.
16. Berkow RL, Dodson RW. Functional analysis of the marginating pool of human polymorphonuclear leukocytes. *Am J Hematol.* 1987;24(1):47–54.
17. Okutsu M, Ishii K, Niu KJ, Nagatomi R. Cortisol-induced CXCR4 augmentation mobilizes T lymphocytes after acute physical stress. *Am J Physiol Regul Integr Comp Physiol.* 2005;288(3):R591–R599.
18. Ortega E, Collazos ME, Maynar M, Barriga C, De la Fuente M. Stimulation of the phagocytic function of neutrophils in sedentary men after acute moderate exercise. *Eur J Appl Physiol Occup Physiol.* 1993;66(1):60–64.
19. Nieman DC, Nehlsen-Cannarella SL, Fagoaga OR, et al. Effects of mode and carbohydrate on the granulocyte and monocyte response to intensive, prolonged exercise. *J Appl Physiol.* 1998;84(4):1252–1259.
20. Ortega E. Neuroendocrine mediators in the modulation of phagocytosis by exercise: physiological implications. *Exerc Immunol Rev.* 2003;9:70–93.
21. Bishop NC, Gleeson M, Nicholas CW, Ali A. Influence of carbohydrate supplementation on plasma cytokine and neutrophil degranulation responses to high intensity intermittent exercise. *Int J Sport Nutr Exerc Metab.* 2002;12(2):145–156.
22. Dziedziak W. The effect of incremental cycling on physiological functions of peripheral blood granulocytes. *Biol Sport.* 1990;7:239–247.
23. Pyne DB. Regulation of neutrophil function during exercise. *Sports Med.* 1994;17(4):245–258.

24. Suzuki K, Nakaji S, Yamada M, et al. Impact of a competitive marathon race on systemic cytokine and neutrophil responses. *Med Sci Sports Exerc.* 2003;35(2):348–355.
25. Nieman DC, Miller AR, Henson DA, et al. Effects of high- vs moderate-intensity exercise on natural killer cell activity. *Med Sci Sports Exerc.* 1993;25(10):1126–1134.
26. Bigley AB, Rezvani K, Chew C, et al. Acute exercise preferentially redeploys NK-cells with a highly-differentiated phenotype and augments cytotoxicity against lymphoma and multiple myeloma target cells. *Brain Behav Immun.* 2014;39:160–171.
27. Bishop NC, Walker GJ, Gleeson M, Wallace FA, Hewitt CR. Human T lymphocyte migration towards the supernatants of human rhinovirus infected airway epithelial cells: influence of exercise and carbohydrate intake. *Exerc Immunol Rev.* 2009;15:127–144.
28. Simpson RJ, Spielmann G, Hanley PJ, Bollard CM. A single bout of exercise augments the expansion of multi-virus specific T-cells in healthy humans. In: *Psychoneuroimmunology Research Society (PNIRS) 21st Annual Scientific Meeting; Philadelphia, PA, USA*; 2014.
29. Nieman DC, Miller AR, Henson DA, et al. Effect of high- versus moderate-intensity exercise on lymphocyte subpopulations and proliferative response. *Int J Sports Med.* 1994;15(4):199–206.
30. Bishop NC, Walker GJ, Bowley LA, et al. Lymphocyte responses to influenza and tetanus toxoid in vitro following intensive exercise and carbohydrate ingestion on consecutive days. *J Appl Physiol.* 2005;99(4):1327–1335.
31. Green KJ, Rowbottom DG, Mackinnon LT. Exercise and T-lymphocyte function: a comparison of proliferation in PBMC and NK cell-depleted PBMC culture. *J Appl Physiol.* 2002;92(6):2390–2395.
32. Bruunsgaard H, Hartkopp A, Mohr T, et al. In vivo cell-mediated immunity and vaccination response following prolonged, intense exercise. *Med Sci Sports Exerc.* 1997;29(9):1176–1181.
33. Harper Smith AD, Coakley SL, Ward MD, Macfarlane AW, Friedmann PS, Walsh NP. Exercise-induced stress inhibits both the induction and elicitation phases of in vivo T-cell-mediated immune responses in humans. *Brain Behav Immun.* 2011;25(6):1136–1142.
34. Pascoe AR, Fiatarone Singh MA, Edwards KM. The effects of exercise on vaccination responses: a review of chronic and acute exercise interventions in humans. *Brain Behav Immun.* 2014;39:33–41.
35. Bishop NC, Gleeson M. Acute and chronic effects of exercise on markers of mucosal immunity. *Front Biosci.* 2009;14:4444–4456.
36. Gleeson M, Bishop NC, Walsh NP. *Exercise Immunology.* Oxford, UK, New York, USA: Routledge; 2013.
37. Simpson RJ, Florida-James GD, Whyte GP, Black JR, Ross JA, Guy K. Apoptosis does not contribute to the blood lymphocytopenia observed after intensive and downhill treadmill running in humans. *Res Sports Med.* 2007;15(3):157–174.
38. Gannon GA, Rhind SG, Suzui M, et al. Beta-endorphin and natural killer cell cytolytic activity during prolonged exercise. Is there a connection? *Am J Physiol.* 1998;275(6 pt 2): R1725–R1734.
39. Diment BC, Fortes MB, Edwards JP, et al. Exercise intensity and duration effects on in vivo immunity. *Med Sci Sports Exerc.* 2014;47:1390–1398.
40. Kunz H, Bishop NC, Spielmann G, et al. Fitness level impacts salivary antimicrobial protein responses to a single bout of cycling exercise. *Eur J Appl Physiol.* 2015;115: 1015–1027.
41. Simpson RJ, Lowder TW, Spielmann G, Bigley AB, Lavoy EC, Kunz H. Exercise and the aging immune system. *Ageing Res Rev.* 2012;11:404–420.
42. Simpson RJ, Cosgrove C, Ingram LA, et al. Senescent T-lymphocytes are mobilised into the peripheral blood compartment in young and older humans after exhaustive exercise. *Brain Behav Immun.* 2008;22(4):544–551.

43. Ceddia MA, Price EA, Kohlmeier CK, et al. Differential leukocytosis and lymphocyte mitogenic response to acute maximal exercise in the young and old. *Med Sci Sports Exerc.* 1999;31(6):829–836.
44. Spielmann G, Bollard CM, Bigley AB, et al. The effects of age and latent cytomegalovirus infection on the redeployment of CD8 + T cell subsets in response to acute exercise in humans. *Brain Behav Immun.* 2014;39:142–151.
45. Feldman RD, Limbird LE, Nadeau J, Robertson D, Wood AJ. Alterations in leukocyte beta-receptor affinity with aging. A potential explanation for altered beta-adrenergic sensitivity in the elderly. *N Engl J Med.* 1984;310(13):815–819.
46. Davison G, Simpson RJ. Immunity. In: Lanham-New SA, Stear SJ, Shirreffs SM, Collins AL, eds. *Sport and Exercise Nutrition.* Oxford, UK: Wiley-Blackwell; 2011:281–303.
47. Walsh NP, Gleeson M, Pyne DB, et al. Position statement. Part two: maintaining immune health. *Exerc Immunol Rev.* 2011;17:64–103.
48. Pistillo M, Bigley AB, Spielmann G, et al. The effects of age and viral serology on gammadelta T-cell numbers and exercise responsiveness in humans. *Cell Immunol.* 2013;284(1–2):91–97.
49. Turner JE, Aldred S, Witard OC, Drayson MT, Moss PM, Bosch JA. Latent cytomegalovirus infection amplifies CD8 T-lymphocyte mobilisation and egress in response to exercise. *Brain Behav Immun.* 2010;24(8):1362–1370.
50. Bigley AB, Lowder TW, Spielmann G, et al. NK-cells have an impaired response to acute exercise and a lower expression of the inhibitory receptors KLRG1 and CD158a in humans with latent cytomegalovirus infection. *Brain Behav Immun.* 2012;26(1):177–186.
51. Bigley AB, Rezvani K, Pistillo M, et al. Acute exercise preferentially redeploys NK-cells with a highly-differentiated phenotype and augments cytotoxicity against lymphoma and multiple myeloma target cells. Part II: impact of latent cytomegalovirus infection and catecholamine sensitivity. *Brain Behav Immun.* 2015. [epub ahead of print].
52. LaVoy EC, Nieman DC, Henson DA, et al. Latent cytomegalovirus infection and innate immune function following a 75 km cycling time trial. *Eur J Appl Physiol.* 2013;113(10):2629–2635.
53. Lavoy EC, Bigley AB, Spielmann G, et al. CMV amplifies T-cell redeployment to acute exercise independently of HSV-1 serostatus. *Med Sci Sports Exerc.* 2014;46:257–267.
54. Pedersen BK, Ullum H. NK cell response to physical activity: possible mechanisms of action. *Med Sci Sports Exerc.* 1994;26(2):140–146.
55. Spence L, Brown WJ, Pyne DB, et al. Incidence, etiology, and symptomatology of upper respiratory illness in elite athletes. *Med Sci Sports Exerc.* 2007;39(4):577–586.
56. Gleeson M. Immune responses to intensified periods of training. In: Gleeson M, Bishop NC, Walsh NP, eds. *Exercise Immunology.* Oxford, UK, New York, USA: Routledge; 2013:186–206.
57. Horn PL, Pyne DB, Hopkins WG, Barnes CJ. Lower white blood cell counts in elite athletes training for highly aerobic sports. *Eur J Appl Physiol.* 2010;110(5):925–932.
58. Baj Z, Kantorski J, Majewska E, et al. Immunological status of competitive cyclists before and after the training season. *Int J Sports Med.* 1994;15(6):319–324.
59. Gleeson M, McDonald WA, Pyne DB, et al. Immune status and respiratory illness for elite swimmers during a 12-week training cycle. *Int J Sports Med.* 2000;21(4):302–307.
60. Suzui M, Kawai T, Kimura H, et al. Natural killer cell lytic activity and CD56(dim) and CD56(bright) cell distributions during and after intensive training. *J Appl Physiol.* 2004;96(6):2167–2173.
61. Lancaster GI, Halson SL, Khan Q, et al. Effects of acute exhaustive exercise and chronic exercise training on type 1 and type 2T lymphocytes. *Exerc Immunol Rev.* 2004;10: 91–106.

62. Verde TJ, Thomas SG, Moore RW, Shek P, Shephard RJ. Immune responses and increased training of the elite athlete. *J Appl Physiol.* 1992;73(4):1494–1499.
63. Prieto-Hinojosa A, Knight A, Compton C, Gleeson M, Travers PJ. Reduced thymic output in elite athletes. *Brain Behav Immun.* 2014;39:75–79.
64. Gleeson M, Pyne DB, Austin JP, et al. Epstein-Barr virus reactivation and upper-respiratory illness in elite swimmers. *Med Sci Sports Exerc.* 2002;34(3):411–417.
65. He CS, Handzlik M, Muhamad A, Gleeson M. Influence of CMV/EBV serostatus on respiratory infection incidence during 4 months of winter training in a student cohort of endurance athletes. *Eur J Appl Physiol.* 2013;113(10):2613–2619.
66. Mehta SK, Crucian BE, Stowe RP, et al. Reactivation of latent viruses is associated with increased plasma cytokines in astronauts. *Cytokine.* 2013;61(1):205–209.
67. Glaser R, Friedman SB, Smyth J, et al. The differential impact of training stress and final examination stress on herpesvirus latency at the United States Military Academy at West Point. *Brain Behav Immun.* 1999;13(3):240–251.
68. Sarid O, Anson O, Yaari A, Margalith M. Human cytomegalovirus salivary antibodies as related to stress. *Clin Lab.* 2002;48(5–6):297–305.
69. Gleeson M, McDonald WA, Pyne DB, et al. Salivary IgA levels and infection risk in elite swimmers. *Med Sci Sports Exerc.* 1999;31(1):67–73.
70. Neville V, Gleeson M, Folland JP. Salivary IgA as a risk factor for upper respiratory infections in elite professional athletes. *Med Sci Sports Exerc.* 2008;40(7):1228–1236.
71. Kohut ML, Arntson BA, Lee W, et al. Moderate exercise improves antibody response to influenza immunization in older adults. *Vaccine.* 2004;22(17–18):2298–2306.
72. Woods JA, Keylock KT, Lowder T, et al. Cardiovascular exercise training extends influenza vaccine seroprotection in sedentary older adults: the immune function intervention trial. *J Am Geriatr Soc.* 2009;57(12):2183–2191.
73. Spielmann G, McFarlin BK, O'Connor DP, Smith PJ, Pircher H, Simpson RJ. Aerobic fitness is associated with lower proportions of senescent blood T-cells in man. *Brain Behav Immun.* 2011;25(8):1521–1529.
74. Shinkai S, Kohno H, Kimura K, et al. Physical activity and immune senescence in men. *Med Sci Sports Exerc.* 1995;27(11):1516–1526.
75. Pedersen BK, Bruunsgaard H. Possible beneficial role of exercise in modulating low-grade inflammation in the elderly. *Scand J Med Sci Sports.* 2003;13(1):56–62.
76. Yan H, Kuroiwa A, Tanaka H, Shindo M, Kiyonaga A, Nagayama A. Effect of moderate exercise on immune senescence in men. *Eur J Appl Physiol.* 2001;86(2):105–111.
77. Phillips MD, Flynn MG, McFarlin BK, Stewart LK, Timmerman KL. Resistance training at eight-repetition maximum reduces the inflammatory milieu in elderly women. *Med Sci Sports Exerc.* 2010;42(2):314–325.
78. Woods JA, Ceddia MA, Wolters BW, Evans JK, Lu Q, McAuley E. Effects of 6 months of moderate aerobic exercise training on immune function in the elderly. *Mech Ageing Dev.* 1999;109(1):1–19.
79. Drela N, Kozdron E, Szczypiorski P. Moderate exercise may attenuate some aspects of immunosenescence. *BMC Geriatr.* 2004;4:8.
80. Simpson RJ, Bosch JA. Special issue on exercise immunology: current perspectives on aging, health and extreme performance. *Brain Behav Immun.* 2014;39:1–7.
81. Lowder T, Padgett DA, Woods JA. Moderate exercise protects mice from death due to influenza virus. *Brain Behav Immun.* 2005;19(5):377–380.
82. Hojman P, Dethlefsen C, Brandt C, Hansen J, Pedersen L, Pedersen BK. Exercise-induced muscle-derived cytokines inhibit mammary cancer cell growth. *Am J Physiol Endocrinol Metab.* 2011;301(3):E504–E510.
83. Simpson RJ. Aging, persistent viral infections, and immunosenescence: can exercise "make space"? *Exerc Sport Sci Rev.* 2011;39(1):23–33.

84. ElKassar N, Gress RE. An overview of IL-7 biology and its use in immunotherapy. *J Immunotoxicol.* 2010;7(1):1–7.
85. Andersson A, Yang SC, Huang M, et al. IL-7 promotes CXCR3 ligand-dependent T cell antitumor reactivity in lung cancer. *J Immunol.* 2009;182(11):6951–6958.
86. Haugen F, Norheim F, Lian H, et al. IL-7 is expressed and secreted by human skeletal muscle cells. *Am J Physiol Cell Physiol.* 2010;298(4):C807–C816.
87. Fairey AS, Courneya KS, Field CJ, Bell GJ, Jones LW, Mackey JR. Randomized controlled trial of exercise and blood immune function in postmenopausal breast cancer survivors. *J Appl Physiol.* 2005;98(4):1534–1540.
88. Nieman DC, Nehlsen-Cannarella SL, Markoff PA, et al. The effects of moderate exercise training on natural killer cells and acute upper respiratory tract infections. *Int J Sports Med.* 1990;11(6):467–473.
89. Perna FM, LaPerriere A, Klimas N, et al. Cardiopulmonary and CD4 cell changes in response to exercise training in early symptomatic HIV infection. *Med Sci Sports Exerc.* 1999;31(7):973–979.
90. Gleeson M, Bishop NC, Stensel DJ, Lindley MR, Mastana SS, Nimmo MA. The anti-inflammatory effects of exercise: mechanisms and implications for the prevention and treatment of disease. *Nat Rev Immunol.* 2011;11(9):607–615.
91. Steensberg A, Fischer CP, Keller C, Moller K, Pedersen BK. IL-6 enhances plasma IL-1ra, IL-10, and cortisol in humans. *Am J Physiol Endocrinol Metab.* 2003;285(2): E433–E437.
92. Cupps TR, Fauci AS. Corticosteroid-mediated immunoregulation in man. *Immunol Rev.* 1982;65:133–155.
93. Bergmann M, Gornikiewicz A, Sautner T, et al. Attenuation of catecholamine-induced immunosuppression in whole blood from patients with sepsis. *Shock.* 1999;12(6):421–427.
94. McFarlin BK, Flynn MG, Phillips MD, Stewart LK, Timmerman KL. Chronic resistance exercise training improves natural killer cell activity in older women. *J Gerontol A Biol Sci Med Sci.* 2005;60(10):1315–1318.
95. Kawanishi N, Yano H, Yokogawa Y, Suzuki K. Exercise training inhibits inflammation in adipose tissue via both suppression of macrophage infiltration and acceleration of phenotypic switching from M1 to M2 macrophages in high-fat-diet-induced obese mice. *Exerc Immunol Rev.* 2010;16:105–118.
96. Shaikh SR, Edidin M. Polyunsaturated fatty acids, membrane organization, T cells, and antigen presentation. *Am J Clin Nutr.* 2006;84(6):1277–1289.
97. Bosch JA, Fischer JE, Fischer JC. Psychologically adverse work conditions are associated with CD8+ T cell differentiation indicative of immunosenescence. *Brain Behav Immun.* 2009;23(4):527–534.
98. Sachdev S, Davies KJ. Production, detection, and adaptive responses to free radicals in exercise. *Free Radic Biol Med.* 2008;44(2):215–223.
99. Mobius-Winkler S, Hilberg T, Menzel K, et al. Time-dependent mobilization of circulating progenitor cells during strenuous exercise in healthy individuals. *J Appl Physiol.* 2009;107(6):1943–1950.

CHAPTER SIXTEEN

Exercise Regulation of Cognitive Function and Neuroplasticity in the Healthy and Diseased Brain

Gilian F. Hamilton[1], Justin S. Rhodes

Department of Psychology, The Beckman Institute, University of Illinois at Urbana-Champaign, Urbana, Illinois, USA

[1]Corresponding author: e-mail address: gillianh@illinois.edu

Contents

Abstract

Regular exercise broadly enhances physical and mental health throughout the lifespan. Animal models have provided us with the tools to gain a better understanding of the underlying biochemical, physiological, and morphological mechanisms through which exercise exerts its beneficial cognitive effects. One brain region in particular, the hippocampus, is especially responsive to exercise. It is critically involved in learning and memory and is one of two regions in the mammalian brain that continues to generate new neurons throughout life. Exercise prevents the decline of the hippocampus from aging and ameliorates many neurodegenerative diseases, in part by increasing adult

Progress in Molecular Biology and Translational Science, Volume 135
ISSN 1877-1173
http://dx.doi.org/10.1016/bs.pmbts.2015.07.004

hippocampal neurogenesis but also by activating a multitude of molecular mechanisms that promote brain health. In this chapter, we first describe some rodent models used to study effects of exercise on the brain. Then we review the rodent work focusing on the mechanisms behind which exercise improves cognition and brain health in both the normal and the diseased brain, with emphasis on the hippocampus.

1. INTRODUCTION

As compared to many other animals, humans have a sustained long-distance running capability.[1–3] This high running ability has evolved in concert with cellular and molecular mechanisms that aid in the proper functioning of the human body.[2,3] This chapter will explore some of these health-enhancing effects of exercise in the brain.

Keeping physically active as you age maintains the speed of central nervous system processing. One of the earliest mentions of this beneficial effect was in a study by Spirduso and Clifford.[4] In their study, older active men (60–70 years old) with an established pattern of playing racket sports or running at least 3 miles four times a week moved and reacted faster to simple stimuli when compared to their sedentary aged-matched counterparts.[4] Since then, positive correlations between physical activity, brain health, and cognitive performance consistently have been demonstrated in humans.[5] Moreover, consistent exercising maintains cardiovascular, immune, and mental health and also delays aging.[6,7] In order to examine how exercise achieves its dramatic influence over body and mind, researchers have developed many different animal models in order to uncover the mechanisms. They can typically be divided into two categories: aerobic or anaerobic. Each model of exercise may uniquely impact the brain and the body. Section 2 will discuss the functionality and use of animal models of exercise.

2. ANIMAL MODELS OF EXERCISE

In humans, aerobic exercise consistently has been associated with improved cognition and increased volume of certain regions of the brain. Aerobic exercise involves low- to high-intensity activities that increase the efficiency and endurance of the cardiovascular system. This includes activities such as walking, jogging, swimming, and cycling. As such, aerobic exercise has been shown to increase both frontal cortex and hippocampal blood flow and oxygen delivery. This increase in blood flow appears to aid in earlier recall for complex spatial objects.[8,9] Further, in elderly subjects, 6 months of aerobic training has been shown to mitigate age-related decline in both verbal and spatial memory,[10] while a low activity, 8-week yoga intervention significantly improved performance on task switching, working memory, and the n-back task.[11] In young males, high-intensity cycling has been associated with better acquisition and retention 7 days later on the visuomotor tracking task.[12] Further, in aging human adults, those that participated in leisure time physical activity at midlife had 60% lower odds of developing Alzheimer's disease compared to their sedentary counterparts.[13] Moreover, a 16-week exercise intervention (treadmill, 30 min, twice a week and moderate intensity of 60% VO_2 max) aids in cognitive functioning of elderly adults suffering from mild dementia.[14] Thus, exercising a few times a week for approximately 30–45 min at a moderate level of intensity improves mental health in humans.

Therefore, the question becomes, can we see the same beneficial effects of aerobic exercise in rodents, given their use in the laboratory setting. Doing so would allow us to identify and study the mechanisms through which exercise exerts its influence. The most commonly used model of aerobic activity in rodents is free access to a running wheel (Fig. 1). Mice enjoy running and they will take advantage of the access to the running wheel, some more than others. In fact, mice have been recorded as running as much as 2–10 km/day.[15] Thus, using the voluntary wheel-running model,

Figure 1 Wheel-running paradigm. (A) A standard laboratory cage where a mouse has 24-h access to a running wheel. (B) Standard laboratory cage without a running wheel. (See the color plate.)

researchers do not have control over how much the animal will run, but the researcher can precisely record wheel-running activity, and then use individual variation in running as a the predictor or dependent variable influencing the neurological outcomes.

Similar to the effects in humans, voluntary wheel running produces a plethora of positive consequences including enhanced cognitive performance and regional changes in brain volume. For example, male and female C57BL/6J mice that were given 40 days voluntary access to running wheels exhibited significantly enhanced performance on the Morris Water Maze when compared to their nonrunning peers.[16] The Morris Water Maze is a standard behavioral test in the laboratory setting that tests whether an animal can remember where a location is based on the surrounding environment (Fig. 2). Further, similar to its influence in humans, voluntary aerobic exercise alters the volume of certain brain regions that are linked to spatial memory performance. In fact, voluntary access to a running wheel in rodents has also been shown to increase hippocampal volume.[17] Therefore, voluntary wheel running produces similar benefits in rodents as aerobic exercise does in humans.

A second model of aerobic activity that is known to be beneficial is forced running. This model is exactly what it sounds: an animal is placed on a treadmill and forced to run (either via mild shocks or via foam pads that

Figure 2 Morris Water Maze. A photo of a mouse in the middle of a trial. For this task, the animal is placed in one of the four cardinal starting points, and then using landmarks around the room, it must navigate the water as quickly as possible to find the platform that is submerged under the water (note the black dot in the bottom left-hand corner). (See the color plate.)

restrict the animal's placement to only the treadmill) for a period of time. Similar to both the human aerobic and rodent voluntary wheel-running literature, forced treadmill running improves cognitive performance and brain health in rodents. For example, rats that are forced to run on a treadmill for 6 weeks (30 min at 8 m/min 5 days/weeks[18]) or 12 weeks (50 min 30 m/min[19]) exhibit improved performance on the Morris Water Maze. Additionally, forced exercise has been shown to improve cognitive impairments that result from amphetamine withdrawal in rats.[20] The benefits of forced exercise are not restricted to cognitive performance. About 16 weeks of forced exercise (60 min $\sim 10.9 \pm 1.6$ m/min 5 days/week) increased the hippocampal volume of mice when compared to their sedentary counterparts.[17] However, rats exposed to forced exercise exhibit signs of stress such as increased anxiety levels.[18] Further, research shows that voluntary exercise produces more prominent plastic changes in the hippocampus when compared to the impact of forced exercise on the hippocampus.[21] This suggests that forced exercise may not be the best way to model aerobic activity in rodents as it unintentionally produces stress-inducing side effects that make it more difficult to tease apart the influence of exercise from psychological stress effects on the brain. Despite this, forced exercise still improves learning and memory along with overall brain health in rodents.

Much less research has explored the influence of anaerobic exercise, such as strength training, on cognitive performance and behavior as compared to aerobic exercise in rodents and humans. The limited evidence suggests cognitive performance is enhanced with strength training. In humans, strength training refers to sets of repetitions of certain exercises (i.e., leg presses, leg or arm-extensions, chest presses, lat-pulldowns, etc.) performed multiple times per week with a gradual increase in weights. Strength training uses resistance to induce muscular contractions and thus increases muscle strength and endurance. Unlike aerobic exercise, however, it does not necessarily enhance cardiovascular endurance. In humans, strength training has been shown to increase gray and white matter volume in multiple brain regions of aging adults, including the hippocampus.[22] Further, in aged adults, a resistance-type exercise program of two sessions per week improved attention and working memory on a variety of tasks.[23] Further, a 24-week strength training intervention significantly reduced scores of the Geriatric Depression Scale in elderly females. It is interesting that strength training failed to impact cognitive function in this study.[24] Finally, in late-middle-age adults, a single bout of resistance exercise (two sets of 10 repetitions of 70% of 10-repetition maximum of seven exercises) led to better performance on the Tower of London task,[25] suggesting that acute strength training has a positive effect on cognition. Together, these data suggest that strength training improves brain function in humans, but more research in this area is needed.

Until recently, most of the rodent literature has focused on aerobic exercise given their natural desire to run, thus leaving a gap in the field for effects of anerobic exercise on cognitive performance in rodent models. However, more recently, the field has begun to develop rodent models of strength training. Strength training in rodents can be achieved through resistance-based training such as a squat-training apparatus,[26] weight pulling on a treadmill,[27] and ladder climbing with weights attached to their tails.[28] Studies examining the beneficial effects of strength training in rodents are not as plentiful as the aerobic exercise studies. However, it has been found that adult male Wistar rats that underwent 8 weeks of resistance training on a vertical ladder showed improved learning and spatial memory on the Morris Water Maze that mirrored their treadmill-running peers, when compared to control animals.[29] Additionally, the resistance training group exhibited enhanced performance on a step-through passive avoidance task, when compared to both a sedentary and a sham-treated group that were left at the top of the vertical ladder for 15 min while resistance training animals exercised on the same apparatus.[30] Still, similar to forced exercise, these models of strength training are riddled with confounding stress variables, as the animals are many times motivated to lift the weights or to move through food deprivation or mild shocks (for review, see Ref. 31). Still, strength training does appear to be a viable new avenue to examine the beneficial role of exercise on brain function.

While we have so far focused on chronic exposures to exercise (i.e., consistent running over a long period of time), it is important to recognize the influence of acute exercise exposure on brain function in both rodents and humans. Acute refers to a single exercise bout. In humans, a single 12-min aerobic exercise exposure (jogging in place while maintaining target heart rate range) improved the selective visual attention performance in adolescents, a benefit that lasted 45 min.[32] Children exposed to an aerobic fitness assessment (PACER task), demonstrated significantly better learning of the word lists and significantly better recall of the words after a brief delay.[33] Further, rodent work demonstrates that a single 3-h bout of wheel running significantly enhanced the acquisition, extinction, and reconsolidation of context-conditioned fear.[34] It is clear that both acute and chronic exercise paradigms influence learning and memory. However, the influence of acute exercise is ephemeral. Benefits from acute exercise are no longer apparent after 24 h.[33] Therefore, the remainder of this chapter will focus on the influence of chronic exercise exposure. Specifically, in section 3, we will examine the mechanisms through which chronic exercise exposure exerts it influence on cognitive function.

3. HOW EXERCISE IMPACTS THE PHYSIOLOGY OF THE BRAIN

Overall, the message of this chapter is clear: exercise is good for the brain. High levels of exercise lead to larger regional brain volumes and better brain functioning. For example, older adults who participated in light aerobic activity for 6 months (walked for 40 min a day once a week) exhibited a 2% increase in hippocampal volume compared to adults that did not.[35] Thus, exercise effectively reversed the natural age-related decline of hippocampal volume. Not only does aerobic fitness appear to prevent cortical decay, but it also improves cognitive performance. In humans, exercise has been shown to enhance spatial learning, pattern separation, executive function, working memory, and processing speed, among others.[5,36] For example, 6 months of aerobic training mitigated age-related decline in both verbal and spatial memory[10] and a low activity, 8-week yoga intervention significantly improved performance on working memory.[11]

Still, how does exercise produce these changes in brain function? To produce both acute and chronic neurological effects on many brain regions and systems, exercise triggers a variety of neurobiological mechanisms. These neurobiological mechanisms produce long-term neurological changes. For example, chronic exposure to exercise increases the total number of granule neurons in the dentate gyrus of the hippocampus.[37,38] It is theorized that the increased production of these neurons may play a substantial role in the improved hippocampal-dependent spatial learning and memory that is seen following chronic exercise. Still, the influence of exercise on the hippocampus is immense, and it will be described in more detail later in Section 5. First, we must focus on the neurobiological mechanisms through which exercise produces such drastic changes both morphologically and functionally in the brain.

3.1 Exercise and Neurotransmitters

Exercise modulates chemicals that communicate information in the brain known as neurotransmitters. This influence over neurotransmitters may play a role in how exercise exerts its neurological effects. Physical activity directly influences the central dopaminergic, noradrenergic, and serotonergic systems (for review, see Ref. 39). Alterations in these systems can cause disorders (i.e., depression). Therefore, it is possible that activation of these systems through exercise could repair brain function. In fact, both administration of a 6-month low-dosage dopamine agonist (ropinirole—Adartrel®, GlaxoSmithKline, UK) and a 6-month exercise training intervention (cycling in a recumbent position at an intensity of 60–65% of the patient's maximal exercise intensity three times per week) in patients suffering from restless legs syndrome were effective in treating symptoms, such as depression.[40] Further, 16 weeks of aerobic training (45 min, 6 km/h, 3 days/week) has been shown to produce transient elevations in plasma tryptophan (a serotonin precursor) availability to the brain.[41] Together, these data illustrate that exercise increases neurotransmitter levels and suggest that in doing so exercise may restore proper brain function.

Still, in order to determine the functional role of neurotransmitters, researchers have utilized rodent models wherein they can directly manipulate and measure levels of neurotransmitters in the brain. Rodent models also demonstrate the link between exercise, neurotransmitters, and improved brain function. For example, exercise has been shown to alter levels of monoamines as a mechanism for alleviating symptoms in rodent models of Parkinson's disease,[42] ADHD,[43] and Huntington's disease.[44] Together, these data illustrate the overt influence exercise has on the neurotransmitter systems in order to promote healthy brain plasticity.

3.2 Exercise and Hormones

An interaction between exercise and hormones may be vital to a properly functioning brain. In regularly menstruating women (24–37 years old), physical activity levels are strongly associated with salivary levels of estradiol. Low levels of physical activity produced high levels of estradiol, while high levels of physical activity produced low levels of estradiol.[45] Alterations in hormonal levels can lead to health problems (i.e., high levels of estradiol are associated with a higher risk of cancer).

Approximately 50% of the world's population will suffer a significant loss in sex steroid hormones due to menopause. Female menopause occurs due to a cessation of estradiol and progesterone production by the ovaries. This causes a variety of symptoms that include short-term memory loss and difficulty concentrating. The most commonly utilized treatment for postmenopausal cognitive decline in women is hormone therapy. However, it has been shown that a combination of exercise and hormone therapy is much more efficient. In fact, higher fitness levels can augment the effects of shorter durations of hormone treatment, which can offset the associated risks of a longer hormone replacement therapy.[46] Through the use of rodent models, researchers have further validated the role of exercise in rescuing behavioral deficits caused by low levels of hormones. For example, Garcia-Mesa and colleagues[47] examined the neuroprotective effect of voluntary running exercise in a mouse model of Alzheimer's disease (3 × Tg-AD female mice). Mice were either ovariectomized or sham operated at 4 months of age and then 2 months later received either voluntary wheel running or remained in standard housing for 3 months. Ovariectomized mice without access to a running wheel exhibited major cognitive impairments on the Morris Water Maze, which was rescued by exposure to a running wheel. Therefore, exercise and estrogen levels may interact in order to maintain proper brain function.

3.3 Exercise and Neurotrophic Factors

Growing support implicates the role of neurotrophic factors in exercise-induced neurological changes. Voluntary wheel running increases the concentration of several different growth and trophic factors that likely support the morphological changes in the brain discussed above including fibroblast growth factor-2 (FGF-2),[48] insulin-like growth factor-1 (IGF-1),[49] brain-derived neurotrophic factor (BDNF),[50] and vascular endothelial growth factor (VEGF)[18] among others. BDNF in particular appears to be especially susceptible to regulation by exercise.[50,51] Physical activity increases levels of BDNF in the lumbar spinal cord, cerebellum, cortex, and hippocampus.[51,52] BDNF promotes the differentiation, neurite extension, and survival of a variety of neuronal populations, and it potentiates synaptic transmission, participates in gene transcription, modifies synaptic morphology, and enhances neuronal resilience,[53] implicating BDNF as a prime candidate behind exercise-induced neuronal plasticity and learning enhancement. Adult humans participating in intense rowing exercise and treadmill exercise exhibit increased levels of plasma BDNF.[54] Further, forced treadmill exercise, voluntary exercise, and strength training paradigms have all proven to increase serum levels of BDNF in rodents.[55–57] However, in the human studies, the source of the BDNF in the plasma is not clear as it could come from many different tissues including brain or muscle.

The activity-dependent enhancement of BDNF in the brain may provide a mechanism through which exercise improves learning and memory. About 1 week of forced treadmill exercise significantly enhanced BDNF levels throughout the hippocampus and resulted in both spatial and non-spatial learning improvements in rodents.[56] Further, Vaynman, Ying, and Gomez-Pinilla[51] showed that in rats, 1 week of voluntary exercise enhanced spatial memory performance on the Morris Water Maze along with increased hippocampal BDNF levels. Interestingly, blockade of hippocampal TrkB receptors during the pretesting exercise period eliminated the behavioral performance enhancement from exercise, suggesting the necessity of BDNF for the exercise-induced memory enhancements.

3.4 Exercise, Blood Flow, and Microvasculature

Physical activity may produce neurological changes by altering cerebral blood flow, microvasculature, and VEGF expression. In humans, aerobic exercise has been shown to increase frontal cortex blood flow and oxygen delivery with more intense exercise creating a greater enhancement.[9] Additionally, Sato and colleagues[58] reported increased blood flow through the carotid and vertebral arteries in an intensity-graded manner in adults involved in graded cycling exercise.

In animal models, angiogenesis has been shown to be a prerequisite for many forms of neural and behavioral plasticity.[59] More recently, the beneficial effects of exercise-enhanced microvasculature have been shown in macaque monkeys given 1 h a day of treadmill exercise for 5 months. This exercise paradigm increased vascular density in the motor cortex and resulted in the monkeys needing fewer trials to reach criterion on a spatial learning task.[60] Interestingly, when the monkeys were left sedentary for 3 months, their performance diminished. This suggests exercise must remain continuous in order to gain benefit. This is further evidenced in mice given access to a running wheel for either 1, 3 or 10 days. Although both blood vessel density increased after only 3 days and levels of hippocampal neurogenesis increased after 10 days, both measures returned to baseline following 24 h removal of the running wheel.[61] Thus, it appears that the continuous exercise exposure is essential to produce substantial changes in microvasculature and blood flow.

3.5 Exercise and Oxidative Stress

Physical activity may produce neurological changes in the brain, in part, through a reduction in oxidative stress levels. Oxidative stress occurs when an organism cannot eliminate chemically reactive molecules (produced from metabolism) fast enough before they start attacking and degrading important molecules for biological functions. In rodents, exercise reduces levels of reactive oxygen species[62,63] and oxidative protein damage[64] and increases levels of some endogenous antioxidants.[62] For example, voluntary access to a running wheel reduced levels of reactive oxygen species and increased endogenous antioxidants in the hippocampus of 12-month-old female rats.[65] Further, in 12-month-old male and female mice, moderate treadmill exercise significantly extended lifespan and behavioral performance on spatial and nonspatial tasks. This was associated with decreased oxidative stress markers in the brain, heart liver, and kidney.[66] These data illustrate the beneficial influence of exercise on reducing oxidative stress.

3.6 Exercise and Apoptosis

Exercise has consistently proven to alter levels of apoptosis in the brain. For example, during the first week of voluntary wheel running, there is a rapid increase of apoptosis that is evident in the hippocampus[67]; however, this effect is transient and signs of apoptosis are no longer present following 2[68] or 8 weeks[69] of voluntary exercise. This temporary increase in apoptosis at the onset of exercise may trigger a cascade of events that promotes the birth of new cells and enhanced functioning in the hippocampus. One example of this can be found in a study that utilized aged animals. Approximately 24-month-old aged rats were given 30 min of forced treadmill running once a day for 6 weeks. This exercise exposure significantly reduced the number of hippocampal apoptotic cells, while it significantly increased the number of newly born hippocampal cells. Moreover, the treadmill exercise exposure prevented age-related impairments on a short-term memory task.[70] Still, it is important to note that alongside these aged animals, Kim and colleagues[70] also tested a cohort of young 5-month-old animals and although they saw an increase in the number of newly born hippocampal cells, there was no impact on apoptosis levels. Therefore, it appears that the influence of exercise on apoptosis is more apparent in the aged brain. Exercise appears to cause a transient increase in apoptosis that triggers neuroprotective mechanisms (e.g., the birth of new cells).

4. EXERCISE-INDUCED SIGNALING PATHWAYS IN THE BRAIN

It is through the modulation of many different signaling pathways that exercise ultimately influences behavior and cognition (Fig. 3). In fact, the

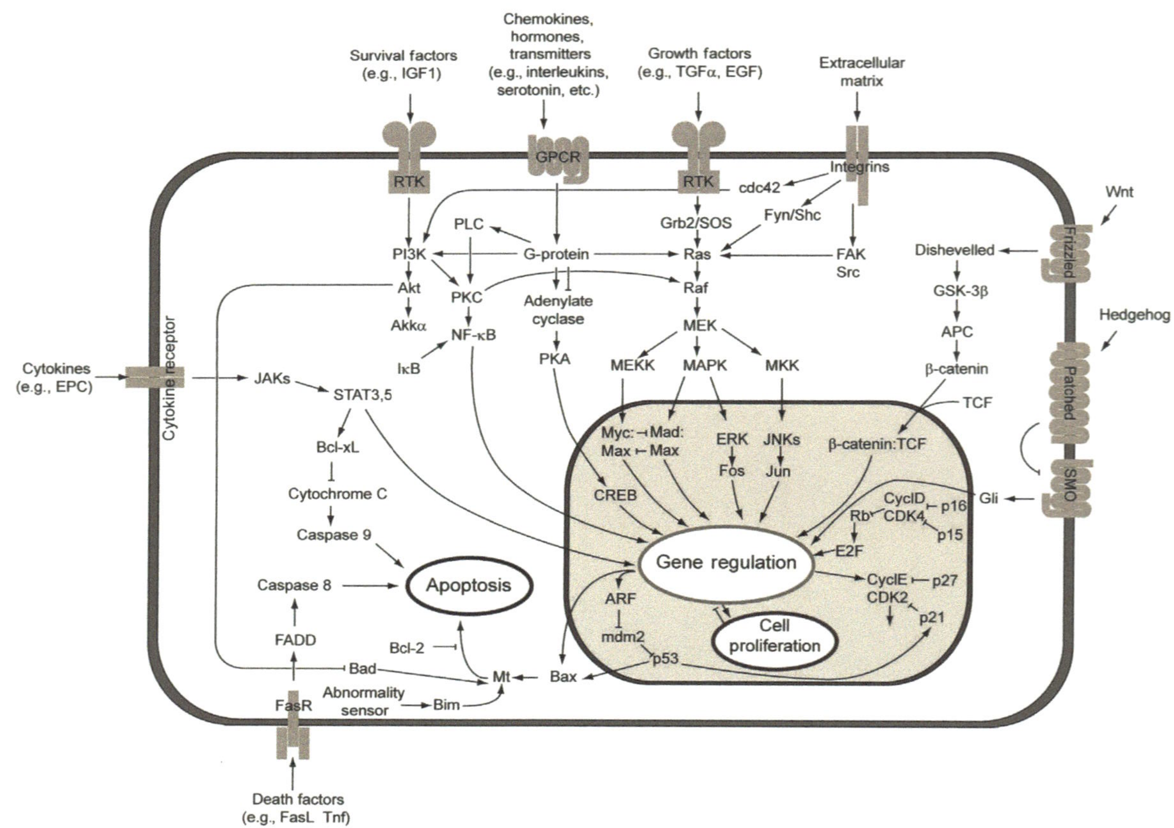

Figure 3 An overview of the major signal transduction pathways modulated by exercise in mammals. Please note that this still does not include all of the pathways.

response to exercise is so massive that the ability to pinpoint a single signaling pathway through which exercise exerts its effect is next to impossible. Rather, exercise appears to activate various components of different signaling pathways that promote cellular growth and brain plasticity concurrently. Cassilhas and colleagues[29] alluded to this vast influence by exposing 90-day-old male rats to either an aerobic treadmill intervention (week 1: 15 m/min for 25 min and increased to week 8: 20 m/min for 30 min) or an anaerobic strength training intervention (vertical ladder training: eight climbing series with a progressively heavier load). The results demonstrated that not only did both forms of exercise enhance spatial learning and memory but, interestingly, they activated divergent intracellular signal transduction pathways. For example, aerobic exercise increased levels of IGF-1, BDNF, TrkB, and calcium/calmodulin-dependent kinase II (β-CaMKII) in the hippocampus, while strength training acted on the central nervous system through induced IGF-1 with concomitant activation of receptor for IGF-1 (IGF-1R) and AKT in the hippocampus. Still, the end result was the same: increased synapsin 1 and synaptophysin protein expression, which are important for synaptic transmission and neurotransmitter release (i.e., greater neuronal function). This study illustrates that depending on the form of exercise; different signaling pathways may be activated in order to enhance cellular plasticity and, resultantly, improve spatial learning and memory. There are many different intracellular signaling pathways that influence cellular plasticity and proliferation, and through its ability to modulate them, exercise may produce both neurobiological and behavioral improvements.

Exercise also exerts its influence by causing stable long-term change of gene expression (in which genes are turned on or off), i.e., epigenetic changes. For example, adolescent male mice exposed to 1 week of wheel running exhibited increased global acetylation of histone 3 in the hippocampus that was correlated with increased hippocampal BDNF.[71] Further, in 3-month-old male Sprague–Dawley rats, 1 week of voluntary exercise increased DNA demethylation, specifically in the **BDNF** promotor IV, which is known to be highly responsive to neuronal activity. Further, exercise increased levels of chromatin remodeling through elevated levels of phosphorylated *CAMKII* and *cAMP response element-binding protein*.[72] Exercise exerts a similar influence on *Wnt* signaling, proteins known to play an important role in adult hippocampal neurogenesis. About 5-week-old Sprague–Dawley rats exposed to 4–5 days of forced treadmill exercise (start speed of 4.2 m/min, increased progressively by 1 m/min every 30 s to a speed of 12 m/min) for 36 weeks exhibited significantly lower levels of *Dkk-1*, a *Wnt* antagonist that has been associated with neuronal death in Alzheimer's disease and Parkinson's disease, when compared to levels exhibited by sedentary animals. In comparison to their sedentary counterparts, exercising animals also expressed higher protein levels of the anti-apoptotic protein *Bcl-2*.[73] Exercise may activate the **Wnt** pathway in order to prevent neuronal degeneration in both the aging and the diseased brain. These studies illustrate that changes in gene expression are occurring alongside long-term continued activation of the signaling pathways, which produce the neurobiological changes. Overall, the impact of exercise on molecular mechanisms is complex and vast (Fig. 3; for review, see Ref. 5). Perhaps, the reason exercise is such a powerful therapeutic tool is that it can modulate a variety of pathways in order to promote enhanced brain health and function.

5. THE HIPPOCAMPUS IS THE BRAIN REGION MOST INFLUENCED BY EXERCISE

The hippocampus is located in the medial temporal lobe (Fig. 4), and it plays an integral role in learning and memory. This was notably exemplified by H.M., one of the most well-known case studies in the neuroscience field. In order to cure his epilepsy, H.M. underwent a bilateral medial temporal lobectomy, which resulted in severe anterograde amnesia, and he was unable to commit new events to his explicit memory. Still, over the years, H.M. retained his short-term working memory and intellect, and he was left with residual learning capabilities.[74] For example, he could still perform many

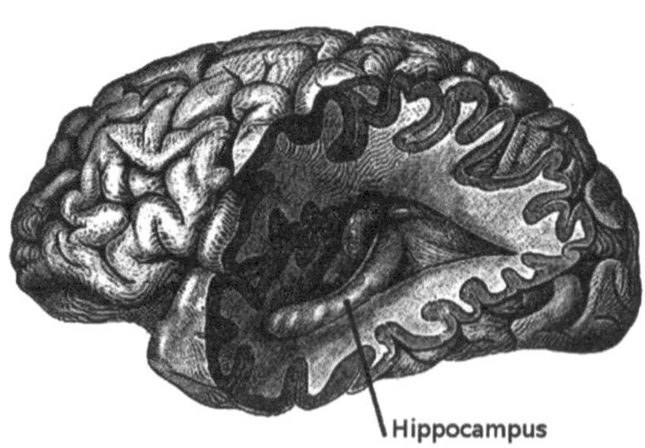

Figure 4 Gross anatomy of the hippocampus. The hippocampus is located in the medial temporal lobe of the brain. (See the color plate.)

types of motor learning tasks, though he could not remember learning them. The case of H.M. was pivotal in demonstrating the important role of the hippocampus in long-term memory formation, given that he was unable to create new memories following the removal of his hippocampus.

5.1 The Influence of Exercise on Hippocampal Adult Neurogenesis

One important region of the hippocampus is the dentate gyrus. The dentate gyrus is the first region where all sensory modalities merge together to form unique representations and memories that bind stimuli together, and thus, it plays a critical role in learning and memory. Furthermore, the dentate gyrus is one of two known regions in which adult neurogenesis—the continuous birth of new neurons throughout adult hood—occurs.[75] The process of hippocampal adult neurogenesis begins with the proliferation of a precursor progenitor cell and ends with the integration of a functional cell into the preexisting hippocampal network. Exercise not only increases levels of hippocampal adult neurogenesis, the continuous birth of new neurons in adulthood[37,38,76] but it also increases the dendritic length and complexity of dentate granule cells.[76] Thus, exercise appears to modulate both the structure and function of the hippocampal dentate gyrus.

Adult neurogenesis can be modulated by many different extrinsic and intrinsic factors, but none is greater than the influence of exercise over adult neurogenesis. In fact, many of the intrinsic factors governed by exercise (i.e., hormones, neurotrophic factors) are known to regulate levels of adult neurogenesis. Thus, as a result, when it comes to modulation of adult neurogenesis, exercise is a powerful source. Exercise triggers a multitude of intrinsic factors and in doing so, it substantially increases levels of adult-born neurons, more so than a single intrinsic factor alone.

In 1999, van Praag hypothesized that exercise increased levels of adult neurogenesis by increasing the survival of proliferating cells in the dentate gyrus. This effect has been countlessly replicated.[15,16,37] In fact, the distance an animal runs strongly predicts the level of neurogenesis. This is evident in a study by Clark and colleagues[15] in which running activity was studied in 12 different strains of mice that varied greatly with respect to voluntary running levels. For example, 129S1/SvlmJ mice ran approximately 2 km/day, whereas B6129SF1/J mice ran approximately 10 km/day. As a result, B6129SF1/J mice had a four- to fivefold increase, while 129S1/SvlmJ mice exhibited a modest two- to threefold increase in levels of adult neurogenesis. Additionally, in genetically identical mice, levels of adult neurogenesis were positively correlated with the level of their physical activity in an environmental enrichment cage.[77] Thus, physical activity modulates levels of adult neurogenesis, and this effect is dependent on the amount of physical activity an animal undertakes.

Not only does exercise consistently increase levels of hippocampal adult neurogenesis, but it also enhances cognitive performance. In particular, it improves spatial memory: the ability to remember the location of an object relative to other objects in the environment.[15,37] For example, allowing mice access to a running wheel for either 1 or 6 months improves performance on the Morris Water Maze and increased levels of adult neurogenesis when compared to aged-matched controls.[78]

More recently, adult-born neurons have been implicated in pattern separation. Pattern separation is the ability to recognize and encode a subtle difference between two very similar stimuli (e.g., objects, textures, patterns, etc.). Creer and colleagues[79] examined the relationship between levels of adult neurogenesis and pattern separation. In their study, 3- and 22-month-old male mice were given either housed with a running wheel or in standard cages and then tested on a spatial discrimination task. In the young mice, exercise significantly increased levels of adult neurogenesis, and these levels were tightly correlated with their improved performance on the spatial pattern separation task. In contrast, the aged mice were impaired on the spatial discrimination task regardless of exercise condition coincident with greatly reduced cell genesis that was refractory to running. Together, these data imply a causal link between exercise-induced levels of adult neurogenesis and spatial learning and memory. In fact, when levels of adult neurogenesis are ablated through focal irradiation, the procognitive effects of exercise are no longer present on Morris Water Maze performance in C57BL/6J mice.[16] Recently, Groves and colleagues[80] performed a meta-analysis that explored the relationship between newly generated neurons and performance on spatial tasks. The results showed that the new neurons were not required for baseline performance. Although this meta-analysis did not take into account the role of exercise, the results are still important, as they indicate a possible limitation of the neurogenesis lesion method. It is possible that the brain is able to compensate from the loss of neurons. Therefore, one hypothesis may be that new neurons are preferentially recruited when an animal is learning a task; however, if they are not present, then there are redundant mechanisms in the brain to compensate for their loss.

Exercise-enhanced levels of adult neurogenesis also are correlated with improved performance on a multitude of nonspatial tasks. For example,

Wojtowicz and colleagues (2008) found that a 68% reduction in neurogenesis produced a striking 68% reduction in freezing in the contextual fear conditioning task. Moreover, performance on the motor performance task, the rotarod, is consistently enhanced from wheel running and hence associated with increased levels of adult neurogenesis.[16,78] Therefore, although exercise-enhanced levels of adult neurogenesis correlate with improved performance on spatial memory tasks, a causal role remains unclear.

6. THERAPEUTIC ROLE OF EXERCISE

The beneficial influence of exercise appears to be even greater in the damaged brain. The hippocampus is a brain region severely impacted by fetal alcohol spectrum disorders (FASD),[81–83] Alzheimer's disease,[84–86] and aging.[6,35,86] If we can understand how to grow new neurons in this region, perhaps through exercise, we may be able to treat these diseases and others. Researchers have therefore begun to explore the therapeutic role of exercise on the diseased brain. In this section, we will examine how exercise protects against some brain-related diseases.

6.1 Exercise as an Intervention for FASD

Despite growing awareness of the dangers of prenatal alcohol exposure, between 2.4% and 4.8% of children have some form of FASD. Furthermore, this number is not decreasing and is a conservative estimate.[87] From an economic standpoint, annual cost estimates for the United States were 75 million dollars in 1984. However, as of 1998, this cost had blossomed to be as high as 4 billion dollars.[88] Therefore, the need for interventions is overdue. Thus, research has focused on the therapeutic role of exercise.

Developmental alcohol exposure can disrupt proper functioning of the cerebellum. For example, postnatal day (PD) 4–9 alcohol exposure (4.5 g/kg/day), in rats, reduced cerebellar Purkinje and granule cell number, and also impaired performance on a complex motor task. However, a 10-day exposure to a rehabilitative motor skill training paradigm, but not pure running alone, enhanced cerebellar plasticity along with motor skill performance.[89] Further, pure running alone was sufficient to reverse increased levels of oxidative stress in the cerebellum of rats exposed to alcohol during all three trimester equivalents.[90] These studies suggest that exercise may protect against some alcohol damage in the cerebellum.

In addition, postnatal alcohol exposure impairs hippocampal-associated learning and memory, an effect that is mitigated by exercise. Christie and colleagues[81] administered a liquid diet of ethanol (35.5% ethanol) to pregnant rat dams, provided a running wheel intervention to half the pups beginning on PD28 and then tested the pups at PD60 on the Morris Water Maze. Alcohol exposure produced pronounced deficits in both reference and working memory; however, this was attenuated by exercise. Further, exercise also rescued deficits on Morris Water Maze performance along with open field performance that resulted from a third trimester equivalent (PD4–9; 5.25 g/kg/day) alcohol exposure.[83] In addition, exercise can enhance long-term potentiation in the perforant pathway in prenatally exposed animals,[81] and it can also increase hippocampal expression of immediate early gene c-Fos, which is a marker of neuronal activation,[91] suggesting that anatomical changes may be correlated with behavioral changes.

In fact, exercise has been found to increase levels of adult neurogenesis in rodent models of FASD. Sprague–Dawley rats exposed to alcohol throughout gestation that were provided access to a running wheel from PD35–50 exhibited enhanced levels of cell proliferation and cell survival.[92] While, in Long–Evan rats administered alcohol on PD4–9 (5.25 g/kg/day), a 12-day exposure to a running wheel was sufficient to increase cell proliferation but not cell survival.[82] In contrast, Sprague–Dawley rats administered alcohol through all three trimesters equivalents receiving 12 days of voluntary exercise exhibited a robust benefit in both cell proliferation and cell survival.[93] Reasons for these conflicting results could be the sex or strain of the animals used, as well as differing developmental time points for wheel-running access (early vs. late adolescence), timing of BrdU administration and tissue analysis. Together, these data suggest that exercise is a potential behavioral therapy for individuals with FASD.

6.2 Exercise as an Intervention for Alzheimer's Disease

Physical activity is a strong candidate for treatment against Alzheimer's disease (for review, see Ref. 84), a disease from which approximately 5.2 million Americans suffer.[94] In aging human adults, those that participated in leisure time physical activity at midlife had 60% lower odds of Alzheimer's disease compared to their sedentary counterparts.[13] Further, elderly adults suffering from mild dementia presented with significant improvement on the Cambridge Cognitive Examination following a 16-week exercise intervention (treadmill, 30 min, twice a week and moderate intensity of 60% VO_2 max[14]).

Rodent work has attempted to understand the mechanisms through which exercise inhibits the progression of Alzheimer's disease. Through the use of a transgenic mouse model of AD (APP/PS1), Liu and colleagues[85] investigated the long-term influence of treadmill exercise. Results showed that a 5-month exposure to exercise significantly reduced hippocampal β-amyloid (Aβ) deposition and tau phosphorylation along with APP phosphorylation and PS1 expression. Further, inhibition of the GSK3-dependent signaling pathway was observed, suggesting this may be a potential target to attenuate Alzheimer's disease-like neuropathology. In addition, transgenic Alzheimer's disease mice given 4 months of voluntary exercise exhibited improved spatial learning in the Barnes maze along with lowered soluble Aβ1–42 levels in the cortex and hippocampus.[86] It is possible that the neuroprotective effect of physical activity is achieved through BDNF mechanisms[47] or through lowered levels of apoptosis.[95] Overall, it appears that in patients with dementia or Alzheimer's disease, simply being more active may stimulate brain plasticity, and provide neuroprotection.

6.3 Exercise as an Intervention for Aging

Aerobic fitness appears effective as an intervention to prevent age-related cortical decay and cognitive impairments. In older adults (58–77 years old), there is a positive association between cardiovascular fitness and executive functioning[36] and cognition.[96] Further, continued exercise performance in older adults is associated with larger left and right hippocampi,[6,10] increased prefrontal cortical volume,[53] and increased prefrontal and cingulated gray matter.[97] A 3-month aerobic exercise intervention produced changes in hippocampal perfusion of older adults along earlier recall for complex spatial objects.[8] Further, 6 months of exercise mitigated age-related decline in both verbal and spatial memory.[98] Finally, in older adults, a low activity, 8-week yoga intervention significantly improved performance on task switching, running memory, and the n-back task.[11] Again, these data illustrate that simply being physically active will promote brain plasticity and serve to protect against aging.

6.4 Exercise as an Intervention for Stroke

Exercise has proven to be an effective treatment in stroke recovery. Rand and colleagues[99] investigated the effectiveness of a 6-month program of exercise for 2 h and recreation for 1 h weekly in aged adults suffering from chronic stroke. The exercise intervention improved performance on the dual task (walking while talking), response inhibition (Stroop test), and memory (Rey Auditory Verbal Learning Test-long delay). In rodent models, mice subjected to focal cerebral ischemia have impaired performance on the Morris Water Maze; however, voluntary exercise mitigated these effects, increased levels of adult neurogenesis, and upregulated phosphorylation of cAMP response element-binding protein.[100] Similarly, in rats, exercise rescued stroke-induced deficits on the Morris Water Maze; however, this was accompanied by a reduction in oxidative stress[101] along with increased hippocampal dendritic complexity, levels of BDNF and PSD-95.[102] In fact, exercise-induced levels of BDNF appear to play a key role in stroke recovery.[103–105] Thus, physical activity induces neuroplasticity, which aids in the recovery from stroke.

7. CONCLUSION

This chapter illustrates that exercise is critical for physical and mental health throughout the lifespan. Despite the many different forms of exercise, the overall result appears to be that any type of activity be it acute or chronic, forced or voluntary improves cognitive functions and enhances brain plasticity. In fact, through the use of a variety of animal models of exercise in both humans and rodents, research has demonstrated that exercise increases longevity and improves cognitive and spatial memory performance in both the healthy and the damaged brain. Further, given its role in learning and memory, this chapter focused on the influence of exercise on the hippocampus. The hippocampus is a highly plastic brain region, due, in part, to the continuous birth of adult neurons that occur in the dentate gyrus. As such, it is highly susceptible to beneficial factors such as exercise and detrimental factors such as disease and decay. The ability of exercise to not only improve hippocampal learning and memory but also increase levels of hippocampal neurogenesis is remarkable. Further, exercise activates a multitude of signaling pathways that promote neuroplasticity and that are associated with enhanced learning and memory, which indicates the therapeutic role of exercise in both a healthy and a diseased brain. Exercise also produces beneficial long-term epigenetic changes that have the potential to be passed on to future generations. The evidence justifies serious consideration of a role for exercise in therapeutics. Overall, the data presented in this chapter provide evidence in support of exercise as a means to promote enhanced cognitive function and brain plasticity, both in the healthy and in the diseased brain.

REFERENCES

1. Bramble DM, Lieberman DE. Endurance running and the evolution of Homo. *Nature.* 2004;432:345–352.
2. Marino FE. The evolutionary basis of thermoregulation and exercise performance. *Med Sport Sci.* 2008;53:1–13.
3. Mattson MP. Evolutionary aspects of human exercise—born to run purposefully. *Ageing Res Rev.* 2012;11:347–352.
4. Spirduso WW, Clifford P. Replication of age and physical activity effects on reaction and movement time. *J Gerontol.* 1978;33:26–30.
5. Voss MW, Vivar C, Kramer AF, van Praag H. Bridging animal and human models of exercise-induced brain plasticity. *Trends Cogn Sci.* 2013;17:525–544.
6. Erickson KI, Kramer AF. Aerobic exercise effects on cognitive and neural plasticity in older adults. *Br J Sports Med.* 2009;43:22–24.
7. Erickson KI, Miller DL, Roecklein KA. The aging hippocampus: interactions between exercise, depression, and BDNF. *Neuroscientist.* 2012;18:82–97.
8. Maass A, Duzel S, Goerke M, et al. Vascular hippocampal plasticity after aerobic exercise in older adults. *Mol Psychiatry.* 2015;20:585–593.
9. Vogiatzis I, Louvaris Z, Habazettl H, et al. Frontal cerebral cortex blood flow, oxygen delivery and oxygenation during normoxic and hypoxic exercise in athletes. *J Physiol.* 2011;589:4027–4039.
10. Ten Brinke LF, Bolandzadeh N, Nagamatsu LS, et al. Aerobic exercise increases hippocampal volume in older women with probable mild cognitive impairment: a 6-month randomised controlled trial. *Br J Sports Med.* 2015;49(4):248–254.
11. Gothe NP, Kramer AF, McAuley E. The effects of an 8-week Hatha yoga intervention on executive function in older adults. *J Gerontol A Biol Sci Med Sci.* 2014;69:1109–1116.
12. Skriver K, Roig M, Lundbye-Jensen J, et al. Acute exercise improves motor memory: exploring potential biomarkers. *Neurobiol Learn Mem.* 2014;116:46–58.
13. Rovio S, Kareholt I, Helkala EL, et al. Leisure-time physical activity at midlife and the risk of dementia and Alzheimer's disease. *Lancet Neurol.* 2005;4:705–711.
14. Arcoverde C, Deslandes A, Moraes H, et al. Treadmill training as an augmentation treatment for Alzheimer's disease: a pilot randomized controlled study. *Arq Neuropsiquiatr.* 2014;72:190–196.
15. Clark PJ, Kohman RA, Miller DS, et al. Genetic influences on exercise-induced adult hippocampal neurogenesis across 12 divergent mouse strains. *Genes Brain Behav.* 2011;10:345–353.
16. Clark PJ, Brzezinska WJ, Thomas MW, et al. Intact neurogenesis is required for benefits of exercise on spatial memory but not motor performance or contextual fear conditioning in C57BL/6J mice. *Neuroscience.* 2008;155:1048–1058.
17. Yuede CM, Zimmerman SD, Dong H, et al. Effects of voluntary and forced exercise on plaque deposition, hippocampal volume, and behavior in the Tg2576 mouse model of Alzheimer's disease. *Neurobiol Dis.* 2009;35:426–432.
18. Uysal N, Kiray M, Sisman A, et al. Effects of voluntary and involuntary exercise on cognitive functions, and VEGF and BDNF levels in adolescent rats. *Biotech Histochem.* 2015;90:55–68.
19. Ang ET, Dawe GS, Wong PT, Moochhala S, Ng YK. Alterations in spatial learning and memory after forced exercise. *Brain Res.* 2006;1113:186–193.
20. Mokhtari-Zaer A, Ghodrati-Jaldbakhan S, Vafaei AA, et al. Effects of voluntary and treadmill exercise on spontaneous withdrawal signs, cognitive deficits and alterations in apoptosis-associated proteins in morphine-dependent rats. *Behav Brain Res.* 2014;271:160–170.
21. Arida RM, Scorza CA, da Silva AV, Scorza FA, Cavalheiro EA. Differential effects of spontaneous versus forced exercise in rats on the staining of parvalbumin-positive neurons in the hippocampal formation. *Neurosci Lett.* 2004;364:135–138.

22. Niemann C, Godde B, Voelcker-Rehage C. Not only cardiovascular, but also coordinative exercise increases hippocampal volume in older adults. *Front Aging Neurosci*. 2014;6:170.
23. van de Rest O, van der Zwaluw NL, Tieland M, et al. Effect of resistance-type exercise training with or without protein supplementation on cognitive functioning in frail and pre-frail elderly: secondary analysis of a randomized, double-blind, placebo-controlled trial. *Mech Ageing Dev*. 2014;136–137:85–93.
24. Alves CR, Merege Filho CA, Benatti FB, et al. Creatine supplementation associated or not with strength training upon emotional and cognitive measures in older women: a randomized double-blind study. *PLoS One*. 2013;8:e76301.
25. Chang YK, Ku PW, Tomporowski PD, Chen FT, Huang CC. Effects of acute resistance exercise on late-middle-age adults' goal planning. *Med Sci Sports Exerc*. 2012;44:1773–1779.
26. Tamaki T, Uchiyama S, Nakano S. A weight-lifting exercise model for inducing hypertrophy in the hindlimb muscles of rats. *Med Sci Sports Exerc*. 1992;24:881–886.
27. Aparicio VA, Nebot E, Porres JM, et al. Effects of high-whey-protein intake and resistance training on renal, bone and metabolic parameters in rats. *Br J Nutr*. 2011;105:836–845.
28. Cassilhas RC, Reis IT, Venancio D, et al. Animal model for progressive resistance exercise: a detailed description of model and its implications for basic research in exercise. *Motriz-Revista De Educacao Fisica*. 2013;19:178–184.
29. Cassilhas RC, Lee KS, Fernandes J, et al. Spatial memory is improved by aerobic and resistance exercise through divergent molecular mechanisms. *Neuroscience*. 2012;202:309–317.
30. Cassilhas RC, Lee KS, Venancio DP, et al. Resistance exercise improves hippocampus-dependent memory. *Braz J Med Biol Res*. 2012;45:1215–1220.
31. Seo DY, Lee SR, Kim N, et al. Humanized animal exercise model for clinical implication. *Pflugers Arch*. 2014;466:1673–1687.
32. Tine M. Acute aerobic exercise: an intervention for the selective visual attention and reading comprehension of low-income adolescents. *Front Psychol*. 2014;5:575.
33. Etnier J, Labban JD, Piepmeier A, Davis ME, Henning DA. Effects of an acute bout of exercise on memory in 6th grade children. *Pediatr Exerc Sci*. 2014;26:250–258.
34. Siette J, Reichelt AC, Westbrook RF. A bout of voluntary running enhances context conditioned fear, its extinction, and its reconsolidation. *Learn Mem*. 2014;21:73–81.
35. Erickson KI, Voss MW, Prakash RS, et al. Exercise training increases size of hippocampus and improves memory. *Proc Natl Acad Sci USA*. 2011;108:3017–3022.
36. Colcombe S, Kramer AF. Fitness effects on the cognitive function of older adults: a meta-analytic study. *Psychol Sci*. 2003;14:125–130.
37. van Praag H, Christie BR, Sejnowski TJ, Gage FH. Running enhances neurogenesis, learning, and long-term potentiation in mice. *Proc Natl Acad Sci USA*. 1999;96:13427–13431.
38. van Praag H, Kempermann G, Gage FH. Running increases cell proliferation and neurogenesis in the adult mouse dentate gyrus. *Nat Neurosci*. 1999;2:266–270.
39. Meeusen R, De Meirleir K. Exercise and brain neurotransmission. *Sports Med*. 1995;20:160–188.
40. Giannaki CD, Sakkas GK, Karatzaferi C, et al. Effect of exercise training and dopamine agonists in patients with uremic restless legs syndrome: a six-month randomized, partially double-blind, placebo-controlled comparative study. *BMC Nephrol*. 2013;14:194.
41. Melancon MO, Lorrain D, Dionne IJ. Changes in markers of brain serotonin activity in response to chronic exercise in senior men. *Appl Physiol Nutr Metab*. 2014;39:1–7.
42. Mabandla MV, Kellaway LA, Daniels WM, Russell VA. Effect of exercise on dopamine neuron survival in prenatally stressed rats. *Metab Brain Dis*. 2009;24:525–539.

43. Kim H, Heo HI, Kim DH, et al. Treadmill exercise and methylphenidate ameliorate symptoms of attention deficit/hyperactivity disorder through enhancing dopamine synthesis and brain-derived neurotrophic factor expression in spontaneous hypertensive rats. *Neurosci Lett*. 2011;504:35–39.
44. Renoir T, Chevarin C, Lanfumey L, Hannan AJ. Effect of enhanced voluntary physical exercise on brain levels of monoamines in Huntington disease mice. *PLoS Curr*. 2011;3:RRN1281. http://dx.doi.org/10.1371/currents.RRN1281.
45. Jasienska G, Ziomkiewicz A, Thune I, Lipson SF, Ellison PT. Habitual physical activity and estradiol levels in women of reproductive age. *Eur J Cancer Prev*. 2006;15:439–445.
46. Erickson KI, Colcombe SJ, Elavsky S, et al. Interactive effects of fitness and hormone treatment on brain health in postmenopausal women. *Neurobiol Aging*. 2007;28:179–185.
47. Garcia-Mesa Y, Pareja-Galeano H, Bonet-Costa V, et al. Physical exercise neuroprotects ovariectomized 3× Tg-AD mice through BDNF mechanisms. *Psychoneuroendocrinology*. 2014;45:154–166.
48. Gomez-Pinilla F, Dao L, So V. Physical exercise induces FGF-2 and its mRNA in the hippocampus. *Brain Res*. 1997;764:1–8.
49. Ding Q, Vaynman S, Akhavan M, Ying Z, Gomez-Pinilla F. Insulin-like growth factor I interfaces with brain-derived neurotrophic factor-mediated synaptic plasticity to modulate aspects of exercise-induced cognitive function. *Neuroscience*. 2006;140:823–833.
50. Neeper SA, Gomez-Pinilla F, Choi J, Cotman CW. Physical activity increases mRNA for brain-derived neurotrophic factor and nerve growth factor in rat brain. *Brain Res*. 1996;726:49–56.
51. Vaynman S, Gomez-Pinilla F. License to run: exercise impacts functional plasticity in the intact and injured central nervous system by using neurotrophins. *Neurorehabil Neural Repair*. 2005;19:283–295.
52. Cotman CW, Berchtold NC. Exercise: a behavioral intervention to enhance brain health and plasticity. *Trends Neurosci*. 2002;25:295–301.
53. de Lange FP, Koers A, Kalkman JS, et al. Increase in prefrontal cortical volume following cognitive behavioural therapy in patients with chronic fatigue syndrome. *Brain*. 2008;131:2172–2180.
54. Rasmussen P, Brassard P, Adser H, et al. Evidence for a release of brain-derived neurotrophic factor from the brain during exercise. *Exp Physiol*. 2009;94:1062–1069.
55. Ding Q, Ying Z, Gomez-Pinilla F. Exercise influences hippocampal plasticity by modulating brain-derived neurotrophic factor processing. *Neuroscience*. 2011;192:773–780.
56. Griffin EW, Mullally S, Foley C, et al. Aerobic exercise improves hippocampal function and increases BDNF in the serum of young adult males. *Physiol Behav*. 2011;104:934–941.
57. Novaes Gomes FG, Fernandes J, Vannucci Campos D, et al. The beneficial effects of strength exercise on hippocampal cell proliferation and apoptotic signaling is impaired by anabolic androgenic steroids. *Psychoneuroendocrinology*. 2014;50:106–117.
58. Sato K, Ogoh S, Hirasawa A, Oue A, Sadamoto T. The distribution of blood flow in the carotid and vertebral arteries during dynamic exercise in humans. *J Physiol*. 2011;589:2847–2856.
59. Berggren KL, Kay JJ, Swain RA. Examining cerebral angiogenesis in response to physical exercise. *Methods Mol Biol*. 2014;1135:139–154.
60. Rhyu IJ, Bytheway JA, Kohler SJ, et al. Effects of aerobic exercise training on cognitive function and cortical vascularity in monkeys. *Neuroscience*. 2010;167:1239–1248.
61. Van der Borght K, Kobor-Nyakas DE, Klauke K, et al. Physical exercise leads to rapid adaptations in hippocampal vasculature: temporal dynamics and relationship to cell proliferation and neurogenesis. *Hippocampus*. 2009;19:928–936.

62. Liu JK, Yeo HC, Overvik-Douki E, et al. Chronically and acutely exercised rats: biomarkers of oxidative stress and endogenous antioxidants. *J Appl Physiol.* 2000;89:21–28.
63. Radak Z, Toldy A, Szabo Z, et al. The effects of training and detraining on memory, neurotrophins and oxidative stress markers in rat brain. *Neurochem Int.* 2006;49:387–392.
64. Ogonovszky H, Berkes I, Kumagai S, et al. The effects of moderate-, strenuous- and over-training on oxidative stress markers, DNA repair, and memory, in rat brain. *Neurochem Int.* 2005;46:635–640.
65. Marosi K, Bori Z, Hart N, et al. Long-term exercise treatment reduces oxidative stress in the hippocampus of aging rats. *Neuroscience.* 2012;226:21–28.
66. Navarro A, Gomez C, Lopez-Cepero JM, Boveris A. Beneficial effects of moderate exercise on mice aging: survival, behavior, oxidative stress, and mitochondrial electron transfer. *Am J Physiol Regul Integr Comp Physiol.* 2004;286:R505–R511.
67. Kerr AL, Swain RA. Rapid cellular genesis and apoptosis: effects of exercise in the adult rat. *Behav Neurosci.* 2011;125:1–9.
68. Hamilton GF, Boschen KE, Goodlett CR, Greenough WT, Klintsova AY. Housing in environmental complexity following wheel running augments survival of newly generated hippocampal neurons in a rat model of binge alcohol exposure during the third trimester equivalent. *Alcohol Clin Exp Res.* 2012;36:1196–1204.
69. Uysal N, Tugyan K, Kayatekin BM, et al. The effects of regular aerobic exercise in adolescent period on hippocampal neuron density, apoptosis and spatial memory. *Neurosci Lett.* 2005;383:241–245.
70. Kim DH, Ko IG, Kim BK, et al. Treadmill exercise inhibits traumatic brain injury-induced hippocampal apoptosis. *Physiol Behav.* 2010;101:660–665.
71. Abel JL, Rissman EF. Running-induced epigenetic and gene expression changes in the adolescent brain. *Int J Dev Neurosci.* 2013;31:382–390.
72. Gomez-Pinilla F, Zhuang Y, Feng J, Ying Z, Fan G. Exercise impacts brain-derived neurotrophic factor plasticity by engaging mechanisms of epigenetic regulation. *Eur J Neurosci.* 2011;33:383–390.
73. Bayod S, Menella I, Sanchez-Roige S, et al. Wnt pathway regulation by long-term moderate exercise in rat hippocampus. *Brain Res.* 2014;1543:38–48.
74. Augustinack JC, van der Kouwe AJ, Salat DH, et al. H.M.'s contributions to neuroscience: a review and autopsy studies. *Hippocampus.* 2014;24:1267–1286.
75. Altman J. Are new neurons formed in the brains of adult mammals? *Science.* 1962;135:1127–1128.
76. Eadie BD, Redila VA, Christie BR. Voluntary exercise alters the cytoarchitecture of the adult dentate gyrus by increasing cellular proliferation, dendritic complexity, and spine density. *J Comp Neurol.* 2005;486:39–47.
77. Freund J, Brandmaier AM, Lewejohann L, et al. Emergence of individuality in genetically identical mice. *Science.* 2013;340:756–759.
78. Marlatt MW, Potter MC, Lucassen PJ, van Praag H. Running throughout middle-age improves memory function, hippocampal neurogenesis, and BDNF levels in female C57BL/6J mice. *Dev Neurobiol.* 2012;72:943–952.
79. Creer DJ, Romberg C, Saksida LM, van Praag H, Bussey TJ. Running enhances spatial pattern separation in mice. *Proc Natl Acad Sci USA.* 2010;107:2367–2372.
80. Groves JO, Leslie I, Huang GJ, et al. Ablating adult neurogenesis in the rat has no effect on spatial processing: evidence from a novel pharmacogenetic model. *PLoS Genet.* 2013;9:e1003718.
81. Christie BR, Swann SE, Fox CJ, et al. Voluntary exercise rescues deficits in spatial memory and long-term potentiation in prenatal ethanol-exposed male rats. *Eur J Neurosci.* 2005;21:1719–1726.

82. Helfer JL, Goodlett CR, Greenough WT, Klintsova AY. The effects of exercise on adolescent hippocampal neurogenesis in a rat model of binge alcohol exposure during the brain growth spurt. *Brain Res.* 2009;1294:1–11.
83. Thomas JD, Sather TM, Whinery LA. Voluntary exercise influences behavioral development in rats exposed to alcohol during the neonatal brain growth spurt. *Behav Neurosci.* 2008;122:1264–1273.
84. Radak Z, Hart N, Sarga L, et al. Exercise plays a preventive role against Alzheimer's disease. *J Alzheimers Dis.* 2010;20:777–783.
85. Liu HL, Zhao G, Zhang H, Shi LD. Long-term treadmill exercise inhibits the progression of Alzheimer's disease-like neuropathology in the hippocampus of APP/PS1 transgenic mice. *Behav Brain Res.* 2013;256:261–272.
86. Walker JM, Klakotskaia D, Ajit D, et al. Beneficial effects of dietary EGCG and voluntary exercise on behavior in an Alzheimer's disease mouse model. *J Alzheimers Dis.* 2015;44:561–572.
87. May PA, Baete A, Russo J, et al. Prevalence and characteristics of fetal alcohol spectrum disorders. *Pediatrics.* 2014;134:855–866.
88. Lupton C, Burd L, Harwood R. Cost of fetal alcohol spectrum disorders. *Am J Med Genet C Semin Med Genet.* 2004;127C:42–50.
89. Klintsova AY, Goodlett CR, Greenough WT. Therapeutic motor training ameliorates cerebellar effects of postnatal binge alcohol. *Neurotoxicol Teratol.* 2000;22: 125–132.
90. Brocardo PS, Boehme F, Patten A, et al. Anxiety- and depression-like behaviors are accompanied by an increase in oxidative stress in a rat model of fetal alcohol spectrum disorders: protective effects of voluntary physical exercise. *Neuropharmacology.* 2012;62:1607–1618.
91. Sim YJ, Kim H, Shin MS, et al. Effect of postnatal treadmill exercise on c-Fos expression in the hippocampus of rat pups born from the alcohol-intoxicated mothers. *Brain Dev.* 2008;30:118–125.
92. Redila VA, Christie BR. Exercise-induced changes in dendritic structure and complexity in the adult hippocampal dentate gyrus. *Neuroscience.* 2006;137:1299–1307.
93. Boehme F, Gil-Mohapel J, Cox A, et al. Voluntary exercise induces adult hippocampal neurogenesis and BDNF expression in a rodent model of fetal alcohol spectrum disorders. *Eur J Neurosci.* 2011;33:1799–1811.
94. Alzheimer's Association. 2014 Alzheimer's disease facts and figures. *Alzheimers Dement.* 2014;10:e47–e92.
95. Um HS, Kang EB, Koo JH, et al. Treadmill exercise represses neuronal cell death in an aged transgenic mouse model of Alzheimer's disease. *Neurosci Res.* 2011;69: 161–173.
96. Churchill JD, Galvez R, Colcombe S, et al. Exercise, experience and the aging brain. *Neurobiol Aging.* 2002;23:941–955.
97. Floel A, Ruscheweyh R, Kruger K, et al. Physical activity and memory functions: are neurotrophins and cerebral gray matter volume the missing link? *Neuroimage.* 2010;49:2756–2763.
98. Nagamatsu LS, Chan A, Davis JC, et al. Physical activity improves verbal and spatial memory in older adults with probable mild cognitive impairment: a 6-month randomized controlled trial. *J Aging Res.* 2013;2013:861893.
99. Rand D, Eng JJ, Liu-Ambrose T, Tawashy AE. Feasibility of a 6-month exercise and recreation program to improve executive functioning and memory in individuals with chronic stroke. *Neurorehabil Neural Repair.* 2010;24:722–729.
100. Luo CX, Jiang J, Zhou QG, et al. Voluntary exercise-induced neurogenesis in the post-ischemic dentate gyrus is associated with spatial memory recovery from stroke. *J Neurosci Res.* 2007;85:1637–1646.

101. Cechetti F, Worm PV, Elsner VR, et al. Forced treadmill exercise prevents oxidative stress and memory deficits following chronic cerebral hypoperfusion in the rat. *Neurobiol Learn Mem*. 2012;97:90–96.
102. Shih PC, Yang YR, Wang RY. Effects of exercise intensity on spatial memory performance and hippocampal synaptic plasticity in transient brain ischemic rats. *PLoS One*. 2013;8:e78163.
103. Chang HC, Yang YR, Wang SG, Wang RY. Effects of treadmill training on motor performance and extracellular glutamate level in striatum in rats with or without transient middle cerebral artery occlusion. *Behav Brain Res*. 2009;205:450–455.
104. Mang CS, Campbell KL, Ross CJ, Boyd LA. Promoting neuroplasticity for motor rehabilitation after stroke: considering the effects of aerobic exercise and genetic variation on brain-derived neurotrophic factor. *Phys Ther*. 2013;93:1707–1716.
105. Ploughman M, Windle V, MacLellan CL, et al. Brain-derived neurotrophic factor contributes to recovery of skilled reaching after focal ischemia in rats. *Stroke*. 2009;40:1490–1495.
106. Wojtowicz JM, Askew ML, Winocur G. The effects of running and inhibiting adult neurogenesis on learning and memory in rats. *Eur J Neurosci*. 2008;27:1494–1502.

CHAPTER SEVENTEEN

Exercise, Autophagy, and Apoptosis

Frank C. Mooren[1], Karsten Krüger
Department of Sports Medicine, University of Giessen, Giessen, Germany
[1]Corresponding author: e-mail address: Frank-Christoph.Mooren@sport.uni-giessen.de

Contents

Abstract

Exercise is a form of physiological stress which is known to induce an adaptational response.

It is proposed that both apoptosis and autophagy are processes which are necessary for adaptation to exercise. Apoptosis and autophagy are induced during exercise to limit tissue damage, restore tissue integrity, terminate inflammatory responses, or induce direct signals for adaptation. Apoptosis is induced by specific mediators like reactive oxygen species, cytokines, and hormones. Autophagic pathways are activated by altered proteins/organelles with the aim to conserve and recycle the cellular resources. In this case, the cell is flooded with damaged molecules, the repairing mechanisms are overtaxed, and apoptosis is induced. In conclusion, autophagy seems to be necessary for adaptation by providing locally the conditions for muscle plasticity and apoptosis systemically by mobilizing progenitor cells.

Progress in Molecular Biology and Translational Science, Volume 135
ISSN 1877-1173
http://dx.doi.org/10.1016/bs.pmbts.2015.07.023

1. APOPTOSIS

The term apoptosis describes a special process of cell death that allows cells to die and be lysed in a well-controlled fashion. In general, apoptotic processes play an important role during development and in maintaining tissue homeostasis by guaranteeing a balance between the generation of new cells and the removal of damaged or aged cells. In case of apoptosis induction, the cell is eliminated without necrotic processes which are known to result in local inflammatory conditions.[1] Therefore, apoptosis offers the opportunity to remove cells without any collateral damage in adjacent tissues. Besides physiological cell turnover, apoptosis has also an important role in a variety of pathophysiological conditions like inflammatory responses, cell stress or cell damage, or in preventing neoplastic diseases. The process of apoptosis is highly controlled and orchestrated, and is characterized by a specific morphology.[2,3]

1.1 Morphology of Apoptosis

Early processes of apoptosis comprise cell shrinkage and pyknosis.[4] During cell shrinkage, the cells get smaller in size and the cytoplasm gets more dense because the organelles are tightly packed. Pyknosis, which represent the most characteristic feature of apoptosis, is the result of chromatin condensation. During chromatin condensation, nuclear material aggregates under the nuclear membrane. Extensive plasma membrane blebbing occurs followed by karyorrhexis and separation of cell fragments into apoptotic bodies. Apoptotic bodies consist of cytoplasm with tightly packed organelles with or without a nuclear fragment. The organelle integrity is still maintained and all of this is enclosed within an intact plasma membrane. These bodies are subsequently phagocytosed by macrophages, parenchymal cells, or neoplastic cells.[5,6] However, currently, their role in signal transduction is not fully understood.

1.2 Pathways of Apoptosis

Cells have specific intracellular signaling pathways which control the balance between surviving or dying signals. In many cell types, apoptosis depends on the presence or absence of specific death-promoting or cell-stabilizing signals. Based on the nature of the apoptotic stimuli, two main pathways for apoptosis induction have been differentiated: first, an extrinsic pathway which is initiated by the binding of a ligand. These ligands are more frequently expressed after specific physiologic stimuli and bind to a death-promoting receptor. The second main pathway is known as the intrinsic pathway, which is mainly triggered by damage to the cell or single components.[1,2]

1.2.1 The Extrinsic Signaling Pathway

The extrinsic signaling pathway leading to apoptosis is initiated by ligands and transmembrane death receptors. The receptors are members of the tumor necrosis factor (TNF) receptor gene superfamily. In case of apoptosis induction, members of this receptor family bind to extrinsic ligands and transduce intracellular signals that result in the process of apoptosis. Some of the ligands like Fas ligand (FasL), TNF-alpha, Apo3L, and Apo2L are well characterized. These ligands bind to their corresponding receptors known as Fas receptor (FasR), TNFR1, DR3, and DR4/DR5. The signal transduction of the extrinsic pathway involves specific proteins like caspases which are proteases with specific intracellular targets. After activation, caspases affect several cellular functions as part of a process that results in the death of the cells. A major signaling pathway for the extrinsic induction of apoptosis is the FasR/FasL pathway.[1,7] After binding, the receptor is trimerized and the Fas adaptor protein, Fas-associated death domain protein (FADD), binds to the trimerized Fas cytoplasmatic region. After recruitment of procaspase-8 to FADD, the FasR, FADD, and procaspase-8 form a functional death-inducing signaling complex. This process induces self-activation of caspase-8, which is released into the cytosol and activates the downstream effector caspase-3, which cleaves cellular targets and induces cell death.[3]

1.2.2 The Intrinsic Pathway of Apoptosis

The intrinsic signaling pathways that initiate apoptosis involve a diverse array of nonreceptor-mediated stimuli that produce intracellular signals. These stimuli act directly on targets within the cell and are mitochondrial-initiated events. The stimuli that initiate the intrinsic pathway produce intracellular signals that may act in either a positive or negative fashion. Cellular stress stimuli like heat shock, oxidative stress, growth factor withdrawal, and DNA damage are known to activate the intrinsic apoptosis pathway. Mitochondria are involved in both caspase-dependent and caspase-independent apoptosis pathways.

There are several links between the receptor and the mitochondrial pathways. In case of death receptor triggering, activation of caspase-8 may result in cleavage of Bid, a Bcl-2 family protein with a BH3 domain only, which translocates to mitochondria inducing the release of cytochrome *c*. Thereby, a mitochondrial amplification loop is initiated.[8] The cleavage of caspase-6 downstream of mitochondria may feedback to the receptor pathway by cleaving caspase-8.[5,8]

2. EXERCISE AND APOPTOSIS

Exercise is a type of physiological stress which affects concentration of several cytokines, hormones, growth factors, and the oxidative status. Additionally, exercise affects energy balance by mobilizing and metabolizing high amounts of substrates like carbohydrates and free fatty acids. All these factors are known to potentially mediate either accelerated death or prolonged cellular survival. Thereby, exercise-induced apoptosis signaling depends on exercise intensity and duration, which affect the critical balance between prosurvival and proapoptotic factors as well as the intracellular protection systems contributing to apoptosis resistance.[9]

2.1 Exercise Leukocytes Apoptosis

Because in the field of exercise, apoptosis is mainly investigated in leukocytes and muscle cells, this chapter focuses on these cell types. In leukocytes, apoptosis is of special importance because these cells have high rates of cell turnover, and apoptosis plays an important role in inflammatory processes. For instance, in lymphocytes, apoptosis is necessary during development and differentiation in thymus. Thereby, immature thymocytes expressing either self-reactive or nonfunctional T-cell receptors are eliminated. In case of an inflammatory response, cells develop a certain apoptosis resistance to increase cell population (clonal expansion), while after termination of the immune reaction, these cells have to be deleted effectively through controlled cell death (clonal contraction).[1] In neutrophils, a special form of apoptosis is called NETosis. NETosis is a process of generation of neutrophil extracellular traps (NETs), which represent chromatin structures associated with antimicrobial molecules. They are produced to trap and kill nonspecific bacterial, fungal, and protozoal pathogens. NETs production is suggested to be an essential immune response to infection. In addition to the antimicrobial function, NETosis is involved in many inflammatory and autoimmune disorders and participates in the regulation of noninfectious processes.[10]

Several studies demonstrated that exercise has a marked impact on numbers and functions of leukocytes in blood and tissues. Postexercise blood lymphopenia is suggested to be the result of either redistribution into various tissues and organs or apoptotic cell death. In this regard, exercise intensity is assumed to be a main effector of apoptosis induction. An increase of apoptosis was observed after endurance as well as resistance exercise. In particular, an increase of apoptosis was observed after ultramarathon,[11] marathon run,[12] intensive treadmill running and ergometer cycling,[13,14] and triathlon.[14] In contrast, moderate exercise did not or only marginally affected lymphocyte apoptosis.[12] There are some reports indicating that an increase of lymphocyte apoptosis occurs after exceeding a threshold of 40–60% of VO_{2max}.[15] It can be speculated that the concentration of potential apoptosis mediators is gradiently expressed with increasing exercise intensity and that they induce apoptosis after exceeding a specific threshold.[9] Moreover, after the athlete exceeds a specific duration of exercise, expression of several potential mediators is amplified and might exceed a death-inducing threshold.[9] In addition, it was also shown that lymphocyte apoptosis increased also after intensive resistance exercise.[16] Because the blood compartment represents only a small fraction of total lymphocytes, some studies investigated lymphocyte apoptosis inside organs or tissues. Using mouse models, increased lymphocyte apoptosis was investigated for several lymphatic organs. Here, it was shown that intensive treadmill exercise increased lymphocyte apoptosis in several other organs like gut tissue, thymus, spleen, bone marrow, and others in mice.[17,18] Thus, the extent of apoptosis, the kinetics, and the inducing mechanisms seemed to have a certain tissue specificity.

Regarding neutrophil apoptosis and exercise, results are conflicting. While Syu *et al.* demonstrated that acute incremental exercise induced an oxidative state in neutrophils which resulted in an increase of spontaneous neutrophil apoptosis others found a significant delay of neutrophil cell death after different exercise types.[19,20] Thus, the delay of apoptosis seemed to be related to exercise intensity because it was not observed after moderate exercise types. It is suggested that the delay is related to inflammatory processes after exercise because G-CSF signaling seems to be involved.[20] Recently, first evidence was provided that exercise induces an increased NET formation. It is suggested that these structures might contribute to exercise-induced liberation of circulation-free DNA (cf-DNA).[21] It was further shown that an excess cf-DNA in response to intense exercise is immediately and efficiently counterbalanced in healthy individuals by a transient increase in serum deoxyribonuclease activity. The emergence of NETs in the "exercising vasculature" raises important questions considering short-term and long-term effects of exercise on immune homeostasis (Fig. 1).

2.2 Exercise and Apoptosis in Skeletal Muscle

Apoptosis in multinucleated cells, like myocytes, and as a mechanism for myonuclear loss with atrophy is controversial. Some studies have analyzed molecular signals which displayed unique features of apoptosis. Here, the activation of the apoptotic cascade was shown to result from the removal of myonuclei and a relative portion of sarcoplasm, a process known as myonuclear apoptosis. This process is followed by fiber atrophy rather than wholesale cell death. In addition, apoptotic signaling may stimulate muscle protein degradation through the activation of the ubiquitin–proteasome system. Accordingly, muscle proteolysis and apoptosis are strongly connected, and it is suggested that apoptotic signaling is required for and precedes protein degradation during muscle atrophy.[22,23]

Current data indicate that exercise affects apoptotic signals in myocytes differently. First apoptotic signals are stimulated after acute prolonged or eccentric exercise which are known to cause skeletal muscle damage.[24] A significant increase in DNA fragmentation, myofibril and cytoskeletal

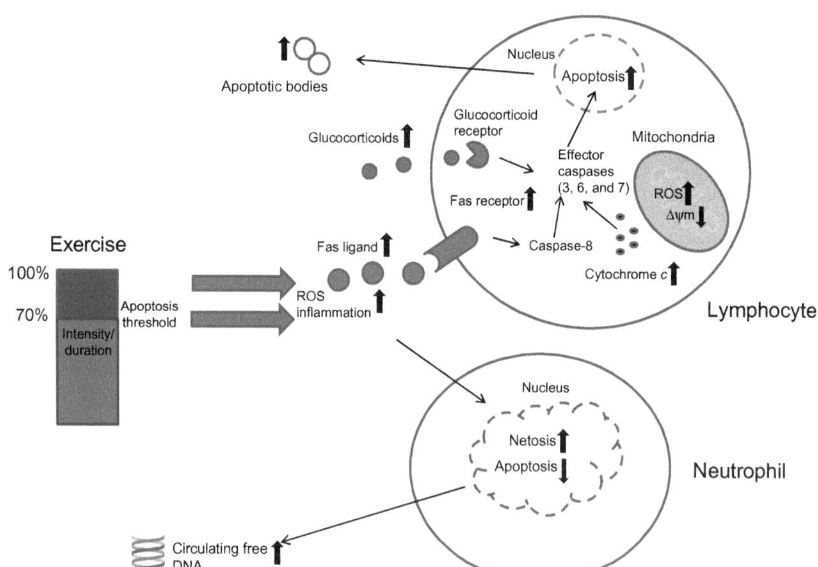

Figure 1 Suggested mediators and signaling pathways on lymphocyte and neutrophil apoptosis during exercise. (See the color plate.)

damage, TUNEL-positive nuclei, increased expression of Bax proteins, and activation of caspase-3 were found. Accordingly, it is suggested that apoptosis might be involved during muscle remodeling and repair in response to certain types of functional demand. Second, regular physical exercise is considered to attenuate myocyte apoptosis.[25,26] In this regard, Song *et al.* showed that a 12-week treadmill exercise program reduced the expression of Bax in the gastrocnemius muscle of old rats.[27] Levels of Bcl-2 were increased in exercised rodents, resulting in a decrease in the Bax-to-Bcl-2 ratio. Furthermore, cleavage of caspase-3 was lowered in old rats after exercise training. Consequently, the extent of apoptotic DNA fragmentation was significantly attenuated by exercise training, resulting in a similar apoptosis level as in young animals.[27] These changes were accompanied by an increased fiber cross-sectional area and diminished extramyocyte space. Similarly, it was found that the age-related increase of the death receptor pathway was reversed by exercise. In this regard, Marzetti *et al.* demonstrated that exercise training downregulated the expression of TNF-R1, active caspase-8, and cleaved caspase-3 in the extensor digitorum longus muscle of old rats. In parallel, treadmill running reduced levels of apoptotic DNA fragmentation. These adaptations were accompanied by improvements in exercise capacity and grip strength.[25,26]

2.3 Apoptosis Signaling and Mediators During Exercise

Regarding signaling pathways, there is evidence that exercise induces apoptosis by the intrinsic as well as the extrinsic pathways. For peripheral human lymphocytes, it was shown that intensive treadmill running, marathon running, and intensive resistance training lead to an upregulation of CD95 receptors and ligands.[13] Experiments with intestinal lymphocytes suggest that the intrinsic pathway of apoptosis is also engaged by exercise. Here, apoptosis after exercise was accompanied by mitochondrial membrane depolarization, an increase of cytosolic cytochrome *c*, and a significant reduction of Bcl-2 protein content.[18] The mediators of exercise-induced apoptosis are intensively discussed. Intensive exercise is known to affect the balance between the production of free radicals and their depletion by antioxidant defense mechanisms. Reactive oxygen species (ROS) are known to affect certain pathways of apoptosis. On the one hand, ROS reduce cellular Bcl-2 and depolarize the outer layer of the mitochondrial membrane. On the other hand, ROS crosslink the intrinsic with the extrinsic apoptosis pathway by increasing Fas expression.[1,9] Glucocorticoids are known to induce apoptosis in monocytes, macrophages, and T cells. They act via binding to their intracellular receptors known as the glucocorticoid receptors. A first step in glucocorticoid apoptosis induction is mitochondrial dysfunction followed by the release of cytochrome *c* into the cytosol. Further steps in glucocorticoid-induced apoptosis involved the activation of the caspase cascade (Fig. 1).[16]

2.4 Potential Role of Apoptosis in Adaptation

In early studies, lymphocyte apoptosis was often considered to contribute to postexercise lymphopenia. In this regard, leukocyte apoptosis was interpreted as loss of immunological competence. This idea was supported by studies which demonstrated a transient increase of upper respiratory tract infections after prolonged intensive exercise.[28] However, a potential role of apoptosis in a clinical context of immunosuppression has never been shown. Beside detrimental effects on immunity, researchers also considered that lymphocyte apoptosis is a regulatory mechanism involved in adaptation to regular training. In this regard, leukocyte apoptosis is suggested to be a mechanism to remove senescent, activated, or autoreactive lymphocytes.[29] Other studies demonstrated that exercise stimulates both apoptosis and progenitor cell mobilization.[30] Therefore, exercise promotes both cell death and cell production. It can be speculated that an increased turnover of cells might be part of exercise-induced adaptation processes.

3. AUTOPHAGY

Under physiological conditions, autophagy is an intracellular recycling system, which is involved in the clearance of damaged cell components. For proper function, proteins need to be folded into a correct three-dimensional structure, which is challenged by internal processing errors like transcriptional and translational errors as well as external stressors such as high temperature, osmotic stress, and hypoxia. Cells employ several systems to guarantee protein quality control. Heat-shock proteins and other chaperone-like molecules modulate protein folding and repair, while the ubiquitin–proteasome system and lysosome-dependent systems, such as autophagy, degrade dysfunctional proteins or organelles. So far, three main forms of autophagy have been described—macroautophagy, microautophagy, and chaperone-mediated autophagy.[31] The major distinctions between these forms are the substrate selection and delivery mechanisms involved. While microautophagy and chaperone-mediated autophagy degrade small portions of cytosol and selected proteins, macroautophagy (autophagy) is a more complex process which includes the genesis of double-membrane vesicles which engulf either large portions of cytosol or entire organelles for degradation after fusion with cellular lysosomes.

As part of the cell's catabolic machinery, autophagy is responsible for the degradation of both long-lived structural proteins and organelles. Therefore, basal autophagy is involved in the normal protein turnover. On specific stress stimuli, such as starvation or exercise, autophagy is enhanced in order to meet the altered nutritional and energy demands. In this context, autophagy takes on the responsibility for cell homeostasis, development, and function. Autophagy was long considered a coarse nonselective degradation system, but recent data indicate that autophagy represents a specific recycling machinery for the selective elimination of defective organelles.[32]

In contrast, autophagic dysfunction is associated with cellular malfunction in a wide range of diseases including neurodegenerative disorders, cancer, infections, inflammatory diseases, and insulin resistance. Thus, deviations of autophagic functions in both directions may be detrimental for the cell's fate. With respect to muscle tissue, excessive autophagy induces muscle atrophy, while impaired autophagic function results in muscle weakness and muscle degeneration.[33,34]

3.1 Autophagic Process and Its Regulation

The autophagic process comprises several steps such as initiation, vesicle nucleation, vesicle expansion, cargo recognition, fusion with lysosomes, and finally cargo degradation.[35] It occurs in every cell type albeit the exact mechanism and regulation may vary. Therefore, the processes described below may not be valid for every tissue. Everything starts with the formation of a phagophore, which most likely originates from existing membrane material from endoplasmic reticulum, golgi apparatus, and mitochondria.[36] A central switch for autophagy activation is mammalian target of rapamycin (MTOR). It is part of the IGF-1/Akt/MTOR axis which is the main signaling pathway for protein synthesis in skeletal muscle.[37,38] In contrast, upon amino acid starvation, low cellular ATP content, and the accumulation of ROS, mTOR is inhibited, which results in activation of a multiprotein complex comprising unc-51-like-kinase (ULK1), autophagy-specific gene (Atg) number 13, Atg101, and FAK-family-interacting protein (FIP200). Together with the Beclin-1/VPS34/VPS15/Atg14 protein complex, nucleation of vesicles is promoted. Beclin-1 is a substrate of ULK1 and activated upon phosphorylation on a serine residue.[39] In contrast, the anti-apoptotic protein BCL-2 (B-cell leukemia/lymphoma-2) inhibits autophagy by binding to Beclin-1. Many other intracellular stress signals and signaling pathways, e.g., C-jun N-terminalkinase 1, seem to converge on these two protein complexes thereby affecting autophagy. Expansion of autophagosomal membranes is regulated by two other complexes, the Atg12/Atg5/Atg16 complex and a conjugation of the microtubule-associated protein 1 light chain 3 (LC3) with phosphatidylethanolamine referred to as LC3-II.[39] Their formation depends on the interplay of several different Atgs, such as 4, 7, and 10. LC3-II is also important for cargo recognition as it binds to adaptor proteins such as p62 which in turn selectively recognize autophagy substrates through an ubiquitin-associated domain.[40,41] Finally, the growing autophagosome fuses with the lysosome. This process seems to require a number of different molecules such as microtubules, presenilins, SNAP29, VAMP7/8, MFN2, and Tecpr1, while tubulin polymerization-promoting protein, two-pore channel 2 (TPC2), and vacuolin-1 inhibit the formation of the autophagosome–lysosome complex. After protein degradation, the released amino acids are exported to the cytosol through lysosomal amino acid transporters. Together, the complete autophagic machinery comprises more than 30 proteins and regulators.

Another major regulator of autophagy is adenosine monophosphate-activated protein kinase (AMPK) known to be a master player in skeletal muscle homeostasis which is activated by energy deprivation during various conditions such as exercise and starvation. In this regard, AMPK is known to enhance catabolic metabolism of carbohydrates and fatty acids for the generation of ATP, while glyconeogenesis and fatty acid synthesis are inhibited. Recently, the role of AMPK in activation of autophagy has been described in more detail. First, AMPK phosphorylates and activates Ulk1 directly and via inhibition of MTOR.[42] Second, AMPK also induces the transcription of several Atgs via activation of the Forkhead box class O protein 3 (FOXO3). Moreover, FOXO3 is responsible for the transcription of several E3 ligases, such as muscle ring finger protein 1 (MuRF1) and muscle atrophy F-box (MAFbx/atrogin-1), which promote the breakdown of both muscle-specific transcription factors, such as myogenin and sarcomeric proteins.[43]

4. EFFECTS OF EXERCISE ON AUTOPHAGY

Exercise is a newly defined stimulus that induces autophagy *in vivo*. Although a first description of exercise-induced autophagy dates back to

1984, this topic has attracted more attention just recently. In 2011, Grumati and coworkers showed that physical exercise activated autophagy in skeletal muscle as indicated by an enhanced conversion of LC3-I to LC3-II and the presence of autophagosomes. In collagen VI null mice, which show an impaired autophagic flux, the acute exercise stimulus exacerbated a dystrophic phenotype as indicated by muscle-fiber apoptosis and degeneration.[44] In humans, expression of autophagy-related genes and proteins has been found upregulated in vastus lateralis muscle samples after ultraendurance exercise.[45,46] Acute exercise is a challenge to cell and tissue integrity. Alterations of intracellular calcium homeostasis may result in the activation of intracellular proteases. Increased mitochondrial activity is accompanied by enhanced generation of ROS leading to lipid peroxidation and protein carbonylation. Electrolyte and osmotic changes have the ability to perturb cell metabolism. Together, these exercise-associated factors can substantially affect cell viability and function. By the removal of damaged molecules and structures, autophagy plays an important role in maintaining and restoring cell health after exercise as indicated by the above-mentioned exercise studies. But it can be speculated that autophagy supports also cellular energy homeostasis during and after exercise as the released amino acids may serve as alternative energy substrates. The role of autophagy in mediating metabolic effects during exercise is further supported by a study of He *et al.*. They used transgenic mice containing knock-in mutations in BCL-2 phosphorylation sites (BCL2AAA), which showed normal levels of basal autophagy but were deficient in stimulus-induced autophagy.[47] During acute exercise, BCL2AAA mice showed an altered glucose metabolism as indicated by less of a decline in plasma glucose and insulin than in wild-type animals. Moreover, BCL2AAA mice failed to increase Glut4 glucose transporter insertion into the plasma membrane of the vastus lateralis and soleus muscles. These molecular events were accompanied by functional impairments such as a lower maximal exercise capacity. Furthermore, in a model of diet-induced obesity, they found evidence that exercise-induced autophagy mediates the improvements of insulin sensitivity by regular exercise.[47] However, conflicting results have been presented about exercise-induced autophagy and its role in metabolism. Kim *et al.* demonstrated that, in the recovery period after acute exercise, the expression of some autophagic proteins such as LC3-II and Atg7 was decreased, while components of the ubiquitin–proteasome machinery like MuRF1 were enhanced.[48] Furthermore, autophagy was shown to improve fat metabolism, glucose control, and insulin sensitivity in diet-induced obesity. To overcome these discrepancies, it has been suggested that the induction of autophagy by exercise stimulus is regulated in an exercise duration and/or intensity-dependent manner. It was demonstrated recently by Schwalm *et al.* that increasing exercise intensity resulted in higher autophagic activity.[49] Nutrient availability may also be a factor to consider as. Jamart *et al.* found that exercise performed in the fasted state induced higher autophagic flux, as indicated by higher increases of LC3B-II levels, than in the fed state.[43] There are only limited data available about the effects of resistance exercise on autophagy. Fry *et al.* showed that autophagy was reduced after an acute bout of resistance exercise in skeletal muscle of both young and old adults.[50]

The effects of chronic exercise training on autophagy-related proteins showed also considerable variability. After training rodents from 5 days to 3 months, no significant changes were found for autophagy-related proteins such as Atg5/12 or LC3-II/LC3-I ratio.[44,51] Others found an upregulation of LC3, Atg7, Beclin-1, and FOXO3 after 4 or 8 weeks of training.[52,53] However, the response of the autophagic machinery was fiber type dependent with a dichotomous regulation of basal autophagy and autophagy protein expression.[53] While an upregulation of autophagy protein expression occurs after increased contractile activity, the increase of autophagic flux depended on an enhanced oxidative phenotype. In contrast, in young mice, the expression of Beclin-1 and Atg7 was downregulated following 8 weeks of aerobic exercise training.[42] In the same study, an upregulation of these autophagy-related proteins was shown for aged mice after endurance training. Nonetheless, training experiments with Atg6-deficient mice suggest that autophagy is required for metabolic and structural adaptation of skeletal muscle to endurance exercise as well as for improving aerobic performance.[53]

Finally, the effects of chronic resistance training on autophagy have also been investigated. Luo *et al.* found that this type of exercise resulted in an increased autophagic activity as suggested by increased levels of autophagy regulatory proteins including Beclin-1, Atg5, 12, and 7, while levels of p62 and the ratio of LC3-II/LC3-I were reduced.[54] Moreover, the upstream signaling pathways including total AMPK, phosphorylated AMPK, and FOXO3 were upregulated. Interestingly, the enhanced autophagic activity was accompanied by reduced muscle apoptosis.

In summary, the available data suggest an important role of autophagy for the regulation of the acute response to exercise stimulus as well as the adaptive response to exercise training. However, more studies are needed to clarify the effects of different exercise modalities with respect to intensity, duration, and type of exercise and to determine the effects of cofactors such as nutritional and prior training status.

5. AUTOPHAGY AND APOPTOSIS: CONCLUDING REMARKS

The "self-destroying" ability of cells displayed during apoptosis shows some similarities with the cellular "self-eating" mechanisms during autophagy. Therefore, the relationships or interactions between both degrading mechanisms are of considerable interest. Both autophagy and apoptosis can be activated by similar stress signals. Both processes show similar regulation by BCL-2 which exhibits both antiautophagic and antiapoptotic characteristics. However, the antiautophagic effects of BCL-2 are dependent on an intact apoptotic signaling chain including proapoptotic BAX and BAK.[55] Otherwise, autophagy-related proteins such as Beclin-1, Atg 5, 7, and 12 are involved in modulating apoptosis.[56] It can be proposed that during exercise both processes act together in the pursuit of similar aims—to limit tissue damage and to restore its integrity—but might be activated at different thresholds. Autophagic pathways will be activated early after the appearance of exercise induced by altered proteins/organelles with the aim to conserve and recycle the resources. When the cell is flooded with damaged molecules, the repairing mechanisms are overtaxed and apoptosis is induced. However, both processes seem to be involved in the adaptive response to exercise—autophagy by providing locally the conditions for muscle plasticity and apoptosis systemically by mobilizing stem/progenitor cells. After exposure to exercise training, the upregulation of ROS-defending systems as well as of the autophagic machinery seems to be adaptive responses which limit the release of damaged proteins/organelles leading eventually to decreased exercise-induced apoptosis as demonstrated recently.[9,30]

REFERENCES

1. Krammer PH, Arnold R, Lavrik IN. Life and death in peripheral T cells. *Nat Rev Immunol.* 2007;7(7):532–542.
2. Korsmeyer SJ. Programmed cell death and the regulation of homeostasis. *Harvey Lect.* 1999–2000;95:21–41.
3. Curtin JF, Cotter TG. Live and let die: regulatory mechanisms in Fas-mediated apoptosis. *Cell Signal.* 2003;15(11):983–992.
4. Kerr JF, Wyllie AH, Currie AR. Apoptosis: a basic biological phenomenon with wide-ranging implications in tissue kinetics. *Br J Cancer.* 1972;26:239–257.

5. Blander JM. A long-awaited merger of the pathways mediating host defence and programmed cell death. *Nat Rev Immunol.* 2014;14(9):601–618.
6. Ishizaki Y, Cheng L, Mudge AW, Raff MC. Programmed cell death by default in embryonic cells, fibroblasts, and cancer cells. *Mol Biol Cell.* 1995;6(11):1443–1458.
7. Wajant H. Principles and mechanisms of CD95 activation. *Biol Chem.* 2014;395(12):1401–1416.
8. Cory S, Adams JM. The Bcl2 family: regulators of the cellular life-or-death switch. *Nat Rev Cancer.* 2002;2:647–656.
9. Krüger K, Mooren FC. Exercise-induced leukocyte apoptosis. *Exerc Immunol Rev.* 2014;20:117–134.
10. Vorobjeva NV, Pinegin BV. Neutrophil extracellular traps: mechanisms of formation and role in health and disease. *Biochemistry (Mosc).* 2014;79(12):1286–1296.
11. Atamaniuk J, Stuhlmeier KM, Vidotto C, Tschan H, Dossenbach-Glaninger A, Mueller M. Effects of ultra-marathon on circulating DNA and mRNA expression of pro- and anti-apoptotic genes in mononuclear cells. *Eur J Appl Physiol.* 2008;104(4):711–717.
12. Mooren FC, Lechtermann A, Völker K. Exercise-induced apoptosis of lymphocytes depends on training status. *Med Sci Sports Exerc.* 2004;36(9):1476–1483.
13. Mooren FC, Bloming D, Lechtermann A, Lerch MM, Volker K. Lymphocyte apoptosis after exhaustive and moderate exercise. *J Appl Physiol.* 2002;93(1):147–153.
14. Levada-Pires AC, Cury-Boaventura MF, Gorjão R, et al. Induction of lymphocyte death by short- and long-duration triathlon competitions. *Med Sci Sports Exerc.* 2009;41(10):1896–1901.
15. Navalta JW, Sedlock DA, Park KS. Effect of exercise intensity on exercise induced lymphocyte apoptosis. *Int J Sports Med.* 2007;28(6):539–542.
16. Krüger K, Agnischock S, Lechtermann A, et al. Intensive resistance exercise induces lymphocyte apoptosis via cortisol and glucocorticoid receptor-dependent pathways. *J Appl Physiol.* 2011;110(5):1226–1232.
17. Krüger K, Frost S, Most E, Völker K, Pallauf J, Mooren FC. Exercise affects tissue lymphocyte apoptosis via redox-sensitive and Fas-dependent signaling pathways. *Am J Physiol Regul Integr Comp Physiol.* 2009;296(5):R1518–R1527.
18. Quadrilatero J, Hoffman-Goetz L. N-Acetyl-L-cysteine prevents exercise-induced intestinal lymphocyte apoptosis by maintaining intracellular glutathione levels and reducing mitochondrial membrane depolarization. *Biochem Biophys Res Commun.* 2004;319(3):894–901.
19. Syu GD, Chen HI, Jen CJ. Severe exercise and exercise training exert opposite effects on human neutrophil apoptosis via altering the redox status. *PLoS One.* 2011;6(9): e24385.
20. Mooren FC, Völker K, Klocke R, Nikol S, Waltenberger J, Krüger K. Exercise delays neutrophil apoptosis by a G-CSF-dependent mechanism. *J Appl Physiol.* 2012;113(7):1082–1090.
21. Beiter T, Fragasso A, Hartl D, Nieß AM. Neutrophil extracellular traps: a walk on the wild side of exercise immunology. *Sports Med.* 2015;45(5):625–640.
22. Tower J. Programmed cell death in aging. *Ageing Res Rev.* 2015;23(pt A):90–100.
23. Brooks NE, Myburgh KH. Skeletal muscle wasting with disuse atrophy is multi-dimensional: the response and interaction of myonuclei, satellite cells and signaling pathways. *Front Physiol.* 2014;5:99.
24. Paulsen G, Mikkelsen UR, Raastad T, Peake JM. Leucocytes, cytokines and satellite cells: what role do they play in muscle damage and regeneration following eccentric exercise? *Exerc Immunol Rev.* 2012;18:42–97.
25. Marzetti E, Privitera G, Simili V, et al. Multiple pathways to the same end: mechanisms of myonuclear apoptosis in sarcopenia of aging. *Sci World J.* 2010;10:340–349.

26. Marzetti E, Calvani R, Bernabei R, Leeuwenburgh C. Apoptosis in skeletal myocytes: a potential target for interventions against sarcopenia and physical frailty—a mini-review. *Gerontology.* 2012;58(2):99–106.
27. Song W, Kwak HB, Lawler JM. Exercise training attenuates age-induced changes in apoptotic signaling in rat skeletal muscle. *Antioxid Redox Signal.* 2006;8(3–4): 517–528.
28. Nieman DC. Exercise, infection, and immunity. *Int J Sports Med.* 1994;15(suppl 3): S131–S141.
29. Simpson RJ. Aging, persistent viral infections, and immunosenescence: can exercise "make space"? *Exerc Sport Sci Rev.* 2011;39(1):23–33.
30. Mooren FC, Krüger K. Apoptotic lymphocytes induce progenitor cell mobilization after exercise. *J Appl Physiol (1985).* 2015;119(2):135–139. http://dx.doi.org/10.1152/japplphysiol.00287.2015.
31. Vainshtein A, Grumati P, Sandri M, Bonaldo P. Skeletal muscle, autophagy, and physical activity: the ménage à trois of metabolic regulation in health and disease. *J Mol Med (Berl).* 2014;92(2):127–137.
32. Lim J, Lachenmayer ML, Wu S, et al. Proteotoxic stress induces phosphorylation of p62/SQSTM1 by ULK1 to regulate selective autophagic clearance of protein aggregates. *PLoS Genet.* 2015;11(2):e1004987.
33. Masiero E, Agatea L, Mammucari C, et al. Autophagy is required to maintain muscle mass. *Cell Metab.* 2009;10(6):507–515.
34. Sandri M. Autophagy in skeletal muscle. *FEBS Lett.* 2010;584(7):1411–1416.
35. Zhang J, Kim J, Alexander A, et al. A tuberous sclerosis complex signalling node at the peroxisome regulates mTORC1 and autophagy in response to ROS. *Nat Cell Biol.* 2013;15(10):1186–1196.
36. Rubinsztein DC, Shpilka T, Elazar Z. Mechanisms of autophagosome biogenesis. *Curr Biol.* 2012;22(1):R29–R34.
37. Zhang J. Teaching the basics of autophagy and mitophagy to redox biologists—mechanisms and experimental approaches. *Redox Biol.* 2015;4:242–259.
38. Dunlop EA, Tee AR. mTOR and autophagy: a dynamic relationship governed by nutrients and energy. *Semin Cell Dev Biol.* 2014;36:121–129.
39. Russell RC, Tian Y, Yuan H, et al. ULK1 induces autophagy by phosphorylating Beclin-1 and activating VPS34 lipid kinase. *Nat Cell Biol.* 2013;15(7):741–750.
40. Johansen T, Lamark T. Selective autophagy mediated by autophagic adapter proteins. *Autophagy.* 2011;7(3):279–296.
41. Mostowy S, Sancho-Shimizu V, Hamon MA, et al. p62 and NDP52 proteins target intracytosolic Shigella and Listeria to different autophagy pathways. *J Biol Chem.* 2011;286(30):26987–26995.
42. Kim J, Kim YC, Fang C, et al. Differential regulation of distinct Vps34 complexes by AMPK in nutrient stress and autophagy. *Cell.* 2013;152(1–2):290–303.
43. Mammucari C, Milan G, Romanello V, et al. FoxO3 controls autophagy in skeletal muscle in vivo. *Cell Metab.* 2007;6(6):458–471.
44. Grumati P, Coletto L, Schiavinato A, et al. Physical exercise stimulates autophagy in normal skeletal muscles but is detrimental for collagen VI-deficient muscles. *Autophagy.* 2011;7(12):1415–1423.
45. Jamart C, Francaux M, Millet GY, Deldicque L, Frère D, Féasson L. Modulation of autophagy and ubiquitin-proteasome pathways during ultra-endurance running. *J Appl Physiol (1985).* 2012;112(9):1529–1537.
46. Jamart C, Benoit N, Raymackers JM, Kim HJ, Kim CK, Francaux M. Autophagy-related and autophagy-regulatory genes are induced in human muscle after ultraendurance exercise. *Eur J Appl Physiol.* 2012;112(8):3173–3177.
47. He C, Bassik MC, Moresi V, et al. Exercise-induced BCL2-regulated autophagy is required for muscle glucose homeostasis. *Nature.* 2012;481(7382):511–515.

48. Kim YA, Kim YS, Song W. Autophagic response to a single bout of moderate exercise in murine skeletal muscle. *J Physiol Biochem*. 2012;68(2):229–235.
49. Schwalm C, Jamart C, Benoit N, et al. Activation of autophagy in human skeletal muscle is dependent on exercise intensity and AMPK activation. *FASEB J*. 2015;29(8):3515–3526.
50. Fry CS, Drummond MJ, Glynn EL, et al. Skeletal muscle autophagy and protein breakdown following resistance exercise are similar in younger and older adults. *J Gerontol A Biol Sci Med Sci*. 2013;68(5):599–607.
51. Smuder AJ, Kavazis AN, Min K, Powers SK. Exercise protects against doxorubicin-induced markers of autophagy signaling in skeletal muscle. *J Appl Physiol (1985)*. 2011;111(4):1190–1198.
52. Feng Z, Bai L, Yan J, et al. Mitochondrial dynamic remodeling in strenuous exercise-induced muscle and mitochondrial dysfunction: regulatory effects of hydroxytyrosol. *Free Radic Biol Med*. 2011;50(10):1437–1446.
53. Lira VA, Okutsu M, Zhang M, et al. Autophagy is required for exercise training-induced skeletal muscle adaptation and improvement of physical performance. *FASEB J*. 2013;27(10):4184–4193.
54. Luo L, Lu AM, Wang Y, et al. Chronic resistance training activates autophagy and reduces apoptosis of muscle cells by modulating IGF-1 and its receptors, Akt/mTOR and Akt/FOXO3a signaling in aged rats. *Exp Gerontol*. 2013;48(4):427–436.
55. Lindqvist LM, Heinlein M, Huang DC, Vaux DL. Prosurvival Bcl-2 family members affect autophagy only indirectly, by inhibiting Bax and Bak. *Proc Natl Acad Sci USA*. 2014;111(23):8512–8517.
56. Rubinstein AD, Eisenstein M, Ber Y, Bialik S, Kimchi A. The autophagy protein Atg12 associates with antiapoptotic Bcl-2 family members to promote mitochondrial apoptosis. *Mol Cell*. 2011;44(5):698–709.

CHAPTER EIGHTEEN

Exercise and Stem Cells

Marni D. Boppart*,†,1, Michael De Lisio*, Sarah Witkowski‡

*Department of Kinesiology and Community Health, University of Illinois, Urbana, Illinois, USA
†Beckman Institute for Advanced Science and Technology, University of Illinois, Urbana, Illinois, USA
‡Department of Kinesiology, School of Public Health and Health Sciences, University of Massachusetts, Amherst, Massachusetts, USA
[1]Corresponding author: e-mail address: mboppart@illinois.edu

Contents

Abstract

Stem cells are traditionally studied in the context of embryonic development, yet studies confirm that a fraction remains in the adult organism for the purpose of daily remodeling and rejuvenation of multiple tissues following injury. Adult stem cells (ASCs) are found in close proximity to vessels and respond to tissue-specific cues in the microenvironment that dictate their fate and function. Exercise can dramatically alter strain sensing, extracellular matrix composition, and inflammation, and such changes in the niche likely alter ASC quantity and function postexercise. The field of stem cell biology is still in its infancy and identification and terminology of ASCs continues to evolve; thus, current information regarding exercise and stem cells is lacking. This chapter summarizes the literature that reports on the ASC response to acute exercise and exercise training, with particular emphasis on hematopoietic stem cells, endothelial progenitor cells, and mesenchymal stem cells.

Progress in Molecular Biology and Translational Science, Volume 135
ISSN 1877-1173
http://dx.doi.org/10.1016/bs.pmbts.2015.07.005

1. STEM CELL DEFINITION

Stem cells are unspecialized cells that have the potential to self-renew and differentiate into a specialized cell type for the purpose of developing an embryo (embryonic stem cell, ESC) or replacing tissue in the adult organism (adult stem cell, ASC). The property of self-renewal ensures maintenance of a stem cell pool. Cells that appear in a developing embryo postfertilization are totipotent or "totally potent" and have the capacity to differentiate into one of over two hundred different cell types that compose the human body (Fig. 1). Upon gastrulation, or the creation of the three germ layers that provide the basis for body structure, ESCs become increasingly lineage restricted and give rise to a subset of cell types. Pluripotent or multipotent stem cells can be extracted as the inner cell mass from the blastocyst of a developing embryo and provide the basis for ESC therapies for the treatment of disease or tissue injury, yet risk of teratoma and ethical concerns has largely diminished enthusiasm for use of these cells in the clinic.

2. INTRODUCTION TO STEM CELLS AND EXERCISE

The discovery of stem cells in multiple tissues of the adult organism for the purpose of tissue replacement and turnover has generated new and exciting questions regarding the role for the ASC in organ and whole system health and has generated significant interest in the potential use of these cells to treat a wide variety of degenerative conditions throughout the lifespan. The fact that physical activity is a potent stimulator of tissue remodeling, as well as an essential component of maintaining health and wellness, suggests that ASCs may provide a basis for some of the positive outcomes associated with exercise.

Hematopoietic and mesenchymal stem cells are two populations of ASCs that retain multilineage potential and are distributed in small concentrations in the bone marrow, blood, and multiple tissues in the adult human.[2,3] Relative to multipotent stem cells, the majority of ASCs in the human body have been described as unipotent stem cells, or resident progenitor cells, with limited capacity for differentiation. Examples of these lineage committed stem cells include endothelial cells (vasculature), satellite cells (skeletal muscle), neural stem cells, intestinal stem cells, bulge cells (skin), germ cells, oval cells (liver), cardiac progenitors, and bronchioloalveolar cells (lung).[4–13] Endothelial, satellite, and neural stem cells aside, minimal information exists

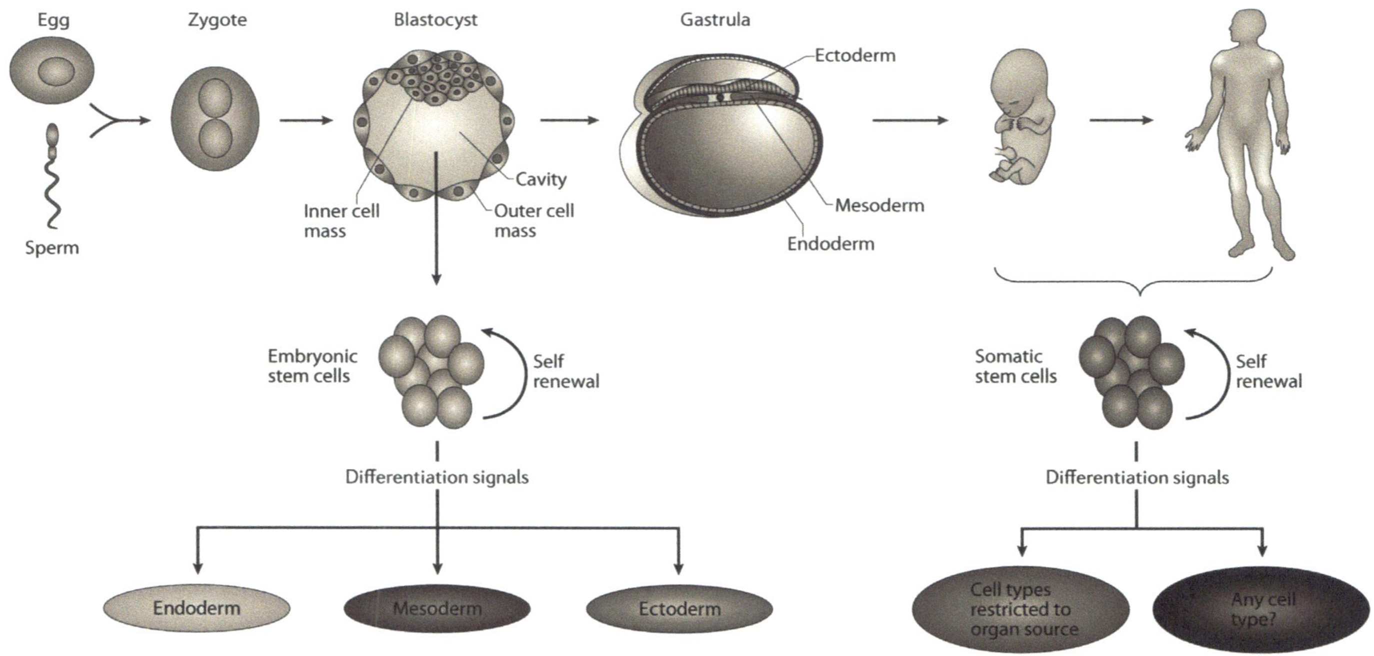

Figure 1 Introduction to stem cells. Embryonic stem cells (ESCs) are important for development of a fetus and may be extracted from the inner cell mass of the blastocyst for the purpose of ESC-based therapies. Adult or somatic stem cells (ASCs) reside in the body after birth for the purpose of routine remodeling or repair and regeneration of tissues following injury. While some ASCs retain multilineage potential, the majority are unipotent and differentiation capacity is limited to the tissue type in which they reside. Positive health outcomes associated with exercise, including enhanced repair and improvements in function, may occur as a result of alterations in ASC quantity and/or function in a variety of tissues. *Reprinted with permission from O'Connor and Crystal.*[1] (See the color plate.)

regarding the multipotent and resident progenitor cell responses to exercise. Studies that have evaluated the effects of exercise on neural stem cells will be reviewed in a separate chapter of this volume and limited information regarding satellite cells will be provided in this chapter. The purpose of this chapter is to provide an overview of ASCs, with particular emphasis on hematopoietic stem cells (HSCs), endothelial progenitor cells (EPCs), and mesenchymal stem cells (MSCs), and of our current understanding of the response of these cells to acute or repeated bouts of exercise.

3. HEMATOPOIETIC STEM CELLS

HSCs are the most primitive cells of the hematopoietic system from which all other blood cells are derived. First functionally characterized in mice in 1961 by Drs. Till and McCulloch,[14] and later characterized phenotypically and functionally in humans by the lab of John Dick,[15] HSCs are among the longest and most studied ASC population. HSCs are extremely rare, comprising only $\sim 1/10^4 - 10^5$ cells in whole bone marrow.[15] The function of HSCs is to ensure continual maintenance of peripheral blood cell populations through replenishment of mature hematopoietic cells lost through naturally occurring cell turnover, or as a result of stress. HSCs carry out this function by continuous activation, proliferation, and differentiation throughout the lifespan. During times of hematopoietic stress, such as with an acute infection, blood loss, or loss of hematopoietic cells due to chemotherapy or radiation for cancer treatment, expansion of HSCs is rapidly increased to quickly regenerate the hematopoietic system. This impressive capacity for regeneration has made HSCs the stem cell of choice for cell therapy interventions as is evidenced by their use in bone marrow transplantation since the late 1950s.[16,17]

The initial assays used to examine HSCs relied on their ability to form colonies *in vitro* and to regenerate all hematopoietic lineages *in vivo*. The "gold standard" assay for evaluating HSC quantity and function is the hematopoietic cell transplantation (HCT) assay.[18] In this assay, HSCs or whole bone marrow from mice or humans are transplanted into lethally irradiated recipient mice and hematopoiesis can be evaluated *in vivo*.[19] Due to the large quantity of mice and extended period of time required to perform the HCT assay, *in vitro* colony-forming unit (CFU) assays were developed to evaluate hematopoiesis *in vitro*. Progenitors from the macrophage, monocyte, granulocyte, and erythrocyte lineages as well as more primitive long-term culture initiating cells can be evaluated using these *in vitro* assays over a relatively short period of time. Recently, much effort has been devoted toward defining the cell surface phenotype of mouse and human HSCs for direct quantification and isolation by flow cytometry or fluorescence-activated cell sorting. In both humans and mice, the markers presently available allow for the isolation of a heterogeneous population of HSCs and not necessarily a completely pure population. In mice, HSCs are enriched within the cell population coexpressing c-kit and Sca-1 in the lineage negative population, the so-called LSK population.[20] When combined with the signaling lymphocyte activation molecule family of markers, LSK cells can be further enriched for populations of HSCs (LSK $CD150^{+}CD244^{-}CD48^{-}$) and multipotent progenitors (LSK $CD150^{-}CD244^{+}CD48^{-}$).[21,22] In humans, HSCs are enriched in the population of cells from the bone marrow, circulating in peripheral blood and in cord blood expressing the CD34 antigen.[23] CD34-positive cells represent a heterogeneous population consisting of a variety of stem/progenitor cells including endothelial progenitors and mesenchymal progenitors with the most primitive HSCs found within the rare $CD34^{+}CD38^{-}$ population.[15] Additionally, in both mice and humans, HSCs have been found to be highly enriched within the side population (SP) of cells in bone marrow and peripheral blood.[24,25] The term SP denotes a subpopulation that is distinct from the main population of cells extracted from a tissue as analyzed via flow cytometry.

3.1 Effect of Acute Exercise on HSCs

The effects of exercise on HSCs have been recently reviewed.[26] The following section will provide an updated discussion of the relevant literature only with regard to HSCs defined by CFUs, LSK, CD34, $CD34^{+}/CD38^{-}$, or SP as well as speculate on their physiological function following exercise-stimulated mobilization. Due to the heterogeneity of HSC populations, these cells will be described as hematopoietic stem/progenitor cells (HSPCs). This section will only review the literature that defined human $CD34^{+}$ or SP cells by *in vitro* colony assays, or referred to them directly as HSPCs. Studies that added additional phenotypic markers to their CD34 evaluation to examine endothelial progenitors will be discussed in Section 4.

It was first determined that an acute bout of exercise could increase the quantity of HSPCs in peripheral blood in 1978 when Barrett and colleagues detected a fourfold increase in the number of peripheral blood CFUs following an acute exercise session.[27] This early study generated a great deal of excitement as it was suggested that exercise may enhance HSPC mobilization for stem cell transplantation. Two follow-up studies attempted to better define the time course of changes in HSPC quantity in peripheral blood to determine if HSPCs were elevated for enough time to be clinically useful in stem cell transplant. Both of these studies found that HSPC quantity, as determined by CFU-GM, was increased immediately following cessation of exercise, but returned to baseline after 15 min.[28,29] This brief period of elevation was considered to be too short to be clinically useful for aiding in HSPC harvest for clinical bone marrow transplantation as normal apheresis can last many hours. Importantly, the increase in HSPC quantity was only observed with maximal exercise, stair climbing in this case, and not with submaximal exercise lasting only 1 h.[28,29] These data have been replicated in many subsequent studies using maximal exercise in the form of rowing,[30] running,[25,31–33] and cycling.[34,35] In all of these subsequent studies, the increase in peripheral blood HSPCs was transient, increasing within the first 30 min following short-duration, maximal-intensity exercise, and quickly returning to baseline values thereafter (Fig. 2). Few studies have examined the more primitive $CD34^{+}CD38^{-}$ cells population in response to exercise. These studies demonstrate that primitive HSPCs are not responsive to acute exercise,[30,36] suggesting that exercise may be mobilizing a more differentiated population of progenitors and not the most primitive HSPCs.

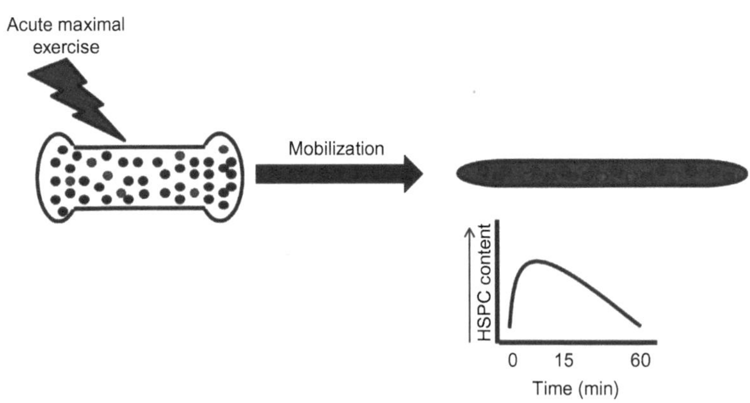

Figure 2 HSPC response to acute exercise. HSPCs (open circles), located in the bone marrow and enriched within the LSK (mouse) and CD34 (human) populations, are mobilized from their bone marrow niche to peripheral blood peaking at around 15 min following an acute, maximal exercise bout. HSPC quantity in peripheral blood returns to preexercise levels rapidly following acute exercise. It is believed that mobilized HSPCs are involved in tissue repair. (See the color plate.)

The kinetics of HSPC mobilization to long-term, lower intensity exercise has been less well examined. In a study examining the quantity of HSPCs in peripheral blood over 4 h of continuous cycling at 70% of $\dot{V}O_2max$, HSPC quantity did not increase until very late into the exercise session (3 h) and quickly returned to basal levels upon cessation of exercise.[37] Bonsignore and colleagues[36] examined the effects of running a marathon on HSPC quantity. In their study, HSPC quantity was not altered immediately after the marathon, but was significantly decreased 24 h post,[36] whereas Adams and colleagues[38] observed a significant decrease in $CD34^+$ cell quantity immediately upon completion of a marathon in elderly individuals. It was suggested that the increased tissue damage induced by long-duration exercise provides a signal for removal of HSPCs from circulation to aid in tissue repair.[38] In summary, HSPCs are transiently mobilized into peripheral blood by a maximal acute exercise stimulus, and the transient increase is due to mobilization of more differentiated progenitors than the most primitive HSPC population (Table 1). This response is consistent between both men and women[25] and young and elderly populations.[34] The transient increase may be due to rapid removal of HSPCs from circulation to aid in repairing tissue damage induced by exercise.

3.2 Effect of Exercise Training on HSCs

Studies examining the effects of exercise training on HSPC quantity are few and mostly equivocal. In young participants, trained marathon runners have a three- to fourfold increase in circulating $CD34^+$ cell quantity versus sedentary controls; however, the magnitude of increase was not related to volume of training.[36] Thijssen and colleagues[34] examined $CD34^+$ cell content in a cross-sectional study where sedentary and trained individuals were defined as <1 h of exercise per week and >8 h of exercise per week, respectively. In this study, HSPC content did not differ between sedentary and trained individuals.[34] Similarly, Wardyn and colleagues[25] did not observe any difference in peripheral HSPC quantity in trained (10 METS for at least 250 min/week over preceding 12 weeks) versus sedentary individuals, and Witkowski and colleagues[43] also did not observe any differences in $CD34^+$ cell quantity in trained versus sedentary individuals. Finally, young adult athletes exhibited lower levels of circulating HSPCs compared to age-matched sedentary controls.[42] The equivocal findings from these exercise training studies may be related to their use of a cross-sectional design, or timing of analysis as athletes in season experience higher training volumes, thus greater hematopoietic stress.[42]

Table 1 HSC Response to Exercise

Cell Type	Response to Exercise	References
Acute exercise		
CFU	4 × increase immediately after stair climbing	Barrett *et al.*[27]
CFU-GM	Increase immediately after stair climbing	Heal and Brightman[28]
CFU-GM	Increase immediately after running	Jootar *et al.*[29]
$CD34^+$ and $CD34^+/CD38^-$	No change immediately after marathon, decrease $CD34^+$ 24 h post	Bonsignore *et al.*[36]
$CD34^+$ and $CD34^+/CD38^-$	~2 × increase in $CD34^+$, no change in $CD34^+/CD38^-$ after 1000 m sprint on rowing ergometer	Morici *et al.*[30]
$CD34^+/CD45^{low}$	Increase in sedentary and trained healthy older and young adults after maximal cycling test	Thijssen *et al.*[34]
$CD34^+$	Increase in pre- and late pubertal boys postintermittent cycling	Zaldivar *et al.*[35]
SP	1.5 × increase after treadmill test to exhaustion	Wardyn *et al.*[25]
$CD34^+$	3 × increase over 4 h cycling at 70% IAT	Möbius-Winkler *et al.*[37]
$CD34^+$	No change after a marathon, but increase after 1500 m run	Bonsignore *et al.*[32]
$CD34^+/CD45^+$	Increase after incremental cycling test	Kroepfl *et al.*[33]
Exercise training		
$CD34^+$	3–4 × increase in trained marathon runners	Bonsignore *et al.*[36]
SP	No difference in active young adults	Wardyn *et al.*[25]
$CD34^+/CD45^{low}$	No change after 8 weeks of cycling in healthy older adults and no difference between sedentary and trained young	Thijssen *et al.*[34]
Secondary CFU	Restored to level of young sedentary in aged rats exposed to running wheel	Stelzer *et al.*[39]
CFU-GM, CFU-GEMM, BFU-E	Significant increases in CFUs in mice exposed to forced treadmill exercise	Baker *et al.*[40]
LSK	20% increase in mice exposed to forced treadmill exercise	De Lisio and Parise[41]
$CD34^+/CD45^{dim}$	~50% decrease in football players measured in season	D'Ascenzi *et al.*[42]

Longitudinal training studies evaluating HSPC content have been conducted in rodents. HSPC content increased ~20% within the bone marrow in mice as determined by flow cytometry,[40,41] while extramedullary hematopoiesis was increased with training as well as determined by CFU assays.[41] Despite this increased quantity, HSPCs from exercise trained donors did not enhance engraftment in ablated recipient mice.[41] In rats, lifelong exposure to a running wheel restored secondary colony-forming capacity to that of young, sedentary mice,[39] and light aerobic exercise combined with intermittent hypoxia increased circulating HSPC quantity.[44] Overall, these animal data suggest that a well-designed training program under controlled conditions can increase the quantity of HSPCs both within the bone marrow and in extramedullary sites such as peripheral blood and spleen. A few studies in humans have conducted longitudinal training studies and examined changes in peripheral blood HSPC content. In elderly individuals trained by cycle ergometry for 8 weeks HSPC content did not increase; however, the training protocol used in this study also did not induce an increase in $\dot{V}O_2$max,[34] suggesting that it may not have been of sufficient intensity. Two studies have examined the effects of long-term physical activity interventions in children and both demonstrated an increase in HSPC quantity with increased physical activity.[45,46] Additionally, in young healthy adults, exercise training increased HSPC quantity, and the effects of exercise were augmented by training under hypoxic conditions.[47] These data suggest that in response to a properly designed training program under well-controlled conditions, HSPC content may indeed increase with exercise training. The increase in response to training may depend on the age of participants, as well as whether HSPC quantity is evaluated in the bone marrow or extramedullary sites of hematopoiesis. To summarize longitudinal training studies conducted in rodents, elderly and children have demonstrated an increase in HSPC quantity; however, the cross-sectional data obtained from young adults is equivocal (Table 1). It is important to consider the timing of HSPC analysis as athletes studied in season, during the period of intense, high volume training, demonstrated a decrease in HSPC quantity.

4. ENDOTHELIAL PROGENITOR CELLS

The disruption of vascular integrity is a central component of diabetes, peripheral artery disease, coronary heart disease, and other ischemic conditions. Endothelial cells line the luminal surface of blood vessels, chiefly comprise capillaries of the microcirculation, and are important for vascular homeostasis. Modulating blood vessel repair, growth, and maintenance are therefore important for prevention and treatment of cardiovascular diseases (CVDs). Asahara and colleagues[48] first identified circulating cells with angiogenic potential in 1979, a discovery that led to the use of the term EPCs and later to the identification of other circulating and resident cell populations with angiogenic capacity.

4.1 The Struggle to Define EPCs—Overlap with HSCs

EPCs have been defined as bone marrow-derived cells with the capacity to home to sites of vascular injury and participate in endothelial repair and angiogenesis via differentiation into endothelial cells (Fig. 3); however, the precise cellular definition of the EPC remains unclear and highly controversial. This controversy led to identification of diverse cell types defined as "EPC," as reflected in the literature on EPC and physical activity. In addition, progenitor cells resident to the blood vessel wall are now considered

	CFU-Hill	CAC	ECFC
Clonal proliferative status	–	–	+
Replating ability	–	–	+
In vitro tube formation	+/–	+/–	+
In vivo de novo vessel formation	–	–	+
Homing to ischemic sites *in vivo*	+	+	+
Paracrine augmentation of angiogenesis	+	+	+
Phenotypic appearance	$CD34^{+/-}$ $CD133^{+}$ $VEGFR2^{+}$ $CD45^{+/-}$ $CD146^{+/-}$ $CD115^{+}$ $CD31^{+}$ $ALDH^{bright}$ acLDL uptake	$CD34^{+/-}$ $CD133^{+}$ $VEGFR2^{+}$ $CD45^{+/-}$ $CD146^{+/-}$ $CD115^{+}$ $CD31^{+}$ $ALDH^{bright}$ acLDL uptake	$CD34^{+/-}$ $CD133^{-}$ $VEGFR2^{+}$ $CD45^{-}$ $CD146^{+}$ $CD115^{-}$ $CD31^{+}$ $ALDH^{bright}$ acLDL uptake

Figure 3 Comparison of the phenotypic and functional characteristics of putative EPC determined by different assays. The cells isolated by the CFU-Hill and CAC assays identify essentially similar proangiogenic hematopoietic subsets. In contrast, the cells isolated in the ECFC assay differ from the proangiogenic cells by displaying clonal proliferative potential, replating ability, and *in vivo* vessel forming ability. In addition, the phenotype of the ECFC can be distinguished from the proangiogenic cells by the lack of CD133, CD45, and CD115 expression by the ECFC. Abbreviations: ALDH, aldehyde dehydrogenase; acLDL, acetylated low density lipoprotein; VEGFR2, vascular endothelial growth factor receptor 2. *Reprinted with permission from Basile and Yoder.*[49]

important for vessel growth and repair via differentiation into endothelial cells,[50,51] suggesting that our knowledge regarding EPC biology continues to evolve.

EPCs are defined with cell surface markers via flow cytometry, and/or characteristics from *in vitro* culture methods.[52] Asahara *et al.*[48] first defined putative endothelial progenitors as cells positive for expression of CD34 and vascular endothelial growth factor 2 receptor (i.e., VEGFR2, KDR for human or Flk-1 for rodents). This definition was widely accepted until Peichev *et al.*[53] added evidence to suggest the HSC marker AC133 constituted endothelial precursors as mature endothelial cells do not contain AC133 and $AC133^{+}VEGFR2^{+}$ cells colonized the surface of left ventricular assist devices in humans. Although later, use of CD133 to define EPCs became controversial when $CD34^{+}AC133^{+}KDR^{+}$ cells did not form capillary-like structures *in vitro* or integrate to human vascular lumen.[54,55] The relationship of EPCs to hematopoietic stem and progenitor cells is still disputed, yet many EPC populations coexpress hematopoietic markers including CD34, CD133, CD117, and CD105.

Three types of *in vitro* culture assays have been used to isolate EPCs and as a result let to the emergence of three distinct EPC populations: (1) endothelial colony-forming cells (ECFC or "late" EPC), (2) circulating angiogenic cells (CAC or "early EPC"), and (3) colony-forming unit-Hill (CFU-Hill or CFU-CAC). These populations display different phenotypic and functional characteristics (Fig. 3). ECFC form colonies and cobblestone appearance after extended time in culture (~2–4 weeks), have robust proliferative capacity, form cells that express mature endothelial surface markers such as $CD31^{+}$, and contribute directly to revascularization. CAC or early EPCs form colonies in culture after ~1 week, have a low capacity to self-replicate, and do not directly contribute to the formation of new vessels *in vitro* or *in vivo* but promote vascular repair by secretion of proangiogenic growth factors. Transcriptomic and proteomic analyses of the ECFC and early EPC revealed that early EPCs have a molecular phenotype similar to monocytes, whereas late EPCs had a phenotype similar to endothelial cells.[56] The third population, CFU-CAC or CFU-Hill CAC, were determined to be related to endothelial function and CVD risk.[57] These cells were subsequently identified as a mixed population of cells of hematopoietic lineage including T cells and monocytes with few endothelial progenitor, which contribute to angiogenesis via secretion of paracrine factors.[58,59]

As definitions of EPC populations have been refined, many cells originally defined as EPCs are no longer considered endothelial progenitors but share characteristics with HSCs, monocytes, and lymphocytes[58–61] and are now considered important regulators of angiogenesis.[62] In fact, it has been recently proposed that EPCs should be defined according to whether they are hematopoietic or nonhematopoietic lineage EPCs.[63] This categorization scheme reconciles observations that EPCs with and without hematopoietic markers contribute to vascular maintenance, repair, and angiogenesis: some that directly form vessels through differentiation to endothelial cells and others that support endothelial health and vascular growth via the secretion of soluble factors such as growth factors and cytokines but do not directly differentiate into endothelial cells.[15] Importantly, hematopoietic and nonhematopoietic "EPCs" have been associated with CVD such that low appearance of EPCs in circulation and EPC dysfunction are related to elevated CVD risk and poor outcomes.[64–71]

4.2 Exercise and EPCs

Exercise can restore endothelial function and decrease CVD risk. The maintenance of endothelial integrity and function is paramount as CVD risk is strongly influenced by endothelial function.[72] Increased EPC appearance in circulation and EPC functional improvements via exercise are hypothesized to constitute novel mechanisms by which physical activity promotes cardiovascular health (Fig. 4). However, the controversies in EPC definition described above are reflected in the literature on exercise and many published reports describing EPCs may in fact be HSCs or other CACs as very few studies in the exercise literature have included experiments confirming that the cells described differentiate into endothelial cells and integrate into vessels or contribute to endothelial homeostasis via the secretion of paracrine factors. Therefore, the term CAC will be used to collectively refer to the EPC populations and will be specifically defined where appropriate in sections 4.3 and 4.4 on physical activity. As the definition of EPCs evolves, a reevaluation of the literature on exercise will be necessary. The effect of acute exercise and exercise training on CACs has been recently reviewed elsewhere.[73–75] The focus of the following is a review of the effect of exercise on CACs, highlighting studies that lend novel insight into the CAC response to exercise or potential mechanisms to explain the role of exercise to change CAC characteristics.

4.3 Effect of Acute Exercise on EPCs

In general, acute exercise is associated with an increase in appearance of CACs in the blood of healthy volunteers,[37] patients with CVD risk

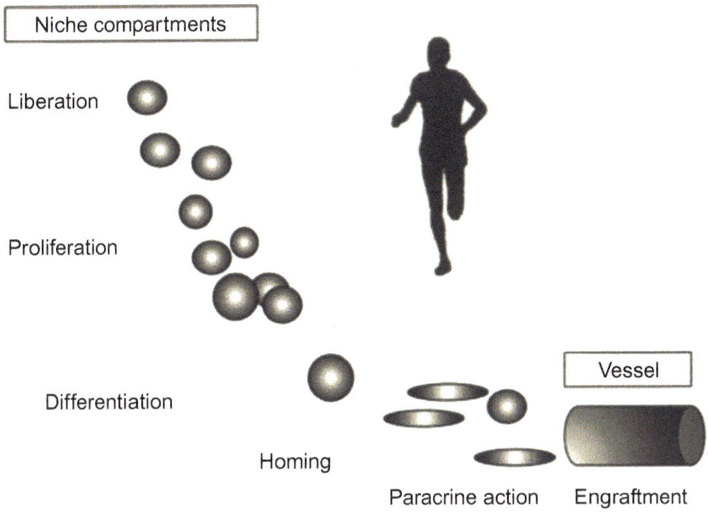

Figure 4 The efficacy of CAC to support vascular growth and regeneration is dependent on the successful participation in several functions. Exercise has the potential to affect each separately, in combination, or all of these functions. Characterization of the mechanisms by which exercise may influence CAC function may help optimize cardiovascular regenerative medicine. *Reprinted with permission from Witkowski* et al.[73]

factors,[74] and patients with established CVD[75] although the acute response varies according to CAC subtype, and exercise type, duration, and intensity as well as disease status (Table 2).[43,75] With a standard acute bout of moderate to intense exercise (30–45 min), maximal, or symptom-limited treadmill or bicycle tests, CACs are mobilized quickly, demonstrating significant increases from rest within approximately 10–30 min following exercise completion and returning to resting levels within 24 h. In patients with heart failure, the mobilization of $CD34^+KDR^+$ was smaller and returned to baseline more quickly than healthy subjects.[79] Under ischemic conditions, the CAC response to acute exercise may be delayed. One study used blood flow restricted acute treadmill exercise (via tourniquet) in healthy individuals and patients with peripheral artery occlusive disease and exercise-induced ischemia, to show that under artificial and diseased limb ischemia, $CD34^+KDR^+$ and $CD133^+KDR^+$ CACs peaked 24 h postexercise and returned to baseline within 72 h in both groups.[78] In this study, the healthy participants increased CACs fourfold compared with threefold in the patients. It is unclear whether the magnitude of the response varies with ischemic versus nonischemic exercise. Longer duration exercise also leads to a different CAC response. For example, with 240 min of high-intensity cycling exercise, many CACs including $CD34^+$, $CD34^+KDR^+$, $CD133^+KDR^+$, and

Table 2 EPC Response to Exercise

Cell Type	Response to Exercise	References
Acute exercise		
$CD34^+/KDR^+$ $AC133^+/KDR^+$	Population: healthy men Result: increased progressively over 240′ of cycling	Mobius-Winkler *et al.*[37]
$AC133^+/CD144^+$	Population: men with CVD risk Result: increased ~fourfold with symptom-limited treadmill or bicycle test	Rehman *et al.*[76]
$CD34^+/KDR^+$	Population: three groups (1) patients with exercise-induced ischemia, (2) no exercise-induced ischemia, and (3) healthy Result: increased 24 and 48 h following symptom-limited bicycle test only in exercise-induced ischemia group	Adams *et al.*[77]
$CD34^+/KDR^+$ $AC133^+/KDR^+$	Population: male and female, two groups (1) PAOD patients with exercise-induced limb ischemia and (2) healthy individuals—ischemic exercise (tourniquet) Result: both groups increased 12 and 24 h with ischemic treadmill exercise. Healthy ischemic group had greater response	Sandri *et al.*[78]
$CD34^+/KDR^+$	Population: CHF and healthy individuals Result: cells increased in healthy subjects within 10 min following exercise test and returned to baseline within 2 h, response in CHF patients was attenuated	Van Craenenbroeck *et al.*[79]
$CD34^+/KDR^+$ $CD34^+$ "early" EPC	Population: mild and severe CHF patients and healthy participants Result: Cardiopulmonary exercise test resulted in increased CAC migration in CHF groups only. No change in $CD34^+/KDR^+$ or $CD34^+$ cells with exercise	Van Craenenbroeck *et al.*[80]

Table 2 EPC Response to Exercise—cont'd

Cell Type	Response to Exercise	References
CFU-CAC	Population: healthy active and sedentary men Result: CFU-CAC increased with acute treadmill exercise in active men only	Jenkins *et al.*[81]
Exercise training		
$CD34^+/KDR^+$ CFU-CAC	Population: stable CHF Training: 8 weeks, 3×/week, 45 min cycling 60% HRR Result: $CD34^+/KDR^+$ and CFU-CAC increased. After 8 weeks discontinued training $CD34^+/KDR^+$ decreased CFU-CAC remained elevated	Sarto *et al.*[82]
$CD133^+/CD34^+/KDR^+$	Population: CVD or CVD risk factors Training: 12 weeks, 3×/week, 30–60 min running Result: threefold increase in CAC	Steiner *et al.*[91]
$CD34^+/KDR^+$	Population: stable CAD Training: 4 weeks, cycling 60–80% VO_2max, 20 min, moderate muscle strength training, and walking Result: 78±34% increase in cell number, 41±11% reduction in EPC apoptosis	Laufs *et al.*[83]
$CD34^+/KDR^+$ "early" EPC	Population: CHF (EF ≤40%), control group, and healthy group Training: 6 months, 3 days/week, 60 min Result: CHF group increased early EPC migration and $CD34^+/KDR^+$	Van Craenenbroeck *et al.*[84]
$CD34^+/KDR^+$	Population: young and older men Training: 8 weeks older men only, cross-sectional for training status in other groups Result: older men had lower resting and exercise-induced EPC, and training status did not alter baseline or exercise-induced EPC	Thijssen *et al.*[34]

$CD133^+$ progressively increased in the blood with continued exercise and decreased within 1–2 h postexercise.[37] In the latter study, $CD146^+$ circulating endothelial cells, a marker associated with damaged endothelium, also increased during cycling and progressively declined after exercise in a fashion similar to the CAC response. It is possible that the mobilization of CACs assists the replacement of normal endothelial cell turnover that results from enhanced shear stress as seen with acute, particularly long-duration exercise. However, there are currently no mechanistic studies evaluating this hypothesis.

Due to overlapping definitions of EPC and hematopoietic progenitors in the literature, the exercise-induced responses of $CD45^+CD34^+$ hematopoietic progenitors and $CD45^-CD34^+KDR^+$ endothelial progenitors were recently compared.[85] Differences in the kinetics and magnitude of appearance of the two populations were revealed, which may explain some of the discrepancies in the literature that stem from the presence of hematopoietic markers such as CD45 on cell populations defined as EPCs. In the same study, the authors also reported cell-type differences in the time course of response to three forms of exercise: concentric endurance exercise, resistance exercise, and eccentric exercise. Concentric endurance exercise mobilized HPCs to a greater degree than EPCs although both cell populations were not significantly different from resting levels by 24 h. Interestingly, when all exercise types were combined, the HPC, but not EPC response, was significantly associated with creatine kinase, a marker of muscle damage. The HPC response is likely part of the normal regenerative response to muscle damage, whereas EPCs may not play a role in this process.

Acute exercise studies have aided in the discovery of events that mobilize CAC. The mechanisms responsible for the increase in CAC with exercise include changes in blood flow in bones, ischemia, signaling proteins, and angiogenic cytokines. Endothelial nitric oxide synthase (eNOS) is the enzyme responsible for the production of the vasodilator nitric oxide (NO) by endothelial cells. Increased shear stress that accompanies greater blood flow with exercise is known to lead to the release of NO by endothelial cells causing smooth muscle cell relaxation and blood vessel dilation.[86] The NO system is also a main regulator of CAC biology. eNOS is an essential component of the bone marrow stromal microenvironment and mobilization of CAC through its regulation of matrix metalloproteinase (MMP)-9.[87] Cubborn *et al.*[88] reported that mobilization of $CD34^+KDR^+$, $CD133^+CD34^+KDR^+$, and $CD34^+CD45^-$ CAC in response to 30 min of high-intensity cycling was blocked with administration of the eNOS inhibitor L-NMMA, indicating that eNOS is likely the main mechanism responsible for exercise-induced CAC mobilization. Data from a mouse model corroborate the importance of eNOS wherein eNOS knockout mice ($NOS3^{-/-}$) and wild-type mice that were delivered the eNOS inhibitor (L-NAME) had lower $Sca\text{-}1^{+}FLK\text{-}1^{+}$ CAC after only 7 days of exercise training.[83]

Ischemia mobilizes CACs from the bone marrow in patient populations but does not appear to be a mechanism of mobilization in healthy individuals during exercise. Adams *et al.*[77] quantified the response of $CD34^{+}/KDR^{+}$ CACs to an acute bout of maximal exercise in three groups of patients: (1) CAD patients with exercise-induced ischemia, (2) asymptomatic CAD patients with no exercise-induced ischemia, and (3) healthy age-matched controls. In response to a symptom-limited exercise test, CACs increased only in CAD patients with exercise-induced myocardial ischemia and the increases were correlated with the changes in plasma vascular endothelial growth factor (VEGF) levels. Others have confirmed an association between VEGF and CAC mobilization with ischemic exercise.[31] Acute exercise may also be associated with the release of cytokines and growth factors that mobilize CAC. Granulocyte colony-stimulating factor (G-CSF), a cytokine produced by various tissues, induces the production and release of progenitors from the bone marrow, has been used to mobilize progenitors for use in stem cell therapy, and is associated with HPC and EPC mobilization with acute exercise.[85]

CACs from patients with disease are not only reduced in circulation but are dysfunctional compared with those from healthy individuals (i.e., they show increased senescence and apoptosis with decreased migratory capacity), which may contribute to poor vascular outcomes and prognosis. Studies with acute exercise in patients with disease reveal that exercise can improve CAC function. For example, acute exercise improved CAC migration in severe and mild CHF patients to a level similar to that evident in healthy controls,[80] demonstrating that as little as one bout of exercise can influence CAC intracellular function. CAC function with acute exercise is influenced by intracellular NO production, which may be one of the mechanisms of CAC dysfunction with disease and chronic inactivity.[89] Jenkins *et al.*[81] demonstrated that, in sedentary men, acute exercise increased CAC intracellular NO (NOi), which was in part due to reduced NADPH oxidase, the main enzyme that produces oxidative stress in endothelial cells.

4.4 Effect of Exercise Training on EPCs

The effect of exercise training in patients with CVD (heart failure, coronary artery disease, and peripheral artery disease) on CAC has been systematically reviewed.[75] Although the nature of the exercise programs of the studies varied, most showed significant increases in resting CACs (defined mainly as $CD133^+CD34^+KDR^+$ or $CD34^+KDR^+$) with training. Most of the programs involved aerobic exercise (70–85% of max heart rate or 60–70% $\dot{V}O_2max$) and in many, but not all, blood was drawn at least 24 h following the final exercise session to eliminate the effect of acute exercise. In patients with heart failure, Sarto *et al.*[82] reported a significant increase in CAC number and CAC function with 8 weeks of aerobic training. CAC number returned to baseline values 8 weeks after discontinuation of aerobic training demonstrating that changes in physical activity level are related to CACs in this population. These data support the hypothesis that CACs may contribute to improvements in cardiovascular health observed in this patient population with exercise regimens. However, additional studies are necessary to verify these results and establish an exercise prescription necessary for positive outcomes.

Exercise training appears to improve CAC more consistently in patient populations than healthy individuals, as both training studies[34] and cross-sectional comparisons[43] in healthy participants have reported no effect of exercise on CACs. Since exercise has a wide range of effects, it is likely that the overall benefits of exercise for patients with disease may be both direct (i.e., on CACs) and indirect (i.e., improvement of individual CVD risk factors associated with CAC impairment). Further, many studies in healthy populations include young active individuals with low CVD risk and the absence of CAC improvements with training may be due to a CAC system that is already optimal. In heart failure patients, Van Craenenbroeck *et al.*[84] reported that the improvement in CAC function with acute exercise prior to training was absent following training, indicating that 6 months of training may have been sufficient to elicit sustained CAC improvements in these patients to levels comparable to healthy controls. A similar finding was reported in a comparison of regularly active and inactive young but healthy men where CAC NOi increased with an acute bout of exercise in the inactive but not active men.[81] However, in this study, the inactive group did not improve CAC NOi to the level of the regularly active men. The repeated acute exercise bouts that accumulate with training are important for resting CAC number and function. When highly active older men with a long-term history of endurance exercise stopped training for 10 days, the number of CACs fell to the level of sedentary men of the same age.[43] In younger active men, 10 days of reduced physical activity resulted in diminished CAC colony-forming capacity and CAC NOi.[90] Overall, these studies demonstrate flexibility of CACs, along with increased quantity with repeated bouts of activity in both healthy and diseased populations and reversal with detraining.

One of the most important questions is whether exercise improves the capacity of CAC to home to and either incorporate into existing endothelium or release factors that assist angiogenesis and endothelialization. Indirect evidence is available that demonstrates an association between CACs and improved vascular function with exercise training. Steiner *et al.*[91] found that 12 weeks of exercise training in patients with asymptomatic CAD and/or CVD risk factors resulted in an approximately threefold increase in $CD133^+CD34^+KDR^+$ CACs, an increase that was correlated with an enhanced flow-mediated dilation and plasma NO synthesis. In chronic heart failure patients, exercise training increased $CD34^+KDR^+$ CAC and CAC migratory activity as well as flow-mediated dilation.[84] In healthy highly trained older men, the reduction in $CD34^+KDR^+$ cells was related to the reduction in forearm blood flow response with 10 days of detraining.[43] Animal models have provided evidence for the improvement of angiogenesis with exercise. Using a disk model of neoangiogenesis in mice, exercise enhanced CACs and new vessel formation compared to sedentary animals.[83] In aged mice, swim training improved the capillary/myofibril ratio, recovery of blood flow, and CAC homing to neovascularization following hindlimb ischemia compared with nontrained mice.[92]

Early EPCs and CFU-CACs (CFU-Hill) are two populations hypothesized to participate in angiogenesis and reendothelialization via angiogenic growth factor and cytokine secretion. Pula *et al.*[93] used a proteomics approach to evaluate the secretome of early EPC and CFU-CAC. They identified peptides known for their angiogenic potential[94,95] and novel peptides not previously recognized as CAC derived including interleukin (IL)-8, MMP-9, monocyte chemotactic protein-1, pre-B cell-enhancing factor, S100 protein family, and thymidine phosphorylase. Currently, mechanistic studies evaluating the effect of exercise on the release of angiogenic factors related to angiogenesis and endothelialization are lacking. This is an important area for future research in order to optimize cell-based therapies for disease.

The contribution of CACs to the maintenance and repair of the vasculature has been investigated for the past 20 years. It has become clear that physical activity greatly influences mobilization and function of CAC populations, particularly in patients with CVD risk or disease, and they may be important mechanisms by which exercise promotes vascular health. There are many unresolved questions regarding CAC biology and the role of physical activity. For example, as CVDs continue to be highly prevalent and level of activity declines with age while the older segment of the population grows, it is particularly important to evaluate the effect of age on CAC and whether the benefits of physical activity on CAC persist with aging and age-related conditions (i.e., menopause). Women are especially not well represented in the current body of literature. Additionally, the optimal exercise prescription to improve CAC characteristics for various populations is still unclear. Finally, it will be important to delineate cell function *in vivo* using novel evaluative tools that are yet to be created.

5. MESENCHYMAL STEM CELLS

MSCs, also currently referred to as multipotent stromal cells, originate from embryonic connective tissue, or mesenchyme, and give rise to hematopoietic and connective tissue in the developing fetus. The MSC was first identified in bone marrow in the adult organism by Friedenstein and colleagues.[96] The cells were identified as fibroblast-like cells with colony-forming capacity that demonstrate osteogenic potential both *in vitro* and following heterotopic transplantation. The ability for these fibroblast-like cells to give rise to multiple mesodermal tissues, such as bone, adipose, and cartilage, has been verified for decades.[97] However, similar to HSCs and EPCs, confusion exists regarding the homogeneity of the population isolated from bone marrow and agreement on the procedures used for identification of MSCs from a variety of tissues.[98–101] A single, unique biomarker does not exist for identification and numerous environmental cues can rapidly alter the transcriptome and cell surface protein expression *in vitro* and *in vivo*. In addition, MSCs have long been recognized to secrete factors, such as IL-1, IL-6, colony-stimulating factor (CSF-1), and granulocyte macrophage-CSF (GM-CSF) that sustain HSC survival and function in bone marrow.[102] The observation that MSCs synthesize and secrete numerous growth factors, cytokines, and extracellular matrix (ECM) proteins in the context of multiple tissue types suggests a predominant supportive or stromal role in repair, growth, and remodeling. The dynamic nature of this cell type, both with regard to identification and function, presents a challenge that will need to be addressed with sophisticated lineage tracing studies. At present, MSCs are identified and isolated from multiple tissues based on one or more of the following criteria: (1) cell surface expression of CD73, CD90, and CD105 and absence of hematopoietic markers CD14 and CD45; (2) evidence of clonal expansion; and (3) capacity for differentiation into multiple cell types, including bone, cartilage, adipose, and tendon.[98]

5.1 Effect of Acute Exercise and Exercise Training on MSCs

MSCs have been identified in peripheral blood and localized to the vascular niche in numerous adult human tissues, including bone marrow, adipose, skeletal muscle, lung, heart, cartilage, tendon, and dental pulp.[3] MSCs represent a very small fraction of the total mononuclear cell in bone marrow (0.001–0.01%).[103] Current studies have investigated the effect of injury on MSC migration, proliferation, and function, yet to our knowledge, few studies exist which have evaluated the MSC response to either acute exercise or exercise training. Ocarino and colleagues[104] isolated MSCs from bone marrow (~90% $CD45^{-}CD73^{+}CD90^{+}$) of healthy and osteopenic rats following 3 months of moderate intensity treadmill running (15 m/min, 30 min/day, 5 days/week). Physical activity restored MSC capacity for osteoblast differentiation in osteopenic rats as detected by alkaline phosphatase expression and formation of mineralized nodules. Bone marrow-derived MSCs from adult (6 months) rats also exhibit greater ALP expression to an extent comparable to young (1 month) rats following 7, 14, and 21 days of treadmill exercise.[105] Physical exercise can also increase bone formation rate and decrease marrow adipocyte volume in the proximal tibia metaphysis following 5 weeks of progressive treadmill running exercise, likely due to suppression of MSC adipogenesis as a result of mechanical loading as demonstrated *in vitro*.[104,106] In a similar study, 10 weeks of progressive treadmill running (3 days/week) reduced marrow cavity adipose content by 78% in mice.[40] Despite these intriguing results, studies that evaluate the tissue-specific MSC response to exercise do not exist outside of skeletal muscle. Therefore, the remaining section on MSCs will focus on current knowledge of the effect of exercise on skeletal muscle resident MSCs and their subsequent influence on satellite cells.

5.1.1 *The Resident Progenitor Cell in Skeletal Muscle and Its Response to Exercise*

Cellular remodeling, including destruction and regrowth, is a predominant outcome of acute or repeated bouts of exercise in multiple tissues. Identified in 1961 by Mauro,[107] the satellite cell is the predominant resident progenitor cell in skeletal muscle responsible for repair and/or regeneration of the myofiber in response to injury. The quiescent $Pax7^+$ satellite cell residing in the interstitial space between the sarcolemma and the basal lamina can become "activated" upon injury, an event that includes upregulation of myogenic transcription factors *Myf5* and *MyoD*, migration to the site of injury, and asymmetric cell division (Fig. 5). Asymmetric cell division allows the satellite cell to give rise to one daughter cell that will return to quiescence and one daughter cell that will express *myogenin* and terminally differentiate into a myoblast with capacity for myofiber fusion (repair) or new fiber synthesis (regeneration).[5] Verification that the satellite cell is a *bona fide* stem cell

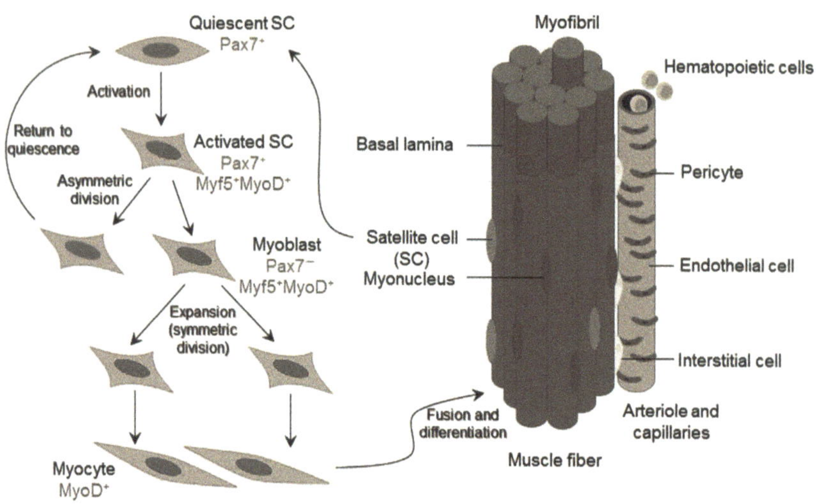

Figure 5 Skeletal muscle resident stem cells. Satellite cells ($Pax7^+$) reside in the space between the sarcolemma and the basal lamina and are required for repair and regeneration of the myofiber following injury. Satellite cell activation in response to injury includes upregulation of the myogenic regulatory factors Myf5 and MyoD and asymmetric cell division. Newly formed myoblasts or myocytes can fuse with damaged fibers or fuse together to form new fibers in the muscle environment. $Pax7^-$ (nonsatellite) stem cells (interstitial cells) and pericytes are located outside the basal lamina in close proximity to vessels. The majority of nonsatellite stem cells in muscle contribute to repair via secretion of factors that allow for satellite cell activation. *Reprinted with permission from Dr. Péter Balogh, University of Pecs, Hungary.* (See the color plate.)

has been demonstrated, as a single luciferase-expressing $Pax7^+$ satellite cell can give rise to 20,000–80,000 progeny upon transplantation.[108] Studies that allow for the conditional ablation of the $Pax7^+$ satellite cell have clearly delineated an important role for the satellite cell in muscle repair and regeneration in response to injury following myotoxin exposure or mechanical overload.[109,110] The satellite cell indeed exhibits limited potential for differentiation aside from myogenesis,[111] and nearly all studies to date have focused on the satellite cell in the context of muscle repair and adaptation to injury.

$Pax7^+$ satellite cells are increased in skeletal muscle following acute and repeated bouts of strength training, with preferential enhancement observed in the niche surrounding Type II fibers, the fiber type most susceptible to contraction-induced injury and age-related atrophy. Type II fiber-associated enhancement of satellite cell content and activation (as reflected by DLK1 expression) has been observed in young recreationally active men 1–3 days following an acute bout of maximal unilateral isolated eccentric exercise or traditional resistance-type exercise.[112–115] Resistance exercise training can also increase satellite cell content in young and aged men and women,[113,116–119] and the increase has been demonstrated to be specific to type II fibers in elderly men and correlated with fiber size or muscle volume in young and aged men.[120,121] The strong relationship between satellite cell content and muscle fiber size postexercise would suggest a role for the satellite cell in growth, yet myonuclear accretion and/or new fiber synthesis are not consistently observed posttraining.[117,118,120,122,123] Conditional ablation of the $Pax7^+$ cell using genetically altered mice ($Pax7^{iCE/+}$: $R26R^{DTA/+}$) also does not compromise growth in response to chronic loading in mice,[110] suggesting that the satellite cell may fulfill an alternative or additional role in muscle postexercise. Satellite cells are enriched in the motor endplate and around capillaries,[124,125] yet the extent to which they protect or repair these structures postexercise is not known. The $Pax7^+$ satellite cell also can suppress fibroblast-mediated collagen deposition.[126] Thus, the satellite cell may regulate cell–cell communication in muscle and its role postexercise may extend beyond myofiber repair and regeneration.

Skeletal muscle fiber injury associated with high-intensity or eccentric (lengthening) contractions and the subsequent requirement for repair are hypothesized to provide the underlying basis for satellite cell activation and differentiation postexercise. However, satellite cell accumulation has been observed postexercise in the absence of injury and/or inflammation[117,127,128] and current studies demonstrate that less intense contractions associated with endurance exercise training may be sufficient for satellite cell activation in rodents and humans.[122,123,129] Multiple intrinsic and extrinsic factors, including growth factors, mechanical strain, hypoxia, inflammation, ECM composition, and topography, are altered by exercise and ultimately dictate the satellite cell response to muscle contraction. The majority of studies that evaluate satellite cell activation and differentiation are conducted either *in vitro* or in response to stimuli that do not mimic the exercise condition, such as myotoxin injection or injection of isolated growth factors in mice. Despite this limitation, it is clear that satellite cells proliferate in response to a wide variety of growth factors, cytokines, and hormones in sterile culture conditions that are also present in muscle during exercise, including fibroblast growth factor, hepatocyte growth factor, insulin-like growth factor (IGF-I), IGF-II, transforming growth factor-β, leukemia inhibitory factor, interleukin-6 (IL-6), platelet-derived growth factor (PDGF), epidermal growth factor, tumor necrosis factor-α, and testosterone.[130,131] While satellite cell expression of myostatin (a satellite cell cycle inhibitor) is decreased at 12 h, IGF-1 and phosphorylated STAT3 (a target of IL-6 receptor activation) are transiently localized to satellite cells 24 h following an acute bout of eccentric or resistance exercise in young men.[115,132,133] Guerci *et al.*[134] demonstrated that serum response factor, a ubiquitous transcription factor expressed in muscle, can regulate the synthesis of IL-6, an event necessary for satellite cell proliferation in response to mechanical load in rodents. Together, these studies suggest an important role for the paracrine factor milieu in satellite cell activation postexercise.

5.1.2 Mesenchymal-Like Stem Cells in Skeletal Muscle and Their Response to Exercise

Muscle-derived stem cells, SP cells, mesenchymal progenitor cells, pericytes, fibro/adipogenic progenitor cells (FAPs), and $PW1^+$ cells have been identified and isolated from skeletal muscle and further characterized as $Pax7^-$ (nonsatellite) stem cells with multilineage capacity (Fig. 5).[135–141] A review of the history and characterization of each cell type has been previously presented, with particular emphasis on the response of each type to injury as a result of myotoxin injection.[142] Briefly, Gussoni *et al.*[143] identified the first $Pax7^-$ stem cell in skeletal muscle as an SP cell (based on Hoechst 33342 dye exclusion) that could reconstitute the entire hematopoietic system in lethally irradiated mice. While only the $CD45^+$ subfraction was found capable of hematopoiesis, the $CD45^-$ subfraction did not express hematopoietic lineage markers and demonstrated myogenic capacity following transplantation into regenerating tibialis anterior muscle.[143,144] Enthusiasm for the discovery of a nonsatellite stem cell with myogenic capacity was diminished when engraftment was found to be limited. However, Motohashi and colleagues[136] later characterized a $CD45^-CD31^-$ SP cell in muscle that could synthesize regenerative growth factors which indirectly enhance satellite cell expansion and fusion postinjury. The majority of studies completed to date suggest the existence of a multipotent, nonmyogenic $Pax7^-Sca\text{-}1^+CD45^-$ $CD31^-CD34^{+/-}$, $PDGFR\alpha^+$ population in muscle that can proliferate in response to injury and provide stromal support that allows for satellite cell activation.[136,137,139,140,145] Expression of PDGFRα appears to be a key cell surface marker that determines myogenic potential, with $PDGFR\alpha^-$ nonsatellite cells exhibiting some potential for myogenesis and $PDGFR\alpha^+$ nonsatellite cells demonstrating MSC-like stromal capacity and indirect capacity for muscle repair.[139,145,146]

The extent to which $Sca\text{-}1^+CD45^-$ nonsatellite stem cells can become enriched and contribute to muscle repair following eccentric exercise was recently addressed in a series of studies in mice. Mice were subjected to a single bout of eccentric exercise (20° decline, 17 m/min, and 30 min) and flow cytometry was used to evaluate the percentage of $Sca\text{-}1^+CD45^-$ cells.[141] $Sca\text{-}1^+CD45^-$ cells increased twofold at 24 h postexercise and cells were localized to large vessels and nerves in the interstitium. Following isolation from muscle and short-term culture, the cells were characterized by flow cytometry and found to express both MSC (CD29, CD73, CD90, and CD105) and pericyte markers (NG2, CD146, and PDGFRβ) and were negative for endothelial cell markers (CD31 and CD34). In addition, $Sca\text{-}1^+CD45^-$ cells were fibroblastic in morphology, adherent to plastic, and exhibited multilineage potential, including capacity for differentiation into osteoblasts, chondrocytes, and adipocytes. The predominance of MSC markers and differentiation capacity lead to the use of the term "mMSC" to denote the isolation of stem cell from muscle with MSC properties. mMSCs are most likely FAPs or type 1 pericytes previously described.[137,146] Interestingly, transplantation of young (5 week) muscle-derived mMSCs (1×10^4) into recipient muscle 1 h following an acute bout of exercise (protocol described above) significantly increased satellite cell number and new fiber synthesis[141] In addition, transplantation increased myonuclear accumulation, myofiber growth, and strength following repeated bouts of eccentric exercise (3×/week, 4 weeks) in adult mice.[147] In both studies, mMSCs did not express Pax7, form new fibers, or fuse with existing fibers. The fact that transplanted cells localized to areas of CLN^+ fibers and vessels and contributed to vessel remodeling following an acute bout of exercise[148] suggests the ability of mMSCs to repair multiple tissues via release of local factors. Finally, mMSCs are highly sensitive to mechanical strain and release a wide variety of growth and neurotrophic factors in a substrate-dependent manner that likely dictates their functional capacity and the ability to stimulate satellite cells throughout the lifespan.[141,148,149]

The human homologue for Sca-1 has not been determined, thus limiting the ability to use this marker for identification of mMSCs in samples of human muscle. The increase in NF-κB expression in $NG2^{+}$ and ALP^{+} pericytes 3 h posteccentric exercise suggests a role for the pericyte in the early adaptive response to exercise in human skeletal muscle.[150] However, it is clear that studies have not fully delineated the pericyte response to acute exercise or exercise training. Future studies will need to optimize immunohistochemistry and flow cytometry methods to identify pericyte (NG2 and ALP), as well as traditional MSC markers (CD90, PDGFRα), for identification of the interstitial multipotent stem cells in adult skeletal muscle. Given the hypothesis that pericytes become detached from vessels in response to injury and express MSC markers, it will be important to evaluate the potential for pericyte–MSC transition using a combination of markers over an extensive time course postexercise. In addition, consideration should be given to the nonsatellite stem cell response to different types of exercise (endurance, resistance, isolated concentric/eccentric contractions, and flexibility), age of individuals, and disease status.

6. CONCLUSION

HSC, EPC (CAC), and MSC quantity and function are critical for tissue and whole body health. The extent to which these stem cells change in response to exercise and are involved in the commonly recognized beneficial adaptations to exercise is not fully understood. The field of stem cell biology is still in its infancy and many questions remain that will be addressed pending the development of new tools necessary for identification and examination of ASCs. Such information may provide the impetus for future stem cell-based therapies perhaps offering opportunities to optimally repair tissues, resist disease, and improve healthspan.

REFERENCES

1. O'Connor TP, Crystal RG. Genetic medicines: treatment strategies for hereditary disorders. *Nat Rev Genet*. 2006;7:261–276.
2. Ratajczak MZ. A novel view of the adult bone marrow stem cell hierarchy and stem cell trafficking. *Leukemia*. 2015;29:776–782.

3. Hass R, Kasper C, Bohm S, Jacobs R. Different populations and sources of human mesenchymal stem cells (MSC): a comparison of adult and neonatal tissue-derived MSC. *Cell Commun Signal*. 2011;9:12.
4. Timmermans F, Plum J, Yoder MC, Ingram DA, Vandkerckhove B, Case J. Endothelial progenitor cells: identity defined? *J Cell Mol Med*. 2009;13:87–102.
5. Charge SB, Rudnicki MA. Cellular and molecular regulation of muscle regeneration. *Physiol Rev*. 2004;84:209–238.
6. Bentzinger CF, Wang YX, von Maltzahn J, Rudnicki MA. The emerging biology of muscle stem cells: implications for cell-based therapies. *Bioessays*. 2012;35:231–241.
7. Li L, Clevers H. Coexistence of quiescent and active adult stem cells in mammals. *Science*. 2010;327:542–545.
8. Seib DR, Martin-Villalba A. Neurogenesis in the normal ageing hippocampus: a mini-review. *Gerontology*. 2015;61:327–335.
9. Barker N, van Es JH, Kuipers J, et al. Identification of stem cells in small intestine and colon by marker gene *Lgr5*. *Science*. 2007;449:1003–1007.
10. Kretzschmar K, Watt FM. Markers of epidermal stem cell subpopulations in adult mammalian skin. *Cold Spring Harb Perspect Med*. 2014;4:10.
11. Shin S, Kaestner KH. The origin, biology, and therapeutic potential of facultative adult hepatic progenitor cells. *Curr Top Dev Biol*. 2014;107:269–292.
12. Leri A, Rota M, Hosoda T, Goichberg P, Anversa P. Cardiac stem cell niches. *Stem Cell Res*. 2014;13:631–646.
13. Wansleeben C, Barkauskas CE, Rock JR, Hogan BL. Stem cells of the adult lung: their development and role in homeostasis, regeneration, and disease. *Wiley Interdiscip Rev Dev Biol*. 2013;2:131–148.
14. Till JE, McCulloch EA. A direct measurement of the radiation sensitivity of normal mouse bone marrow cells. *Radiat Res*. 1961;14:213–222.
15. Bhatia M, Wang JC, Kapp U, Bonnet D, Dick JE. Purification of primitive human hematopoietic cells capable of repopulating immune-deficient mice. *Proc Natl Acad Sci USA*. 1997;94:5320–5325.
16. McGovern Jr JJ, Russell PS, Atkins L, Webster EW. Treatment of terminal leukemic relapse by total-body irradiation and intravenous infusion of stored autologous bone marrow obtained during remission. *N Engl J Med*. 1959;260:675–683.
17. Thomas ED, Lochte Jr HL, Cannon JH, Sahler OD, Ferrebee JW. Supralethal whole body irradiation and isologous marrow transplantation in man. *J Clin Invest*. 1959;38:1709–1716.
18. Purton LE, Scadden DT. Limiting factors in murine hematopoietic stem cell assays. *Cell Stem Cell*. 2007;1:263–270.
19. Dick JE, Bhatia M, Gan O, Kapp U, Wang JC. Assay of human stem cells by repopulation of NOD/SCID mice. *Stem Cells*. 1997;15(Suppl 1):199–203. discussion 204–7.
20. Ema H, Morita Y, Yamazaki S, et al. Adult mouse hematopoietic stem cells: purification and single-cell assays. *Nat Protoc*. 2007;1:2979–2987.
21. Kiel MJ, Yilmaz ÖH, Iwashita T, et al. SLAM family receptors distinguish hematopoietic stem and progenitor cells and reveal endothelial niches for stem cells. *Cell*. 2005;121:1109–1121.
22. Oguro H, Ding L, Morrison SJ. SLAM family markers resolve functionally distinct subpopulations of hematopoietic stem cells and multipotent progenitors. *Cell Stem Cell*. 2013;13:102–116.
23. Sutherland DR, Anderson L, Keeney M, Nayar R, Chin-Yee I. The ISHAGE guidelines for CD34+ cell determination by flow cytometry. International Society of Hematotherapy and Graft Engineering. *J Hematother*. 1996;5:213–226.
24. Challen GA, Boles N, Lin K-YK, Goodell MA. Mouse hematopoietic stem cell identification and analysis. *Cytometry A*. 2009;75A:14–24.

25. Wardyn GG, Rennard SI, Brusnahan SK, et al. Effects of exercise on hematological parameters, circulating side population cells, and cytokines. *Exp Hematol.* 2008;36:216–223.
26. De Lisio M, Parise G. Exercise and hematopoietic stem and progenitor cells: protection, quantity, and function. *Exerc Sport Sci Rev.* 2013;1(2):116–122.
27. Barrett AJ, Longhurst P, Sneath P, Watson JG. Mobilization of CFU-C by exercise and ACTH induced stress in man. *Exp Hematol.* 1978;6:590–594.
28. Heal JM, Brightman A. Exercise and circulating hematopoietic progenitor cells (CFU-GM) in humans. *Transfusion (Paris).* 1987;27:155–158.
29. Jootar S, Chaisiripoomkere W, Thaikla O, Kaewborworn M. Effect of running exercise on hematological changes, hematopoietic progenitor cells (CFU-GM) and fibrinolytic system in humans. *J Med Assoc Thai.* 1992;75:94–98.
30. Morici G, Zangla D, Santoro A, et al. Supramaximal exercise mobilizes hematopoietic progenitors and reticulocytes in athletes. *Am J Physiol Regul Integr Comp Physiol.* 2005;289:R1496–R1503.
31. Sandri M, Adams V, Gielen S, et al. Effects of exercise and ischemia on mobilization and functional activation of blood-derived progenitor cells in patients with ischemic syndromes: results of 3 randomized studies. *Circulation.* 2005;111:3391–3399.
32. Bonsignore MR, Morici G, Riccioni R, et al. Hemopoietic and angiogenetic progenitors in healthy athletes: different responses to endurance and maximal exercise. *J Appl Physiol.* 2010;109:60–67.
33. Kroepfl JM, Pekovits K, Stelzer I, et al. Exercise increases the frequency of circulating hematopoietic progenitor cells, but reduces hematopoietic colony-forming capacity. *Stem Cells Dev.* 2012;21:2915–2925.
34. Thijssen DHJ, Vos JB, Verseyden C, et al. Haematopoietic stem cells and endothelial progenitor cells in healthy men: effect of aging and training. *Aging Cell.* 2006;5:495–503.
35. Zaldivar F, Eliakim A, Radom-Aizik S, Leu S-Y, Cooper DM. The effect of brief exercise on circulating CD34+ stem cells in early and late pubertal boys. *Pediatr Res.* 2007;61:491–495.
36. Bonsignore MR, Morici G, Santoro A, et al. Circulating hematopoietic progenitor cells in runners. *J Appl Physiol.* 2002;93:1691–1697.
37. Möbius-Winkler S, Hilberg T, Menzel K, et al. Time-dependent mobilization of circulating progenitor cells during strenuous exercise in healthy individuals. *J Appl Physiol.* 2009;107:1943–1950.
38. Adams V, Linke A, Breuckmann F, et al. Circulating progenitor cells decrease immediately after marathon race in advanced-age marathon runners. *Eur J Cardiovasc Prev Rehabil.* 2008;15:602–607.
39. Stelzer I, Fuchs R, Schraml E, et al. Decline of bone marrow-derived hematopoietic progenitor cell quality during aging in the rat. *Exp Aging Res.* 2010;36:359–370.
40. Baker JM, De Lisio M, Parise G. Endurance exercise training promotes medullary hematopoiesis. *FASEB J.* 2011;25:4348–4357.
41. De Lisio M, Parise G. Characterization of the effects of exercise training on hematopoietic stem cell quantity and function. *J Appl Physiol.* 2012;113(10):1576–1584.
42. D'Ascenzi F, Zacà V, Maiorca S, et al. A quantitative assessment of circulating progenitor cells in competitive athletes and in sedentary subjects. *J Sports Med Phys Fitness.* 2015;55:241–248.
43. Witkowski S, Lockard MM, Jenkins NT, et al. Relationship between circulating progenitor cells, vascular function and oxidative stress with long-term training and short-term detraining in older men. *Clin Sci (Lond).* 2010;118:303–311.
44. Núñez-Espinosa C, Ferreira I, Ríos-Kristjánsson JG, et al. Effects of intermittent hypoxia and light aerobic exercise on circulating stem cells and side population,

after strenuous eccentric exercise in trained rats. *Curr Stem Cell Res Ther.* 2015;10:132–139.
45. Park J-H, Miyashita M, Kwon Y-C, et al. A 12-week after-school physical activity programme improves endothelial cell function in overweight and obese children: a randomised controlled study. *BMC Pediatr.* 2012;12:111.
46. Walther C, Gaede L, Adams V, et al. Effect of increased exercise in school children on physical fitness and endothelial progenitor cells: a prospective randomized trial. *Circulation.* 2009;120:2251–2259.
47. Wang J-S, Lee M-Y, Lien H-Y, Weng T-P. Hypoxic exercise training improves cardiac/muscular hemodynamics and is associated with modulated circulating progenitor cells in sedentary men. *Int J Cardiol.* 2014;170:315–323.
48. Asahara T, Murohara T, Sullivan A, et al. Isolation of putative progenitor endothelial cells for angiogenesis. *Science.* 1997;275(5302):964–967.
49. Basile DP, Yoder MC. Circulating and tissue resident endothelial progenitor cells. *J Cell Physiol.* 2014;229:10–16.
50. Goligorsky MS, Salven P. Concise review: endothelial stem and progenitor cells and their habitats. *Stem Cells Transl Med.* 2013;2(7):499–504.
51. Yoder MC. Is endothelium the origin of endothelial progenitor cells? *Arterioscler Thromb Vasc Biol.* 2010;30(6):1094–1103.
52. Yoder MC. Defining human endothelial progenitor cells. *J Thromb Haemost.* 2009;7:49–52.
53. Peichev M, Naiyer AJ, Pereira D, et al. Expression of VEGFR-2 and AC133 by circulating human CD34(+) cells identifies a population of functional endothelial precursors. *Blood.* 2000;95(3):952–958.
54. Timmermans F, Van Hauwermeiren F, De Smedt M, et al. Endothelial outgrowth cells are not derived from CD133+ cells or CD45+ hematopoietic precursors. *Arterioscler Thromb Vasc Biol.* 2007;27(7):1572–1579.
55. Case J, Mead LE, Bessler WK, et al. Human CD34+AC133+VEGFR-2+ cells are not endothelial progenitor cells but distinct, primitive hematopoietic progenitors. *Exp Hematol.* 2007;35(7):1109–1118.
56. Medina R, O'Neill C, Sweeney M, et al. Molecular analysis of endothelial progenitor cell (EPC) subtypes reveals two distinct cell populations with different identities. *BMC Med Genomics.* 2010;3(1):18.
57. Hill JM, Zalos G, Halcox JP, et al. Circulating endothelial progenitor cells, vascular function, and cardiovascular risk. *N Engl J Med.* 2003;348(7):593–600.
58. Rohde E, Bartmann C, Schallmoser K, et al. Immune cells mimic the morphology of endothelial progenitor colonies in vitro. *Stem Cells.* 2007;25(7):1746–1752.
59. Rohde E, Malischnik C, Thaler D, et al. Blood monocytes mimic endothelial progenitor cells. *Stem Cells.* 2006;24(2):357–367.
60. Richardson MR, Yoder MC. Endothelial progenitor cells: quo vadis? *J Mol Cell Cardiol.* 2011;50(2):266–272.
61. Hur J, Yang H-M, Yoon C-H, et al. Identification of a novel role of T cells in postnatal vasculogenesis: characterization of endothelial progenitor cell colonies. *Circulation.* 2007;116(15):1671–1682.
62. Yoder MC. Endothelial progenitor cell: a blood cell by many other names may serve similar functions. *J Mol Med (Berl).* 2013;91(3):285–295.
63. Asahara T, Kawamoto A, Masuda H. Concise review: circulating endothelial progenitor cells for vascular medicine. *Stem Cells.* 2011;29(11):1650–1655.
64. Schmidt-Lucke C, Rossig L, Fichtlscherer S, et al. Reduced number of circulating endothelial progenitor cells predicts future cardiovascular events: proof of concept for the clinical importance of endogenous vascular repair. *Circulation.* 2005;111(22):2981–2987.

65. Werner N, Kosiol S, Schiegl T, et al. Circulating endothelial progenitor cells and cardiovascular outcomes. *N Engl J Med.* 2005;353(10):999–1007.
66. Fadini GP, Coracina A, Baesso I, et al. Peripheral blood CD34+KDR+ endothelial progenitor cells are determinants of subclinical atherosclerosis in a middle-aged general population. *Stroke.* 2006;37(9):2277–2282.
67. Fadini GP, de Kreutzenberg S, Agostini C, et al. Low CD34+ cell count and metabolic syndrome synergistically increase the risk of adverse outcomes. *Atherosclerosis.* 2009;207(1):213–219.
68. Fadini GP, de Kreutzenberg SV, Coracina A, et al. Circulating CD34+ cells, metabolic syndrome, and cardiovascular risk. *Eur Heart J.* 2006;27(18):2247–2255.
69. Fadini GP, Maruyama S, Ozaki T, et al. Circulating progenitor cell count for cardiovascular risk stratification: a pooled analysis. *PLoS One.* 2010;5(7):e11488.
70. Fadini GP, Miorin M, Facco M, et al. Circulating endothelial progenitor cells are reduced in peripheral vascular complications of type 2 diabetes mellitus. *J Am Coll Cardiol.* 2005;45(9):1449–1457.
71. Fadini GP, Schiavon M, Cantini M, Avogaro A, Agostini C. Circulating CD34+ cells, pulmonary hypertension, and myelofibrosis. *Blood.* 2006;108(5):1776–1777.
72. Mora S, Cook N, Buring JE, Ridker PM, Lee I-M. Physical activity and reduced risk of cardiovascular events: potential mediating mechanisms. *Circulation.* 2007;116(19):2110–2118.
73. Witkowski S, Jenkins NT, Hagberg JM. Enhancing treatment for cardiovascular disease: exercise and circulating angiogenic cells. *Exerc Sport Sci Rev.* 2011;39(2):93–101.
74. De Biase C, De Rosa R, Luciano R, et al. Effects of physical activity on endothelial progenitor cells (EPCs). *Front Physiol.* 2013;4:414.
75. Ribeiro F, Ribeiro IP, Alves AJ, et al. Effects of exercise training on endothelial progenitor cells in cardiovascular disease: a systematic review. *Am J Phys Med Rehabil.* 2013;92(11):1020–1030.
76. Rehman J, Li J, Parvathaneni L, et al. Exercise acutely increases circulating endothelial progenitor cells and monocyte-/macrophage-derived angiogenic cells. *J Am Coll Cardiol.* 2004;43(12):2314–2318.
77. Adams V, Lenk K, Linke A, et al. Increase of circulating endothelial progenitor cells in patients with coronary artery disease after exercise-induced ischemia. *Arterioscler Thromb Vasc Biol.* 2004;24(4):684–690.
78. Sandri M, Bernhard Beck E, Adams V, et al. Maximal exercise, limb ischemia, and endothelial progenitor cells. *Eur J Cardiovasc Prev Rehabil.* 2011;18(1):55–64.
79. Van Craenenbroeck EM, Bruyndonckx L, Van Berckelaer C, Hoymans VY, Vrints CJ, Conraads VM. The effect of acute exercise on endothelial progenitor cells is attenuated in chronic heart failure. *Eur J Appl Physiol.* 2011;111(9):2375–2379.
80. Van Craenenbroeck EM, Beckers PJ, Possemiers NM, et al. Exercise acutely reverses dysfunction of circulating angiogenic cells in chronic heart failure. *Eur Heart J.* 2010;31(15):1924–1934.
81. Jenkins NT, Witkowski S, Spangenburg EE, Hagberg JM. Effects of acute and chronic endurance exercise on intracellular nitric oxide in putative endothelial progenitor cells: role of NAPDH oxidase. *Am J Physiol Heart Circ Physiol.* 2009;297(5):H1798–H1805.
82. Sarto P, Balducci E, Balconi G, et al. Effects of exercise training on endothelial progenitor cells in patients with chronic heart failure. *J Card Fail.* 2007;13(9):701–708.
83. Laufs U, Werner N, Link A, et al. Physical training increases endothelial progenitor cells, inhibits neointima formation, and enhances angiogenesis. *Circulation.* 2004;109(2):220–226.
84. Van Craenenbroeck EM, Hoymans VY, Beckers PJ, et al. Exercise training improves function of circulating angiogenic cells in patients with chronic heart failure. *Basic Res Cardiol.* 2010;105(5):665–676.

85. Kruger K, Pilat C, Schild M, et al. Progenitor cell mobilization after exercise is related to systemic levels of G-CSF and muscle damage. *Scand J Med Sci Sports*. 2015;25: e283–e291.
86. Goto C, Nishioka K, Umemura T, et al. Acute moderate-intensity exercise induces vasodilation through an increase in nitric oxide bioavailability in humans. *Am J Hypertens*. 2007;20(8):825–830.
87. Aicher A, Heeschen C, Mildner-Rihm C, et al. Essential role of endothelial nitric oxide synthase for mobilization of stem and progenitor cells. *Nat Med*. 2003;9(11):1370–1376.
88. Cubbon RM, Murgatroyd SR, Ferguson C, et al. Human exercise-induced circulating progenitor cell mobilization is nitric oxide-dependent and is blunted in South Asian men. *Arterioscler Thromb Vasc Biol*. 2010;30(4):878–884.
89. Fleissner F, Thum T. Critical role of the nitric oxide/reactive oxygen species balance in endothelial progenitor dysfunction. *Antioxid Redox Signal*. 2011;15(4):933–948.
90. Guhanarayan G, Jablonski J, Witkowski S. Circulating angiogenic cell population responses to 10 days of reduced physical activity. *J Appl Physiol*. 2014;117:500–506.
91. Steiner S, Niessner A, Ziegler S, et al. Endurance training increases the number of endothelial progenitor cells in patients with cardiovascular risk and coronary artery disease. *Atherosclerosis*. 2005;181(2):305–310.
92. Cheng XW, Kuzuya M, Kim W, et al. Exercise training stimulates ischemia-induced neovascularization via phosphatidylinositol 3-kinase/Akt-dependent hypoxia-induced factor-1{alpha} reactivation in mice of advanced age. *Circulation*. 2010;122:707–716.
93. Pula G, Mayr U, Evans C, et al. Proteomics identifies thymidine phosphorylase as a key regulator of the angiogenic potential of colony-forming units and endothelial progenitor cell cultures. *Circ Res*. 2009;104(1):32–40.
94. Urbich C, Aicher A, Heeschen C, et al. Soluble factors released by endothelial progenitor cells promote migration of endothelial cells and cardiac resident progenitor cells. *J Mol Cell Cardiol*. 2005;39(5):733–742.
95. Yang Z, von Ballmoos MW, Faessler D, et al. Paracrine factors secreted by endothelial progenitor cells prevent oxidative stress-induced apoptosis of mature endothelial cells. *Atherosclerosis*. 2010;211(1):103–109.
96. Friedenstein AJ, Petrakova KV, Kurolesova AI, Frolova GP. Heterotopic of bone marrow. Analysis of precursor cells for osteogenic and hematopoietic tissues. *Transplantation*. 1968;6:230–247.
97. Owen M, Friedenstein AJ. Stromal stem cells: marrow-derived osteogenic precursors. *Ciba Found Symp*. 1988;136:42–60.
98. Dominici M, Le Blanc K, Mueller I, et al. Minimal criteria for defining multipotent mesenchymal stromal cells. The International Society for Cellular Therapy position statement. *Cytotherapy*. 2006;8:315–317.
99. Bianco P, Robey PG, Simmons PJ. Mesenchymal stem cells: revisiting history, concepts, and assays. *Cell Stem Cell*. 2008;2:313–319.
100. Bianco P, Cao X, Frenette PS, et al. The meaning, the sense and the significance: translating the science of mesenchymal stem cells into medicine. *Nat Med*. 2013;19:35–42.
101. Phinney DG, Galipeau J, Krampera M, Martin I, Shi Y, Sensebe L. MSCs: science and trials. *Nat Med*. 2013;19:812.
102. Kittler EL, McGrath H, Temeles D, Crittenden RB, Kister VK, Quesenberry PJ. Biologic significance of constitutive and subliminal growth factor production by bone marrow stroma. *Blood*. 1992;79:3168–3178.
103. Pittenger MF, Mackay AM, Veck SC, et al. Multilineage potential of adult human mesenchymal stem cells. *Science*. 1999;284:143–147.
104. Ocarino NM, Boeloni JN, Goes AM, Silva JF, Marubayashi U, Serakides R. Osteogenic differentiation of mesenchymal stem cells from osteopenic rats subjected to

physical activity with and without nitric oxide synthase inhibition. *Nitric Oxide.* 2008;19:320–325.
105. Hell RC, Ocarino NM, Boeloni JN, et al. Physical activity improves age-related decline in the osteogenic potential of rats' bone marrow-derived mesenchymal stem cells. *Acta Physiol (Oxf).* 2012;205:292–301.
106. David V, Martin A, Lafage-Proust MH, et al. Mechanical loading down-regulates peroxisome proliferator-activated receptor gamma in bone marrow stromal cells and favors osteoblastogenesis at the expense of adipogenesis. *Endocrinology.* 2007;148:2553–2562.
107. Mauro A. Satellite cell of skeletal muscle fibers. *J Biophys Biochem Cytol.* 1961;9:493–495.
108. Sacco A, Doyonnas R, Kraft P, Vitorovic S, Blau HM. Self-renewal and expansion of single transplanted muscle stem cells. *Nature.* 2008;456:502–506.
109. Sambasivan R, Yao R, Kissenpfennig A, et al. Pax7-expressing satellite cells are indispensable for adult skeletal muscle regeneration. *Development.* 2011;138:3647–3656.
110. McCarthy JJ, Mula J, Miyazaki M, et al. Effective fiber hypertrophy in satellite cell-depleted skeletal muscle. *Development.* 2011;138:3657–3666.
111. Starkey JD, Yamamoto M, Yamamoto S, Goldhamer DJ. Skeletal muscle cells are committed to myogenesis and do not spontaneously adopt nonmyogenic fates. *J Histochem Cytochem.* 2011;59:33–46.
112. Cermak NM, Snijders T, McKay BR, et al. Eccentric exercise increases satellite cell content in type II muscle fibers. *Med Sci Sports Exerc.* 2013;45:230–237.
113. Bellamy LM, Joanisse S, Grubb A, et al. The acute satellite cell response and skeletal muscle hypertrophy following resistance training. *PLoS One.* 2014;9:e109739.
114. Farup J, Rahbek SK, Knudsen IS, de Paoli F, Mackey AL, Vissing K. Whey protein supplementation accelerates satellite cell proliferation during recovery from eccentric exercise. *Amino Acids.* 2014;46:2503–2516.
115. Snijders T, Verdijk LB, Smeets JS, et al. The skeletal muscle satellite cell response to a single bout of resistance-type exercise is delayed with aging in men. *Age.* 2014;36:9699.
116. Roth SM, Martel GF, Ivey FM, et al. Skeletal muscle satellite cell characteristics in young and older men and women after heavy resistance strength training. *J Gerontol A Biol Sci Med Sci.* 2001;56:B240–B247.
117. Crameri RM, Langberg H, Magnusson P, et al. Changes in satellite cells in human skeletal muscle after a single bout of high intensity exercise. *J Physiol.* 2004;558:333–340.
118. Mackey AL, Esmarck B, Kadi F, et al. Enhanced satellite cell proliferation with resistance training in elderly men and women. *Scand J Med Sci Sports.* 2007;17:34–42.
119. Petrella JK, Kim JS, Mayhew DL, Cross JM, Bamman MM. Potent myofiber hypertrophy during resistance training in humans is associated with satellite cell-mediated myonuclear addition. *J Appl Physiol.* 2008;104:1736–1742.
120. Verdijk LB, Gleeson BG, Jonkers RA, et al. Skeletal muscle hypertrophy following resistance training is accompanied by a fiber type-specific increase in satellite cell content in elderly men. *J Gerontol A Biol Sci Med Sci.* 2009;64:332–339.
121. Verdijk LB, Snijders T, Drost M, Delhaas T, Kadi F, van Loon LJ. Satellite cells in human skeletal muscle; from birth to old age. *Age.* 2014;36:545–547.
122. Charifi N, Kadi F, Feasson L, Denis C. Effects of endurance training on satellite cell frequency in skeletal muscle of old men. *Muscle Nerve.* 2003;28:87–92.
123. Kadi F, Schjerling P, Andersen LL, et al. The effects of heavy resistance training and detraining on satellite cells in human skeletal muscles. *J Physiol.* 2004;558: 1005–1012.
124. Umnova MM, Seene TP. The effect of increased functional load on the activation of satellite cells in the skeletal muscle of adult rats. *Int J Sports Med.* 1991;12:501–504.
125. Mounier R, Chretien F, Chazaud B. Blood vessels and the satellite cell niche. *Curr Top Dev Biol.* 2011;96:121–138.

126. Fry CS, Lee JD, Jackson JR, et al. Regulation of the muscle fiber microenvironment by activated satellite cells during hypertrophy. *FASEB J.* 2014;28:1654–1665.
127. Malm C, Sjodin TL, Sjoberg B, et al. Leukocytes, cytokines, growth factors and hormones in human skeletal muscle and blood after uphill or downhill running. *J Physiol.* 2004;556:983–1000.
128. Mikkelsen UR, Langberg H, Helmark IC, et al. Local NSAID infusion inhibits satellite cell proliferation in human skeletal muscle after eccentric exercise. *J Appl Physiol.* 2009;107:1600–1611.
129. Shefer G, Rauner G, Stuelsatz P, Benayahu D, Yablonka-Reuveni Z. Moderate-intensity treadmill running promotes expansion of the satellite cell pool in young and old mice. *FEBS J.* 2013;280:4063–4073.
130. Hawke TJ, Garry DJ. Myogenic satellite cells: physiology to molecular biology. *J Appl Physiol.* 2001;91:534–551.
131. Macaluso F, Myburgh KH. Current evidence that exercise can increase the number of adult stem cells. *J Muscle Res Cell Motil.* 2012;33:187–198.
132. McKay BR, O'Reilly CE, Phillips SM, Tarnopolsky MA, Parise G. Co-expression of IGF-1 family members with myogenic regulatory factors following acute damaging muscle-lengthening contractions in humans. *J Physiol.* 2008;586:5549–5560.
133. Toth KG, McKay BR, De Lisio M, Little JP, Tarnopolsky MA, Parise G. IL-6 induced STAT3 signaling is associated with the proliferation of human muscle satellite cells following acute muscle damage. *PLoS One.* 2011;6:e17392.
134. Guerci A, Lahoute C, Hebrard S, et al. *Cell Metab.* 2012;15:25–37.
135. Qu-Petersen Z, Deasy B, Jankowski R, et al. Identification of a novel population of muscle stem cells in mice: potential for muscle regeneration. *J Cell Biol.* 2002;157:851–864.
136. Motohashi N, Uezumi A, Yada E, et al. Muscle CD31(−)CD45(−) side population cells promote muscle regeneration by stimulating proliferation and migration of myoblasts. *Am J Pathol.* 2008;173:781–791.
137. Joe AW, Yi L, Natarajan A, et al. Muscle injury activates resident fibro/adipogenic progenitors that facilitate myogenesis. *Nat Cell Biol.* 2010;12:153–163.
138. Mitchell KJ, Pannerec A, Cadot B, et al. Identification and characterization of a non-satellite cell muscle resident progenitor during postnatal development. *Nat Cell Biol.* 2010;12:257–266.
139. Uezumi A, Fukada S, Yamamoto N, Takeda S, Tsuchida K. Mesenchymal progenitors distinct from satellite cells contribute to ectopic fat cell formation in skeletal muscle. *Nat Cell Biol.* 2010;12:143–152.
140. Doyle MJ, Zhou S, Tanaka KK, et al. Abcg2 labels multiple cell types in skeletal muscle and participates in muscle regeneration. *J Cell Biol.* 2011;195:147–163.
141. Valero MC, Huntsman HD, Liu J, Zou K, Boppart MD. Eccentric exercise facilitates mesenchymal stem cell appearance in skeletal muscle. *PLoS One.* 2012;7:e29760.
142. Boppart MD, De Lisio M, Zou K, Huntsman HD. Defining a role for non-satellite stem cells in the regulation of muscle repair following exercise. *Front Physiol.* 2013;4:310.
143. Gussoni E, Soneoka Y, Strickland CD, et al. Dystrophin expression in the mdx mouse restored by stem cell transplantation. *Nature.* 1999;401:390–394.
144. Asakura A, Seale P, Girgis-Gabardo A, Rudnicki MA. Myogenic specification of side population cells in skeletal muscle. *J Cell Biol.* 2002;159:123–134.
145. Pannerec A, Formicola L, Besson V, Marazzi G, Sassoon DA. Defining skeletal muscle resident progenitors and their cell fate potentials. *Development.* 2013;140:2879–2891.
146. Birbrair A, Zhang T, Wang ZM, et al. Role of pericytes in skeletal muscle regeneration and fat accumulation. *Stem Cells Dev.* 2013;22:2298–2314.
147. Zou K, Huntsman HD, Valero MC, et al. Mesenchymal stem cells augment the adaptive response to eccentric exercise. *Med Sci Sports Exerc.* 2015;47:315–325.

148. Huntsman HD, Zachwieja N, Zou K, et al. Mesenchymal stem cells contribute to vascular growth in skeletal muscle in response to eccentric exercise. *Am J Physiol Heart Circ Physiol*. 2013;304:H72–H81.
149. De Lisio MD, Jensen T, Sukiennik RA, Huntsman HD, Boppart MD. Substrate and strain alter the muscle-derived mesenchymal stem cell secretome to promote myogenesis. *Stem Cell Res Ther*. 2014;5:74.
150. Hyldahl RD, O'Fallon KS, Schwartz LM, Clarkson PM. Activation of nuclear factor-kB following muscle eccentric contractions in humans is localized primarily to skeletal muscle-residing pericytes. *FASEB J*. 2011;25:2956–2966.

CHAPTER NINETEEN

Exercise and Gene Expression

Mark Hargreaves[1]
Department of Physiology, The University of Melbourne, Melbourne, Australia
[1]Corresponding author: e-mail address: m.hargreaves@unimelb.edu.au

Contents

Abstract

Acute and transient changes in gene transcription following a single exercise bout, if reinforced by repeated exercise stimuli, result in the longer lasting effects on protein expression and function that form the basis of skeletal muscle training adaptations. Changes in skeletal muscle gene expression occur in response to multiple stimuli associated with skeletal muscle contraction, various signaling kinases that respond to these stimuli, and numerous downstream pathways and targets of these kinases. In addition, DNA methylation, histone acetylation and phosphorylation, and micro-RNAs can alter gene expression via epigenetic mechanisms. Contemporary studies rely upon "big omics data," in combination with computational and systems biology, to interrogate, and make sense of, the complex interactions underpinning exercise adaptations. The exciting potential is a greater understanding of the integrative biology of exercise.

1. EXERCISE ADAPTATIONS

Physical exercise results in profound changes in virtually all of the physiological systems of the body to ensure that contracting skeletal muscles are appropriately supplied with fuels and oxygen, that metabolic waste products are removed, and that key whole body functions are preserved in the face of major challenges to homeostasis. Numerous chapters in this volume describe these various changes in more detail, as do recent reviews.[1,2] Repeated bouts of exercise (i.e., training) elicit adaptive responses that enhance functional capacity and improve health and well-being. Skeletal

Progress in Molecular Biology and Translational Science, Volume 135
ISSN 1877-1173
http://dx.doi.org/10.1016/bs.pmbts.2015.07.006

muscle, by virtue of its mass, capacity for substrate utilization and storage, and remarkable plasticity, has attracted much attention in terms of studying exercise responses. This has been amplified by observations in the last 10–15 years that contracting skeletal muscles release biologically active molecules (myokines) that influence the function and characteristics of other tissues and organs and, in so doing, may partly mediate the beneficial effects of regular exercise on whole body function (see chapter Exercise and regulation of adipokine and myokine production by Eckel et al. of this volume). For these reasons, the focus of this chapter will be on adaptive responses in skeletal muscle and will draw primarily on studies in human volunteers whenever possible.

The molecular bases of skeletal muscle adaptations to exercise, be they increased mitochondrial mass, altered substrate metabolism, enhanced angiogenesis, or myofiber hypertrophy, fundamentally involve the function and/or mass of key proteins. This is mediated by an array of signaling events, pre- and posttranscriptional processes, regulation of translation and protein expression, and modulation of protein (enzyme) activities and/or intracellular localization. There are multiple stimuli associated with skeletal muscle contraction, various signaling kinases that respond to these stimuli, and numerous downstream pathways and targets of these kinases. Some of these are listed in Table 1 and more detailed summaries can be found in recent reviews.[3,4] There are complex spatial and temporal interactions between these various elements that combine to produce the integrated response

Table 1 Selected List of Putative Stimuli, Sensors, and Signal Transduction Mechanisms That Potentially Mediate Skeletal Muscle Adaptive Responses to Exercise

Stimuli
$[Ca^{2+}]$, energy charge, [CP], [glycogen], amino acids, redox state, oxygenation, muscle temperature, $[NAD^{+}]$, muscle tension and length, hormones
Sensors
AMPK, calcineurin, CaMK, FAK, MAPK, mTOR, SIRT
Signal transduction
ATF2, CREB, FoxO, HDAC, HIF1-α, MEF2, NFAT, PGC-1α, SRF

Abbreviations: AMPK, AMP-activated protein kinase; ATF2, activating transcription factor 2; CaMK, calcium/calmodulin-dependent protein kinase; CP, creatine phosphate; CREB, cAMP response element-binding protein; FAK, focal adhesion kinase; FoxO, forkhead box O transcription factors; HDAC, histone deacetylase; HIF1-α, hypoxia-inducible factor 1-alpha; MAPK, mitogen-activated protein kinase; MEF2, myocyte enhancer factor 2; mTOR, mammalian target of rapamycin; NAD^{+}, nicotinamide adenine dinucleotide; NFAT, nuclear factor of activated T cells; PGC-1α, peroxisome proliferator-activated gamma coactivator-1 alpha; SIRT, sirtuin; SRF, serum response factor.

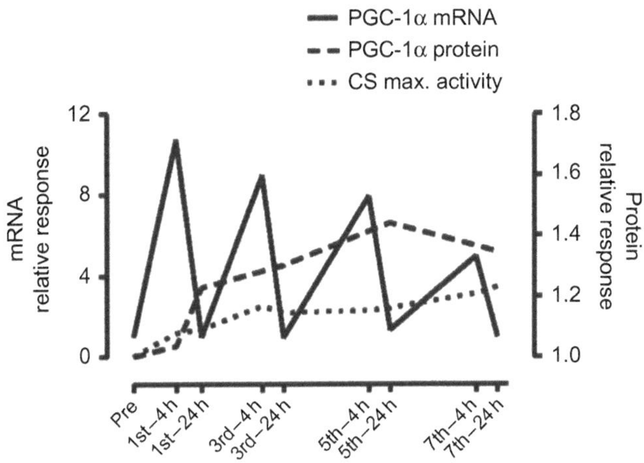

Figure 1 Temporal relationships between PGC-1α mRNA and protein and citrate synthase (CS) maximal activity during 2 weeks of high-intensity exercise training. Muscle samples obtained before training and 4 and 24 h after the 1st, 3rd, 5th, and 7th exercise training sessions. *Reproduced from Perry* et al.[5] *with permission.*

to an exercise challenge. They may also account for the observed differences in skeletal muscle responses to exercise of differing intensity, duration, and mode.

An important concept in the past ~20 years has been that the responses to exercise training reflect the cumulative effects of the responses to repeated, single exercise bouts. Acute and transient changes in gene transcription following a single exercise bout (see Section 2), if reinforced by repeated exercise stimuli, result in the longer lasting effects on protein expression and function that form the basis of skeletal muscle training adaptations (Fig. 1).[5–7] Similarly, removal of the regular, exercise stimulus during a period of "detraining" results in the rapid reversal of the training adaptation.[8] Hence, there has been a great interest in understanding the molecular regulation of altered skeletal muscle gene transcription in response to a single exercise bout.

2. EXERCISE AND GENE TRANSCRIPTION

An increase in skeletal muscle oxidative capacity and the associated alteration in substrate metabolism is a hallmark of endurance exercise adaptation.[9] The regulation of mitochondrial biogenesis and increased mitochondrial protein expression involves the interplay between the nuclear and mitochondrial genomes. One of the first studies to probe the molecular basis of this adaptation was from Williams *et al.*,[10] who observed increased cytochrome *c* mRNA and mitochondrial DNA following chronic electrical stimulation of rabbit tibialis anterior muscle. Although changes in mRNA stability can influence mRNA levels, it is generally believed that an increase in mRNA largely reflects an increase in gene transcription. Indeed, regulation at the transcriptional level was confirmed when it was observed that a single bout of exercise transiently increased GLUT4 and citrate synthase gene transcription in rat skeletal muscle.[11] Subsequently, a number of groups observed transient increases in GLUT4 and glycogenin,[12] hexokinase,[12,13] and lipoprotein lipase[14] mRNA levels in human skeletal muscle following a single bout of exercise. Modification of the method for assessing gene transcriptional activity enabled measurements in small samples of human skeletal muscle and it was demonstrated that a single bout of exercise increased transcription of a number of metabolic genes[15] and the IL-6[16] and PGC-1α (peroxisome proliferator-activated gamma coactivator-1 alpha)[17] genes. Measurements of mRNA were extended to various transcription factors involved in the metabolic adaptations to endurance exercise,[18] metabolic and mitogenic gene responses to resistance and endurance exercise,[19] and gene expression changes following high-intensity interval exercise.[20] In parallel with advances in analytical technologies, measurement of skeletal muscle gene expression (as reflected by mRNA levels) soon became a regular component of many exercise studies. The mRNA responses to acute exercise were shown to be influenced by training status,[15,17,21] preceding diet[22] and muscle glycogen,[16,23] blood glucose[24,25] and amino acid[26] availability. The use of gene microarrays enabled analysis of acute exercise[27,28] and training[29] effects on skeletal muscle gene expression. These studies demonstrated that acute and chronic exercise altered transcription of genes involved in a diverse range of cellular functions, reinforcing the power of exercise to dramatically influence skeletal muscle phenotype. Interestingly, it was also observed that gene expression in inactive muscle exercising individuals was also altered,[28] most likely as a consequence of changes in circulating hormones (catecholamines) and free fatty acids.

To elaborate on the interactions between exercise stimuli, sensors, and signal transduction that mediate exercise-induced alterations in skeletal muscle gene expression, our own work on the regulation of GLUT4 expression is a useful example. As mentioned above, a single bout of exercise increases GLUT4 mRNA in human skeletal muscle for several hours after exercise.[12]

This has been associated with an increase in GLUT4 protein,[30] although this is not always observed.[31,32] Our current understanding of the regulation of exercise-induced changes in GLUT4 expression is summarized in Fig. 2.[33]

It is known that the skeletal muscle expression of GLUT4 is dependent upon two transcription factors: myocyte enhancer factor 2 (MEF2) and GLUT4 enhancer factor (GEF).[34] MEF2 is subject to transcriptional repression by the class II histone deacetylases (HDAC), notably HDAC4 and HDAC5. Phosphorylation of these HDACs results in their nuclear export and removal of the inhibition of MEF2 transcriptional activity. Upstream kinases are AMP-activated protein kinase (AMPK)[35] and calcium/calmodulin-dependent protein kinase (CaMK).[36] The increase in transcription of the human GLUT4 gene by exercise is dependent upon response elements within −895 bp of the promoter,[37] implicating both MEF2 and GEF. In human skeletal muscle, a single bout of exercise reduces nuclear abundance of HDAC4,[38] HDAC5,[38,39] and MEF2-associated HDAC5,[39] with a concomitant increase in GLUT4 mRNA.[39] Exercise also increased

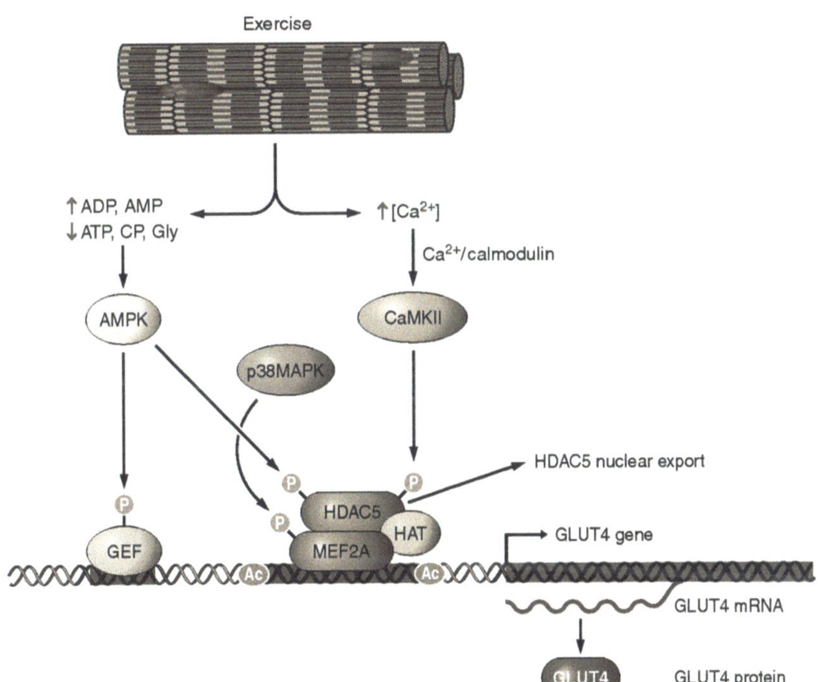

Figure 2 Overview of the molecular events involved in the exercise-induced increase in GLUT4 expression in human skeletal muscle. *Reproduced from Richter and Hargreaves[33] with permission.* (See the color plate.)

MEF2-associated PGC-1α,[39] known to recruit histone acetyltransferases (HATs) such as p300, and p38 mitogen-activated protein kinase-mediated phosphorylation of MEF2[39] which, together with the removal of HDAC5 repression, increases MEF2 transcriptional activity. These changes were associated with increased nuclear abundance of the AMPK α2 isoform,[40] activation of both AMPK and CaMK,[38] and increased MEF2 and GEF DNA binding.[41] As for resting muscle,[42] there is redundancy in the control of GLUT4 expression by exercise, which is preserved in the absence of AMPK[43] and with inhibition of calcineurin activity.[44] Interestingly, exercise in mice lacking AMPK activity results in compensatory activation of protein kinase D, another known HDAC5 kinase, and preservation of the metabolic responses to energetic stress.[45]

As mentioned, there are complex interactions between various kinases and transcription factors and transcriptional coactivators that mediate exercise effects on gene transcription. The activity of these factors and coactivators can be modified by their expression, their phosphorylation, and/or acetylation status and by their intracellular localization. The transcriptional coactivator PGC-1α has attracted considerable attention due to its role in mediating exercise-induced alterations in skeletal muscle mitochondrial biogenesis and angiogenesis.[46] A single bout of exercise increases PGC-1α gene expression[17,47] and protein level[47,48]; however, there is evidence that mitochondrial biogenesis commences before the increase in PGC-1α protein levels,[48] suggesting that changes in PGC-1α activity and/or localization may be important initial responses to exercise. Indeed, it has been observed that exercise results in translocation of PGC-1α to the nucleus and mitochondria in skeletal muscle,[49–52] thereby promoting nuclear–mitochondrial communication[51] and increased nuclear MEF2–PGC-1α association.[39]

3. EXERCISE AND EPIGENETIC MODIFICATIONS

Exercise effects on gene transcription and the transcriptome have not involved fundamental changes in the gene sequence; rather, they arise through epigenetic modifications on DNA. There has been debate on the exact definition of epigenetics, but that proposed by Bird[53] is useful: "the structural adaptation of chromosomal regions so as to register, signal or perpetuate altered activity states". Such epigenetic modifications link changes in the environment, notably exercise or activity level, with alterations in gene expression and function. The two major epigenetic modifications are DNA methylation and posttranslational histone modifications, including acetylation, methylation, phosphorylation, and ubiquitination.[54] The modulation of gene transcription and translation by noncoding RNAs, including microRNAs, is another form of epigenetic modification and is covered in more detail in the next chapter of this volume. In relation to GLUT4 expression, it has recently been shown that the micro-RNA miR-199a is elevated in the plasma of patients with type 2 diabetes and is associated with reduced GLUT4 expression and insulin resistance.[55]

DNA methylation is thought to suppress gene expression by restricting access of the transcriptional machinery to chromatin. Relatively, few studies have examined exercise effects on DNA methylation in skeletal muscle. Recently, it was demonstrated that increased gene expression following a single bout of exercise was associated with transient DNA hypomethylation.[56] Further experiments in L6 myotubes implicated increased Ca^{2+} as necessary, but not sufficient, for DNA promoter hypomethylation.[56] Increased DNA methylation of various genes was seen in skeletal muscle from individuals with a family history of type 2 diabetes, an effect that was reduced by 6 months of exercise training.[57] In contrast, reduced physical activity due to a bed rest intervention resulted in hypermethylation of the PGC-1α gene.[58] Finally, an interesting observation is that elite athletes harbor polymorphisms in DNA methylation enzymes that generally favor DNA hypomethylation and enhanced gene expression.[59] In terms of the intergenerational transmission of such epigenetic modifications, it has recently been shown that maternal exercise prevented PGC-1α gene hypermethylation and the transmission of the negative effects of maternal high fat feeding to aging offspring.[60] This is an exciting observation that demonstrates the potential long-term benefits of physical activity in pregnancy.

Chromatin consists of double-stranded DNA wrapped around a core of histone proteins. The histone core is positively charged and forms a tight structure with the negatively charged DNA backbone, thereby restricting access of transcriptional regulators to DNA and suppressing gene expression. Modifications to the amino acid residues in the histone proteins, notably histones 3 (H3) and 4 (H4), have major effects on chromatin structure and gene expression by altering the histone–DNA association and facilitating recruitment and access of key transcriptional regulators to DNA. Such modifications include lysine acetylation, methylation and ubiquitination, arginine methylation, and serine phosphorylation. We have shown that exercise does not change global H3 lysine (K)9/14 acetylation, but increases global

H3-K36 acetylation,[38] epigenetic marks that are associated with transcriptional initiation and elongation, respectively. Given that global acetylation will reflect hyper- and hypoacetylation on various residues, the significance of such changes requires further investigation. Notably, in follow-up studies on exercise effects on genome-wide histone modifications, we have observed increased acetylation of the H3 K9/14 and K36 residues on the PGC-1α and GLUT4 genes (M. Hargreaves and A. El Osta, unpublished observations). Histone modifications are controlled by various enzymes that are also subject to regulation. These include HATs and deacetylases (HDACs), histone methyltransferases and demethylases, histone kinases, and histone ligases. Exercise activates AMPK and CaMK, resulting in phosphorylation and nuclear export of the class II HDACs HDAC4,[38] and HDAC5,[38,39] thereby removing repression of MEF2 and activating transcription of MEF2-dependent genes. Given the importance of acetyl-CoA for these acetylation reactions,[61] it is likely that it provides the link between increased substrate metabolism and histone acetylation during exercise, although this has not been directly studied.

Exercise has also been shown to increase H3 serine phosphorylation[62] and this may be another modification with implications for exercise-induced changes in gene expression. We have also observed an increase in ubiquitinated HDAC5 following exercise,[38] which may be relevant for longer term adaptations to exercise. It has been shown that the class II HDACs can be ubiquitinated and degraded by the proteasome system, increasing MEF2 activation, and increasing muscle oxidative capacity and endurance performance.[63] It was suggested that lower steady state levels of the class II HDACs accounted for the oxidative phenotype in slow twitch fibers. In our studies,[38,39] we have not seen changes in skeletal muscle HDAC5 expression following a single bout of exercise; however, an increase in ubiquitinated HDAC5 after exercise[38] may be the precursor to HDAC5 degradation later in recovery or following repeated exercise bouts within a longer training program.

4. EXERCISE, "OMICS," AND SYSTEMS BIOLOGY

The remarkable advances in molecular and cell biology, and analytical technologies and methodologies have resulted in a much greater understanding of the molecular bases of skeletal muscle adaptations to acute and chronic exercise. Contemporary studies rely upon "big data" generated from genomics, transcriptomics, epigenomics, proteomics, phosphoproteomics, and metabolomics, in combination with computational expertise, to interrogate, and make sense of, the complex interactions underpinning exercise adaptations.[64–72] The power of these approaches lies in the identification of molecular networks and pathways that interact to ultimately affect physiological function. Investigation of the skeletal muscle responses to resistance and endurance exercise training,[64–68,71] aging,[67] diabetic obesity,[69] and altered dietary protein intake[72] are examples of how these techniques can provide novel insights on common and/or distinct mechanisms underpinning changes in skeletal muscle adaptation or maladaptation, identify the molecular bases of the variation in responses between individuals and inform the development of interventions that target specific molecular networks and pathways. Over the next few years, there is likely to an increase in studies on the "systems biology of exercise," taking advantage of the emerging and rapidly changing "omics" approaches, in partnership with computational and quantitative scientists. As exercise scientists, we should embrace the potential of such approaches to enhance our ability to investigate the "exercise response," while recognizing their limitations.[73] The exciting potential is a greater understanding of the integrative biology of exercise and the ability to optimize individualized exercise interventions and to identify novel therapeutic strategies that enhance health and well-being.

ACKNOWLEDGMENTS

The author acknowledges Dr. Sean McGee, Deakin University, for ongoing support and collaboration in studies on molecular responses to exercise and the Diabetes Australia Research Trust for financial support of our original investigations.

REFERENCES

1. Hawley JA, Hargreaves M, Joyner MJ, Zierath JR. Integrative biology of exercise. *Cell*. 2014;159:738–749.
2. Heinonen I, Kalliokoski KK, Hannukainen JC, Duncker DJ, Nuutila P, Knuuti J. Organ-specific physiological responses to acute physical exercise and long-term training in humans. *Physiology*. 2014;29:421–436.
3. Bassel-Duby R, Olson EN. Signaling pathways in skeletal muscle remodeling. *Annu Rev Biochem*. 2006;75:19–37.
4. Egan B, Zierath JR. Exercise metabolism and the molecular regulation of skeletal muscle adaptation. *Cell Metab*. 2013;17:162–184.
5. Perry CGR, Lally J, Holloway GP, Heigenhauser GJF, Bonen A, Spriet LL. Repeated transient mRNA bursts precede increases in transcriptional and mitochondrial proteins during training in human skeletal muscle. *J Physiol*. 2010;588:4795–4810.
6. Egan B, O'Connor PL, Zierath JR, O'Gorman DJ. Time course analysis reveals gene-specific transcript and protein kinetics of adaptation to short-term aerobic exercise training in human skeletal muscle. *PLoS One*. 2013;8:e74098.

7. Kraniou GN, Cameron-Smith D, Hargreaves M. Effect of short-term training on GLUT4 mRNA and protein expression in human skeletal muscle. *Exp Physiol.* 2004;89:559–563.
8. McCoy M, Proietto J, Hargreaves M. Effect of detraining on GLUT4 protein in human skeletal muscle. *J Appl Physiol.* 1994;77:1532–1536.
9. Holloszy JO, Coyle EF. Adaptations of skeletal muscle to endurance exercise and their metabolic consequences. *J Appl Physiol.* 1984;56:831–838.
10. Williams RS, Salmons S, Newsholme EA, Kaufman RE, Mellor J. Regulation of nuclear and mitochondrial gene expression by contractile activity in skeletal muscle. *J Biol Chem.* 1986;261:376–380.
11. Neufer PD, Dohm GL. Exercise induces a transient increase in transcription of the GLUT4 gene in skeletal muscle. *Am J Physiol.* 1993;265:C1597–C1603.
12. Kraniou Y, Cameron-Smith D, Misso M, Collier G, Hargreaves M. Effects of exercise on GLUT4 and glycogenin gene expression in human skeletal muscle. *J Appl Physiol.* 2000;88:794–796.
13. Koval JA, DeFronzo RA, O'Doherty RM, et al. Regulation of hexokinase II activity and expression in human skeletal muscle by moderate exercise. *Am J Physiol.* 1998;274:E304–E308.
14. Seip RL, Mair K, Cole TG, Semenkovich CF. Induction of human skeletal muscle lipoprotein lipase gene expression by short-term exercise is transient. *Am J Physiol.* 1997;272: E255–E261.
15. Pilegaard H, Ordway GA, Saltin B, Neufer PD. Transcriptional regulation of gene expression in human skeletal muscle during recovery from exercise. *Am J Physiol.* 2000;279:E806–E814.
16. Keller C, Steensberg A, Pilegaard H, et al. Transcriptional activation of the IL-6 gene in human contracting skeletal muscle: influence of muscle glycogen content. *FASEB J.* 2001;15:2748–2750.
17. Pilegaard H, Saltin B, Neufer PD. Exercise induces transient transcriptional activation of the PGC-1α gene in human skeletal muscle. *J Physiol.* 2003;546:851–858.
18. Russell AP, Hesselink MK, Lo SK, Schrauwen P. Regulation of metabolic transcriptional co-activators and transcription factors with acute exercise. *FASEB J.* 2005;19:986–988.
19. Yang Y, Creer A, Jemiolo B, Trappe S. Time course of myogenic and metabolic gene expression in response to acute exercise in human skeletal muscle. *J Appl Physiol.* 2005;98:1745–1752.
20. Gibala MJ, McGee SL, Garnham AP, Howlett KF, Snow RJ, Hargreaves M. Brief intense interval exercise activates AMPK and p38 MAPK signaling and increases the expression of PGC-1α in human skeletal muscle. *J Appl Physiol.* 2009;106:929–934.
21. Coffey VG, Shield A, Canny BJ, Carey KA, Cameron-Smith D, Hawley JA. Interaction of contractile activity and training history on mRNA abundance in skeletal muscle from trained athletes. *Am J Physiol.* 2006;290:E849–E855.
22. Arkinstall MJ, Tunstall RJ, Cameron-Smith D, Hawley JA. Regulation of metabolic genes in human skeletal muscle by short-term exercise and diet manipulation. *Am J Physiol.* 2004;287:E25–E31.
23. Pilegaard H, Keller C, Steensberg A, et al. Influence of pre-exercise muscle glycogen content on exercise-induced transcriptional regulation of metabolic genes. *J Physiol.* 2002;541:261–271.
24. Civitarese AE, Hesselink MKC, Russell AP, Ravussin E, Schrauwen P. Glucose ingestion blunts exercise-induced gene expression of skeletal muscle fat oxidative genes. *Am J Physiol.* 2005;289:E1023–E1029.
25. Cluberton LJ, McGee SL, Murphy RM, Hargreaves M. Effect of carbohydrate ingestion on exercise-induced alterations in metabolic gene expression. *J Appl Physiol.* 2005;99: 1359–1363.

26. Hulmi JJ, Kovanen V, Selänne H, Kraemer WK, Häkkinen K, Mero AA. Acute and long-term effects of resistance exercise with or without protein ingestion on muscle hypertrophy and gene expression. *Amino Acids*. 2009;37:297–308.
27. Mahoney DJ, Parise G, Melov S, Safdar A, Tarnopolsky MA. Analysis of global mRNA expression in human skeletal muscle during recovery from endurance exercise. *FASEB J*. 2005;19:1498–1500.
28. Catoire M, Mensink M, Boekschoten MV, et al. Pronounced effects of acute endurance exercise on gene expression in resting and exercising human skeletal muscle. *PLoS One*. 2012;7:e51066.
29. Stepto NK, Coffey VG, Carey AL, et al. Global gene expression in skeletal muscle from well-trained strength and endurance athletes. *Med Sci Sports Exerc*. 2009;41:546–565.
30. Kraniou GN, Cameron-Smith D, Hargreaves M. Acute exercise and GLUT4 expression in human skeletal muscle: influence of exercise intensity. *J Appl Physiol*. 2006;101:934–937.
31. Leick L, Plomgaard P, Gronlokke L, Al-Abaiji F, Wojtaszewski JF, Pilegaard H. Endurance exercise induces mRNA expression of oxidative enzymes in human skeletal muscle late in recovery. *Scand J Med Sci Sports*. 2010;20:593–599.
32. Steensberg A, van Hall G, Keller C, et al. Muscle glycogen content and glucose uptake during exercise in humans: influence of prior exercise and dietary manipulation. *J Physiol*. 2002;541:273–281.
33. Richter EA, Hargreaves M. Exercise, GLUT4 and skeletal muscle glucose uptake. *Physiol Rev*. 2013;93:993–1017.
34. Knight JB, Eyster CA, Griesel BA, Olson AL. Regulation of the GLUT4 gene promoter: interaction between a transcriptional activator and myocyte enhancer factor 2A. *Proc Natl Acad Sci USA*. 2003;100:14725–14730.
35. McGee SL, van Denderen BJ, Howlett KF, et al. AMP-activated kinase regulates GLUT4 transcription by phosphorylating histone deacetylase 5. *Diabetes*. 2008;57:860–867.
36. Smith JAH, Kohn TA, Chetty AK, Ojuka EO. CaMK activation during exercise is required for histone hyperacetylation and MEF2A binding at the MEF2 site on the Glut4 gene. *Am J Physiol*. 2008;295:E698–E704.
37. MacLean PS, Zhang D, Jones JP, Olson AL, Dohm GL. Exercise-induced transcription of the muscle glucose transporter (GLUT4) gene. *Biochem Biophys Res Commun*. 2002;292:409–414.
38. McGee SL, Fairlie E, Garnham AP, Hargreaves M. Exercise-induced histone modifications in human skeletal muscle. *J Physiol*. 2009;587:5951–5958.
39. McGee SL, Hargreaves M. Exercise and myocyte enhancer factor 2 regulation in human skeletal muscle. *Diabetes*. 2004;53:1208–1214.
40. McGee SL, Howlett KF, Starkie RL, Cameron-Smith D, Kemp BE, Hargreaves M. Exercise increases nuclear AMPK alpha2 in human skeletal muscle. *Diabetes*. 2003;52:926–928.
41. McGee SL, Sparling D, Olson AL, Hargreaves M. Exercise increases MEF2 and GEF DNA binding activity in human skeletal muscle. *FASEB J*. 2006;20:348–349.
42. Murgia M, Elbenhardt JT, Cusinato M, Garcia M, Richter EA, Schiaffino S. Multiple signaling pathways redundantly control GLUT4 gene transcription in skeletal muscle. *J Physiol*. 2009;587:4319–4327.
43. Holmes BF, Lang DB, Birnbaum MJ, Mu J, Dohm GL. AMP kinase is not required for the GLUT4 response to exercise and denervation in skeletal muscle. *Am J Physiol*. 2004;287:E739–E743.
44. Garcia-Roves PM, Jones TE, Otani K, Han DH, Holloszy JO. Calcineurin does not mediate exercise-induced increase in muscle GLUT4. *Diabetes*. 2005;54:624–628.
45. McGee SL, Swinton C, Morrison S, et al. Compensatory regulation of HDAC5 in muscle maintains metabolic adaptive responses and metabolism in response to energetic stress. *FASEB J*. 2014;28:3384–3395.

46. Yan Z, Okutsu M, Akhtar YN, Lira VA. Regulation of exercise-induced fiber type transformation, mitochondrial biogenesis, and angiogenesis in skeletal muscle. *J Appl Physiol.* 2011;110:264–274.
47. Baar K, Wende AR, Jones TE, et al. Adaptations of skeletal muscle to exercise: rapid increase in the transcriptional coactivator PGC-1. *FASEB J.* 2002;16:1879–1886.
48. Wright DC, Han DH, Garcia-Roves PM, Geiger PC, Jones TE, Holloszy JO. Exercise-induced mitochondrial biogenesis begins before the increase in muscle PGC-1α expression. *J Biol Chem.* 2007;282:194–199.
49. Little JP, Safdar A, Cermak N, Tarnopolsky MA, Gibala MJ. Acute endurance exercise increases the nuclear abundance of PGC-1α in trained human skeletal muscle. *Am J Physiol.* 2010;298:R912–R917.
50. Little JP, Safdar A, Bishop D, Tarnopolsky MA, Gibala MJ. An acute bout of high-intensity interval training increases the nuclear abundance of PGC-1α and activates mitochondrial biogenesis in human skeletal muscle. *Am J Physiol.* 2011;300:R1303–R1310.
51. Safdar A, Little JP, Stokl AJ, Hettinga BP, Akhtar M, Tarnopolsky MA. Exercise increases mitochondrial PGC-1α content and promotes nuclear-mitochondrial cross-talk to coordinate mitochondrial biogenesis. *J Biol Chem.* 2011;286:10605–10617.
52. Smith BK, Mukai K, Lally JS, et al. AMP-activated protein kinase is required for exercise-induced peroxisome proliferator-activated receptor γ co-activator 1α translocation to subsarcolemmal mitochondria in skeletal muscle. *J Physiol.* 2013;591:1551–1561.
53. Bird A. Perceptions of epigenetics. *Nature.* 2007;447:396–398.
54. Evertts AG, Zee BM, Garcia BA. Modern approaches for investigating epigenetic signaling pathways. *J Appl Physiol.* 2010;109:927–933.
55. Yan ST, Li CL, Tian H, et al. MiR-199a is overexpressed in plasma of type 2 diabetes patients which contributes to type 2 diabetes by targeting GLUT4. *Mol Cell Biochem.* 2014;397:45–51.
56. Barres R, Yan J, Egan B, et al. Acute exercise remodels promoter methylation in human skeletal muscle. *Cell Metab.* 2012;15:405–411.
57. Nitert MK, Dayeh T, Volkov P, et al. Impact of an exercise intervention on DNA methylation in skeletal muscle from first-degree relatives of patients with type 2 diabetes. *Diabetes.* 2012;61:3322–3332.
58. Alibegovic AL, Sonne MP, Højbjerre L, et al. Insulin resistance induced by physical inactivity is associated with multiple transcriptional changes in skeletal muscle in young men. *Am J Physiol.* 2010;299:E752–E763.
59. Terruzzi I, Senesi P, Montesano A, et al. Genetic polymorphisms of the enzymes involved in DNA methylation and synthesis in elite athletes. *Physiol Genomics.* 2011;43:965–973.
60. Laker RC, Lillard TS, Okutsu M, et al. Exercise prevents maternal high-fat diet-induced hypermethylation of the Pgc-1α gene and age-dependent metabolic dysfunction in the offspring. *Diabetes.* 2014;63:1605–1611.
61. Lu C, Thompson CB. Metabolic regulation of epigenetics. *Cell Metab.* 2012;16:9–17.
62. Yu M, Stepto NK, Chibalin AV, et al. Metabolic and mitogenic signal transduction in human skeletal muscle after intense cycling exercise. *J Physiol.* 2003;546:327–335.
63. Potthoff MJ, Wu H, Arnold MA, et al. Histone deacetylase degradation and MEF2 activation promote the formation of slow-twitch myofibers. *J Clin Invest.* 2007;117:2459–2467.
64. Timmons JA, Knudsen S, Rankinen T, et al. Using molecular classification to predict gains in maximal aerobic capacity following endurance exercise training in humans. *J Appl Physiol.* 2010;108:1487–1496.

65. Bouchard C, Sarzynski MA, Rice TK, et al. Genomic predictors of the maximal O_2 uptake response to standardized exercise training programs. *J Appl Physiol.* 2011;110: 1160–1170.
66. Keller P, Vollaard NB, Gustafsson T, et al. A transcriptional map of the impact of endurance exercise training on skeletal muscle phenotype. *J Appl Physiol.* 2011;110:46–59.
67. Phillips BE, Atherton JP, Gustafsson T, et al. Molecular networks of human muscle adaptation to exercise and age. *PLoS Genet.* 2013;9:e1003389.
68. Ghosh S, Vivar JC, Sarzynski MA, et al. Integrative pathway analysis of a genome-wide association study of VO_2 max response to exercise training. *J Appl Physiol.* 2013;115:1343–1359.
69. Rowlands DS, Page RA, Sukala WR, et al. Multi-omic integrated networks connect DNA methylation and miRNA with skeletal muscle plasticity to chronic exercise in type 2 diabetic obesity. *Physiol Genomics.* 2014;46:747–765.
70. Philp A, Rowland T, Perez-Schindler J, Schenk S. Understanding the acetylome: translating target proteomics into meaningful physiology. *Am J Physiol.* 2014;307: C763–C773.
71. Thalacker-Mercer A, Stec M, Cui X, Cross J, Windham S, Bamman M. Cluster analysis reveals differential transcript profiles associated with resistance training-induced human skeletal muscle hypertrophy. *Physiol Genomics.* 2013;45:499–507.
72. Rowlands DS, Thomson JS, Timmons BW, et al. Transcriptome and translational signaling following endurance exercise in trained skeletal muscle: impact of dietary protein. *Physiol Genomics.* 2011;43:1004–1020.
73. Greenhaff PL, Hargreaves M. 'Systems biology' in human exercise physiology: is it something different from integrative physiology? *J Physiol.* 2011;589:1031–1036.

CHAPTER TWENTY

Exercise, Skeletal Muscle and Circulating microRNAs

Aaron P. Russell[1], **Séverine Lamon**
Centre for Physical Activity and Nutrition (C-PAN) Research, School of Exercise and Nutrition Sciences, Deakin University, Burwood, Victoria, Australia
[1]Corresponding author: e-mail address: aaron.russell@deakin.edu.au

Contents

Abstract

Regular exercise stimulates numerous structural, metabolic, and morphological adaptations in skeletal muscle. These adaptations are vital to maintain human health over the life span. Exercise is therefore seen as a primary intervention to reduce the risk of chronic disease. Advances in molecular biology, biochemistry, and bioinformatics, combined with exercise physiology, have identified many key signaling pathways as well as transcriptional and translational processes responsible for exercise-induced adaptations. Noncoding RNAs, and specifically microRNAs (miRNAs), constitute a new regulatory component that may play a role in these adaptations. The short single-stranded miRNA sequences bind to the 3′ untranslated region of specific messenger RNAs (mRNAs) on the basis of sequence homology. This results in the degradation of the target mRNA or the inhibition of protein translation causing repression of the corresponding protein. While tissue specificity or enrichment of certain miRNAs makes them ideal targets to manipulate and understand tissue development, function, health, and disease, other

Progress in Molecular Biology and Translational Science, Volume 135
ISSN 1877-1173
http://dx.doi.org/10.1016/bs.pmbts.2015.07.018

miRNAs are ubiquitously expressed; however, it is uncertain whether their mRNA/protein targets are conserved across different tissues. miRNAs are stable in plasma and serum and their altered circulating expression levels in disease conditions may provide important biomarker information. The emerging research into the role that miRNAs play in exercise-induced adaptations has predominantly focused on the miRNA species that are regulated in skeletal muscle or in circulation. This chapter provides an overview of these current research findings, highlights the strengths and weaknesses identified to date, and suggests where the exercise-miRNA field may move into the future.

1. INTRODUCTION

Physical exercise typically results in the activation of multiple intra- and extracellular signals that positively or negatively influence the transcription and translation of genes and proteins controlling skeletal muscle structure and function.[1,2] The transcription and translation processes are conjointly controlled by transcription factors,[3,4] histone modification,[5] and DNA methylation.[6,7] microRNAs (miRNAs) were first identified 20 years ago and have generated an additional level of complexity in our understanding of transcriptional and translational regulation.[8,9] miRNAs are short, single-stranded RNA molecules (20–30 nucleotides) that bind to specific messenger RNAs (mRNAs) on the basis of sequence homology[10,11] and repress the corresponding protein expression.[12,13] miRNAs usually bind to specific sites of the 3′ untranslated region (UTR) of their target transcripts[11] thanks to a conserved nucleotide sequence called the "seed" region and located at base positions 2–8 on the 5′ end of the miRNA. miRNA/mRNA binding principally results in the degradation of the target mRNA,[12] but can sometimes directly inhibit protein translation. In rare cases, miRNAs can also stabilize mRNA targets.[14] miRNAs are now demonstrated major regulators of biological processes[15] and play an essential role in the cellular response to stress stimuli and the maintenance of healthy cellular function.

Specific miRNAs are found to be particularly enriched in specific tissues.[16] MyomiR is the term referring to the miRNAs highly expressed in skeletal muscle and includes miR-1, miR-133a, miR-133b, miR-206, miR-208, miR-208b, miR-486, and miR-499.[17,18] These miRNAs regulate fundamental biological processes in the muscle, including muscle growth, development, metabolism, and repair.[19] For example, myogenesis is a complex process mediated by several key transcription factors (known as myogenic regulatory factors, MRFs), including MyoD, myogenin, Myf5, and MRF4,[20,21] that closely control and coordinate every aspect of muscle growth. miRNAs in turn regulate the MRFs and therefore play an essential role in the control of muscle development. Similarly, miRNAs have been shown to be involved in muscle fiber-type regulation,[22] muscle protein synthesis,[23,24] and muscle regeneration.[25] In the muscle, physical exercise triggers multiple short- and long-term adaptive physiological responses that essentially depend on these biological processes. Therefore, exercise influences the expression of the miRNAs present in skeletal muscle as well as the members of their biogenesis machinery.

2. THE REGULATION OF miRNA BIOGENESIS MACHINERY WITH EXERCISE

miRNAs originate from the eukaryotic genome.[15] In the nucleus, miRNAs are primarily transcribed in a several kilobases-long double-stranded RNA molecule referred to as primary miRNAs (pri-miRNA).[15] Pri-miRNAs are cleaved by the RNase-III-type endonuclease Drosha associated with Pasha (DGCR8) into 60–70 nt long precursor molecules called pre-miRNAs.[26] Exportin-5 (XPO5) then mediates the transport of pre-miRNAs into the cytoplasm,[27] where Dicer-1 cleaves them into their mature form.[26] Two mature miRNAs can originate from the opposite arms of the same pre-miRNA and are differentiated with a -3p or -5p suffix. Alternatively, the less abundant of the two mature miRNA species is sometimes identified by an asterisk following its name. For example, miR-149 and miR-149* share a same pre-miRNA sequence, but miR-149 is more abundant in the cell. To exert their effects, mature miRNAs are incorporated into the RNA-induced silencing complex (RISC),[15,28] a protein complex that uses the miRNA as a template for identifying and binding the complementary mRNA molecule, binding that eventually results in the repression of the target expression. Very few studies have investigated the influence of exercise on the miRNA biogenesis machinery. To date, an increase in XPO5, Drosha, and Dicer mRNA has been observed 3 h following a single bout of endurance exercise in untrained males.[29] Similarly, XPO5 and Drosha mRNA were increased in old but not young subjects during the 6-h period following a single-bout resistance exercise.[30] Surprisingly, XPO5 protein levels were reduced in both groups. Presently, how the exercise-induced changes in components of the miRNA biogenesis machinery influence miRNA levels has not been determined. In addition to XPO5 exporting pre-miRNA from the nucleus to the cytoplasm, it can also stabilize pre-miRNAs[31] and potentially increase the pool of pre-miRNAs. With this in mind, the exercise-induced upregulation of skeletal muscle XPO5 may assist in processing of new pre-miRNAs that regulate exercise-induced adaptations.

3. THE REGULATION OF SKELETAL MUSCLE miRNAs BY EXERCISE

Exercise plays an important role in maintaining muscle health throughout the lifespan, with resistance exercise being a potent anabolic stimulus that enhances muscle protein synthesis and muscle growth.[32–36] On the other hand, endurance exercise is essential to maintain whole-body energy metabolism, largely via the uptake, storage, and oxidation of metabolic substrates by skeletal muscle. As miRNAs influence numerous intracellular networks, their regulation by exercise and contribution to exercise-induced adaptations is starting to be investigated.

3.1 Resistance Exercise—Single-Bout Exercise

Few studies have investigated the changes in skeletal muscle miRNA species following a single bout of resistance exercise. Six young (29 ± 2 years) and six older (70 ± 2 years) subjects completed 8×10 repetitions of two-legged extension exercises at 70% of 1 repetition maximum (1 RM).[30] One hour postexercise, the subjects consumed 20 g of a leucine-enriched essential amino acid (EAA) solution. Molecular analysis was performed in muscle samples taken preexercise, 3 and 6 h postexercise. Expression of primary (pri-), but not mature, miR-1 and 133a was higher in the older when compared to the younger subjects. At 6 h postexercise, pri-miR-1 and -133a were reduced in the young but not the older subjects. Pri-miR-206 was elevated at 3 and 6 h postexercise in the older subjects only. Compared to basal, the mature miR-1 was decreased only in the young subjects 3 and 6 h post-exercise. The functional significance of elevated basal levels of pri-miRNAs without a concomitant increase in the mature miRNA isoforms is unknown. Furthermore, the biological role of decreasing miR-1 levels post resistance exercise combined with EAA ingestion has not been established. *In vitro*, miR-1 inhibits IGF-1.[18,37] It is plausible that miR-1 down-regulation post resistance exercise combined with EAA ingestion may promote muscle protein synthesis by removing its repression on IGF-1. Therefore, failure to downregulate miR-1 in elderly subjects following an acute bout of resistance exercise and EAA ingestion may be partially responsible for the attenuated muscle protein synthesis in response to the anabolic stimuli.

High-throughput miRNA screening has recently been performed in muscle biopsies taken from young and older subjects following a single-bout of resistance exercise. In a study by Rivas *et al.*, eight young (22 ± 1 years) and eight older (74 ± 2 years) subjects completed 3 × 10 repetitions of two-legged knee extension exercises at 80% of 1 RM.[23] Biopsies were taken pre- and 6 h postexercise from the left and right vastus lateralis muscles, respectively. miRNA analysis was performed on a custom-made PCR array containing 60 assays for miRNAs previously observed to play a role in the regulation of skeletal muscle function. At baseline, 14 miRNAs were significantly lower in the older subjects compared to the younger subjects. Exercise did not alter the expression of any miRNAs in the older subjects but decreased 16 and increased 1 miRNA species in the younger subjects. The lack of miRNA regulation in the older subjects was mirrored at the mRNA level, with 42 genes altered in the older subjects compared with 175 genes in the younger subjects, when measured 6 h postexercise. Of the miRNAs identified, bioinformatics analysis predicted that miR-126 may regulate targets involved in muscle growth. An important addition to this human exercise study was the *in vitro* investigation into the potential growth mechanisms controlled by miR-126. Inhibiting miR-126 in C2C12 myoblasts increased IRS1 protein levels and decreased FOXO1 protein levels, while increasing miR-126 levels in myoblasts reduced MyoD and Myf5 protein levels. In differentiated C2C12 myotubes, inhibiting miR-126, followed by anabolic simulation with IGF-1 for 30 min, increased the phosphorylation of Akt and its downstream target FOXO1 to a greater extent than in the vehicle-treated cells. These *in vitro* observations suggest that downregulating miR-126 following resistance exercise may be an anabolic adaptive response; however, this requires validation *in vivo*. Zacharewicz *et al.* completed an extensive high-throughput miRNA screening following single-bout resistance exercise in 10 young (24 ± 0.9 years) and 10 older (67 ± 1.0 years) males.[24] Here, the subjects completed 3 × 14 repetitions of leg extension exercise at 60% of 1 RM. The expression levels of 754 microRNAs were screened via PCR in the vastus lateralis muscle biopsy samples taken before and 2 h postexercise. Of these, 257 miRNA species were reliably expressed in the muscle samples analyzed. Twenty-six miRNAs were regulated with age, exercise, or a combination of both factors. Bioinformatics pathway analysis predicted nine of these microRNAs to regulate targets within the Akt-mTOR signaling pathway. Specifically, six of these miRNAs, miR-99a, miR-99b, miR-100, miR-149-3p, miR-196, and miR-199a, have been validated to target several mRNAs that encode proteins within the protein synthesis pathways in various cell types.[38–44]

3.2 Resistance Exercise—Training

The relationship between miRNA regulation and muscle growth following resistance training was investigated in 56 young (18–30 years) healthy males. Resistance training used a progressive overload design and was performed 5 days per week, consisting of upper and lower body exercises. Table 1 provides details of the different training protocols used. Selected miRNAs were measured in skeletal muscle samples of subjects defined as "high responders" versus "low responders" ($n=8$–9 per group). High responders had an increase of approximately 4.5 kg of lean body mass (LBM), while low responders were judged as those having little or no muscle hypertrophy following the training intervention.[45] Following training, there was an increase in skeletal muscle miR-451 and a decrease in miR-26a, miR-29a, and miR-378 in the "low responder" group only. A positive correlation was observed between the LBM (kg) and the change in miR-378 levels. A gene ontology analysis of the predicted targets of these miRNAs suggested the regulation of the mTOR pathway, known to positively regulate muscle growth[48], extra-cellular matrix (ECM)-receptor interaction and focal adhesion kinase, both involved in skeletal muscle mechanotransduction.[49] Low muscle hypertrophy response to resistance exercise training in healthy young subjects is referred to as anabolic resistance,[49] and this phenomenon is also linked to age-related muscle wasting. Whether miR-451, miR-26a, miR-29a, and miR-378 contribute to the resistance exercise training-induced muscle hypertrophy and the mechanisms they control still require experimental validation.

3.3 Endurance Exercise—Single-Bout Exercise

Single bout of endurance exercise has also been used to investigate changes in skeletal muscle miRNAs. Skeletal muscle samples were taken before and 1 and 3 h after 60 min of cycling at 60% of peak Watt in 10 untrained healthy men (30.5 ± 5 years).[46] An increase in miR-1 and miR-133a was observed 1 h, but not 3 h postexercise. Interestingly, when the same subjects performed the same single exercise bout in the trained state, there was no increase in these miRNAs. This suggests that unaccustomed endurance exercise induces a stress that rapidly and transiently upregulates several

Table 1 Overview of Studies Focused on Skeletal Muscle miRNAs Following Acute and Long-Term Resistance and Endurance Exercise

References	Type of Exercise	Protocol	Time	miRNA ↑	miRNA ↓	miRNAs Investigated (qPCR)
30	Acute resistance + EAA ingestion Young subjects	8 × 10 repetitions of two-legged extension exercises at 70% of 1 repetition maximum (1 RM) with 3 min rest between sets	3 h 6 h	N/A	miR-1 miR-1	3
30	Acute resistance + EAA ingestion Old subjects	8 × 10 repetitions of two-legged extension exercises at 70% of 1 repetition maximum (1 RM) with 3 min rest between sets	3 h 6 h	N/A	N/A	3
23	Acute resistance Young subjects	3 × 10 repetitions of two-legged knee extension exercises at 80% of 1 repetition maximum (1 RM)	6 h	miR-23b-3p, miR-24-3p, miR-26a-5p, miR-26b-5p, miR-27a-3p, miR-27b-3p, miR-29c-3p, miR-30a-5p, miR-30d-5p, miR-95-3p, miR-126-3p, miR-133a, miR-133b, miR-140-3p, miR-181a-5p, miR-378a-5p	miR-423-5p	60

Continued

Table 1 Overview of Studies Focused on Skeletal Muscle miRNAs Following Acute and Long-Term Resistance and Endurance Exercise—cont'd

References	Type of Exercise	Protocol	Time	miRNA ↑	miRNA ↓	miRNAs Investigated (qPCR)
23	Acute resistance Old subjects	3 × 10 repetitions of two-legged knee extension exercises at 80% of 1 repetition maximum (1 RM)	6 h	N/A	N/A	60
24	Acute resistance Young subjects	3 × 14 repetitions of leg extension exercise at 60% of 1 RM consisting of 3 × 14 repetitions with 2-min rest periods between each set	2 h	miR-486-3p	miR-149-3p, miR-99b-5p, miR-520-5p	754
24	Acute resistance Old subjects	3 × 14 repetitions of leg extension exercise at 60% of 1 RM consisting of 3 × 14 repetitions with 2-min rest periods between each set	2 h	miR-499a-5p, miR-99a-5p	miR-489-5p, miR-196b-5p, miR-628-5p, miR-186-5p, miR-335-5p	754

45	Resistance training Low responders only	5 days/week for 12 weeks including pushing exercises, including military press, bench press, seated chest fly, and seated triceps extension; pulling exercises, including seated lateral pull down, seated wide grip row, seated reverse fly, seated biceps curl, and a series of abdominal exercises without weights; and leg exercises including incline (45°) leg press, 2-leg knee extension, 2-leg hamstring curl, and seated calf raise. All exercises were performed at 80% of 1 RM. The program was progressively overloaded over 12 weeks	Posttraining	miR-451	miR-26a, miR-29a, miR-378	21
46	Acute endurance Young subjects—untrained only	60 min of cycling at 60% of peak Watt	1 h	miR-1, miR-133	N/A	4

Continued

Table 1 Overview of Studies Focused on Skeletal Muscle miRNAs Following Acute and Long-Term Resistance and Endurance Exercise—cont'd

References	Type of Exercise	Protocol	Time	miRNA ↑	miRNA ↓	miRNAs Investigated (qPCR)
46	Acute endurance Young subjects—untrained only	60 min of cycling at 60% of peak Watt	3 h	N/A	N/A	4
29	Acute endurance Young subjects	60 min of cycling at 70% of VO_2 peak	3 h	miR-1, miR-133a, miR-133b, miR-181	N/A	12
46	Endurance training	Cycling training 5 days/week for 12 weeks (75–90% P_{max} for 70–81 min)	Posttraining	N/A	miR-1, miR-133a, miR-133b, miR-206	4
47	Endurance training	Cycling training 4 days/week for 6 weeks (45 min at 70% VO_{2max})	Posttraining	miR-125a, miR-183, miR-189, miR-432*, miR-575, miR-616, miR-637	miR-101, miR-133, miR-144, miR-15b, miR-26b, miR-28, miR-29b, miR-338, miR-455, miR-92, miR-98, miR-451, miR-589, miR-1	1488 (probes) 4 confirmed by qPCR

29	Endurance training	10 days cycling training (75% of VO_{2peak} for 45 min on days 1, 5, 6, and 10, 60 min on day 3, and 90 min on day 8; 6 × 5 min intervals at 90–100% VO_{2peak} with 2 min recovery below 40% VO_{2peak} on days 2, 4, 7, and 9)	Posttraining	miR-1, miR-29b	miR-31	12

muscle-enriched miRNAs. How this regulation influences muscle adaptation to exercise is not known. Similarly, in nine healthy untrained males (23 ± 5 years) miR-1, -133a, -133-b, and miR-181a were increased in skeletal muscle samples taken 3 h following 60 min of cycling at 70% of VO_2 peak.[29] In this study, several miRNAs previously shown to be dysregulated in muscle myopathies, including miR-9, -23a, -23b, and -31, were decreased. Negative correlations were observed between miR-9 and HDAC4 protein, miR-31 and HDAC4 protein, and between miR-31 and NRF1 protein levels when measured 3 h postexercise. *In vitro* reporter assays performed in C2C12 myoblasts indicated that miR-31 directly interacts with HDAC4, a component of the MAPK pathway, as well as with nuclear respiratory factor-1 (NRF1), which is involved in mitochondrial biogenesis and metabolism. The *in vivo* human observations combined with the *in vitro* data suggest that mitochondrial and metabolic adaptations in skeletal muscle following endurance exercise may involve the regulation of miR-31. The direct role of miR-31 in skeletal muscle *in vivo* still requires experimental validation.

3.4 Endurance Exercise—Training

The effect of supervised endurance cycling training over a period of 5 days per week for 12 weeks (55–90% at maximal power for 60–150 min, depending on the intensity) on the expression of muscle-enriched miRNAs in 10 healthy males (30 ± 5 years) was investigated. miR-1, miR-133a, miR-133b, and miR-206 were all significantly downregulated following training. These miRNAs returned to pretraining baseline levels 2 weeks after the cessation of training.[46] While the transient response of these miRNAs supports an endurance training effect, their role in muscle adaptation to exercise is unknown. Following 6 weeks of supervised endurance training (4 × 45 min at 70% VO_{2max} per week) in young sedentary males (23 ± 1 year), a miRNA screening identified 14 miRNAs that were decreased and 7 that were increased in skeletal muscle.[47] Of the 14 miRNAs that were decreased, miR-92, -98, -101, and -104 were predicted to target RUNX1, SOX9, and PAX3. These genes targets were upregulated in the same muscle samples suggesting that the downregulation of miR-92, -98, -101, and -104 during endurance training may permit aerobic adaptation to occur. However, this requires further experimental validation. The effect of 10 days endurance training consisting of cycling at 75% of VO_2 peak or intervals at 90–100% VO_2 peak on skeletal muscle miRNA levels has been investigated. This protocol increased miR-1, concomitantly with an increase in miR-29b and a decrease in miR-31.[29] These studies demonstrate that muscle miRNA expression is sensitive to endurance exercise training and can be rapidly reversed following a period of inactivity. The precise miRNA targets and the molecular processes regulated by changes in these miRNAs remain to be established and should be a priority for future research.

3.5 Role of miRNAs Is the Adaptation to Exercise

Validating the precise role of a single miRNA or miRNA family in exercise-induced adaptations requires a controlled manipulation of the miRNA species of interest. At present, these experiments can only be performed *in vivo* in non-human models. So far, the role of miR-23a in regulating adaptations to endurance exercise has been investigated. miR-23a binds to and downregulates peroxisome proliferator-activated receptor gamma, coactivator 1 alpha (PGC-1α) in a 3′UTR-dependent manner *in vitro*.[50] Endurance exercise in mice and humans increases skeletal muscle PGC-1α expression[1,51,52] and reduces miR-23a expression.[29,53] Transgenic mice overexpressing miR-23a (whole body) have a reduction in skeletal muscle PGC-1α mRNA and protein and in several PGC-1α downstream targets required for efficient mitochondrial function.[54] If the exercise-induced downregulation of miR-23a is required for adaptation, then it might be expected that mice with elevated miR-23a levels may not adapt to exercise as well as wild-type mice. Under basal conditions and when compared to wild-type mice, miR-23a transgenic (miR-23a Tg) mice have lower levels of protein markers of mitochondrial content, including PGC-1α, and cytochrome *c* oxidase complex IV (COX IV), in the slow soleus muscle, but not the fast plantaris muscle. There is also a decrease in type IId/x fibers only in the soleus muscle. Following 4 weeks of voluntary wheel running, there was no difference in the endurance exercise capacity or in several muscle adaptive responses, including an increase in muscle mass, capillary density, or the protein content of myosin heavy-chain (MHC) IIa, PGC-1α, COX IV, and cytochrome *c*, between the miR-23a Tg mice and the wild-type mice. These results show that miR-23a targets PGC-1α and regulates basal metabolic properties of slow-, but not fast-twitch muscles. Elevated levels of miR-23a do not impact whole-body endurance capacity or exercise-induced muscle adaptations in the fast plantaris muscle. This suggests that miR-23a plays a limited role in regulating the skeletal muscle adaptation to endurance exercise.

Future studies investigating the role of other miRNAs, individually or in combination, in exercise adaptation are required. Alternative technologies to manipulate miRNA species *in vivo*, as well as the use of different *in vivo* models, are required to identify the miRNAs that may be responsible for exercise-induced health and performance adaptations.

4. THE REGULATION OF CIRCULATING miRNAs BY EXERCISE

The expression of miRNAs across the body may be tissue specific or tissues may be enriched with certain miRNAs. In addition, miRNAs are present and highly stable in the bloodstream.[55,56] Whereas miRNAs circulating under their free form would be susceptible to ribonucleases (RNases), blood miRNAs exist within exosomes, lipoprotein, and ribonucleoprotein complexes that prevent them from RNAse degradation.[57,58] It is likely that circulating miRNAs originate from tissues, and from hematopoietic cells in particular.[59,60] However, their function in circulation remains mostly unclear. Recent studies suggested a possible role in cell-to-cell communication, where circulating miRNAs might be able to mediate gene expression in target tissues in a way comparable to hormones and cytokines.[59,61,62] Aberrant expression of certain miRNAs in blood has been linked to specific pathologies or health conditions,[55,63–65] and recently, it became apparent that an abnormal blood expression pattern might constitute the signature of a disease burden in a specific tissue. It follows that miRNAs are now considered as promising biomarkers for various diseases, including cancer, type 2 diabetes, and other chronic syndromes. Stress situations such as nutrition, hypoxia,[66,67] and exercise also have the potential to modify the blood miRNA expression profile. Circulating miRNAs may therefore represent important novel indicators of muscle adaptation to exercise.

4.1 Endurance Exercise—Single-Bout Exercise

Studies have shown that various types and durations of exercise induce specific changes in plasma or serum miRNA expression. Table 2 provides a summary of the exercise-induced changes in blood miRNAs following exercise. Single, acute exercise sessions have the ability to elicit rapid modifications of the blood miRNA profile. Early studies investigated the expression of circulating miRNAs immediately after exercise with a focus on muscle-enriched miRNAs (myomiRs), but also on miRNAs involved in angiogenesis, inflammation, ischemia, and hypoxia and therefore,

Table 2 An Overview of Circulating miRNAs Relevant to Exercise and Muscle Function That Are Regulated Immediately After Acute Endurance or Resistance Exercise

miRNA	Roles	Exercise Stimulus ↑miRNA	Exercise Stimulus ↓miRNA	No Effect	Detectability
miR-146a	Inflammation Hypoxia	Cycling VO_{2max} test (pre- and posttraining)[68]	1 h cycling 65% VO_{2max} (trained)[69]		–
miR-221	Angiogenesis Muscle cell differentiation	Cycling VO_{2max} test (pretraining)[68]	1 h cycling 65% VO_{2max} (trained)[69]	Cycling VO_{2max} test (posttraining)[68]	–
miR-222	Angiogenesis Muscle cell differentiation	Cycling VO_{2max} test (pre- and posttraining)[68]		1 h cycling 65% VO_{2max} (trained)[69]	–
miR-21	Inflammation Hypoxia	Cycling VO_{2max} test (pretraining)[68]		Cycling VO_{2max} test (posttraining)[68] 1 h cycling 65% VO_{2max} (trained)[69]	–
miR-133a	Muscle function Muscle cell proliferation	Marathon race (trained) Resistance exercisemax intensity[70–72]		Cycling VO_{2max} test (pre- and posttraining)[68] 4 h cycling 70% VO_{2max} (trained)[71] 1 h cycling 65% VO_{2max} (trained)[69] resistance exercise 70% max[73]	Questioned[74]

Continued

Table 2 An Overview of Circulating miRNAs Relevant to Exercise and Muscle Function That Are Regulated Immediately After Acute Endurance or Resistance Exercise—cont'd

miRNA	Roles	Exercise Stimulus ↑miRNA	Exercise Stimulus ↓miRNA	No Effect	Detectability
miR-126	Vascular endothelium Protein synthesis	Marathon race (trained)[71] 4 h cycling 70% VO_{2max} (trained)[71]		1 h cycling 65% VO_{2max} (trained)[69]	–
miR-1	Muscle function Muscle cell differentiation	Marathon race (trained)[70,72]		1 h cycling 65% VO_{2max} (trained)[69]	Questioned[74]
miR-499	Muscle function MHC expression fiber type	Marathon race (trained)[70,72]		1 h cycling 65% VO_{2max} (trained)[69]	Questioned[74]
miR-206	Muscle function Muscle cell differentiation	Marathon race (trained)[72]		1 h cycling 65% VO_{2max} (trained)[69]	Questioned[74,75]
miR-208b	Muscle function MHC expression fiber type	Marathon race (trained)[72]		1 h cycling 65% VO_{2max} (trained)[69]	Questioned[74]
miR-486	Protein synthesis		1 h cycling 70% VO_{2max} (untrained)[74]	1 h cycling 65% VO_{2max} (trained)[69]	–

potentially regulating biological processes relevant to exercise adaptation. In competitive rowers (19.1 ± 0.6 years), a classical $\dot{V}O_{2max}$ test completed on a cycling ergometer to maximal exertion transiently increased the expression levels of circulating miR-146a, miR-222, miR-21, and miR-221; miRNAs all involved in angiogenesis or in hypoxia-induced stress response.[76–79] Interestingly, miR-146a and miR-222 but not miR-21 and miR-221 were responsive to the same acute exercise bout following 90 days of organized, team-based rowing training period aimed to optimize 5 km performance, suggesting a training-induced habituation specific to some plasma miRNAs.[68] Three studies showed that a marathon race increased the expression levels of circulating miR-133a,[70–72] with additional increases reported for miR-126[71] and for several myomiRs including miR-1, miR-499, miR-206, and miR-208b.[70,72] Four hours of cycling exercise at 70% $\dot{V}O_{2max}$ exclusively increased miR-126 plasma concentration in 12 healthy well-trained men (32.4 ± 2.3 years).[71] miR-133 is a muscle-enriched miRNA that is known to regulate myogenesis *in vitro* via increasing myoblast proliferation.[80] Its expression levels in skeletal muscle are sensitive to muscle contraction in response to various types of endurance exercise,[46,81,82] making it a good candidate for potential postexercise regulation at the plasma level. However, a classical $\dot{V}O_{2max}$ test performed on a cycling ergometer or a 4-h cycling exercise at 70% $\dot{V}O_{2max}$ failed to increase its expression in plasma in young healthy subjects,[68,71] while another study reported that circulating miR-133a and other myomiRs, including miR-1, miR-499, miR-206, and miR-208b, were too lowly expressed at rest and following exercise to be reliably measured in serum,[74] an observation partly shared by others.[75] Interestingly, miR-126, an endothelial-specific miRNA that was originally investigated as a marker for endothelial cell damage,[83,84] was also identified as a regulator of the transcriptional response to resistance exercise able to activate IGF-1/Akt signaling *in vitro*.[23] Whether it plays a different role in endurance exercise-induced adaptation is unknown. In contrast to miR-126, miR-486, a recognized myomiR that also positively regulates the IGF-1/Akt pathway,[85] was significantly decreased following a 60-min cycling exercise at 70% $\dot{V}O_{2max}$ in 11 healthy young men (21.5 ± 4.5 years).[74] Exercise-induced changes in circulating myomiRs are contraction mode dependent. In nine young healthy males (27–36 years), 30 min eccentric (downhill) but not concentric (uphill) walking exercise induced an increase of miR-1, miR-133a, and miR-208b during the early recovery period and were therefore proposed as alternative biomarkers for contraction-induced muscle damage.[75] A more recent study used miRNA array technology to screen the expression levels of circulating miRNAs following moderate intensity cycling exercise.[69] The advantage of this platform is that it allows for the simultaneous and accurate quantitation of hundreds of human miRNAs and importantly contains their own normalization miRNAs and internal controls. It showed that a 60-min cycle ergometer exercise bout at 65% of maximum power ($n=13$ trained men, 28 ± 8 years) decreased the expression levels of eight circulating miRNAs immediately after exercise,[69] including miR-221 and miR-146, two miRNAs for which an increase had previously been reported following a classical VO_{2max} test performed on a cycling ergometer.[68]

4.2 Resistance Exercise—Single-Bout Exercise

The effect of an acute bout of resistance exercise on circulating miRNA expression has been less studied. A miRNA array completed after a resistance exercise session consisting of bench press and bilateral leg press repetitions at 70% of maximum strength revealed no immediate changes in plasma miRNA expression in 12 males aged 29.9 ± 1.2.[73] This is in contrast to another study reporting that in 13 men and women aged 37 ± 2, circulating miR-133a increased immediately after a single bout of resistance exercise constituted of three different resistance exercises (lateral pulldown, leg press, and butterfly) performed at maximal intensity.[71] However, an increase in miR-149*, a miRNA that directly targets the Akt signaling pathway in cancer cells,[43] and a decrease in miR-221 and miR-146a levels were observed 3 days post-resistance exercise.[73] Interestingly, these miRNAs involved in angiogenesis and stress-induced response, respectively, are the same that were reported to decrease immediately after a 60-min cycling exercise, suggesting a dynamic and exercise mode-dependant regulation of the blood miRNA profiles.[69]

4.3 Exercise Training

Little is known about the regulation of circulating miRNAs with long-term exercise training. Trained rowers aged 19.1 ± 0.6 were subjected to plasma sampling prior and following a 3-month sustained training period.[68] Plasma levels of miR-20a, a miRNA mediating angiogenesis,[86] were significantly higher at the completion of the training period. Interestingly, none of the miRNAs influenced by an acute cycling session in the same athletes pre- and posttraining[68] were altered by sustained training. Finally, miR-486 plasma levels decreased following 4 weeks of cycling training

(70% $\dot{V}O_{2max}$ for 30 min three times per week) in 11 healthy young men (21.5 ± 4.5 years).[74]

4.4 Discrepancies in Circulating miRNA Detectability and Regulation

Discrepancies in circulating miRNA detectability at baseline or in miRNA regulation with exercise are not unusual. Differences between exercise protocols, subjects training, or fitness status certainly account for some of these contrasting results. However, these disparities may also be attributable to the various miRNA extraction protocols[87] and normalization approaches used in the different studies. There is therefore an urgent need to establish stringent standards in both domains, a process successfully initiated in a recent study.[69] Moreover, hemolysis is susceptible to occur during blood collection and to induce changes in circulating miRNA levels that will reflect exercise-induced changes in blood cell numbers[88–90] rather than muscle-specific adaptations. Altogether, these factors can potentially generate significant differences in the expression levels of circulating miRNAs and need to be considered interpreting results and comparing studies.

4.5 Circulating miRNAs as Potential Biomarkers of the Adaptation to Exercise

Associations between blood miRNAs profile and aerobic exercise capacity have been made. An early study demonstrated a positive linear correlation between training-induced changes in miR-20a levels and training-induced changes in $\dot{V}O_{2max}$, suggesting for the first time that circulating miRNAs could represent potential biomarkers of cardiorespiratory fitness trainability.[68] Similarly, the magnitude of the decrease in miR-486 following a single bout of steady-state cycling exercise at 70% $\dot{V}O_{2max}$ for 60 min was negatively correlated with the subjects $\dot{V}O_{2max}$[74] and the magnitude of the increase in the plasma concentrations of the myomiRs miR-1, miR-133a, and miR-206 following a marathon race was correlated to several aerobic performance parameters including $\dot{V}O_{2max}$.[72] A screening study confirmed this hypothesis and revealed that the basal expression levels of three miRNAs involved in hypoxia adaptation (miR-21 and miR-210) and/or angiogenesis (miR-210 and miR-222)[77,79,91,92] were increased in healthy subjects with low $\dot{V}O_{2max}$ when compared to subjects with high $\dot{V}O_{2max}$.[93] Altogether, these studies suggest interesting perspectives for circulating miRNAs, which might one day be used as biomarkers to assess adaptation to training, fitness, or blood doping status.[94]

5. LIMITATIONS AND CONCLUSION

While the important role played by miRNAs in health and disease is now well established, the regulation of the different miRNA species with exercise has just started to be investigated. The majority of studies have investigated the influence of exercise stimuli on the expression of skeletal muscle and circulating miRNAs. However, as other systems, including the immune and the cardiovascular system, are also responsive to exercise, it is nonsurprising that changes in the corresponding miRNA expression have been reported. For example, an acute bout of resistance exercise completed by untrained subjects led to significant changes in the expression of 20–40 miRNAs in different populations of circulating leukocytes, including neutrophils, peripheral blood mononuclear cells, and natural killer cells.[95–97] Similarly, in healthy young males, 30 min of treadmill running at 80% VO_{2max} led to the differential regulation of 56 miRNAs expressed in leukocytes postexercise. Of interest, four of these miRNAs were identified to potentially regulate telomeric mRNAs, providing a potential mechanism for the maintenance of longer leukocyte telomere length with exercise.[98] Exercise training also induces cardiac hypertrophy. In rodents, swimming and treadmill running training regulate miRNA expression in the heart.[99] These exercise-induced responses may be able to restore the normal expression of some miRNAs that are aberrantly expressed in cardiac pathologies, such as miR-208a,[99] a negative regulator of the contractile efficiency of the heart. Interestingly, some of the miRNAs identified in leukocytes had a similar response to exercise in other tissues such as skeletal muscle, heart, or plasma, suggesting the existence of individually coordinated regulation processes potentially cross talking between the different systems.

To date, most of the exercise studies are purely descriptive, measuring changes in miRNA expression following acute and long-term exercise in skeletal muscle, serum, or plasma. These studies have contributed to our understanding of which miRNAs are sensitive to the stress of exercise-induced muscle contractions. However, they have not yet identified the role that miRNAs play in skeletal muscle adaptation to exercise. To move the field, forward studies are now required to establish biological cause-and-effect relationships and to identify the specific genes or proteins regulated by these miRNAs during exercise or during the recovery period post-exercise. This is necessary to fully understand the role that specific miRNAs play in the numerous skeletal muscle adaptations, such as muscle structure, function, and metabolism, in response to acute and chronic exercise. *In vitro* experimental validation using miRNA mimics or inhibitors as well as reporter assay experiments are required to investigate the potential mechanisms regulated by miRNAs in skeletal muscle cells and their target genes and proteins. These steps are essential to identify the most promising miRNA candidates and to validate their precise function in exercise-induced adaptations by manipulating their expression *in vivo* in non-human models. Presently, most studies investigate only a limited number of miRNAs and use differing exercise models and protocols. This often fails to provide reliable comparison points between different studies. Implementing high-throughput miRNA screening technologies that allow the accurate quantification of hundreds of miRNAs as well as using consistent endurance and resistance exercise protocols will help to overcome this problem. Ultimately, the use of deep-sequencing technology might allow the discovery of numerous, so far unidentified miRNA species that may be of importance in the regulation of the molecular response of human skeletal muscle to exercise in health and disease conditions.

REFERENCES

1. Russell AP, Hesselink MK, Lo SK, Schrauwen P. Regulation of metabolic transcriptional co-activators and transcription factors with acute exercise. *FASEB J.* 2005;19:986–988.
2. Short KR, Vittone JL, Bigelow ML, et al. Impact of aerobic exercise training on age-related changes in insulin sensitivity and muscle oxidative capacity. *Diabetes.* 2003;52:1888–1896.
3. Keller C, Steensberg A, Pilegaard H, et al. Transcriptional activation of the IL-6 gene in human contracting skeletal muscle: influence of muscle glycogen content. *FASEB J.* 2001;15:2748–2750.
4. McGee SL, Sparling D, Olson AL, Hargreaves M. Exercise increases MEF2- and GEF DNA-binding activity in human skeletal muscle. *FASEB J.* 2006;20:348–349.
5. McGee SL, Fairlie E, Garnham AP, Hargreaves M. Exercise-induced histone modifications in human skeletal muscle. *J Physiol.* 2009;587:5951–5958.
6. Barres R, Yan J, Egan B, et al. Acute exercise remodels promoter methylation in human skeletal muscle. *Cell Metab.* 2012;15:405–411.
7. Nakajima K, Takeoka M, Mori M, et al. Exercise effects on methylation of ASC gene. *Int J Sports Med.* 2010;31:671–675.
8. Lee RC, Feinbaum RL, Ambros V. The *C. elegans* heterochronic gene lin-4 encodes small RNAs with antisense complementarity to lin-14. *Cell.* 1993;75:843–854.
9. Reinhart BJ, Slack FJ, Basson M, et al. The 21-nucleotide let-7 RNA regulates developmental timing in *Caenorhabditis elegans. Nature.* 2000;403:901–906.
10. Brennecke J, Stark A, Russell RB, Cohen SM. Principles of microRNA–target recognition. *PLoS Biol.* 2005;3:e85.
11. Hu Z, Bruno AE. The influence of 3′UTRs on microRNA function inferred from human SNP data. *Comp Funct Genomics.* 2011;2011:1–9.

12. Huili G, Ingolia NT, Weissman JS, Bartel DP. Mammalian microRNAs predominantly act to decrease target mRNA levels. *Nature*. 2010;466:835–840.
13. Pillai RS, Bhattacharyya SN, Artus CG, et al. Inhibition of translational initiation by Let-7 microRNA in human cells. *Science*. 2005;309:1573–1576.
14. Vasudevan S, Tong Y, Steitz JA. Switching from repression to activation: microRNAs can up-regulate translation. *Science*. 2007;318:1931–1934.
15. Bartel DP. MicroRNAs: genomics, biogenesis, mechanism, and function. *Cell*. 2004;116:281–297.
16. Sood P, Krek A, Zavolan M, Macino G, Rajewsky N. Cell-type-specific signatures of microRNAs on target mRNA expression. *Proc Natl Acad Sci USA*. 2006;103:2746–2751.
17. Callis TE, Deng Z, Chen J-F, Wang D-Z. Muscling through the microRNA world. *Exp Biol Med*. 2008;233:131–138.
18. McCarthy JJ, Esser KA. MicroRNA-1 and microRNA-133a expression are decreased during skeletal muscle hypertrophy. *J Appl Physiol*. 2007;102:306–313.
19. Guller I, Russell AP. MicroRNAs in skeletal muscle: their role and regulation in development, disease and function. *J Physiol*. 2010;588:4075–4087.
20. Rao PK, Kumar RM, Farkhondeh M, Baskerville S, Lodish HF. Myogenic factors that regulate expression of muscle-specific microRNAs. *Proc Natl Acad Sci USA*. 2006;103:8721–8726.
21. Rosenberg MI, Georges SA, Asawachaicharn A, Analau E, Tapscott SJ. MyoD inhibits Fstl1 and Utrn expression by inducing transcription of miR-206. *J Cell Biol*. 2006;175:77–85.
22. van Rooij E, Quiat D, Johnson BA, et al. A family of microRNAs encoded by myosin genes governs myosin expression and muscle performance. *Dev Cell*. 2009;17:662–673.
23. Rivas DA, Lessard SJ, Rice NP, et al. Diminished skeletal muscle microRNA expression with aging is associated with attenuated muscle plasticity and inhibition of IGF-1 signaling. *FASEB J*. 2014;28:4133–4147.
24. Zacharewicz E, Della Gatta PA, Reynolds J, et al. Identification of microRNAs linked to regulators of muscle protein synthesis and regeneration in young and old skeletal muscle. *PLoS One*. 2014;9:e114009.
25. Nakasa T, Ishikawa M, Shi M, et al. Acceleration of muscle regeneration by local injection of muscle-specific microRNAs in rat skeletal muscle injury model. *J Cell Mol Med*. 2010;14:2495–2505.
26. Lee Y, Ahn C, Han J, et al. The nuclear RNase III Drosha initiates microRNA processing. *Nature*. 2003;425:415–419.
27. Lund E, Güttinger S, Calado A, Dahlberg JE, Kutay U. Nuclear export of microRNA precursors. *Science*. 2004;303:95–98.
28. Lee I, Ajay SS, Yook JI, et al. New class of microRNA targets containing simultaneous 5′-UTR and 3′-UTR interaction sites. *Genome Res*. 2009;19:1175–1183.
29. Russell AP, Lamon S, Boon H, et al. Regulation of miRNAs in human skeletal muscle following acute endurance exercise and short-term endurance training. *J Physiol*. 2013;591:4637–4653.
30. Drummond MJ, McCarthy JJ, Fry CS, Esser KA, Rasmussen BB. Aging differentially affects human skeletal muscle microRNA expression at rest and after an anabolic stimulus of resistance exercise and essential amino acids. *Am J Physiol Endocrinol Metab*. 2008;58: E1333–E1340.
31. Lund E, Guttinger S, Calado A, Dahlberg JE, Kutay U. Nuclear export of microRNA precursors. *Science*. 2004;303:95–98.
32. Fry AC. The role of resistance exercise intensity on muscle fibre adaptations. *Sports Med*. 2004;34:663–679.

33. Koopman R, Gleeson B, Gijsen AP, et al. Post-exercise protein synthesis rates are only marginally higher in type I compared with type II muscle fibres following resistance-type exercise. *Eur J Appl Physiol.* 2011;111:1871–1878.
34. Kumar V, Selby A, Rankin D, et al. Age-related differences in the dose–response relationship of muscle protein synthesis to resistance exercise in young and old men. *J Physiol.* 2009;587:211–217.
35. Léger B, Cartoni R, Praz M, et al. Akt signalling through GSK-3beta, mTOR and FOXO1 is involved in human skeletal muscle hypertrophy and atrophy. *J Physiol.* 2006;576:923–933.
36. Phillips SM. Physiologic and molecular bases of muscle hypertrophy and atrophy: impact of resistance exercise on human skeletal muscle (protein and exercise dose effects). *Appl Physiol Nutr Metab.* 2009;34:403–410.
37. Elia L, Contu R, Quintavalle M, et al. Reciprocal regulation of microRNA-1 and insulin-like growth factor-1 signal transduction cascade in cardiac and skeletal muscle in physiological and pathological conditions. *Circulation.* 2009;120:2377–2385.
38. Doghman M, El Wakil A, Cardinaud B, et al. Regulation of insulin-like growth factor-mammalian target of rapamycin signaling by microRNA in childhood adrenocortical tumors. *Cancer Res.* 2010;70:4666–4675.
39. Fornari F, Milazzo M, Chieco P, et al. miR-199a-3p regulates mTOR and c-Met to influence the doxorubicin sensitivity of human hepatocarcinoma cells. *Cancer Res.* 2010;70:5184–5193.
40. Hou T, Ou J, Zhao X, et al. MicroRNA-196a promotes cervical cancer proliferation through the regulation of FOXO1 and p27. *Br J Cancer.* 2014;110:1260–1268.
41. Jin Y, Tymen SD, Chen D, et al. MicroRNA-99 family targets AKT/mTOR signaling pathway in dermal wound healing. *PLoS One.* 2013;8:e64434.
42. Lerman G, Avivi C, Mardoukh C, et al. miRNA expression in psoriatic skin: reciprocal regulation of hsa-miR-99a and IGF-1R. *PLoS One.* 2011;6:e20916.
43. Lin RJ, Lin YC, Yu AL. miR-149* induces apoptosis by inhibiting Akt1 and E2F1 in human cancer cells. *Mol Carcinog.* 2010;49:719–727.
44. Wei F, Liu Y, Guo Y, et al. miR-99b-targeted mTOR induction contributes to irradiation resistance in pancreatic cancer. *Mol Cancer.* 2013;12:81.
45. Davidsen PK, Gallagher IJ, Hartman JW, et al. High responders to resistance exercise training demonstrate differential regulation of skeletal muscle microRNA expression. *J Appl Physiol.* 2011;110:309–317.
46. Nielsen S, Scheele C, Yfanti C, et al. Muscle specific microRNAs are regulated by endurance exercise in human skeletal muscle. *J Physiol.* 2010;588:4029–4037.
47. Keller P, Vollaard NB, Gustafsson T, et al. A transcriptional map of the impact of endurance exercise training on skeletal muscle phenotype. *J Appl Physiol.* 2011;110:46–59.
48. Rommel C, Bodine SC, Clarke BA, et al. Mediation of IGF-1-induced skeletal myotube hypertrophy by PI(3)K/Akt/mTOR and PI(3)K/Akt/GSK3 pathways. *Nat Cell Biol.* 2001;3:1009–1013.
49. Durieux AC, Desplanches D, Freyssenet D, Fluck M. Mechanotransduction in striated muscle via focal adhesion kinase. *Biochem Soc Trans.* 2007;35:1312–1313.
50. Russell AP, Wada S, Vergani L, et al. Disruption of skeletal muscle mitochondrial network genes and miRNAs in amyotrophic lateral sclerosis. *Neurobiol Dis.* 2013;49: 107–117.
51. Russell AP, Feilchenfeldt J, Schreiber S, et al. Endurance training in humans leads to fiber type-specific increases in levels of peroxisome proliferator-activated receptor-gamma coactivator-1 and peroxisome proliferator-activated receptor-alpha in skeletal muscle. *Diabetes.* 2003;52:2874–2881.

52. Ikeda S, Kawamoto H, Kasaoka K, et al. Muscle type-specific response of PGC-1 alpha and oxidative enzymes during voluntary wheel running in mouse skeletal muscle. *Acta Physiol (Oxf)*. 2006;188:217–223.
53. Safdar A, Abadi A, Akhtar M, Hettinga BP, Tarnopolsky MA. miRNA in the regulation of skeletal muscle adaptation to acute endurance exercise in C57Bl/6J male mice. *PLoS One*. 2009;4:e5610.
54. Wada S, Kato Y, Sawada S, et al. MicroRNA-23a has minimal effect on endurance exercise-induced adaptation of mouse skeletal muscle. *Pflugers Arch*. 2015;467:389–398 [Epub ahead of print].
55. Chen X, Ba Y, Ma L, et al. Characterization of microRNAs in serum: a novel class of biomarkers for diagnosis of cancer and other diseases. *Cell Res*. 2008;18:997–1006.
56. Gilad S, Meiri E, Yogev Y, et al. Serum microRNAs are promising novel biomarkers. *PLoS One*. 2008;3:e3148.
57. Valadi H, Ekstrom K, Bossios A, et al. Exosome-mediated transfer of mRNAs and microRNAs is a novel mechanism of genetic exchange between cells. *Nat Cell Biol*. 2007;9:654–659.
58. Vickers KC, Palmisano BT, Shoucri BM, Shamburek RD, Remaley AT. MicroRNAs are transported in plasma and delivered to recipient cells by high-density lipoproteins. *Nat Cell Biol*. 2011;13:423–433.
59. Kosaka N, Iguchi H, Yoshioka Y, et al. Secretory mechanisms and intercellular transfer of microRNAs in living cells. *J Biol Chem*. 2010;285:17442–17452.
60. Pritchard CC, Kroh E, Wood B, et al. Blood cell origin of circulating microRNAs: a cautionary note for cancer biomarker studies. *Cancer Prev Res*. 2012;5:492–497.
61. Wang K, Zhang S, Weber J, Baxter D, Galas DJ. Export of microRNAs and microRNA-protective protein by mammalian cells. *Nucleic Acids Res*. 2010;38:7248–7259.
62. Pegtel DM, Cosmopoulos K, Thorley-Lawson DA, et al. Functional delivery of viral miRNAs via exosomes. *Proc Natl Acad Sci USA*. 2010;107:6328–6333.
63. D'Alessandra Y, Devanna P, Limana F, et al. Circulating microRNAs are new and sensitive biomarkers of myocardial infarction. *Eur Heart J*. 2010;31:2765–2773.
64. Freedman J, Ercan B, Morin K, et al. The distribution of circulating microRNA and their relation to coronary disease. *F1000Research*. 2012;1:1–12.
65. Mitchell PS, Parkin RK, Kroh EM, et al. Circulating microRNAs as stable blood-based markers for cancer detection. *Proc Natl Acad Sci USA*. 2008;105:10513–10518.
66. Mouillet JF, Chu T, Hubel CA, et al. The levels of hypoxia-regulated microRNAs in plasma of pregnant women with fetal growth restriction. *Placenta*. 2010;31:781–784.
67. Ho AS, Huang X, Cao H, et al. Circulating miR-210 as a novel hypoxia marker in pancreatic cancer. *Transl Oncol*. 2010;3:109–113.
68. Baggish AL, Hale A, Weiner RB, et al. Dynamic regulation of circulating microRNA during acute exhaustive exercise and sustained aerobic exercise training. *J Physiol*. 2011;589:3983–3994.
69. Nielsen S, Akerstrom T, Rinnov A, et al. The miRNA plasma signature in response to acute aerobic exercise and endurance training. *PLoS One*. 2014;9:e87308.
70. Baggish AL, Park J, Min PK, et al. Rapid upregulation and clearance of distinct circulating microRNAs after prolonged aerobic exercise. *J Appl Physiol*. 2014;116:522–531.
71. Uhlemann M, Möbius-Winkler S, Fikenzer S, et al. Circulating microRNA-126 increases after different forms of endurance exercise in healthy adults. *Eur J Prev Cardiol*. 2012;21:484–491.
72. Mooren FC, Viereck J, Kruger K, Thum T. Circulating microRNAs as potential biomarkers of aerobic exercise capacity. *Am J Physiol Heart Circ Physiol*. 2014;306:H557–H563.

73. Sawada S, Kon M, Wada S, et al. Profiling of circulating microRNAs after a bout of acute resistance exercise in humans. *PLoS One*. 2013;8:e70823.
74. Aoi W, Ichikawa H, Mune K, et al. Muscle-enriched microRNA miR-486 decreases in circulation in response to exercise in young men. *Front Physiol*. 2013;4:80.
75. Banzet S, Chennaoui M, Girard O, et al. Changes in circulating microRNAs levels with exercise modality. *J Appl Physiol*. 2013;115:1237–1244.
76. Kuehbacher A, Urbich C, Zeiher AM, Dimmeler S. Role of Dicer and Drosha for endothelial microRNA expression and angiogenesis. *Circ Res*. 2007;101:59–68.
77. Poliseno L, Tuccoli A, Mariani L, et al. MicroRNAs modulate the angiogenic properties of HUVECs. *Blood*. 2006;108:3068–3071.
78. Chan SY, Zhang YY, Hemann C, et al. MicroRNA-210 controls mitochondrial metabolism during hypoxia by repressing the iron-sulfur cluster assembly proteins ISCU1/2. *Cell Metab*. 2009;10:273–284.
79. Krichevsky AM, Gabriely G. miR-21: a small multi-faceted RNA. *J Cell Mol Med*. 2009;13:39–53.
80. Chen J-F, Mandel EM, Thomson JM, et al. The role of microRNA-1 and microRNA-133 in skeletal muscle proliferation and differentiation. *Nat Genet*. 2006;38:228–233.
81. Aoi W, Naito Y, Mizushima K, et al. The microRNA miR-696 regulates PGC-1α in mouse skeletal muscle in response to physical activity. *Am J Physiol Endocrinol Metab*. 2010;298:E799–E806.
82. Russell AP, Lamon S, Boon H, et al. Regulation of miRNAs in human skeletal muscle following acute endurance exercise and short term endurance training. *J Physiol*. 2013;591:4637–4653.
83. van Solingen C, Seghers L, Bijkerk R, et al. Antagomir-mediated silencing of endothelial cell specific microRNA-126 impairs ischemia-induced angiogenesis. *J Cell Mol Med*. 2009;13:1577–1585.
84. Wang S, Aurora AB, Johnson BA, et al. The endothelial-specific microRNA miR-126 governs vascular integrity and angiogenesis. *Dev Cell*. 2008;15:261–271.
85. Hitachi K, Nakatani M, Tsuchida K. Myostatin signaling regulates Akt activity via the regulation of miR-486 expression. *Int J Biochem Cell Biol*. 2014;47:93–103.
86. Dews M, Homayouni A, Yu D, et al. Augmentation of tumor angiogenesis by a Myc-activated microRNA cluster. *Nat Genet*. 2006;38:1060–1065.
87. McAlexander MA, Phillips MJ, Witwer KW. Comparison of methods for miRNA extraction from plasma and quantitative recovery of RNA from plasma and cerebrospinal fluid. *Front Genet*. 2013;4:83.
88. Connes P, Simmonds MJ, Brun J-F, Baskurt OK. Exercise hemorheology: classical data, recent findings and unresolved issues. *Clin Hemorheol Microcirc*. 2013;53:187–199.
89. Tanimura Y, Kon M, Shimizu K, et al. Effect of 6-day intense Kendo training on lymphocyte counts and its expression of CD95. *Eur J Appl Physiol*. 2009;107:227–233.
90. Tonevitsky A, Maltseva D, Abbasi A, et al. Dynamically regulated miRNA-mRNA networks revealed by exercise. *BMC Physiol*. 2013;13:9.
91. Devlin C, Greco S, Martelli F, Ivan M. miR-210: more than a silent player in hypoxia. *IUBMB Life*. 2011;63:94–100.
92. Chan SY, Loscalzo J. MicroRNA-210: a unique and pleiotropic hypoxamir. *Cell Cycle*. 2010;9:1072–1083.
93. Bye A, Røsjø H, Aspenes ST, et al. Circulating microRNAs and aerobic fitness—the HUNT Study. *PLoS One*. 2013;8:e57496.
94. Leuenberger N, Robinson N, Saugy M. Circulating miRNAs: a new generation of anti-doping biomarkers. *Anal Bioanal Chem*. 2013;405:9617–9623.
95. Radom-Aizik S, Zaldivar F, Haddad F, Cooper DM. Impact of brief exercise on peripheral blood NK cell gene and microRNA expression in young adults. *J Appl Physiol*. 2013;114:628–636.

96. Radom-Aizik S, Zaldivar Jr F, Leu SY, et al. Effects of exercise on microRNA expression in young males peripheral blood mononuclear cells. *Clin Transl Sci*. 2012;5:32–38.
97. Radom-Aizik S, Zaldivar Jr F, Oliver S, Galassetti P, Cooper DM. Evidence for microRNA involvement in exercise-associated neutrophil gene expression changes. *J Appl Physiol*. 2010;109:252–261.
98. Chilton WL, Marques FZ, West J, et al. Acute exercise leads to regulation of telomere-associated genes and microRNA expression in immune cells. *PLoS One*. 2014;9:e92088.
99. van Rooij E, Sutherland LB, Qi X, et al. Control of stress-dependent cardiac growth and gene expression by a microRNA. *Science*. 2007;316:575–579.

CHAPTER TWENTY-ONE

Exercise as a Polypill for Chronic Diseases

Helios Pareja-Galeano*[,1], Nuria Garatachea†, Alejandro Lucia*
*European University and Research Institute of Hospital 12 de Octubre ("i+12"), Madrid, Spain
†Faculty of Health and Sport Science, University of Zaragoza, Huesca, Spain
[1]Corresponding author: e-mail address: helios.pareja@gmail.com

Contents

Abstract

Exercise may be described as a polypill to prevent and/or treat almost every chronic disease, with obvious benefits such as its low cost and practical lack of adverse effects. Implementing physical activity interventions in public health is therefore a goal at the medical, social, and economic levels. This chapter describes the importance of health promotion through physical activity and discusses the impacts of exercise on the most prevalent chronic diseases, namely metabolic syndrome-related disorders, cardiovascular diseases, cancer, and Alzheimer's disease. For each of these chronic conditions, we discuss the epidemiological evidence supporting a beneficial role of exercise, provide guidelines for exercise prescription, and describe the biological mechanisms whereby exercise exerts its modulatory effects.

ABBREVIATIONS

Aβ amyloid-β
AD Alzheimer's disease
BDNF brain-derived neurotrophic factor
CHF Chronic heart failure
CHD Coronary heart disease

Progress in Molecular Biology and Translational Science, Volume 135
ISSN 1877-1173
http://dx.doi.org/10.1016/bs.pmbts.2015.07.019

CRP C-reactive protein
CVD Cardiovascular disease
eNOS endothelial nitric oxide synthase
GDF15 growth differentiation factor 15
GLUT-4 muscle glucose transport protein
HbA_{1c} Glycosylated hemoglobin
HDL High-density lipoprotein
IGF-1 insulin-like growth factor-1
IL Interleukin
LDL Low-density lipoprotein
MS Metabolic syndrome
NK natural killer
$NO^{\bullet}$ nitric oxide
PA Physical activity
PAR Population-attributable risks
RCT Randomized control trial
SPARC secreted protein acidic and rich in cysteine
sTNF-R Soluble TNF receptor
T2D Type 2 diabetes
TC Total cholesterol
TG Triglycerides
TNF-α tumor necrosis factor-α
VEGF vascular endothelial growth factor
VLDL Very low-density lipoproteins
VO_{2max} Maximal oxygen uptake
WC Waist circumference
WHO World Health Organization

1. INTRODUCTION

1.1 Chronic Diseases

According to recent estimates by the World Health Organization (WHO), more than 14 million people aged 30–70 die yearly worldwide of a chronic disease and 85% of these people are from a developing country. The *Non-communicable Diseases Country Profiles 2014* by the WHO[1] reports that mortality due to chronic disease in the United States reaches 31% from cardiovascular diseases (CVDs), 23% from cancer, 8% from chronic respiratory diseases, 6% from injuries, 3% from type 2 diabetes (T2D), and 23% from other chronic diseases.

1.1.1 Metabolic Syndrome-Related Disorders

Metabolic syndrome (MS) is a cluster of risk factors for CVD and T2D.[2] The clinical diagnosis of MS is based on a high waist circumference (WC);

elevated triglycerides (TG; $\geq$150 mg/dL); reduced high-density lipoprotein (HDL; <40 mg/dL in males and <50 mg/dL in females); elevated blood pressure (systolic $\geq$130 mmHg and/or diastolic $\geq$ 85 mmHg); and elevated fasting glucose ($\geq$100 mg/dL).[2] Accordingly, several abnormalities such as glucose intolerance, insulin resistance, central obesity, dyslipidemia, and hypertension are typical of the physiopathology of MS.[3] These conditions are biologically and clinically interconnected as evidenced by the high prevalence of T2D coexisting with risk factors such as overweight (80%), hypertension (60–80%), and dyslipidemia (40–50%).[4]

Insulin resistance is a widely accepted mechanism in the pathophysiology of MS.[3] Its main cause is the overabundance of circulating fatty acids derived mainly from adipose tissue TG stores through lipolysis of TG-rich lipoproteins in different tissues by lipoprotein lipase.[5] Insulin acts as an antilipolytic agent in adipose tissue[6] and as an inducer of lipoprotein lipase.[3] Thus, insulin resistance leads to increased lipolysis which generates more fatty acids, in turn, inducing further lipolysis. The high fatty acid levels generated in this way prevent insulin actions in sensitive tissues, mainly skeletal muscle and liver.[3] This impaired insulin activity produces glucose metabolism dysfunction due to an inability to suppress glucose production or mediate glucose uptake and metabolism in insulin-sensitive tissues. Moreover, insulin resistance in pancreatic islet β-cells affects insulin secretion due to the lipotoxicity of excessive concentrations of fatty acids.[3]

Dyslipidemia is caused in some measure by insulin resistance because increased free fatty acid flux to the liver produces apolipoprotein B-containing TG-rich very low-density lipoproteins (VLDL). Hypertriglyceridemia leads also to a reduction in the total cholesterol (TC) content of HDL due to its transformation into a small, dense molecule and via increased HDL clearance from the circulation. Finally, high TG levels raise LDL and modify its molecular composition.[3] Such modifications make LDL more atherogenic, more toxic to the endothelium, capable of crossing the endothelial basement membrane, able to adhere to glycosaminoglycans, more susceptible to oxidation, and confer LDL particles a capacity to bind to scavenger receptors on monocyte-derived macrophages.[3] Thus, hypertriglyceridemia can be considered an independent risk factor for CVD.[7] Hypertension also contributes to MS through different mechanisms related to insulin and fatty acids.[3]

T2D is caused by abnormal insulin secretion and insulin resistance, which give rise to glucose, protein, and fat metabolism modifications[8] producing hyperglycemia. Diabetes is the clinical consequence of most of the pathophysiological conditions detailed above. Chronic hyperglycemia is associated with long-term complications including damage to the eyes, heart, kidneys, nerves, and blood vessels.[9] T2D usually develops in adulthood and is associated with obesity, sedentary behavior, and an unhealthy diet.[8] Mortality rates in diabetic patients are one to four times higher than in the general population, the main cause of death being CVD.[10]

1.1.2 Cardiovascular Disease

A CVD is any disease that affects blood vessels or the heart and includes conditions such as high blood pressure, subclinical atherosclerosis, coronary heart disease (CHD), acute coronary syndrome (myocardial ischemia), angina pectoris, stroke (cerebrovascular disease), congenital cardiovascular defects, cardiomyopathy, chronic heart failure (CHF), and other less prevalent conditions. In this chapter, we describe the most frequent CVDs, namely hypertension and CHD.

More than 62% of all North American adults suffer from hypertension defined as a systolic blood pressure $\geq$140 mmHg, a diastolic blood pressure $\geq$90 mmHg, or a need for blood pressure-lowering medication. High blood pressure is also an important risk factor for other CVDs such as stroke, acute myocardial infarction, cardiac failure, and sudden death.[4] Conversely, low blood pressure (<115 mmHg of systolic and 75 mmHg of diastolic) is associated with a reduced risk of cardiovascular death.[11]

Responsible for one in every six deaths in the United States in 2006,[12] CHD is a disease of the heart and the coronary arteries characterized by myocardial ischemia due to inadequate local blood supply relative to myocardial oxygen requirements. This process is usually caused by atherosclerotic arterial deposits that block blood flow to the heart. When blockage persists for about 20 min, myocardial ischemia leads to infarction which may manifest as chronic stable angina pectoris, acute and chronic heart failure, acute and chronic cardiac arrhythmia, acute unstable angina pectoris, acute myocardial infarct, or sudden death.[4]

1.1.3 Cancer

Cancer is the term used for diseases in which excessive cell growth results in the compression, invasion, and degradation of adjacent healthy tissue. Malignant cells are transported by the blood or lymph to other organs where they give rise to secondary colonies called metastases. The molecular mechanism common to all cancer types consists of mutations in the DNA of tumor cells. Many factors, including exposure to chemical carcinogens (such as tobacco), physical inactivity, and diet, play major roles in the development of cancer.[13,14] In the United States, cancer is the second cause of death accounting for ~23% of all deaths.[15] Prostate cancer, followed by lung and colon cancer, is the most prevalent in men, while in women breast cancer is the most common, followed by lung and colorectal cancer.[16] The symptoms of cancer are innumerable and depend on the type of tumor, its localization, and its stage. However, there are features common to many forms of cancer and these include cachexia—reduced muscle strength and fitness; and increased fatigue—a feeling of being unwell, poor appetite, and infections, among others.[4]

1.1.4 Neurodegenerative Diseases

Neurodegenerative disorders are characterized by a loss of structure or function and even death of neurons and include diseases such as amyotrophic lateral sclerosis, Parkinson's disease, Huntington's disease, and Alzheimer's disease (AD). AD, the most common chronic neurodegenerative disease, is suffered by 35 million people worldwide and 5.5 million in the United States.[17] Of all causes of dementia, AD is the most prevalent. It accounts for 50–56% of all dementias with 13–17% of all AD characterized by the presence of other cerebrovascular disorders.[17] The risk of AD increases with age. Its prevalence doubles every 5 years from the age of 60 years, such that, for instance, 8% of people aged 75 years suffer from AD. By 2050, the Alzheimer's Association estimates that 11–16 million North Americans will be affected by the disease.

1.2 Health Policy Strategies Against Chronic Diseases: Exercise as an Effective Strategy

The high economic burden of chronic diseases makes their prevention and treatment a first order priority. For example, in 2012 the estimated cost of diagnosed T2D in the United States was $245 billion, including $176 billion in direct medical costs and $69 billion in decreased productivity.[18] Medical costs linked to obesity were estimated at $147 billion in 2008. In 2006, medical costs for obese people were $1429 higher than those for normal-weight persons.[19] The cost of heart disease and stroke has been calculated to reach $315.4 billion,[20] while cancer care cost $157 billion in 2010.[21] AD International estimates that the worldwide cost of dementia in 2010 was $604 billion.

The US Centers for Disease Control and Prevention (CDC) defines physical inactivity as an actual cause of chronic disease.[22] In 2011, 52% of adults aged 18 years or older did not meet recommendations for physical activity (PA) and 76% did not reach the minimum recommendations for muscle-strengthening PA.[23] These statistics worsen from childhood to adulthood and aging[24] and may be aggravated by the presence of several disease states.[24,25] It should be noted that most of these recommendation statements refer to aerobic- and resistance-type exercise.[26] In 2012, one in every three adults worldwide was inactive and it was proposed that this endemic inactivity starts early in life.[27] In its *Noncommunicable Diseases Action Plan*, WHO defined political actions regarding four modifiable risk behaviors—i.e., tobacco use, unhealthy diet, physical inactivity, and harmful use of alcohol—linked to the four main preventable chronic diseases: CVD, cancers, T2D, and chronic respiratory diseases. Besides tobacco use, the most modifiable cancer risk factors are dietary choices, weight control, and levels of exercise.[13,14]

Based on the Booth and Laye review,[28] Table 1 shows the main health problems worsened by physical inactivity.

Table 1 Health Problems Worsened by Physical Inactivity

Brain	**Heart**
Alzheimer's disease	Congestive heart failure
Dementia	Endothelial dysfunction
Depression	Greater damage in an ischemic event
Stroke (incidence and recovery)	Inflammation of vascular wall
Neurogenesis	Lower maximal cardiac output
Executive control in memory	**Liver**
Smaller frontal and superior lobes with aging	Non alcoholic fatty liver disease
Metabolic and lipid profile	**Vascular**
High LDL	Hypertension
Low HDL	Arterial stiffness
T2D	Peripheral artery disease
Visceral obesity	Erectile dysfunction
Reduced glucose uptake/insulin sensitivity	**Musculoskeletal tissues**
Reduced oxidative capacity	Osteoporosis
Cancer	Structure/function of joints, ligaments, and tendons
Breast and colon cancer	Neuromuscular strength

Adapted from Booth and Laye.[28]

One of the first observational studies describing the impact of regular exercise on all-cause mortality was performed by Paffenbarger *et al.*[29] This study examined some 17,000 Harvard University alumni aged 35–74 years and showed that death rates declined steadily as expended energy increased from less than 500 to 3500 kcal/week, beyond which rates increased slightly. The National Health Interview Survey provided data on adult levels of adherence to PA recommendations in the *2008 Physical Activity Guidelines for Adults.*[30] Adherence to recommended PA levels was associated with a substantial survival benefit as indicated by the all-cause mortality risk, which dropped by 27% among people without existing chronic comorbidities, and by almost half in individuals with chronic comorbidities. These results were supported by the findings of Arrieta and Russell,[31] who observed beneficial effects of moderate and leisure PA on survival. However, the positive effects of elite sport on mortality were not easily demonstrated. In a recent meta-analysis, we reviewed 10 cohort studies to assess mortality rates in elite athletes.[32] Results showed that top-level athletes live longer than the general population and carry a lower risk of two major causes of mortality, i.e., CVD and cancer. A meta-epidemiological study by Naci and Ioannidis[33] compared the effectiveness of exercise versus drug interventions on mortality outcomes through the review of 16 meta-analyses of randomized control trials (RCTs). They concluded that exercise and many drug interventions are often similar in terms of mortality benefits in the secondary prevention of CHD, rehabilitation after stroke, treatment of CHF, and prevention of T2D.

2. EXERCISE AS A POLYPILL FOR CHRONIC DISEASES

2.1 Metabolic Syndrome-Related Disorders

2.1.1 Insulin Resistance and Type 2 Diabetes

2.1.1.1 Epidemiological Evidence

For many years, exercise has been an integral component of T2D management. As reported in a recent review by our group,[34] endurance, resistance, or combined exercise training reduces glycosylated hemoglobin (HbA_{1c}) levels compared with controls.[35] In addition, it seems that nondrug interventions (diet or exercise) prevent more T2D than drug approaches.[36] These findings are supported by several meta-analysis studies that have shown that PA reduces insulin resistance and improves glucose intolerance in obese persons,[37] reduces HbA_{1c} levels in T2D patients trained for at least 8 weeks,[38] improves glycemia, and reduces visceral adipose tissue and plasma TG, also in patients with T2D.[39]

In a systematic review, it was determined that aerobic training alone or combined with resistance training reduced HbA_{1c}, systolic blood pressure, TG, and WC.[40] These findings conflict with those of Cauza *et al.*,[41] who observed lower HbA_{1c}, glycemia, and insulin resistance in their T2D-patient strength training group compared with the aerobic training group. In the RCT conducted in 2010 by Church *et al.*,[42] it was found that the aerobic-resistance training combined was the only intervention leading to reduced levels of HbA_{1c} compared to levels in sedentary patients. Neither aerobic nor resistance training alone was able to downregulate HbA_{1c}. The combined exercise group also showed an improved maximal oxygen uptake ($\dot{V}O_{2max}$) and all interventions reduced WC relative to measurements recorded in the control group. This is important, since a lack of fitness seems to be an independent prognostic marker of death in T2D patients.[43]

2.1.1.2 Exercise Prescription

Despite the classic prescription of aerobic training in patients with metabolic diseases, resistance training is an important addition because the combination of both may be equally, if not more, beneficial to health.[42,44,45] Endurance plus resistance training has been noted to be better at reducing HbA_{1c} and improving other MS-related factors than each intervention alone.[44] Recently, Earnest *et al.* compared the effects of three 9-month training programs (aerobic, resistance, or both) in patients with T2D. Training-induced decreases in MS scores were higher with the aerobic program or the combination of the aerobic and resistance programs compared to the control and resistance programs.[45] The meta-analysis by Boule *et al.*[38] revealed that both aerobic and resistance training reduced HbA_{1c} levels in T2D patients compared with control groups. According to the studies reviewed, aerobic training regimens consisted of a mean (standard deviation) of 3.4 (0.9) days/week for 18 (15) weeks; and strength training programs of 10 (0.7) exercises, 2.5 (0.7) sets, 13 (0.7) repetitions, 2.5 (0.4) days/week, and over 15 (10) weeks. In a later study, Boule *et al.*[46] positively correlated postintervention HbA_{1c} with exercise intensity but not with exercise volume. According to Cauza *et al.*,[41] strength training is more effective than aerobic training in terms of improving glycemia, HbA_{1c}, and insulin resistance in T2D patients.

For aerobic exercise in T2D patients, the standard recommendation by the American College of Sports Medicine (ACSM)[47] is a minimum of 150 min/week of moderate-intensity exercise or 60 min/week of vigorous exercise. For resistance exercise training, ACSM recommendations are 2–3 sessions/week at a moderate relative intensity (50% of 1-repetition maximum, 1-RM) or high intensity (75–80% of 1-RM).

Training sessions should include 5–10 exercises involving major muscle groups as 10–15 repetitions. Combining aerobic, resistance, and flexibility training is also recommended.

2.1.1.3 Biological Mechanisms

The acute effects of exercise include fuel mobilization, glucose production, muscle glycogenolysis, and insulin-independent and insulin-dependent muscle glucose uptake.[47] Although these responses reverse rapidly after exercise is stopped, they are followed by a marked increase in the sensitivity to insulin of both muscle glucose transport and glycogen synthesis.[48] There are several biological adaptations to exercise training that are beneficial for the prevention and treatment of insulin resistance, impaired glucose tolerance, and T2D. PA also counteracts the reduced ability of insulin to stimulate muscle blood flow that is a characteristic of insulin-resistant obese and T2D individuals by improving the control of insulin over blood glucose.[48] Exercise enhances insulin sensitivity and muscle contraction-induced glucose uptake in the exercised muscle via different physiologic mechanisms including improved postreceptor insulin signaling,[49] GLUT4 mRNA and protein induction,[50] glycogen synthase activity,[51] reduced release and enhanced clearance of free fatty acids,[52] increased hexokinase activity,[53] glucose influx to muscles due to improved muscle capillarization and blood flow,[54] and enhanced β-cell function.[55] PA also beneficially impacts endothelial dysfunction through endothelial nitric oxide ($NO^{\bullet}$) synthase (eNOS) modulation.[56]

Muscle hypertrophy increases the depot size available for glucose disposal thereby improving glycemic control in T2D patients.[57,58] Changes in muscle morphology due to exercise training are associated with improvements in fasting insulin levels and glucose tolerance since exercise increases the conversion of fast-twitch glycolytic, type IIX fibers to fast-twitch oxidative, type IIA fibers. The latter have a greater capillary density and are more sensitive and responsive to insulin than IIX fibers.[48] Exercise also modulates the over hepatic glucose production by regulating blood free fatty acid levels. Indeed, free fatty acids induce hepatic gluconeogenesis which stimulates hepatic glucose production while inhibiting skeletal muscle uptake and storage. Thus, the exercise-mediated modulation of blood free fatty acids increases peripheral glucose clearance and reduces hepatic glucose formation[48] (see Fig. 1).

2.1.2 Dyslipidemia

2.1.2.1 Epidemiological Evidence

In a recent meta-analysis of RCTs by Kelley *et al.*,[59] it emerged that PA reduced TG but not TC, HDL, and LDL. These same authors had

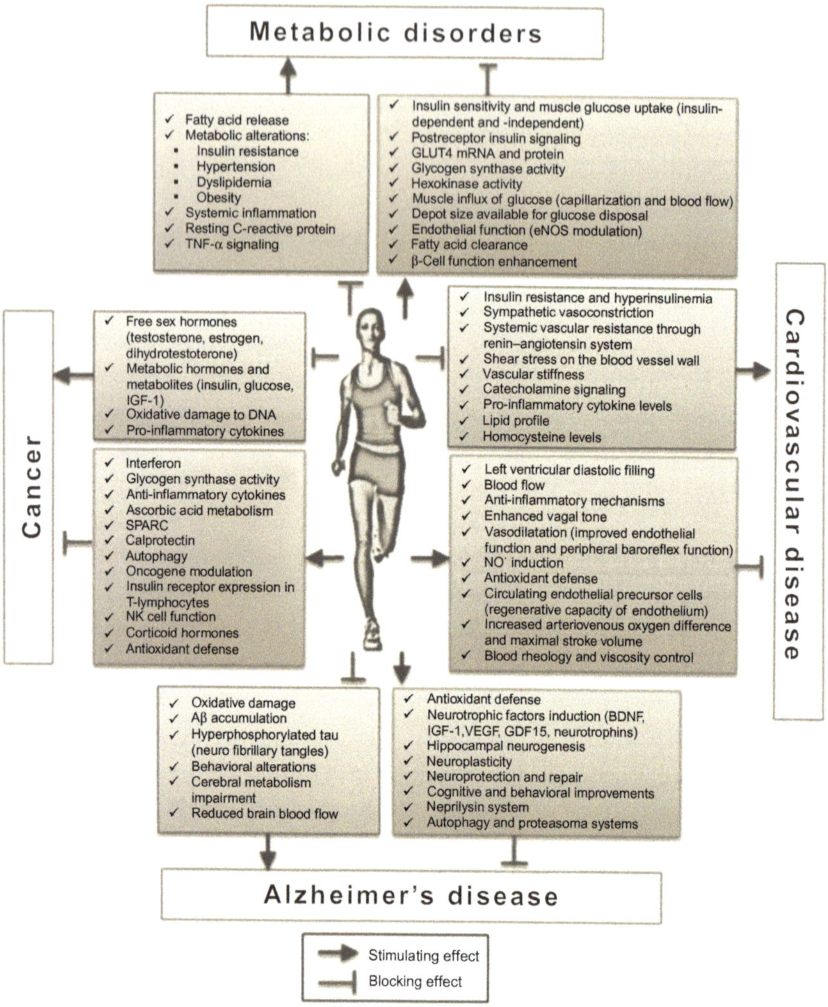

Figure 1 The main biological effects of physical activity on metabolic disorders, cardiovascular disease, Alzheimer's disease, and cancer. (See the color plate.)

previously shown that resistance training reduced TC, the TC:HDL ratio, LDL, and TG, while HDL changes were not significant in adults.[60] Even earlier, Kelley *et al.*[61] had examined the effects of aerobic exercise on lipids and lipoproteins in overweight and obese adults. They found that only TG decreased with PA, while increased HDL levels were correlated with increased $\dot{V}O_{2max}$ and lower body weight, whereas LDL reductions were correlated with lower body weight. In contrast, the US *National Physical Activity Guidelines Advisory Committee Report* (NPAGCR) stated that regular

PA increases blood HDL and decreases TG though this evidence does not clarify whether exercise training reduces LDL.[62] These conclusions are supported by the meta-analysis by Halbert *et al.*,[63] who showed that aerobic exercise training led to significantly reduced TC, LDL, and TG levels and increased HDL levels in both hyperlipidemic and normolipidemic adults. In a meta-analysis centered on T2D patients, Hayashino *et al.*[64] correlated PA with reduced LDL, elevated HDL, and improved blood pressure.

2.1.2.2 Exercise Prescription

The quantitative analysis by Durstine *et al.*[65,66] reveals a dose–response relationship between exercise training volume and blood lipid changes. Thresholds derived from cross-sectional and longitudinal exercise training studies indicate that 15–20 miles/week of brisk walking or jogging (1200–2200 kcal/week) produces TG reductions and HDL increases. As stated by the authors, weekly exercise regimens that match or go beyond this range are more likely to produce the desired lipid changes.

Greene *et al.*[67] compared the impacts of acute exercise and exercise training on the lipid and lipoprotein profiles of obese subjects. These authors found that exercise training (500 kcal/session, 3 sessions/week, and 12 weeks) increased HDL in men and reduced LDL levels in women, whereas acute exercise reduced the TC:HDL ratio in men.

2.1.2.3 Biological Mechanisms

Exercise promotes enzyme activation at the muscle level which enhances lipid metabolism over glycogen metabolism.[4] As mentioned above, exercise improves glycogen synthase activity,[51] reduces the release and enhances the clearance of free fatty acids,[52] increases hexokinase activity,[53] and enhances glucose influx to muscles via improved muscle capillarization and blood flow.[54] The combination of these mechanisms leads to an improvement in blood lipid profile and, as such, exercise leads to a reduction in the risk of atherogenic dyslipidemia caused by the presence of low blood levels of HDL and high levels of LDL and VLDL. PA reduces fasting and nonfasting TG concentrations which, in turn, is attributed to an increase in lipoprotein lipase activity with a concomitant reduction in apolipoprotein C-III levels, a decrease in cholesteryl ester transfer protein activity, and an increased catabolism of TG-rich lipoproteins.[68] The exercise-induced increase in lipoprotein lipase and hepatic lipase activity stimulates TG lipolysis, but the biological signaling of exercise-induced increase in lipoprotein lipase activity remains uncertain[68] (Fig. 1).

2.1.3 *Metabolic Syndrome and Obesity*

2.1.3.1 Epidemiological Evidence

Physical inactivity is a causal mechanism of every MS risk factor including visceral obesity, along with dyslipidemia, hypertension, hyperglycemia, thrombosis, and inflammation.[3,8,12] Exercise training as co-treatment for MS-related disorders such as obesity is thus universally accepted.

Dynamic endurance training reduces WC and blood pressure and increases HDL.[69] In the systematic review by Strasser *et al.*,[70] the impact of resistance training on different MS components was assessed. However, no effects were detected on TC, HDL-C, LDL, TG, and diastolic blood pressure, while impacts were produced on obesity, HbA_{1c} levels, and systolic blood pressure. The CARDIA study found that low fitness predicted the risk of MS as powerfully as conventional risk factors,[71] in agreement with the findings of other studies.[3,72,73] Of note is the cross-sectional study by Lakka *et al.*,[73] in which the relationship between leisure-time PA or $\dot{V}O_{2max}$ and MS was investigated. Their findings revealed that individuals who engaged in moderate-intensity PA $\leq$1.0 h/week were 60% more likely to suffer MS than those who performed $\geq$3.0 h/week of this type of exercise. In addition, those with a $\dot{V}O_{2max}$ $\leq$29.1 mL kg^{-1} min^{-1} were around seven times more likely to have MS than those whose $\dot{V}O_{2max}$ was $\geq$35.5 mL kg^{-1} min^{-1}.

The combination of weight loss plus exercise reduced the incidence of T2D in patients with glucose intolerance.[74] Weight reduction is the consequence of reduced energy intake and increased energy expenditure (i.e., exercise). Thus, weight loss alone will improve all risk factors associated with MS[75] including T2D.[76,77] The effects of exercise on obesity are widely known. In their meta-analysis, Ross and Janssen[78] concluded that short-term training interventions (<16 weeks) in obese persons (body mass index (BMI) <25) reduced total fat in a dose-dependent manner. These results are supported by those emerging from other studies.[79–81] Other studies noted reduced visceral adipose tissue in response to aerobic exercise but not resistance training.[82]

2.1.3.2 Exercise Prescription

ACSM recommendations are similar to those for individuals with insulin resistance and T2D,[47] specifically, at least 150 min/week of moderate-intensity PA. However, a dose effect of exercise is likely, and it has been reported that moderate-intensity PA doses of 250–300 min/week lead to both improved weight loss and better weight regain prevention.[83]

2.1.3.3 Biological Mechanisms

According to Booth *et al.*,[12] physical inactivity is a primary causal mechanism of every risk factor for the MS. Thus, since the latter is a cluster of factors related to insulin resistance, T2D, obesity, hypertension, and dyslipidemia, the biological mechanisms involved in the response to PA are a combination of those described above. These mechanisms include an attenuation of low-grade inflammation, an improvement in glucose and lipid metabolism, and a reduction of both total body fat and visceral adipose tissue. The amelioration of the inflammatory processes due to exercise is mediated by a reduction in resting C-reactive protein (CRP), interleukin-6 (IL-6), IL-8, and tumor necrosis factor-α (TNF-α).[34,84] The metabolic adaptations induced by PA also include increased postreceptor insulin signaling, increased GLUT4 and glycogen synthase activity, and reduced release (as well as enhanced clearance) of free fatty acids. Exercise also enhances muscle capillarization and blood flow, which improves influx of glucose to muscles.[4] In addition, endothelial dysfunction, a critical factor of the MS, is improved by exercise training, independent of changes in BMI, or blood pressure.[85] (Fig. 1).

2.2 Cardiovascular Diseases

2.2.1 Hypertension

2.2.1.1 Epidemiological Evidence

According to Pedersen and Saltin,[4] there is strong evidence in support of the beneficial effects of exercise on blood pressure. Aerobic exercise reduces systolic and diastolic blood pressure in healthy subjects and hypertensive patients, though these effects are more pronounced in the latter.[86] Resistance training, involving either dynamic[87–89] or static exercises,[88–90] also has a blood pressure-lowering effect both in healthy and in hypertensive individuals. However, despite substantial evidence, in a recent meta-analysis by Semlitsch *et al.*,[91] it was concluded that, although blood pressure decreases as a consequence of increased PA, there is still insufficient data from large well-designed RCTs confirming the benefits of exercise in patients with high blood pressure as an isolated health problem.

2.2.1.2 Exercise Prescription

Classically, physicians recommend their hypertensive patients to undertake aerobic physical exercise.[4,91,92] Despite the limited literature addressing the effect of resistance training on hypertension, several studies have highlighted the effectiveness of both dynamic resistance and isometric handgrip training at reducing blood pressure.[88,93] Recently, a meta-analysis by Carlson *et al.*[94] revealed that isometric resistance training lowers systolic and diastolic blood pressure more effectively than dynamic aerobic or resistance training.

According to the ACSM, the training frequencies, intensities, durations, and exercise modalities needed to lower blood pressure remain unclear.[95] Recommendations include 3–5 sessions/week as effective for reducing blood pressure. Physical exercise induces a blood pressure drop that lasts 4–10 h after finishing the bout of exercise,[95] although it may persist for up to 24 h.[96] This means that a hypertensive individual can be normotensive for much of the day.[95] Accordingly, more sessions per week can help control hypertension. The ACSM recommends 30 min or more of continuous or cumulative exercise executed at intensities between 40% and 60% of $\dot{V}O_{2max}$. In addition, it recommends combined endurance-resistance training.

2.2.1.3 Biological Mechanisms

Numerous mechanisms give rise to the blood pressure-lowering effect of exercise.[97] Hypertension often goes hand in hand with insulin resistance and hyperinsulinemia, left ventricular diastolic dysfunction, chronic low-grade inflammation, endothelial dysfunction, and dyslipidemia. Neurohumoral, vascular, and structural adaptations to exercise are responsible for its effects on hypertension.[4] The physiological mechanisms whereby exercise improves insulin resistance and hyperinsulinemia may also regulate hypertension (via the mechanisms described above).[4] Exercise augments left ventricular diastolic filling[98] with favorable effects on hypertension. The antihypertensive effect of exercise is also partially mediated by reduced sympathetic vasoconstriction.[99] Exercise-mediated reduction of systemic vascular resistance involves the sympathetic nervous system and renin–angiotensin system.[100] Exercise increases blood flow reducing the effects of shear stress on the blood vessel wall, and this induces eNOS production which, in turn, causes smooth muscle vasodilatation,[56] preventing vascular stiffness.[101] Exercise also triggers anti-inflammatory mechanisms and decreases catecholamine levels[34] (Fig. 1).

2.2.2 Coronary Heart Disease

2.2.2.1 Epidemiological Evidence

Regular PA delays the development of atherosclerosis and reduces the incidence of CHD events, as has been known for many years.[102–104] The American Heart Association has recognized since 1992 that physical inactivity is a risk factor for CHD and CVD[105] and the US Surgeon General concluded in 1996, that regular PA reduces the risk of CVD and CHD.[106] The *Physical Activity Guidelines Advisory Committee Report* emphasizes that there is a strong inverse relation between a person's usual level of PA and both CHD morbidity and mortality.[62] Individuals who undertake moderate to high volumes of PA lower their CHD risk by 20–30%, compared to sedentary persons.[62]

Tanasescu *et al.*[107] analyzed the relationship between potential CHD risk factors and newly diagnosed cases of CHD and leisure-time PA in 44,452 North American men. Relative risks related to moderate and high exercise intensities were 0.94 and 0.83*R* compared to low exercise intensity. Interestingly, half an hour per day or more of brisk walking was associated with an 18% CHD risk reduction. Total PA, running, weight training, and walking were each correlated with a reduced CHD risk. Average exercise intensity was linked to a reduced risk regardless of the number of metabolic equivalent-hours spent on PA. Similar findings were reported for the cohort study by Lee *et al.*,[108] in which even light to moderate activity was related to lower CHD rates in women. An inverse relationship between CHD development and PA was also detected in women at high risk of CHD. A systematic review by Heran *et al.*[109] analyzed the prevalence of all-cause and cardiac mortality in patients enrolled in cardiac rehabilitation programs and found that exercise-based cardiac rehabilitation reduced overall and cardiovascular mortality along with hospital admissions. Cardiac rehabilitation did not, however, reduce the risk of myocardial infarction, coronary artery bypass graft, or percutaneous transluminal coronary angioplasty.

2.2.2.2 Exercise Prescription

Exercise prescription in CHD patients should include dynamic aerobic exercise involving large muscle groups. Strength training is also beneficial in selected patients. Resistance training sessions may consist of up to 10–12 exercise sets of 10–12 repetitions. The ACSM recommends a minimum exercise frequency and duration of 30–60 min sessions including warm-up and cool-down periods on 3 nonconsecutive days per week. CHD patients can perform exercises at a relative intensity of 40–85% of VO_{2max}, starting slowly and gently and then gradually increasing duration and intensity.[110]

2.2.2.3 Biological Mechanisms

Joyner and Green[111] proposed the following mechanisms for the favorable effects of PA on CHD: (i) enhanced vagal tone via improved peripheral baroreflex function and central nervous system cardiovascular regulation; (ii) enhanced endothelial function through vasodilatation improving peripheral baroreflex function by limiting age- or risk factor-associated increases in vascular stiffness; and (iii) positive interactions between enhanced endothelial function and sympathetic outflow, minimizing the effects of high baseline sympathetic outflow on blood pressure.[12] Further proposed mechanisms included increased $NO^{\bullet}$ and antioxidant levels, decreased circulating pro-inflammatory biomolecule levels (mainly CRP), and enhanced regenerative capacity of the endothelium, manifesting as an increased number of circulating endothelial precursor cells.[112] Exercise training also increases functional capacity and VO_{2max} in CHD patients by increasing the arteriovenous oxygen difference and, in some cases, maximal stroke volume. In addition, some effects of exercise on CHD may be mediated through beneficial effects on the lipid profile (reduced TC, TG, and LDL levels; and increased HDL levels) along with an improvement in blood rheology and viscosity, and reduction in homocysteine levels[113] (Fig. 1).

2.3 Cancer

2.3.1 Epidemiological Evidence

Evidence from cohort and case–control studies indicates that regular PA protects against the risk of cancer, and that the response is dose dependent.[114] Studies have shown that physically active individuals are less prone to develop certain types of cancer, especially colon and breast cancer.[114] This idea is supported by numerous meta-analyses that have examined the effects of PA on cancer recurrence, improvement of cancer-associated side effects, and mortality. As we have observed, even strenuous exercise seems protective of cancer as elite athletes show a lower risk of cancer mortality.[32]

Both prediagnosis and postdiagnosis exercise have been associated with reduced breast cancer-specific mortality.[115] Friedenreich and Orenstein[116] found that 32 of 44 studies analyzed revealed an average breast cancer risk reduction of 30–40% in women who were more physically active. Moreover, they noted a dose–response relationship between PA and breast cancer risk in 20 of 23 studies. These findings are supported by those of another meta-analysis examining the effects of PA on mortality in breast and colorectal cancer survivors, concluding that an active lifestyle before or after cancer diagnosis is associated with a lower mortality risk.[117] Friedenreich and Orenstein[116] also found strong evidence for PA reducing the colon cancer risk. Of 51 studies analyzed, 43 demonstrated a 40–50% reduction in the risk of this cancer among more physically active men and women. Of 29 studies designed to test whether there is a dose–response, 25 detected that increasing levels of exercise were associated with a decreasing colon cancer risk. The risk of colorectal cancer was also found to be reduced by PA in a Japanese population-based cohort study including 65,022 subjects.[118] Indeed, PA is the main lifestyle change proven to reduce an individual's risk of colon cancer.[119] Exercise also reduces the risk of other cancer types including endometrial[120] and lung cancer.[121] Thus, a large study including 81,516 subjects followed for 19 years revealed a 25% reduction in lung cancer risk in men who walked or cycled at least 4 h/week.[122] Based on the studies analyzed by Thune and Furberg,[114] regular PA seems also protective of prostate cancer with risk reduction in the range of 10–70%.

Fatigue and poor fitness are severe problems in cancer patients such that the therapeutic regimen for cancer patients should aim at improving physical performance. Effectively, research has shown an improvement in cancer-related fatigue attributable to exercise both during and after cancer treatment.[123–125]

2.3.2 Exercise Prescription

There is a paucity of data regarding the best exercise prescription for the different phases of carcinogenesis. General recommendations are individualized and progressive exercise prescription targeting specific symptoms. Any type of ambulation is beneficial for most patients with a low fitness level.[126] Despite the scarcity of evidence, McArdle *et al.*[126] recommended exercise to improve movement amplitude and flexibility, programs to improve muscle strength, and global mobility (e.g., static submaximal exercises for antigravitational muscles, deep breathing exercises, and dynamic trunk rotation movements). In addition, they recommend starting with short periods of low-intensity exercises several times per day followed by continuous exercise.

The American Cancer Society (ACS) and ACSM recommend avoiding inactivity and performing moderate regular exercise during and after treatment.[127,128] During treatment, the intensity of exercise may be reduced but patients should try to stay as fit as possible. Sedentary patients may need to start with short periods of low-intensity exercise. When the disease has resolved or stabilized, the ACS[127] advocates a cautious approach to exercise prescription, which should be guided by blood counts during and after treatment. Thus, it is recommended: not to exercise when anemic, or at public gyms when white blood cell counts are low, when under immunosuppression treatment, when levels of minerals in blood (e.g., sodium and potassium) are abnormal, when there is unresolved pain or nausea/vomiting. When a catheter or feeding tube is in place, situations with a risk of infection should be avoided, as should resistance training using muscles in the area of the catheter to prevent dislodging it. To avoid skin irritation, patients receiving radiation treatment should not expose the skin in the treatment area to chlorine in swimming pools. In patients with feet numbness or problems with balance, stationary exercises could be appropriate. Physicians should watch for swollen ankles, unexplained weight gain, or shortness of breath while at rest or when undertaking low volumes of exercise. Bleeding should also be watched for, especially if the patient is taking blood thinners.

According to Pedersen and Saltin,[4] patients who are receiving treatment, even when hospitalized and bedridden, can benefit from exercise training. Patients undergoing chemo- or radiotherapy would be ill-advised to take up exercise training when leukocyte $<0.5 \times 10^9$/L, hemoglobin <6 mmol/L, thrombocyte $<20 \times 10^9$/L, and temperature >38 °C. In addition, patients with bone metastases should not lift heavy weights when strength training.[4]

Davey Smith *et al.*[129] reported that men who walk at a slow pace have a greater cancer risk than those who walk at a fast pace, suggesting a link between exercise intensity and cancer prevention. This link is also affected by the frequency of exercise as well as exercise conditions.[130] Based on the data from epidemiological studies, Lee[121] suggested that the risk of developing colon and breast cancer may be reduced by carrying out 30–60 min/day of moderate-to-vigorous PA.

2.3.3 Biological Mechanisms

Since cancer refers to a heterogeneous cluster of diseases, it has different underlying pathophysiological causes although immune dysfunction seems common to all cancers. At the blood and tissue levels, cancer patients show higher than normal levels of inflammatory cytokines such as IL-1, IL-6, and TNF-α from macrophage or monocyte lineages as well as lower levels of IL-2, interferon-γ, class-II major histocompatibility complex molecules, and natural killer (NK) cell activity.[131–133] Hence, cancer treatment usually includes therapy to improve immunity. The exact mechanism whereby PA benefits cancer patients remains unclear. However, several pathways have been suggested and supported by the findings of intervention studies.

As shown in Fig. 1, exercise enhances immune function through different mechanisms related to enhanced leukocyte function. Physical exercise suppresses tumorigenesis by regulating the activation and proliferation of circulating NK cells. NK cells recognize and destroy carcinoma cells.[134,135]

Antioxidant mechanisms are also antitumorigenic because oxidative stress can lead to DNA damage.[136] Exercise induces various antioxidant defense pathways[137] that mitigate oxidative damage to lipids, proteins, and DNA in a wide range of tissues.[138] Exercise promotes systemic autophagy, a tumor suppressor pathway that is altered in different types of cancer.[139,140] Levels of calprotectin, a protein that induces apoptosis in certain tumor lines,[141] rise in response to an acute exercise bout.[142]

The idea that exercise stimulates antitumorigenic myokines is supported by several lines of evidence.[143] A novel myokine, secreted protein acidic and rich in cysteine (SPARC), which modulates cell proliferation and migration, is released from skeletal muscle into the circulation in response to a single bout of exercise.[144] Preventive effects of SPARC on colon cancer are thought to be mediated by suppression of the formation of aberrant crypt foci, probably via stimulation of apoptosis through caspase-3 and -8.[145] As observed in a recent review, the cumulative effect of regular exercise bouts is protective against cancer through the release of myokines such as IL-4, IL-13, and IL-6 which, in turn, promotes the production of other anti-inflammatory cytokines including IL-1Ra, IL-10, or soluble TNF receptor (sTNF-R), while inhibiting the pro-inflammatory cytokine TNF-α.[34] In addition, exercise downregulates unbound sex hormones (testosterone, estrogen, and dihydrotestosterone) and increases levels of circulating sex hormone-binding globulin, which binds these hormones, thus diminishing their function.[85] Exercise also reduces levels of metabolic hormones, such as insulin[146] and insulin-growth factor (IGF)-1,[147] associated with increased risks of prostate,[148] breast,[149] and colorectal cancers.[150] IGF-1 suppresses the action of the p53 gene (the genome guardian) and it has been shown that both p53 and the downstream p21 genes are upregulated in lysates of serum-stimulated LNCaP cells obtained from exercised individuals.[151]

Other proposed mechanisms whereby exercise exerts positive effects in cancer patients include increased levels of corticoid hormones, enhanced insulin receptor expression on T-lymphocytes, induced interferon production, enhanced glycogen synthase activity, increased ascorbic acid metabolism, and beneficial effects on provirus and oncogene activation.[126] Moreover, since obesity is correlated with the physiopathology of several types of cancer,[152–155] exercise as a weight management tool is receiving growing attention in the fight against cancer.[156–158] Finally, considering the reduced fitness and chronic fatigue of cancer patients, exercise improves muscle strength and fitness which leads to better physical functioning.

2.4 Alzheimer's Disease

2.4.1 *Epidemiological Evidence*

We review here the effects of exercise on AD which globally provide support for its role in both preventing musculoskeletal decline[159] and attenuating disease severity.[160,161]

Recently, Norton *et al.*[162] examined meta-analyses and estimated population-attributable risks (PAR) of AD for seven potentially modifiable risk factors (T2D, midlife hypertension, obesity, physical inactivity, depression, smoking, and poor educational achievement). Physical inactivity was considered a risk factor when adults did neither 20 min of vigorous PA on 3 or more days nor 30 min of moderate PA on 5 or more days per week. They found that, although the highest estimated PAR worldwide was poor educational achievement, the greatest PAR in the United States, Europe, and United Kingdom was physical inactivity. Thus, it was estimated that worldwide, about 13% (nearly 4.3 million) of cases of AD are potentially attributable to physical inactivity. Interestingly, it was also estimated that a 25% reduction in the prevalence of physical inactivity could prevent about 1 million cases of Alzheimer's in the world.

A systematic review and meta-analysis identified 16 prospective studies examining the link between PA and neurodegenerative disease risk.[163] It showed that exercise was inversely associated with risk of dementia. A systematic review by Rolland *et al.*[164] focused on studies published between 1966 and 2007 in which PA and AD were correlated. Out of 24 studies included, 20 showed a significant and independent preventive effect of exercise on cognitive decline and dementia. A Finnish study including 1500 individuals concluded that those who exercised regularly reduced their risk of developing AD by 60% compared with sedentary controls.[165] Similar results were obtained in another study conducted on 2600 older adults.[166] Remarkably, Larson *et al.*[167] noted, during 6 years of follow-up, a reduction of about 40% in the risk of AD in individuals who exercised 3 days/week exercise. Finally, in a recent cohort study by Grande *et al.*,[168] conducted in persons with mild cognitive impairment, higher PA levels were associated with a lower risk of AD.

2.4.2 *Exercise Prescription*

Low-intensity exercise programs are often prescribed in patients with AD, with special attention to the memory loss and depression these patients usually suffer.[169] The cornerstones of any exercise program should be consistency, patience, and enjoyment. Simple repetitive activities can help in the early stages of an exercise program (i.e., walking and cycling). The ACSM specifies that activities should be enjoyable, involve large muscle groups, and have a rate of perceived exertion ranging from 10 to 15 on the 20-point scale. Each session is suggested to last for 40–60 min (although it may be divided into 15–20 min periods of activity). Since AD patients commonly show a higher level of restlessness or agitation at the end of the day, along with high levels of fatigue and tiredness, it is recommended that exercise sessions be performed in the morning. Exercise should include strength, aerobic, and flexibility training. Strength training aims to strengthen postural muscles, especially those in areas of weakness (i.e., quadriceps and hip extensors), with 10–12 repetitions or fewer, as tolerated. Aerobic exercise aims to maintain cardiorespiratory fitness and musculoskeletal function and is based on walking or cycling, while flexibility training targets postural muscle groups and joints.

2.4.3 Biological Mechanisms

Evidence for the beneficial effects of exercise on the prevalence of AD has been provided by human epidemiological studies. However, most biological mechanisms involved have been identified in animal studies. Several mechanisms have been proposed to explain the link between improved Alzheimer's symptoms and PA[170] (Fig. 1). Oxidative damage leads to a neuron death and, in turn, brain atrophy. Exercise reduces oxidative stress in the brain by inducing antioxidant enzymes.[171–175] In addition, exercise increases tissue-specific levels of several neurotrophic factors (IGF-1, brain-derived neurotrophic factor (BDNF), and vascular endothelial growth factor (VEGF)) that promote neurogenesis, neuroprotection, and survival.[176] In humans, lower levels of serum BDNF have been detected in AD patients,[177] while increasing BDNF levels through aerobic exercise may offer protective effects in elderly patients.[178] BDNF mediates several mechanisms that could explain why exercise improves cognitive and behavioral performance as we have shown in a triple transgenic mouse model of AD.[174] In addition, exercise seems to attenuate the physiopathology of Alzheimer's directly by modulating various mechanisms involved in amyloid-β (Aβ) degradation[170] such as the neprilysin[179] and proteasome systems.[180]

3. CONCLUDING REMARKS

Exercise exerts positive effects on the major chronic diseases, i.e., MS-related disorders, CVD, cancer, and AD. The evidence supporting such beneficial effects comes not only from epidemiological research but also from a growing number of mechanistic studies. Indeed, human and animal research has given us some insights into the molecular pathways by which exercise has both preventive and therapeutic (drug-like) effects on the physiopathology of common chronic diseases. These include, among others, gene expression modulation, decreased inflammation, and several proteostatic mechanisms, all of which can counteract, at least partially, disease-induced molecular alterations.

REFERENCES

1. WHO. *Noncommunicable Diseases Country Profiles 2014*. Switzerland: World Health Organization; 2014.
2. Alberti KG, Eckel RH, Grundy SM, et al. Harmonizing the metabolic syndrome: a joint interim statement of the International Diabetes Federation Task Force on Epidemiology and Prevention; National Heart, Lung, and Blood Institute; American Heart Association; World Heart Federation; International Atherosclerosis Society; and International Association for the Study of Obesity. *Circulation*. 2009;120(16):1640–1645.
3. Eckel RH, Grundy SM, Zimmet PZ. The metabolic syndrome. *Lancet*. 2005;365(9468):1415–1428.
4. Pedersen BK, Saltin B. Evidence for prescribing exercise as therapy in chronic disease. *Scand J Med Sci Sports*. 2006;16(suppl 1):3–63.
5. Eckel RH. Lipoprotein lipase. A multifunctional enzyme relevant to common metabolic diseases. *N Engl J Med*. 1989;320(16):1060–1068.
6. Jensen MD, Caruso M, Heiling V, Miles JM. Insulin regulation of lipolysis in nondiabetic and IDDM subjects. *Diabetes*. 1989;38(12):1595–1601.
7. Zambon A, Hokanson JE, Brown BG, Brunzell JD. Evidence for a new pathophysiological mechanism for coronary artery disease regression: hepatic lipase-mediated changes in LDL density. *Circulation*. 1999;99(15):1959–1964.
8. World Health Organization. *About diabetes;* 2014. http://www.who.int/diabetes/action_online/basics/en/. Accessed 16.10.14.
9. Yki-Jarvinen H. Toxicity of hyperglycaemia in type 2 diabetes. *Diabetes Metab Rev*. 1998;14(suppl 1):S45–S50.
10. Kannel WB, McGee DL. Diabetes and cardiovascular disease. The Framingham study. *JAMA*. 1979;241(19):2035–2038.
11. Lewington S, Clarke R, Qizilbash N, Peto R, Collins R. Age-specific relevance of usual blood pressure to vascular mortality: a meta-analysis of individual data for one million adults in 61 prospective studies. *Lancet*. 2002;360(9349):1903–1913.
12. Booth FW, Roberts CK, Laye MJ. Lack of exercise is a major cause of chronic diseases. *Compr Physiol*. 2012;2(2):1143–1211.
13. McGinnis JM, Foege WH. Actual causes of death in the United States. *JAMA*. 1993;270(18):2207–2212.
14. Kushi LH, Doyle C, McCullough M, et al. American Cancer Society Guidelines on nutrition and physical activity for cancer prevention: reducing the risk of cancer with healthy food choices and physical activity. *CA Cancer J Clin*. 2012;62(1):30–67.
15. Arias E, Anderson RN, Kung HC, Murphy SL, Kochanek KD. Deaths: final data for 2001. *Natl Vital Stat Rep*. 2003;52(3):1–115.
16. American Cancer Society. *Cancer Facts and Figures*. Washington, DC: American Cancer Society; 2004.
17. Querfurth HW, LaFerla FM. Alzheimer's disease. *N Engl J Med*. 2010;362(4):329–344.

18. American Diabetes Association. *The cost of diabetes website;* 2014. http://www.diabetes.org/advocate/resources/cost-of-diabetes.html. Accessed September 15.
19. Finkelstein EA, Trogdon JG, Cohen JW, Dietz W. Annual medical spending attributable to obesity: payer- and service-specific estimates. *Health Aff (Millwood)*. 2009;28(5): w822–w831.
20. American Heart Association. *Heart Disease and Stroke Statistics—2014 Update*. AHA Statistical Update Web site; http://circ.ahajournals.org/content/early/2013/12/18/01.cir.0000441139.02102.80.full.pdf; 2014 Accessed September 17.
21. National Cancer Institute. *Cancer prevalence and cost of care projections website;* 2014. http://costprojections.cancer.gov. Accessed September 15.
22. Mokdad AH, Marks JS, Stroup DF, Gerberding JL. Actual causes of death in the United States, 2000. *JAMA*. 2004;291(10):1238–1245.
23. Centers for Disease Control and Prevention. *Exercise or physical activity. NCHS FastStats Web site;* 2014. http://www.cdc.gov/nchs/fastats/exercise.htm. Accessed September 18.
24. Matthews CE, Chen KY, Freedson PS, et al. Amount of time spent in sedentary behaviors in the United States, 2003–2004. *Am J Epidemiol*. 2008;167(7):875–881.
25. King AC, Blair SN, Bild DE, et al. Determinants of physical activity and interventions in adults. *Med Sci Sports Exerc*. 1992;24(6 suppl):S221–S236.
26. American College of Sports Medicine Position Stand. The recommended quantity and quality of exercise for developing and maintaining cardiorespiratory and muscular fitness, and flexibility in healthy adults. *Med Sci Sports Exerc*. 1998;30(6):975–991.
27. Hallal PC, Andersen LB, Bull FC, Guthold R, Haskell W, Ekelund U. Global physical activity levels: surveillance progress, pitfalls, and prospects. *Lancet*. 2012;380(9838): 247–257.
28. Booth FW, Laye MJ. Lack of adequate appreciation of physical exercise's complexities can pre-empt appropriate design and interpretation in scientific discovery. *J Physiol*. 2009;587(pt 23):5527–5539.
29. Paffenbarger Jr RS, Hyde RT, Wing AL, Hsieh CC. Physical activity, all-cause mortality, and longevity of college alumni. *N Engl J Med*. 1986;314(10):605–613.
30. Schoenborn CA, Stommel M. Adherence to the 2008 adult physical activity guidelines and mortality risk. *Am J Prev Med*. 2011;40(5):514–521.
31. Arrieta A, Russell LB. Effects of leisure and non-leisure physical activity on mortality in U.S. adults over two decades. *Ann Epidemiol*. 2008;18(12):889–895.
32. Garatachea N, Santos-Lozano A, Sanchis-Gomar F, et al. Elite athletes live longer than the general population: a meta-analysis. *Mayo Clin Proc*. 2014;89(9):1195–1200.
33. Naci H, Ioannidis JP. Comparative effectiveness of exercise and drug interventions on mortality outcomes: metaepidemiological study. *BMJ*. 2013;347:f5577.
34. Fiuza-Luces C, Garatachea N, Berger NA, Lucia A. Exercise is the real polypill. *Physiology (Bethesda)*. 2013;28(5):330–358.
35. Umpierre D, Ribeiro PA, Kramer CK, et al. Physical activity advice only or structured exercise training and association with HbA1c levels in type 2 diabetes: a systematic review and meta-analysis. *JAMA*. 2011;305(17):1790–1799.
36. Hopper I, Billah B, Skiba M, Krum H. Prevention of diabetes and reduction in major cardiovascular events in studies of subjects with prediabetes: meta-analysis of randomised controlled clinical trials. *Eur J Cardiovasc Prev Rehabil*. 2011;18(6):813–823.
37. Kelley DE, Goodpaster BH. Effects of physical activity on insulin action and glucose tolerance in obesity. *Med Sci Sports Exerc*. 1999;31(11 suppl):S619–S623.
38. Boule NG, Haddad E, Kenny GP, Wells GA, Sigal RJ. Effects of exercise on glycemic control and body mass in type 2 diabetes mellitus: a meta-analysis of controlled clinical trials. *JAMA*. 2001;286(10):1218–1227.

39. Thomas DE, Elliott EJ, Naughton GA. Exercise for type 2 diabetes mellitus. *Cochrane Database Syst Rev*. 2006;3. CD002968.
40. Chudyk A, Petrella RJ. Effects of exercise on cardiovascular risk factors in type 2 diabetes: a meta-analysis. *Diabetes Care*. 2011;34(5):1228–1237.
41. Cauza E, Hanusch-Enserer U, Strasser B, et al. The relative benefits of endurance and strength training on the metabolic factors and muscle function of people with type 2 diabetes mellitus. *Arch Phys Med Rehabil*. 2005;86(8):1527–1533.
42. Church TS, Blair SN, Cocreham S, et al. Effects of aerobic and resistance training on hemoglobin A1c levels in patients with type 2 diabetes: a randomized controlled trial. *JAMA*. 2010;304(20):2253–2262.
43. Myers J, Kaykha A, George S, et al. Fitness versus physical activity patterns in predicting mortality in men. *Am J Med*. 2004;117(12):912–918.
44. Sigal RJ, Kenny GP, Boule NG, et al. Effects of aerobic training, resistance training, or both on glycemic control in type 2 diabetes: a randomized trial. *Ann Intern Med*. 2007;147(6):357–369.
45. Earnest CP, Johannsen NM, Swift DL, et al. Aerobic and strength training in concomitant metabolic syndrome and type 2 diabetes. *Med Sci Sports Exerc*. 2014;46(7): 1293–1301.
46. Boule NG, Kenny GP, Haddad E, Wells GA, Sigal RJ. Meta-analysis of the effect of structured exercise training on cardiorespiratory fitness in type 2 diabetes mellitus. *Diabetologia*. 2003;46(8):1071–1081.
47. Colberg SR, Albright AL, Blissmer BJ, et al. Exercise and type 2 diabetes: American College of Sports Medicine and the American Diabetes Association: joint position statement. Exercise and type 2 diabetes. *Med Sci Sports Exerc*. 2010;42(12):2282–2303.
48. Ivy JL. Role of exercise training in the prevention and treatment of insulin resistance and non-insulin-dependent diabetes mellitus. *Sports Med*. 1997;24(5):321–336.
49. Dela F, Handberg A, Mikines KJ, Vinten J, Galbo H. GLUT 4 and insulin receptor binding and kinase activity in trained human muscle. *J Physiol*. 1993;469:615–624.
50. Dela F, Ploug T, Handberg A, et al. Physical training increases muscle GLUT4 protein and mRNA in patients with NIDDM. *Diabetes*. 1994;43(7):862–865.
51. Ebeling P, Bourey R, Koranyi L, et al. Mechanism of enhanced insulin sensitivity in athletes. Increased blood flow, muscle glucose transport protein (GLUT-4) concentration, and glycogen synthase activity. *J Clin Invest*. 1993;92(4):1623–1631.
52. Ivy JL, Zderic TW, Fogt DL. Prevention and treatment of non-insulin-dependent diabetes mellitus. *Exerc Sport Sci Rev*. 1999;27:1–35.
53. Coggan AR, Spina RJ, Kohrt WM, Holloszy JO. Effect of prolonged exercise on muscle citrate concentration before and after endurance training in men. *Am J Physiol*. 1993;264(2 pt 1):E215–E220.
54. Saltin B, Henriksson J, Nygaard E, Andersen P, Jansson E. Fiber types and metabolic potentials of skeletal muscles in sedentary man and endurance runners. *Ann N Y Acad Sci*. 1977;301:3–29.
55. Dela F, von Linstow ME, Mikines KJ, Galbo H. Physical training may enhance beta-cell function in type 2 diabetes. *Am J Physiol Endocrinol Metab*. 2004;287(5): E1024–E1031.
56. McAllister RM, Hirai T, Musch TI. Contribution of endothelium-derived nitric oxide (EDNO) to the skeletal muscle blood flow response to exercise. *Med Sci Sports Exerc*. 1995;27(8):1145–1151.
57. Eriksson J, Taimela S, Eriksson K, Parviainen S, Peltonen J, Kujala U. Resistance training in the treatment of non-insulin-dependent diabetes mellitus. *Int J Sports Med*. 1997;18(4):242–246.
58. Eriksson JG. Exercise and the treatment of type 2 diabetes mellitus. An update. *Sports Med*. 1999;27(6):381–391.

59. Kelley GA, Kelley KS, Roberts S, Haskell W. Comparison of aerobic exercise, diet or both on lipids and lipoproteins in adults: a meta-analysis of randomized controlled trials. *Clin Nutr.* 2012;31(2):156–167.
60. Kelley GA, Kelley KS. Impact of progressive resistance training on lipids and lipoproteins in adults: a meta-analysis of randomized controlled trials. *Prev Med.* 2009;48(1):9–19.
61. Kelley GA, Kelley KS, Vu Tran Z. Aerobic exercise, lipids and lipoproteins in overweight and obese adults: a meta-analysis of randomized controlled trials. *Int J Obes (Lond).* 2005;29(8):881–893.
62. Physical Activity Guidelines Advisory Committee. *Physical Activity Guidelines Advisory Committee Report.* http://www.health.gov/paguidelines/Report/pdf/CommitteeReport.pdf; 2008.
63. Halbert JA, Silagy CA, Finucane P, Withers RT, Hamdorf PA. Exercise training and blood lipids in hyperlipidemic and normolipidemic adults: a meta-analysis of randomized, controlled trials. *Eur J Clin Nutr.* 1999;53(7):514–522.
64. Hayashino Y, Jackson JL, Fukumori N, Nakamura F, Fukuhara S. Effects of supervised exercise on lipid profiles and blood pressure control in people with type 2 diabetes mellitus: a meta-analysis of randomized controlled trials. *Diabetes Res Clin Pract.* 2012;98(3):349–360.
65. Durstine JL, Grandjean PW, Davis PG, Ferguson MA, Alderson NL, DuBose KD. Blood lipid and lipoprotein adaptations to exercise: a quantitative analysis. *Sports Med.* 2001;31(15):1033–1062.
66. Durstine JL, Grandjean PW, Cox CA, Thompson PD. Lipids, lipoproteins, and exercise. *J Cardiopulm Rehabil.* 2002;22(6):385–398.
67. Greene NP, Martin SE, Crouse SF. Acute exercise and training alter blood lipid and lipoprotein profiles differently in overweight and obese men and women. *Obesity (Silver Spring).* 2012;20(8):1618–1627.
68. Klop B, Elte JW, Cabezas MC. Dyslipidemia in obesity: mechanisms and potential targets. *Nutrients.* 2013;5(4):1218–1240.
69. Pattyn N, Cornelissen VA, Eshghi SR, Vanhees L. The effect of exercise on the cardiovascular risk factors constituting the metabolic syndrome: a meta-analysis of controlled trials. *Sports Med.* 2013;43(2):121–133.
70. Strasser B, Siebert U, Schobersberger W. Resistance training in the treatment of the metabolic syndrome: a systematic review and meta-analysis of the effect of resistance training on metabolic clustering in patients with abnormal glucose metabolism. *Sports Med.* 2010;40(5):397–415.
71. Carnethon MR, Gidding SS, Nehgme R, Sidney S, Jacobs Jr DR, Liu K. Cardiorespiratory fitness in young adulthood and the development of cardiovascular disease risk factors. *JAMA.* 2003;290(23):3092–3100.
72. Cook S, Weitzman M, Auinger P, Nguyen M, Dietz WH. Prevalence of a metabolic syndrome phenotype in adolescents: findings from the third National Health and Nutrition Examination Survey, 1988–1994. *Arch Pediatr Adolesc Med.* 2003;157(8):821–827.
73. Lakka TA, Laaksonen DE, Lakka HM, et al. Sedentary lifestyle, poor cardiorespiratory fitness, and the metabolic syndrome. *Med Sci Sports Exerc.* 2003;35(8):1279–1286.
74. Tuomilehto J, Lindstrom J, Eriksson JG, et al. Prevention of type 2 diabetes mellitus by changes in lifestyle among subjects with impaired glucose tolerance. *N Engl J Med.* 2001;344(18):1343–1350.
75. Clinical Guidelines on the Identification, Evaluation, and Treatment of Overweight and Obesity in Adults—The Evidence Report. National Institutes of Health. *Obes Res.* 1998;6(suppl 2):51S–209S.

76. Knowler WC, Barrett-Connor E, Fowler SE, et al. Reduction in the incidence of type 2 diabetes with lifestyle intervention or metformin. *N Engl J Med.* 2002;346(6): 393–403.
77. Zimmet P, Shaw J, Alberti KG. Preventing type 2 diabetes and the dysmetabolic syndrome in the real world: a realistic view. *Diabet Med.* 2003;20(9):693–702.
78. Ross R, Janssen I. Physical activity, total and regional obesity: dose–response considerations. *Med Sci Sports Exerc.* 2001;33(6 suppl):S521–S527 [discussion S528–529].
79. Ross R, Janssen I. Is abdominal fat preferentially reduced in response to exercise-induced weight loss? *Med Sci Sports Exerc.* 1999;31(11 suppl):S568–S572.
80. Shaw K, Gennat H, O'Rourke P, Del Mar C. Exercise for overweight or obesity. *Cochrane Database Syst Rev.* 2006;4. CD003817.
81. Kelley GA, Kelley KS. Effects of exercise in the treatment of overweight and obese children and adolescents: a systematic review of meta-analyses. *J Obes.* 2013;2013: 783103.
82. Ismail I, Keating SE, Baker MK, Johnson NA. A systematic review and meta-analysis of the effect of aerobic vs. resistance exercise training on visceral fat. *Obes Rev.* 2012;13(1):68–91.
83. Donnelly JE, Blair SN, Jakicic JM, Manore MM, Rankin JW, Smith BK. American College of Sports Medicine Position Stand. Appropriate physical activity intervention strategies for weight loss and prevention of weight regain for adults. *Med Sci Sports Exerc.* 2009;41(2):459–471.
84. Pedersen BK. The diseasome of physical inactivity—and the role of myokines in muscle—fat cross talk. *J Physiol.* 2009;587(pt 23):5559–5568.
85. Roberts CK, Barnard RJ. Effects of exercise and diet on chronic disease. *J Appl Physiol (1985).* 2005;98(1):3–30.
86. Cornelissen VA, Fagard RH. Effects of endurance training on blood pressure, blood pressure-regulating mechanisms, and cardiovascular risk factors. *Hypertension.* 2005;46(4):667–675.
87. Cornelissen VA, Fagard RH. Effect of resistance training on resting blood pressure: a meta-analysis of randomized controlled trials. *J Hypertens.* 2005;23(2):251–259.
88. Cornelissen VA, Fagard RH, Coeckelberghs E, Vanhees L. Impact of resistance training on blood pressure and other cardiovascular risk factors: a meta-analysis of randomized, controlled trials. *Hypertension.* 2011;58(5):950–958.
89. Kelley GA, Kelley KS. Progressive resistance exercise and resting blood pressure: a meta-analysis of randomized controlled trials. *Hypertension.* 2000;35(3):838–843.
90. Owen A, Wiles J, Swaine I. Effect of isometric exercise on resting blood pressure: a meta analysis. *J Hum Hypertens.* 2010;24(12):796–800.
91. Semlitsch T, Jeitler K, Hemkens LG, et al. Increasing physical activity for the treatment of hypertension: a systematic review and meta-analysis. *Sports Med.* 2013;43(10): 1009–1023.
92. Huang G, Shi X, Gibson CA, Huang SC, Coudret NA, Ehlman MC. Controlled aerobic exercise training reduces resting blood pressure in sedentary older adults. *Blood Press.* 2013;22(6):386–394.
93. Cornelissen VA, Smart NA. Exercise training for blood pressure: a systematic review and meta-analysis. *J Am Heart Assoc.* 2013;2(1). e004473.
94. Carlson DJ, Dieberg G, Hess NC, Millar PJ, Smart NA. Isometric exercise training for blood pressure management: a systematic review and meta-analysis. *Mayo Clin Proc.* 2014;89(3):327–334.
95. Pescatello LS, Franklin BA, Fagard R, Farquhar WB, Kelley GA, Ray CA. American College of Sports Medicine position stand. Exercise and hypertension. *Med Sci Sports Exerc.* 2004;36(3):533–553.

96. Park S, Jastremski CA, Wallace JP. Time of day for exercise on blood pressure reduction in dipping and nondipping hypertension. *J Hum Hypertens*. 2005;19(8):597–605.
97. Padilla J, Wallace JP, Park S. Accumulation of physical activity reduces blood pressure in pre- and hypertension. *Med Sci Sports Exerc*. 2005;37(8):1264–1275.
98. Levy WC, Cerqueira MD, Abrass IB, Schwartz RS, Stratton JR. Endurance exercise training augments diastolic filling at rest and during exercise in healthy young and older men. *Circulation*. 1993;88(1):116–126.
99. Esler M, Rumantir M, Kaye D, Lambert G. The sympathetic neurobiology of essential hypertension: disparate influences of obesity, stress, and noradrenaline transporter dysfunction? *Am J Hypertens*. 2001;14(6 pt 2):139S–146S.
100. Fagard RH. Exercise is good for your blood pressure: effects of endurance training and resistance training. *Clin Exp Pharmacol Physiol*. 2006;33(9):853–856.
101. Havlik RJ, Phillips CL, Brock DB, et al. Walking may be related to less vascular stiffness in the activity counseling trial (ACT). *Am Heart J*. 2005;150(2):270–275.
102. Costas Jr P, Garcia-Palmieri MR, Nazario E, Sorlie PD. Relation of lipids, weight and physical activity to incidence of coronary heart disease: the Puerto Rico heart study. *Am J Cardiol*. 1978;42(4):653–658.
103. Powell KE, Thompson PD, Caspersen CJ, Kendrick JS. Physical activity and the incidence of coronary heart disease. *Annu Rev Public Health*. 1987;8:253–287.
104. Thompson PD, Buchner D, Pina IL, et al. Exercise and physical activity in the prevention and treatment of atherosclerotic cardiovascular disease: a statement from the Council on Clinical Cardiology (Subcommittee on Exercise, Rehabilitation, and Prevention) and the Council on Nutrition, Physical Activity, and Metabolism (Subcommittee on Physical Activity). *Circulation*. 2003;107(24):3109–3116.
105. Fletcher GF, Blair SN, Blumenthal J, et al. Statement on exercise. Benefits and recommendations for physical activity programs for all Americans. A statement for health professionals by the Committee on Exercise and Cardiac Rehabilitation of the Council on Clinical Cardiology, American Heart association. *Circulation*. 1992;86(1):340–344.
106. U.S. Department of Health and Human Services. *Physical Activity and Health: A Report of the Surgeon General*. Atlanta, GA: U.S: Centers for Disease Control and Prevention; 1996.
107. Tanasescu M, Leitzmann MF, Rimm EB, Willett WC, Stampfer MJ, Hu FB. Exercise type and intensity in relation to coronary heart disease in men. *JAMA*. 2002;288(16):1994–2000.
108. Lee IM, Rexrode KM, Cook NR, Manson JE, Buring JE. Physical activity and coronary heart disease in women: is "no pain, no gain" passe? *JAMA*. 2001;285(11):1447–1454.
109. Heran BS, Chen JM, Ebrahim S, et al. Exercise-based cardiac rehabilitation for coronary heart disease. *Cochrane Database Syst Rev*. 2011;7. CD001800.
110. American College of Sports Medicine position stand. Exercise for patients with coronary artery disease. *Med Sci Sports Exerc*. 1994;26(3):i–v.
111. Joyner MJ, Green DJ. Exercise protects the cardiovascular system: effects beyond traditional risk factors. *J Physiol*. 2009;587(pt 23):5551–5558.
112. Ribeiro F, Alves AJ, Duarte JA, Oliveira J. Is exercise training an effective therapy targeting endothelial dysfunction and vascular wall inflammation? *Int J Cardiol*. 2010;141(3):214–221.
113. Lavie CJ, Milani RV. Cardiac rehabilitation and exercise training in secondary coronary heart disease prevention. *Prog Cardiovasc Dis*. 2011;53(6):397–403.
114. Thune I, Furberg AS. Physical activity and cancer risk: dose–response and cancer, all sites and site-specific. *Med Sci Sports Exerc*. 2001;33(6 suppl):S530–S550 [discussion S609–510].
115. Zhong S, Jiang T, Ma T, et al. Association between physical activity and mortality in breast cancer: a meta-analysis of cohort studies. *Eur J Epidemiol*. 2014;29(6):391–404.

116. Friedenreich CM, Orenstein MR. Physical activity and cancer prevention: etiologic evidence and biological mechanisms. *J Nutr.* 2002;132(11 suppl):3456S–3464S.
117. Schmid D, Leitzmann MF. Association between physical activity and mortality among breast cancer and colorectal cancer survivors: a systematic review and meta-analysis. *Ann Oncol.* 2014;25(7):1293–1311.
118. Lee KJ, Inoue M, Otani T, Iwasaki M, Sasazuki S, Tsugane S. Physical activity and risk of colorectal cancer in Japanese men and women: the Japan Public Health Center-based prospective study. *Cancer Causes Control.* 2007;18(2):199–209.
119. Word Cancer Research Fund, and American Institute for Cancer Research. *Physical Activity. Food, Nutrition, Physical Activity, and the Prevention of Cancer: A Global Perspective.* Washington DC: World Cancer Research Fund International; 2007.
120. Keum N, Ju W, Lee DH, et al. Leisure-time physical activity and endometrial cancer risk: dose–response meta-analysis of epidemiological studies. *Int J Cancer.* 2014;135(3):682–694.
121. Lee IM. Physical activity and cancer prevention—data from epidemiologic studies. *Med Sci Sports Exerc.* 2003;35(11):1823–1827.
122. Thune I, Lund E. The influence of physical activity on lung-cancer risk: a prospective study of 81,516 men and women. *Int J Cancer.* 1997;70(1):57–62.
123. Cramp F, Byron-Daniel J. Exercise for the management of cancer-related fatigue in adults. *Cochrane Database Syst Rev.* 2012;11. CD006145.
124. Tomlinson D, Diorio C, Beyene J, Sung L. Effect of exercise on cancer-related fatigue: a meta-analysis. *Am J Phys Med Rehabil.* 2014;93(8):675–686.
125. Velthuis MJ, Agasi-Idenburg SC, Aufdemkampe G, Wittink HM. The effect of physical exercise on cancer-related fatigue during cancer treatment: a meta-analysis of randomised controlled trials. *Clin Oncol (R Coll Radiol).* 2010;22(3):208–221.
126. McArdle WD, Katch FI, Katch VL. *Essentials of Exercise Physiology.* 2nd ed. Philadelphia, PA: U.S: Lippincott Williams & Wilkins; 2004.
127. American-Cancer-Society. *Physical Activity and the Cancer Patient.* 2014. http://www.cancer.org/treatment/survivorshipduringandaftertreatment/stayingactive/physical-activity-and-the-cancer-patient. Accessed 23.10.14.
128. Schmitz KH, Courneya KS, Matthews C, et al. American College of Sports Medicine roundtable on exercise guidelines for cancer survivors. *Med Sci Sports Exerc.* 2010;42(7):1409–1426.
129. Davey Smith G, Shipley MJ, Batty GD, Morris JN, Marmot M. Physical activity and cause-specific mortality in the Whitehall study. *Public Health.* 2000;114(5):308–315.
130. Wu AH, Paganini-Hill A, Ross RK, Henderson BE. Alcohol, physical activity and other risk factors for colorectal cancer: a prospective study. *Br J Cancer.* 1987;55(6):687–694.
131. Monteleone G, Pallone F, Stolfi C. The dual role of inflammation in colon carcinogenesis. *Int J Mol Sci.* 2012;13(9):11071–11084.
132. Fantini MC, Pallone F. Cytokines: from gut inflammation to colorectal cancer. *Curr Drug Targets.* 2008;9(5):375–380.
133. Szkaradkiewicz A, Karpinski TM, Drews M, Borejsza-Wysocki M, Majewski P, Andrzejewska E. Natural killer cell cytotoxicity and immunosuppressive cytokines (IL-10, TGF-beta1) in patients with gastric cancer. *J Biomed Biotechnol.* 2010;2010:901564.
134. Smyth MJ, Cretney E, Kelly JM, et al. Activation of NK cell cytotoxicity. *Mol Immunol.* 2005;42(4):501–510.
135. Lanier LL. Up on the tightrope: natural killer cell activation and inhibition. *Nat Immunol.* 2008;9(5):495–502.
136. Klaunig JE, Kamendulis LM, Hocevar BA. Oxidative stress and oxidative damage in carcinogenesis. *Toxicol Pathol.* 2010;38(1):96–109.
137. Ji LL. Exercise-induced modulation of antioxidant defense. *Ann N Y Acad Sci.* 2002;959:82–92.

138. Radak Z, Ihasz F, Koltai E, Goto S, Taylor AW, Boldogh I. The redox-associated adaptive response of brain to physical exercise. *Free Radic Res.* 2014;48(1):84–92.
139. He C, Sumpter Jr R, Levine B. Exercise induces autophagy in peripheral tissues and in the brain. *Autophagy.* 2012;8(10):1548–1551.
140. Levine B. Cell biology: autophagy and cancer. *Nature.* 2007;446(7137):745–747.
141. Yui S, Mikami M, Yamazaki M. Purification and characterization of the cytotoxic factor in rat peritoneal exudate cells: its identification as the calcium binding protein complex, calprotectin. *J Leukoc Biol.* 1995;58(3):307–316.
142. Mortensen OH, Andersen K, Fischer C, et al. Calprotectin is released from human skeletal muscle tissue during exercise. *J Physiol.* 2008;586(14):3551–3562.
143. Hojman P, Dethlefsen C, Brandt C, Hansen J, Pedersen L, Pedersen BK. Exercise-induced muscle-derived cytokines inhibit mammary cancer cell growth. *Am J Physiol Endocrinol Metab.* 2011;301(3):E504–E510.
144. Norheim F, Raastad T, Thiede B, Rustan AC, Drevon CA, Haugen F. Proteomic identification of secreted proteins from human skeletal muscle cells and expression in response to strength training. *Am J Physiol Endocrinol Metab.* 2011;301(5):E1013–E1021.
145. Aoi W, Naito Y, Takagi T, et al. A novel myokine, secreted protein acidic and rich in cysteine (SPARC), suppresses colon tumorigenesis via regular exercise. *Gut.* 2013;62(6):882–889.
146. Giovannucci E. Insulin, insulin-like growth factors and colon cancer: a review of the evidence. *J Nutr.* 2001;131(11 suppl):3109S–3120S.
147. Yu H, Rohan T. Role of the insulin-like growth factor family in cancer development and progression. *J Natl Cancer Inst.* 2000;92(18):1472–1489.
148. Chan JM, Stampfer MJ, Giovannucci E, et al. Plasma insulin-like growth factor-I and prostate cancer risk: a prospective study. *Science.* 1998;279(5350):563–566.
149. Hankinson SE, Willett WC, Colditz GA, et al. Circulating concentrations of insulin-like growth factor-I and risk of breast cancer. *Lancet.* 1998;351(9113):1393–1396.
150. Ma J, Pollak MN, Giovannucci E, et al. Prospective study of colorectal cancer risk in men and plasma levels of insulin-like growth factor (IGF)-I and IGF-binding protein-3. *J Natl Cancer Inst.* 1999;91(7):620–625.
151. Leung PS, Aronson WJ, Ngo TH, Golding LA, Barnard RJ. Exercise alters the IGF axis in vivo and increases p53 protein in prostate tumor cells in vitro. *J Appl Physiol (1985).* 2004;96(2):450–454.
152. Azvolinsky A. Cancer risk: the fat tissue–BMI–obesity connection. *J Natl Cancer Inst.* 2014;106(4). dju100.
153. Bhaskaran K, Douglas I, Forbes H, dos-Santos-Silva I, Leon DA, Smeeth L. Body-mass index and risk of 22 specific cancers: a population-based cohort study of 5.24 million UK adults. *Lancet.* 2014;384(9945):755–765.
154. Kmietowicz Z. Overweight and obesity are linked to 10 common cancers and more than 12,000 UK cases. *BMJ.* 2014;349. g5183.
155. Campbell PT. Obesity: a certain and avoidable cause of cancer. *Lancet.* 2014;384(9945):727–728.
156. World Cancer Research Fund/American Institute for Cancer Research. Continuous Update Project Report. *Food, Nutrition, Physical Activity, and the Prevention of Colorectal Cancer*; 2011.
157. World Cancer Research Fund/American Institute for Cancer Research. Continuous Update Project Report. *Food, Nutrition, Physical Activity, and the Prevention of Breast Cancer*; 2010.
158. Wu Y, Zhang D, Kang S. Physical activity and risk of breast cancer: a meta-analysis of prospective studies. *Breast Cancer Res Treat.* 2013;137(3):869–882.
159. Santana-Sosa E, Barriopedro MI, Lopez-Mojares LM, Perez M, Lucia A. Exercise training is beneficial for Alzheimer's patients. *Int J Sports Med.* 2008;29(10):845–850.

160. Verghese J, Lipton RB, Katz MJ, et al. Leisure activities and the risk of dementia in the elderly. *N Engl J Med.* 2003;348(25):2508–2516.
161. Lucia A, Ruiz JR. Exercise is beneficial for patients with Alzheimer's disease: a call for action. *Br J Sports Med.* 2011;45(6):468–469.
162. Norton S, Matthews FE, Barnes DE, Yaffe K, Brayne C. Potential for primary prevention of Alzheimer's disease: an analysis of population-based data. *Lancet Neurol.* 2014;13(8):788–794.
163. Hamer M, Chida Y. Physical activity and risk of neurodegenerative disease: a systematic review of prospective evidence. *Psychol Med.* 2009;39(1):3–11.
164. Rolland Y, Abellan van Kan G, Vellas B. Physical activity and Alzheimer's disease: from prevention to therapeutic perspectives. *J Am Med Dir Assoc.* 2008;9(6):390–405.
165. Kivipelto M, Solomon A. Alzheimer's disease—the ways of prevention. *J Nutr Health Aging.* 2008;12(1):89S–94S.
166. Laurin D, Verreault R, Lindsay J, MacPherson K, Rockwood K. Physical activity and risk of cognitive impairment and dementia in elderly persons. *Arch Neurol.* 2001;58(3):498–504.
167. Larson EB, Wang L, Bowen JD, et al. Exercise is associated with reduced risk for incident dementia among persons 65 years of age and older. *Ann Intern Med.* 2006;144(2):73–81.
168. Grande G, Vanacore N, Maggiore L, et al. Physical activity reduces the risk of dementia in mild cognitive impairment subjects: a cohort study. *J Alzheimers Dis.* 2014;39(4):833–839.
169. Durstine JL. American College of Sports Medicine. *ACSM's Exercise Management for Persons with Chronic Diseases and Disabilities.* 3rd ed. Champaign, IL: Human Kinetics; 2009.
170. Radak Z, Hart N, Sarga L, et al. Exercise plays a preventive role against Alzheimer's disease. *J Alzheimers Dis.* 2010;20(3):777–783.
171. Ogonovszky H, Berkes I, Kumagai S, et al. The effects of moderate-, strenuous- and over-training on oxidative stress markers, DNA repair, and memory, in rat brain. *Neurochem Int.* 2005;46(8):635–640.
172. Gimenez-Llort L, Garcia Y, Buccieri K, et al. Gender-specific neuroimmunoendocrine response to treadmill exercise in 3xTg-AD mice. *Int J Alzheimers Dis.* 2010;2010:128354.
173. Radak Z, Kaneko T, Tahara S, et al. Regular exercise improves cognitive function and decreases oxidative damage in rat brain. *Neurochem Int.* 2001;38(1):17–23.
174. Garcia-Mesa Y, Pareja-Galeano H, Bonet-Costa V, et al. Physical exercise neuroprotects ovariectomized 3xTg-AD mice through BDNF mechanisms. *Psychoneuroendocrinology.* 2014;45:154–166.
175. Pareja-Galeano H, Brioche T, Sanchis-Gomar F, et al. Effects of physical exercise on cognitive alterations and oxidative stress in an APP/PSN1 transgenic model of Alzheimer's disease. *Rev Esp Geriatr Gerontol.* 2012;47(5):198–204.
176. Neeper SA, Gomez-Pinilla F, Choi J, Cotman C. Exercise and brain neurotrophins. *Nature.* 1995;373(6510):109.
177. Aisen PS. Serum brain-derived neurotrophic factor and the risk for dementia. *JAMA.* 2014;311(16):1684–1685.
178. Coelho FG, Vital TM, Stein AM, et al. Acute aerobic exercise increases brain-derived neurotrophic factor levels in elderly with Alzheimer's disease. *J Alzheimers Dis.* 2014;39(2):401–408.
179. Adlard PA, Perreau VM, Pop V, Cotman CW. Voluntary exercise decreases amyloid load in a transgenic model of Alzheimer's disease. *J Neurosci.* 2005;25(17):4217–4221.
180. Almeida CG, Tampellini D, Takahashi RH, et al. Beta-amyloid accumulation in APP mutant neurons reduces PSD-95 and GluR1 in synapses. *Neurobiol Dis.* 2005;20(2):187–198.

INDEX

Note: Page numbers followed by "*f*" indicate figures and "*t*" indicate tables.

B

G

H

L

M

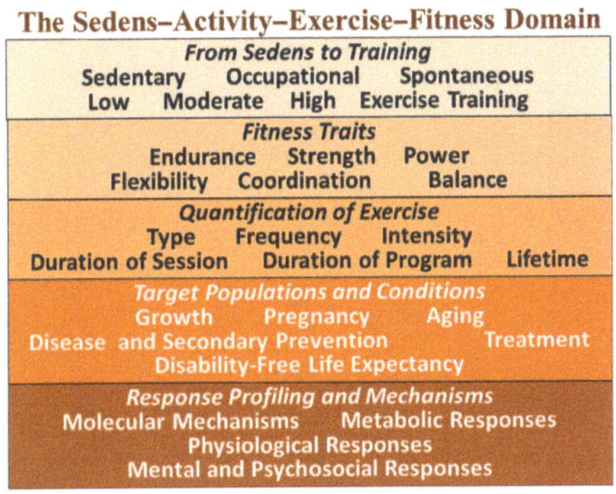

Claude Bouchard, Figure 1 Schematic description of the sedentary behavior, physical activity level, exercise training, and fitness domain with its multiple dimensions and some of its implications.

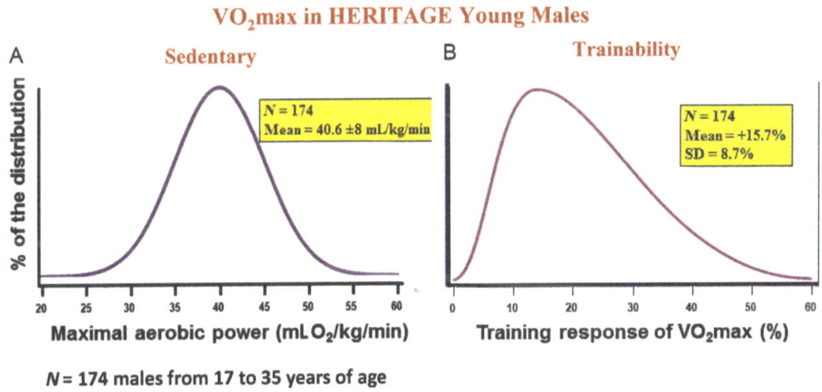

Claude Bouchard, Figure 2 Distribution of VO_2max/kg body weight values in 174 sedentary men, 17–35 years of age, from the HERITAGE Family Study (A). Distribution of the VO_2max changes in % of baseline levels in response to a standardized endurance training program of 20 weeks in the same sedentary subjects (B).

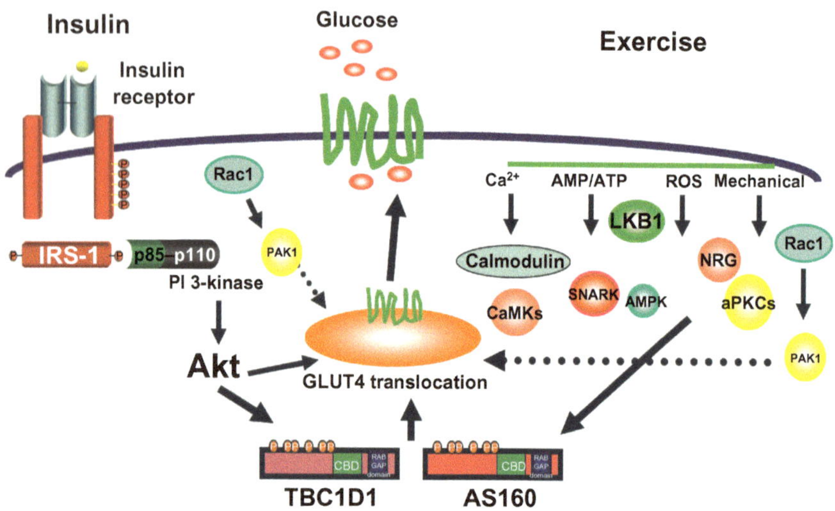

Joram D. Mul *et al.*, Figure 1 *Exercise and insulin regulation of glucose transport.* Proposed model for the signaling pathways mediating exercise- and insulin-induced skeletal muscle glucose transport. Insulin is initiated by binding to its cell service receptor leading to a cascade of phosphorylation reactions involving IRS-1, PI 3-kinase, and Akt among other proteins. Exercise works through a proximal signaling mechanism from that is distinct form that of insulin and is less well defined. It is likely that the proximal exercise signaling mechanism has redundancy as number of stimuli have been implicated in this process including changes in intracellular Ca^{2+}, the AMP:ATP ratio, generation of reactive oxygen species, and mechanical stresses. The insulin and exercise signaling pathways are thought to converge at the level of the Rab GAP proteins TBC1D1 and AS160, which allow for the release of the GLUT4-containing vesicles from intracellular stores, translocation to the transverse tubules and sarcolemma, and an increase in glucose uptake. *Adapted from Ref. 50.*

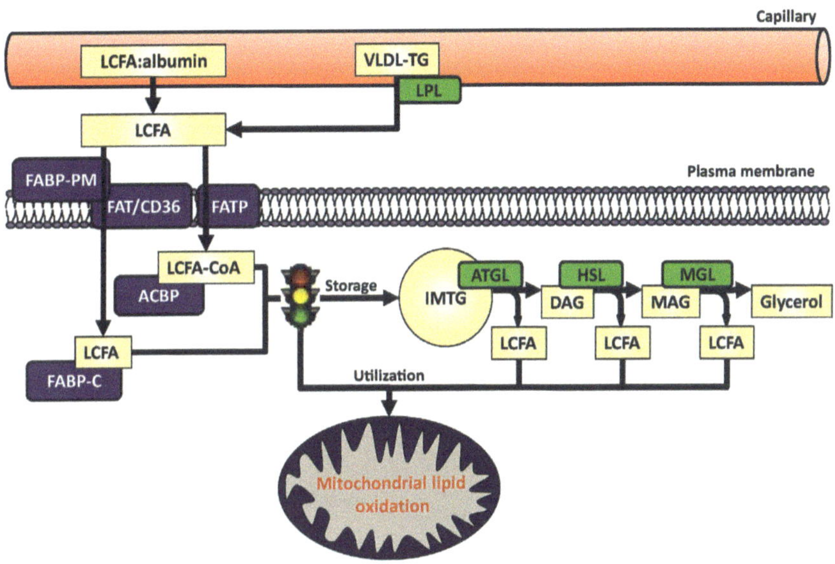

Robert C. Noland, Figure 1 Cellular fatty acid uptake and intracellular lipid handling in skeletal muscle. During moderate-intensity exercise, delivery of lipid (LCFA:albumin and VLDL-TG) through the circulatory system to skeletal muscle increases. In response, upregulation of fatty acid transporters at the plasma membrane during exercise increases fatty acid uptake into skeletal muscle. Additionally, exercise increases lipolysis of intramuscular triglycerides. Collectively, these actions significantly increase fatty acid delivery to the mitochondria for use as metabolic fuel during moderate-intensity endurance exercise. *Abbreviations*: LCFA, long-chain fatty acid; VLDL-TG, very-low-density lipoprotein triglyceride; FABP-PM, plasma membrane fatty acid-binding protein; FAT/CD36, fatty acid translocase, cluster of differentiation 36; FATP, fatty acid transport protein; FABP-C, cytosolic fatty acid-binding protein; ACBP, acyl-CoA-binding protein; IMTG, intramuscular triglyceride; ATGL, adipose triglyceride lipase; HSL, hormone-sensitive lipase; MGL, monoacylglycerol lipase; DAG, diacylglycerol; MAG, monoacylglycerol.

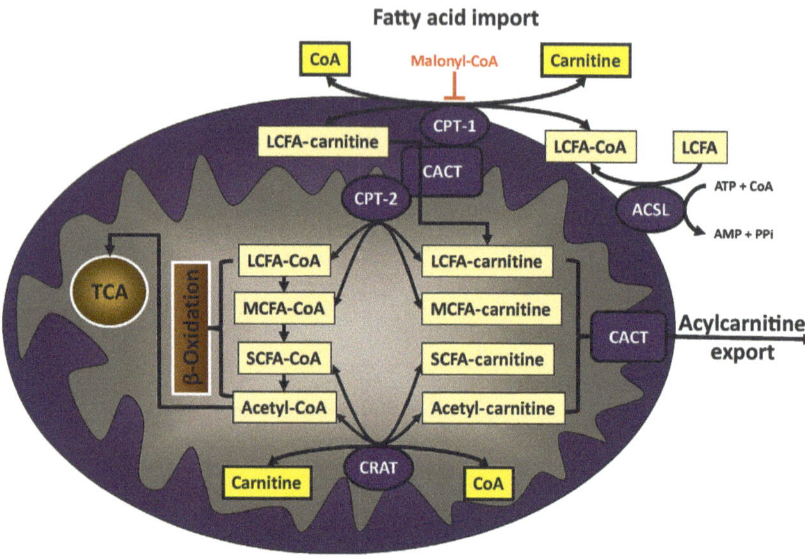

Robert C. Noland, Figure 2 Role of the carnitine shuttle in mitochondrial import and export of fatty acids. Carnitine palmitoyltransferase-1 (CPT-1) converts long-chain acyl-CoA (LCFA-CoA) into long-chain acylcarnitine (LCFA-carnitine), which subsequently enters the mitochondria through carnitine–acylcarnitine translocase (CACT). Within the mitochondrial matrix, carnitine palmitoyltransferase-2 (CPT-2) converts LCFA-carnitine back into LCFA-CoA, which then enters the β-oxidation pathway for use as metabolic fuel. The ability of carnitine to facilitate mitochondrial fatty acid import is essential and supports lipid metabolism during exercise. In the event that mitochondrial lipid entry overwhelms the requirement for metabolic fuel, excess acyl-CoAs generated within the β-oxidative pathway can be converted into acylcarnitine esters by CPT-2 or carnitine acetyltransferase (CRAT). The resultant acylcarnitines can then be exported from the mitochondrial matrix. The ability of carnitine to facilitate the export of excess acyl moieties from the mitochondrial matrix likely preserves optimal mitochondrial performance during exercise. *Abbreviations*: ACSL, long-chain acyl-CoA synthetase; PPi, inorganic pyrophosphate.

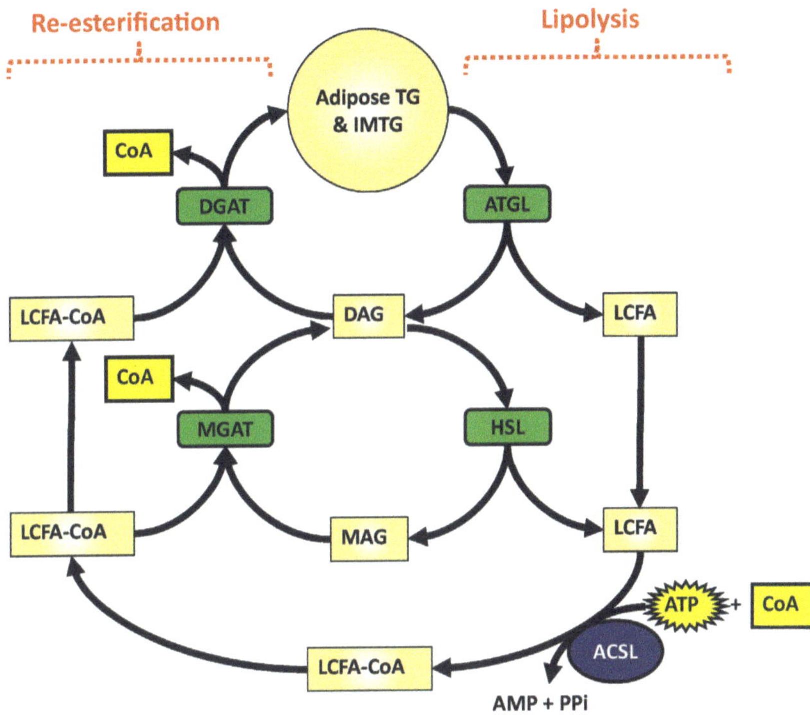

Robert C. Noland, Figure 3 Energy cost of triglyceride/fatty acid (TG/FA) cycling. During exercise, adipose triglyceride lipase (ATGL) and hormone-sensitive lipase (HSL) activity are heightened, which increases mobilization of lipids via lipolysis of triglyceride stores (adipose tissue and IMTG). Heightened lipolysis can be maintained for prolonged periods in the postexercise recovery phase when energy demand drops substantially. Since the mobilized long-chain fatty acids (LCFA) are not utilized as rapidly during the recovery period, they undergo re-esterification. Together, this phenomenon is referred to as triglyceride/fatty acid (TG/FA) cycling. Re-esterification of lipids into the TG pool requires conversion of LCFA to LCFA-CoA via the action of long-chain acyl-CoA synthetase (ACSL). This reaction consumes two ATP equivalents; therefore, TG/FA cycling can be considered an energy-dependent futile cycle. *Abbreviations*: IMTG, intramuscular triglyceride; DAG, diacylglycerol; MAG, monoacylglycerol; MGAT, monoacylglycerol acyltransferase; DGAT, diacylglycerol acyltransferase.

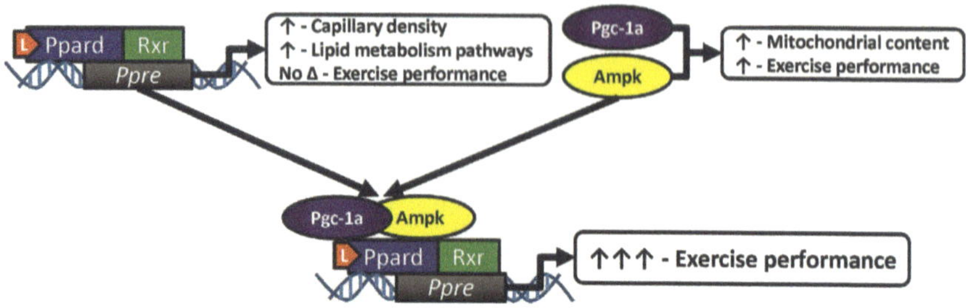

Robert C. Noland, Figure 4 Synergistic coactivation of Ppard by Pgc-1a and Ampk to enhance lipid utilization and exercise performance. Ligand (L)-activated Ppard heterodimerizes with the retinoid X receptor (Rxr) and the resultant complex translocates to the nucleus where it binds to a Ppar response element (*Ppre*) within the DNA. This increases capillary density and upregulates lipid metabolism pathways, but has no effect on exercise performance. Pgc-1a and Ampk activation induce mitochondrial biogenesis, which modestly enhances exercise performance. During exercise, Pgc-1a and Ampk bind to the Ppard:Rxr heterodimer complex resulting in a synergistic improvement in exercise performance. *Abbreviations*: L, ligand—synthetic agonist or endogenous fatty acid; Ppard, peroxisome proliferator-activated receptor delta; Pgc-1a, Ppar-gamma coactivator-1alpha; Ampk, AMP-activated protein kinase alpha; △, change.

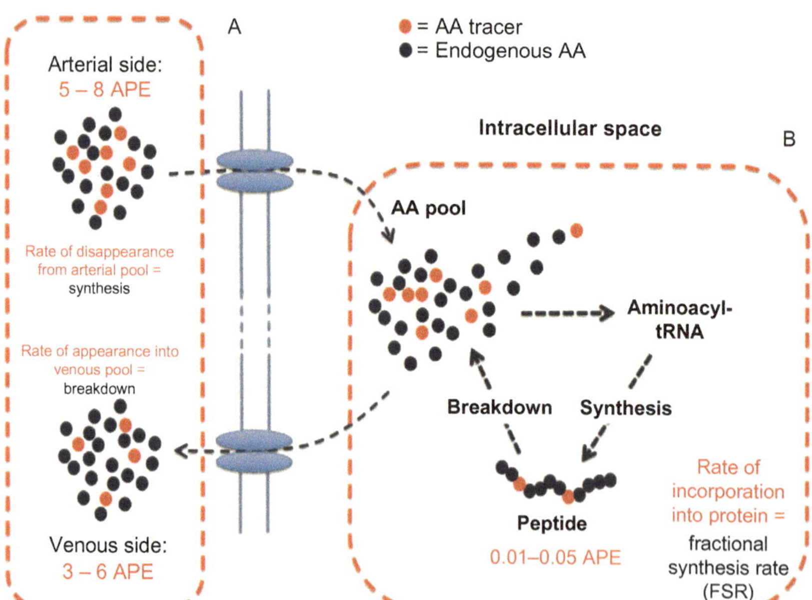

Philip J. Atherton *et al.*, Figure 1 Schematic showing the use of stable isotope tracers for the calculation of protein synthesis and breakdown. (A) represents arterio-venous balance measures, (B) fractional synthetic rate measures. *Abbreviation*: APE, atom percent excess.

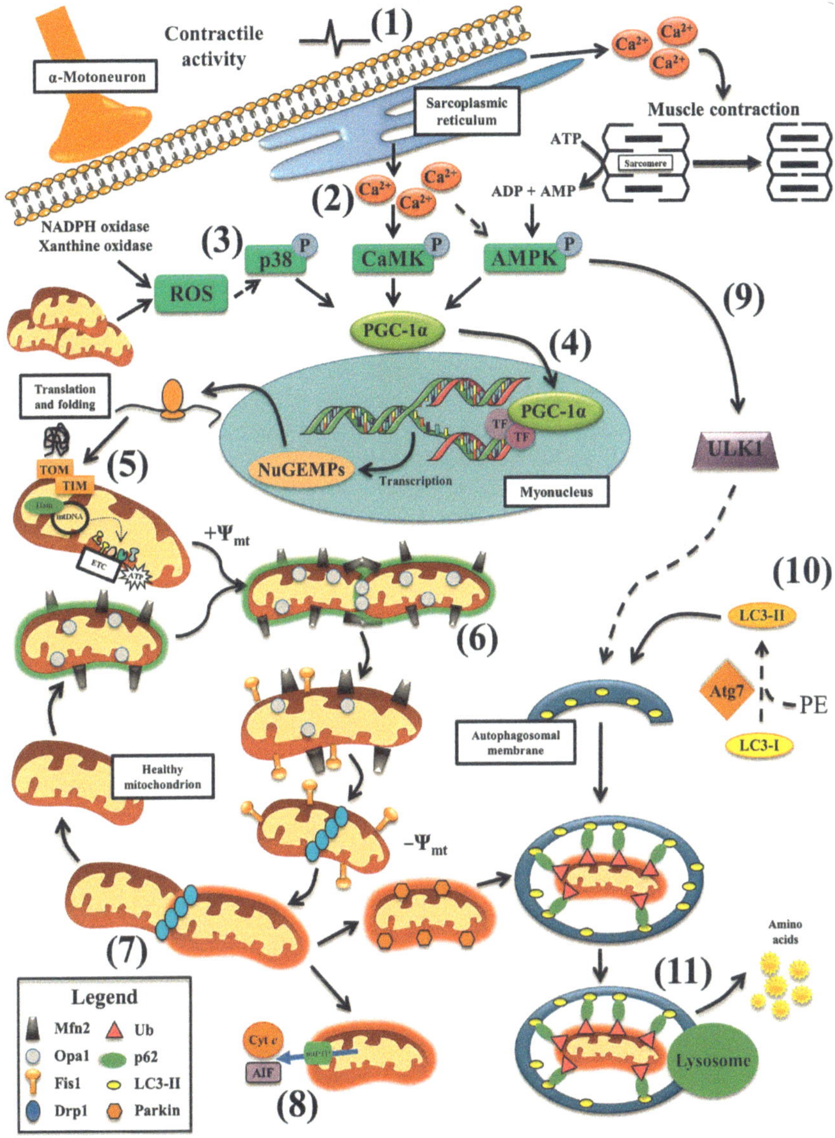

David A. Hood *et al.*, Figure 1 Effect of exercise and training on mitochondrial turnover. During muscle contraction, action potentials propagate down α-motoneurons which innervate muscle fibers (1). Electrical signals are transmitted along the sarcolemma of skeletal muscle and are coupled with the release of Ca^{2+} from the sarcoplasmic reticulum (2). Increases in intracellular Ca^{2+} levels allow for muscle contraction to occur, while also activating Ca^{2+}-sensitive signaling pathways. Contractile activity also consumes cellular ATP, causing a decrease in ATP/ADP ratio and an increase in the formation of AMP, which can activate AMP-activated protein kinase (AMPK). Reactive oxygen species (ROS)

(Continued)

David A. Hood *et al.*, Figure 1—Cont'd production from the mitochondrial electron transport chain (ETC) and other intracellular sources are also enhanced during muscle activity, likely leading to the phosphorylation of kinases such as p38 MAPK (3). These signal transduction pathways target transcription factors (TFs), as well as the transcriptional coactivator peroxisome proliferator-activated receptor γ coactivator-1α (PGC-1α), which stimulates mitochondrial biogenesis (4), among other beneficial adaptations to skeletal muscle. PGC-1α is especially critical in augmenting the transcription of NuGEMPs, which need to be translated and imported into the mitochondrion. Additionally, other nuclear-encoded factors such as mitochondrial transcription factor A (Tfam) must be imported into the mitochondrion, where they can interact with mitochondrial DNA (mtDNA) to assist the expression of mtDNA-encoded subunits of the ETC (5). This coordinated expression of both nuclear and mitochondrial genomes is key to the expansion of the mitochondrial pool. Chronic muscle activity also increases the ratio of fusion-to-fission proteins, promoting the fusion of mitochondria to form a reticular network (6). Essential mammalian skeletal muscle fusion proteins involved in these processes include Mfn2 and the inner mitochondrial membrane protein Opa1, which are necessary for fusion of the outer and inner mitochondrial membranes, respectively. Under conditions of cellular stress, mitochondrial fission can occur, allowing for the isolation of dysfunctional components of the organelle (7). Mitochondrial fission is executed by the formation of a Drp1 oligomer and can occur due to a reduction in mitochondrial membrane potential (Ψ_{mt}) to a portion of the organelle. Mitochondrial fission precedes apoptosis, and exercise training can reduce the susceptibility of both skeletal and cardiac muscle to this process, likely by reducing release of proapoptotic factors, such as cytochrome *c* and apoptosis-inducing factor (AIF) from the mitochondrion (8). Mitochondrial fission is also coupled with mitophagy, the specific degradation, and recycling of dysfunctional mitochondria through the autophagosome–lysosomal system (9). This process is thought to occur through AMPK/ULK1 signaling, in cooperation with the activation of other mitochondrial-specific kinases and ubiquitin ligases, such as Parkin. The activation of this pathway is also crucial for the formation of the autophagosomal membrane, which occurs through the Atg7-mediated lipidation of LC3-I with phosphatidylethanolamine (PE), to form LC3-II (10). Specific interactions between ubiquitinated proteins (Ub) on the mitochondrion and autophagosomal membrane are facilitated by p62, ensuring the precise engulfment of the malfunctioning component of the mitochondrial network. The autophagosome subsequently fuses with a lysosome, and its cargo is degraded and recycled within the cell (11).

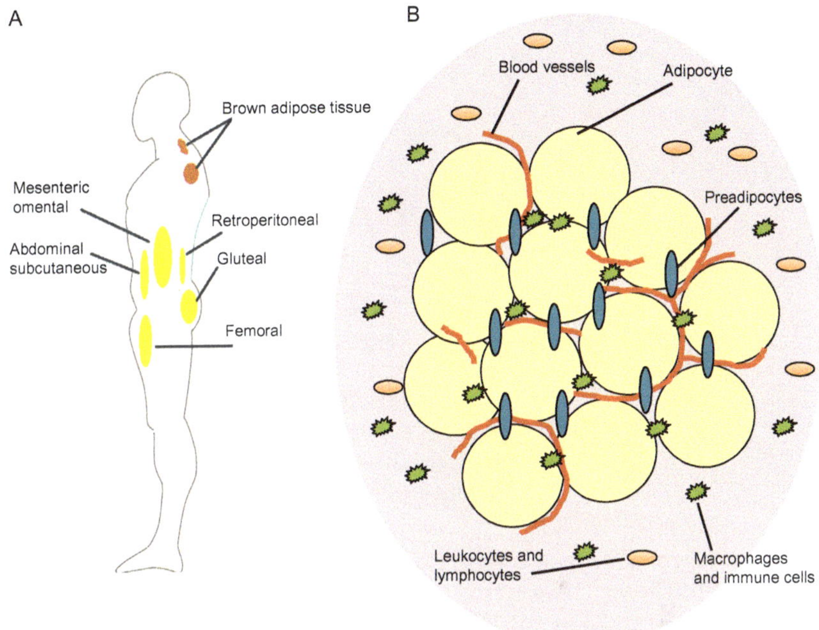

Thomas Tsiloulis and Matthew J. Watt, Figure 1 Adipose tissue localization and composition in humans. (A) White and brown adipose tissue is located in various anatomical locations in humans. Subcutaneous adipose tissue is located under the skin and includes the abdominal and gluteal–femoral depots. Visceral adipose tissue is located near the digestive organs and includes the omental and mesenteric adipose tissue depots. (B) The vast majority of the adipose tissue mass is composed of adipocytes. There are many other cell types that constitute adipose tissue. It is highly vascularized and contains a number of immune cells* such as B cells, mast cells, Tregs (T regulatory cells), macrophages, leukocytes, and lymphocytes.

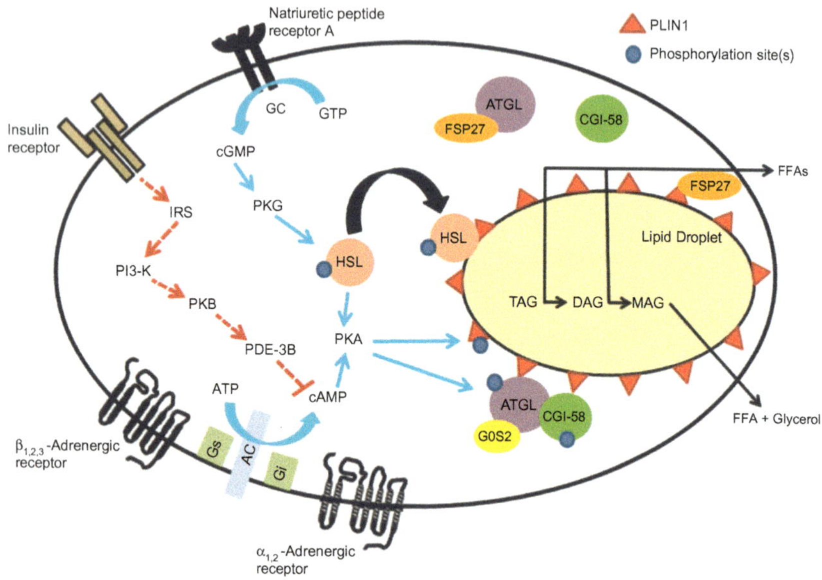

Thomas Tsiloulis and Matthew J. Watt, Figure 2 Regulatory control of lipolysis. Lipolytic and antilipolytic hormones act through their receptors to activate signaling pathways. Thereafter, changes in the cellular localization of key lipolytic proteins, their phosphorylation state, and protein–protein interactions dictate the breakdown of triglyceride stored in lipid droplets. The red arrows denote inhibitory pathways, and the blue arrows stimulatory pathways of lipolysis. Abbreviations: AC, adenylyl cyclase; ATP, adenosine triphosphate; ATGL, adipose triglyceride lipase; CGI-58, comparative gene identification 58; cAMP, cyclic adenosine monophosphate; cGMP, cyclic guanine monophosphate; DAG, diacylglycerol; FSP27, fat-specific protein 27; FFA, free fatty acid; G0S2, G0/G1 switch gene 2; GC, guanylyl cyclase; Gi, inhibitory GTP-binding protein; Gs, stimulatory GTP-binding protein; GTP, guanine triphosphate; HSL, hormone-sensitive lipase; IRS, insulin receptor substrate; MAG, monoacylglycerol; PDE-3B, phosphodiesterase 3B; PI3-K, phosphoinositide 3-kinase; PKA, protein kinase A; PKB, protein kinase B; PKG, protein kinase G; TAG, triacylglycerol.

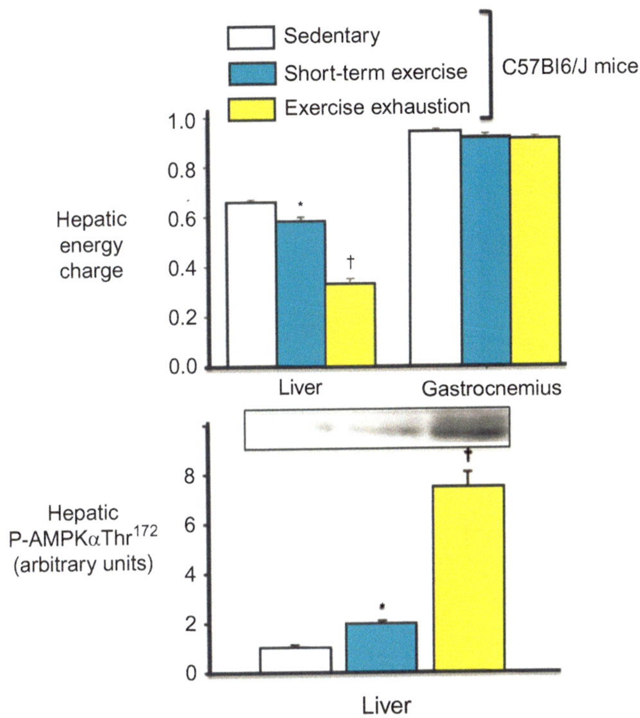

Elijah Trefts *et al.*, Figure 1 Hepatic energy discharge in response to short term and exhaustive exercise in C57Bl/6J mice. Hepatic energy charge decreases with exercise and becomes critically low with exhaustive exercise (top). Skeletal muscle energy charge is not affected by exercise. The decrease in energy charge is associated with an increase in hepatic AMPK activation (bottom). Energy charge is calculated by the equation (ATP + 0.5ADP)/(ATP + ADP + AMP). Data are mean ± SE. *Significantly different from SED ($P < 0.05$). †Significantly different from SED and ST ($P < 0.05$). *Modified from Camacho* et al.[8]

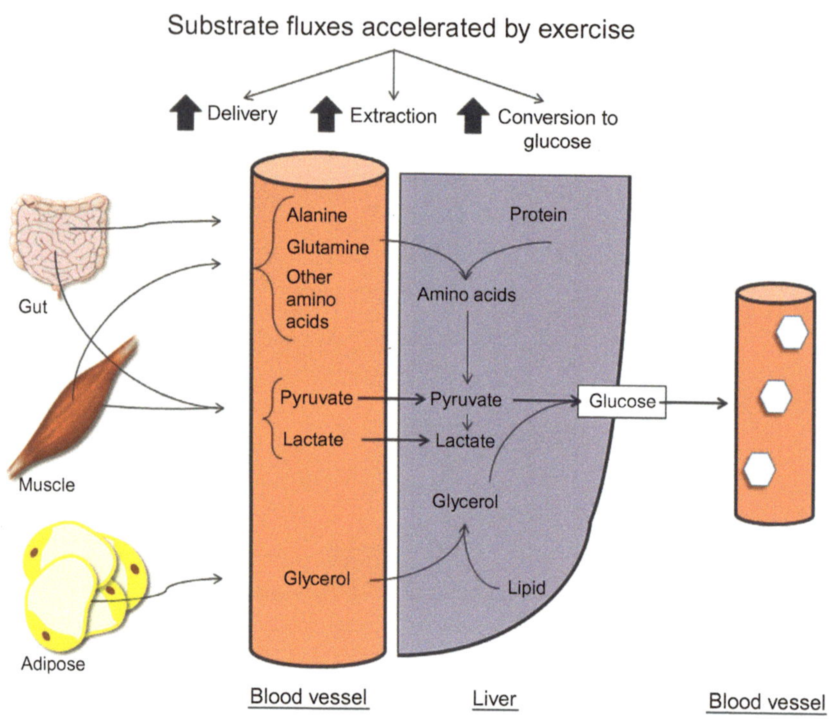

Elijah Trefts *et al.*, Figure 2 Gluconeogenesis is regulated by gluconeogenic precursor supply to the liver, extraction by the liver, and conversion to glucose within the liver. All these processes are accelerated by physical exercise.

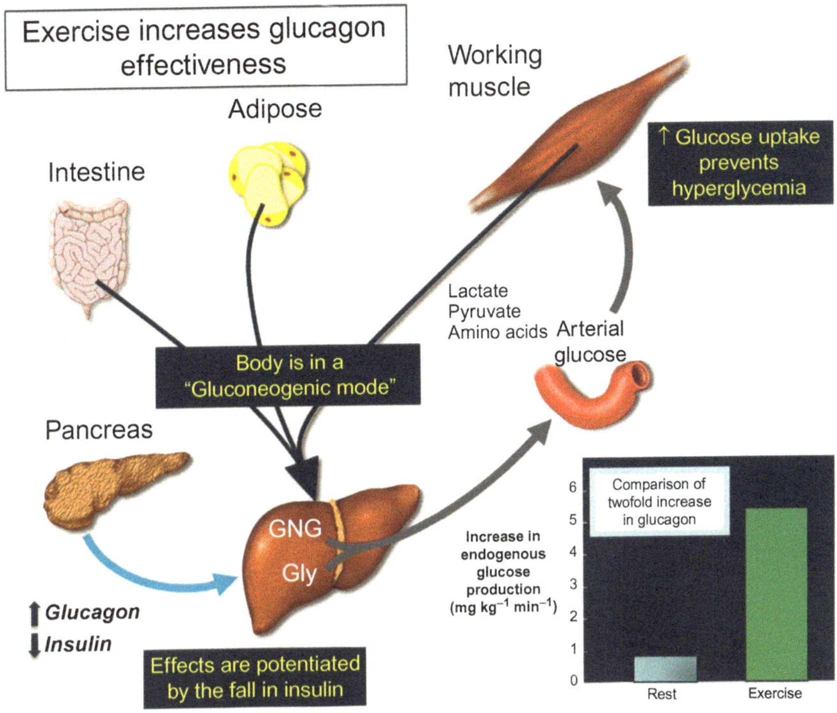

Elijah Trefts *et al.*, Figure 3 The liver is a metabolic hub where pathways for amino acid, fat, and glucose metabolism are integrated. The integration of these pathways serves to provide energy in the form of glucose, recycle carbon-based metabolites, and prevent nitrogen toxicity. The demand on these pathways is accentuated during exercise and driven by glucagon secreted from pancreatic alpha-cells.

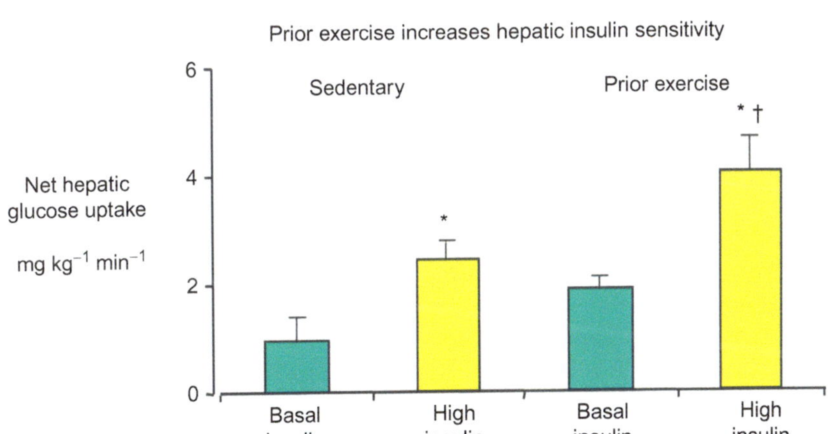

Elijah Trefts *et al.*, Figure 4 Net hepatic glucose uptake in dogs following 150 min of exercise or an equivalent sedentary period. Glucose was infused into a portal vein to increase the liver glucose supply by twofold, and arterial glucose was clamped at 180 mg dl^{-1}. Insulin was infused at basal rates (0.2 mU kg^{-1} min^{-1}) or at rates designed to simulate a meal (1.2 mU kg^{-1} min^{-1}). $^{*}p \leq 0.05$ versus basal insulin. $^{\dagger}p \leq 0.05$ versus sedentary high insulin. Data are mean ± SE. *Modified from Pencek* et al.[51]

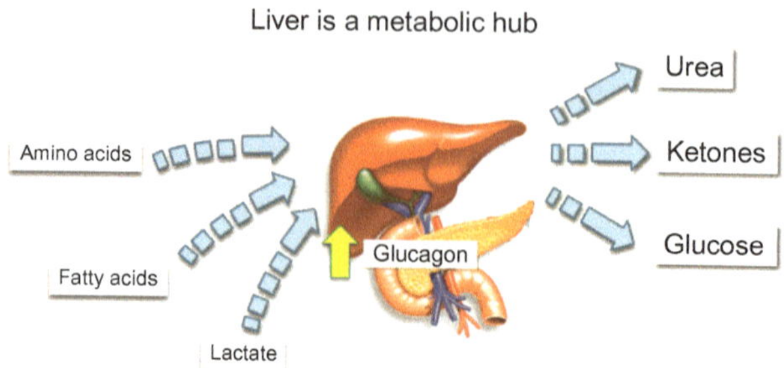

Elijah Trefts *et al.*, Figure 5 The liver is exquisitely sensitive to the actions of glucagon, exceeding glucagon sensitivity in the sedentary state (inset). Glucagon action is fully manifested during exercise because (i) the increased glucose utilization of working muscle prevents the hyperglycemia that accompanies an experimental increase in glucagon; (ii) prolonged exercise creates a physiological environment that supports gluconeogenesis as gluconeogenic substrates are mobilized from muscle, adipose, and intestine; and (iii) exercise causes a fall in insulin that sensitizes the liver to glucagon.

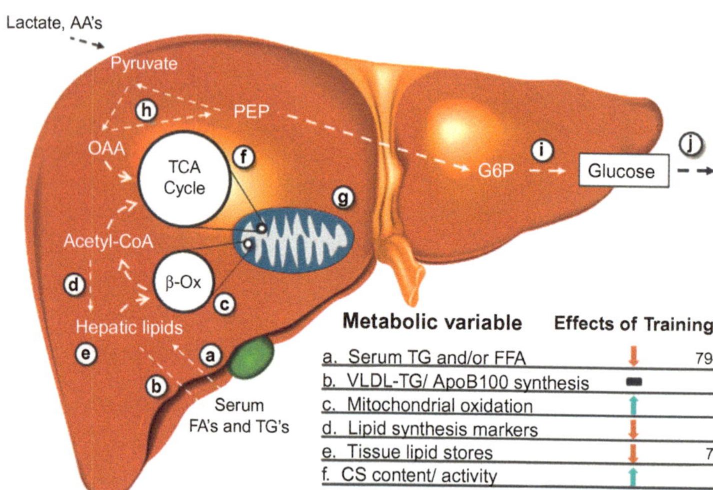

Metabolic variable	Effects of Training	Refs.
a. Serum TG and/or FFA	↓	79–83,90,91,93–97
b. VLDL-TG/ ApoB100 synthesis	–	93
c. Mitochondrial oxidation	↑	79–82
d. Lipid synthesis markers	↓	79–83
e. Tissue lipid stores	↓	79–82,90–93,96
f. CS content/ activity	↑	79,81,82
g. Mitochondrial content markers	↑	79–82,85,86
h. PEPCK	↑	86
i. G6Pase	–	86
j. Suppression of EndoRa by insulin	↑	98

Elijah Trefts *et al.*, Figure 6 Exercise training elicits complex adaptations in liver metabolic processes.[79–83,85,86,90,93–97] Decreased delivery of substrates (a), constant rates of VLDL-TG synthesis (b), increased mitochondrial oxidation of lipid (c), and decreased lipid anabolic processes (d) may contribute to a net decrease in hepatic lipid stores (e). Increased mitochondrial function (f) and/or content (g) may serve to elevate lipid oxidation rates. An increase in PEPCK levels (h), but not G6Pase (i), may indicate elevated gluconeogenic potential elicited through training. An improved ability of the liver to respond to hormonal stimuli, such as insulin (j), may underlie some of the metabolic adaptations seen. Nuclear transcription and protein translation alterations may underlie adaptive changes of this organ.

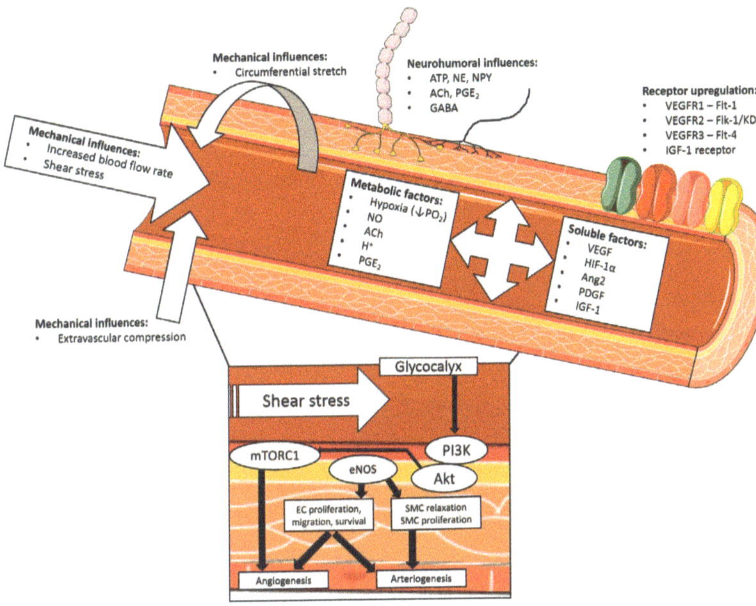

T. Dylan Olver *et al.*, Figure 1 Mechanical and metabolic factors involved in exercise training-induced vascular adaptations. Schematic representation of vascular adaptation in skeletal muscle, coronary, and cerebral vascular tissues. Stimuli include mechanical influences, interluminal metabolites, soluble release factors, and neurohumoral factors. Specific receptor upregulation associated with vascular remodeling and a generalized signaling mechanism are included (inset). Abbreviations: ATP, adenosine triphosphate; NE, norepinephrine; NPY, neuropeptide Y; Ach, acetylcholine, PGE_2, prostaglandin E_2; GABA, gamma aminobutyric acid; VEGFR1/Flt-1, vascular endothelial growth factor receptor 1; VEGFR2/Flk-1, vascular endothelial growth factor receptor 2; VEGFR3/Flt-4, vascular endothelial growth factor receptor 3; IGF-1 receptor, insulin-like growth factor receptor; NO, nitric oxide; H^+, hydrogen ion; VEGF, vascular endothelial growth factor; Ang2, angiopoietin; PDGF, platelet-derived growth factor; IGF-1, insulin-like growth factor; PI3K, phosphatidylinositol-4,5-bisphosphate 3-kinase; Akt, protein kinase B; eNOS, endothelial nitric oxide synthase; mTORC1, mammalian target of rapamycin complex 1; EC, endothelial cell; SMC, smooth muscle cell.

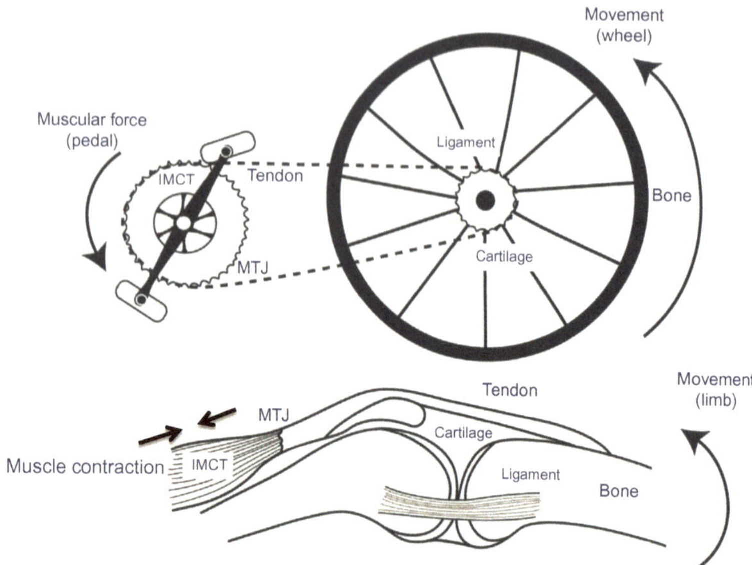

Michael Kjaer *et al.*, Figure 1 A schematic representation of the connective tissue in the musculoskeletal system. The analogy to the transformation of force applied to bicycle pedals, through the chain and the resulting wheel turn, is similar to the muscle cell contraction, the force transmission through the intramuscular connective tissue (IMCT), the myotendinous junction (MTJ), the tendon, over the ligament-stabilized joint with its cartilage, and to the bone tissue, resulting in limb movement.

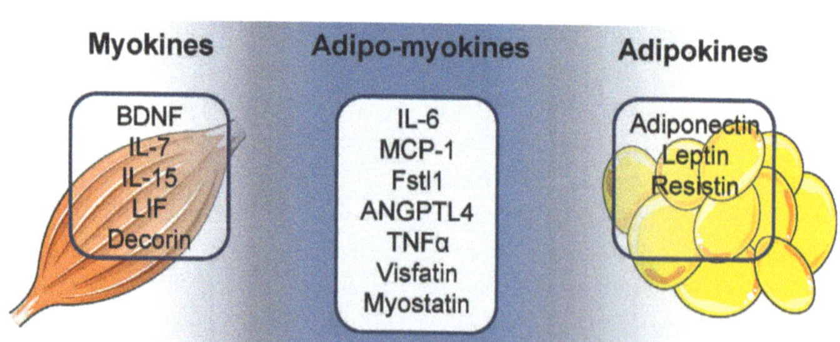

Sven W. Görgens *et al.*, Figure 1 The adipo-myokine concept. A search of original articles in PubMed was performed for the major exercise-regulated myokines and adipokines to identify molecules that were produced and secreted in both tissues. The term adipo-myokines was used for proteins fulfilling both of these criteria. The search terms we used were “skeletal muscle” or “adipose tissue,” “myokine” or “adipokine,” and “exercise.”

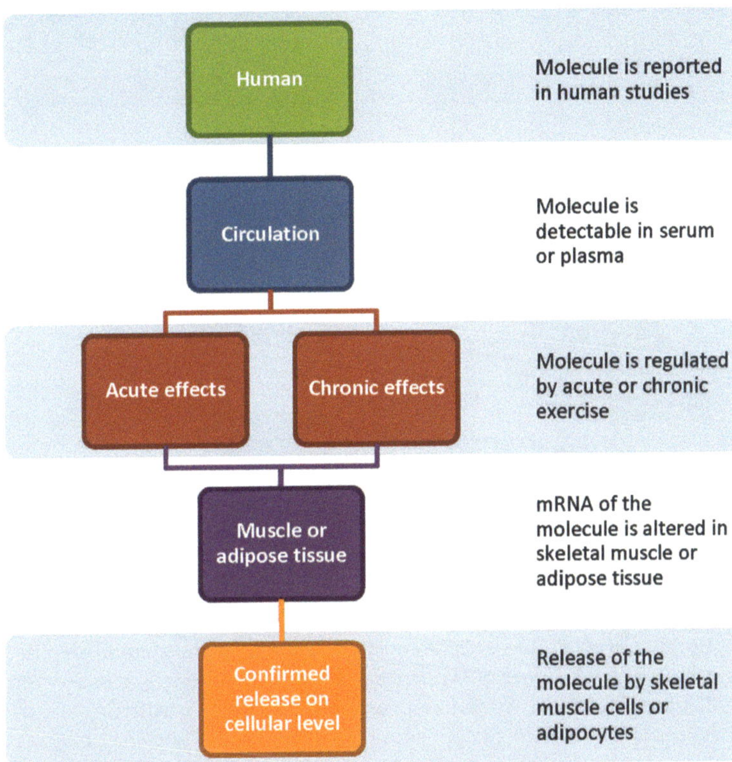

Sven W. Görgens *et al.*, Figure 2 Scheme for the approach to selecting published studies. A search of original articles in PubMed was performed to identify myokines, adipokines, or adipo-myokines, which satisfied all of the following criteria: The molecule must be human and detectable in serum or plasma. Furthermore, the molecule must be regulated by either acute or chronic exercise and expressed in skeletal muscle or adipose tissue. The protein has to be secreted by adipocytes or myocytes. The search terms we used were "skeletal muscle" or "adipose tissue," "myokine" or "adipokine," and "exercise."

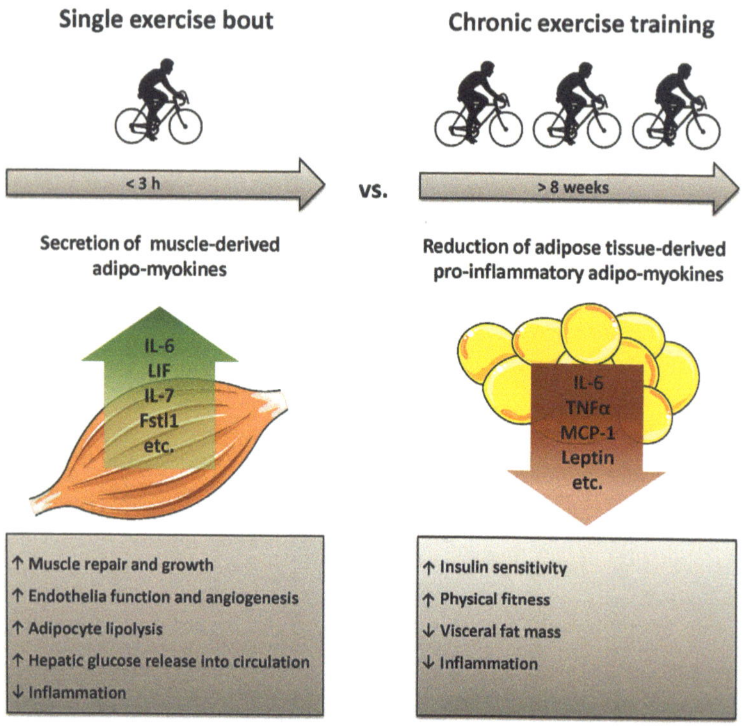

Sven W. Görgens *et al.*, Figure 3 Differential effects of acute and chronic exercise. After acute exercise, a high number of myokines are secreted by skeletal muscle exerting a variety of endocrine effects. Acute induction of myokines like myostatin, IL-7, decorin, and LIF is involved in the regulation of muscle hypertrophy and may play a role in exercise-related restructuring of skeletal muscle. The high level of circulating IL-6 after exercise induces an anti-inflammatory environment by inducing the production of IL-1ra and IL-10, and also inhibits TNFα production. Furthermore, IL-6 has metabolic effects, by affecting insulin-stimulated glucose disposal and fatty acid oxidation. The myokine FSTL1 has protective effects on ischemia–reperfusion injury in muscle and heart tissue. On the other hand, regular exercise training induces a reduction of adipose tissue-derived pro-inflammatory cytokines like IL-6, TNFα, and MCP-1, which are associated with low-grade systemic inflammation and a reduction of whole-body insulin sensitivity. Exercise training has been reported to reduce central adiposity independent of overall weight changes, and this may represent an additional mechanism of the anti-inflammatory action of chronic exercise training.

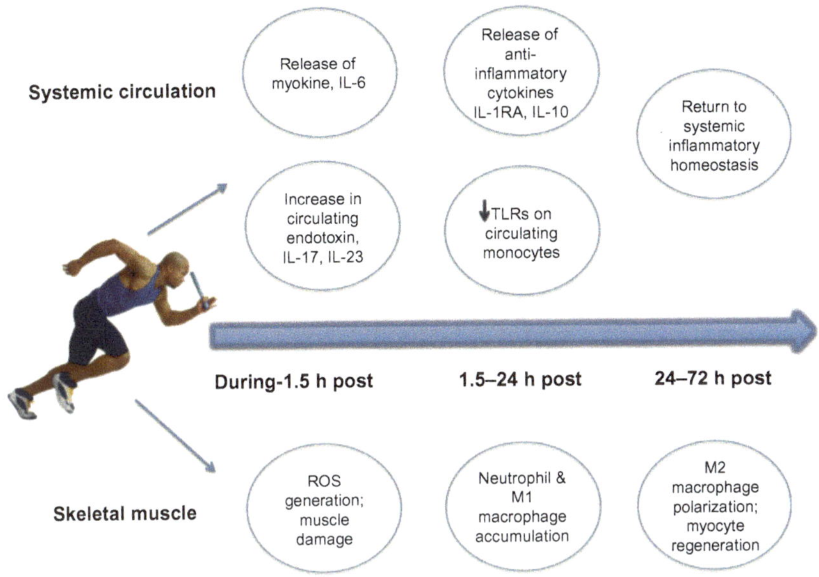

Jacob Allen *et al.*, Figure 1 The effects of acute exercise on skeletal muscle and systemic inflammation. *Abbreviations*: IL, interleukin; ROS, reactive oxygen species; TLR, toll-like receptor; ra, receptor antagonist.

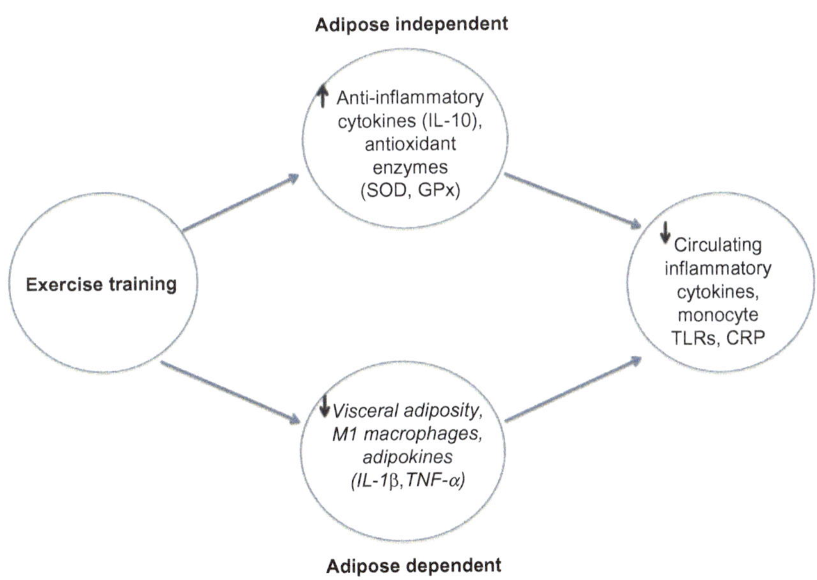

Jacob Allen *et al.*, Figure 2 The effects of long-term, moderate intensity exercise training on inflammatory markers and immune mediators. The anti-inflammatory effects may manifest from both adipose-dependent and -independent mechanism. *Abbreviations*: IL, interleukin; SOD, superoxide dismutase; GPx, glutathione peroxidase.

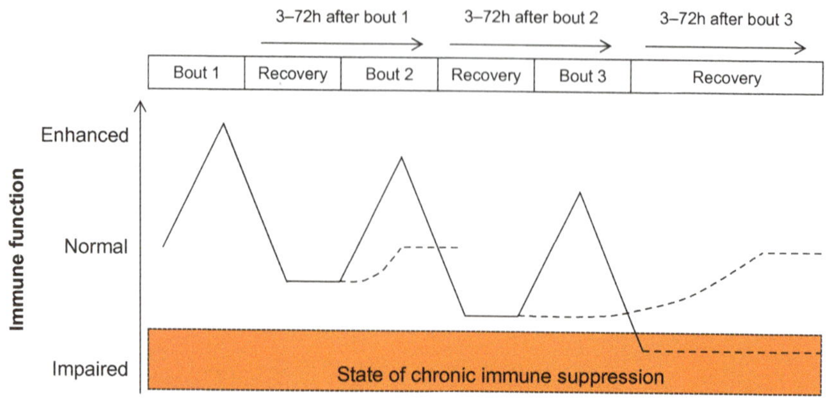

Richard J. Simpson *et al.*, Figure 1 The "open-window" hypotheses adapted from the original model proposed by Pedersen and Ullum.[54] A single bout of exercise is associated with an initial enhancement in immune function that is quickly followed by a transient period of immune depression (the "open window") that can last for 3–72 h after the initial exercise bout (although in most cases immune function is restored to normal levels within 24 h). It is believed that this "open window" leaves the host susceptible to opportunistic infections. If a second bout of exercise is performed during the window (i.e., without adequate recovery), then the exercise-induced enhancement in immunity is blunted and the postexercise immune depression is more severe and prolonged (i.e., the window is opened wider and for longer), rendering the athlete more susceptible to infection. The response is exacerbated with subsequent exercise bouts that, if completed without adequate recovery, may result in a state of chronic immune suppression (a third bout is used here for illustrative purposes, but in reality, it is likely to be many more).

Gilian F. Hamilton and Justin S. Rhodes, Figure 1 Wheel-running paradigm. (A) A standard laboratory cage where a mouse has 24-h access to a running wheel. (B) Standard laboratory cage without a running wheel.

Gilian F. Hamilton and Justin S. Rhodes, Figure 2 Morris Water Maze. A photo of a mouse in the middle of a trial. For this task, the animal is placed in one of the four cardinal starting points, and then using landmarks around the room, it must navigate the water as quickly as possible to find the platform that is submerged under the water (note the black dot in the bottom left-hand corner).

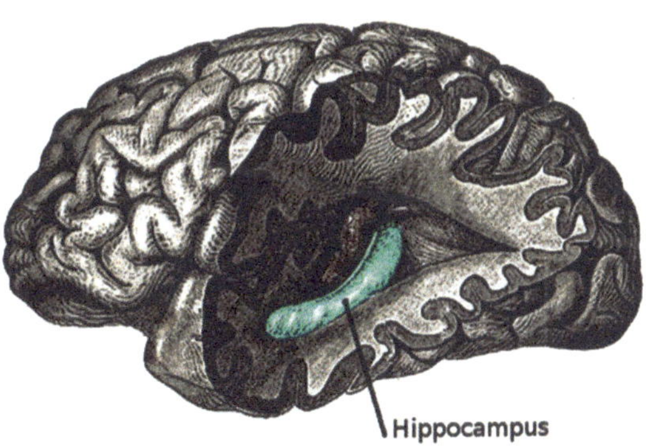

Gilian F. Hamilton and Justin S. Rhodes, Figure 4 Gross anatomy of the hippocampus. The hippocampus is located in the medial temporal lobe of the brain.

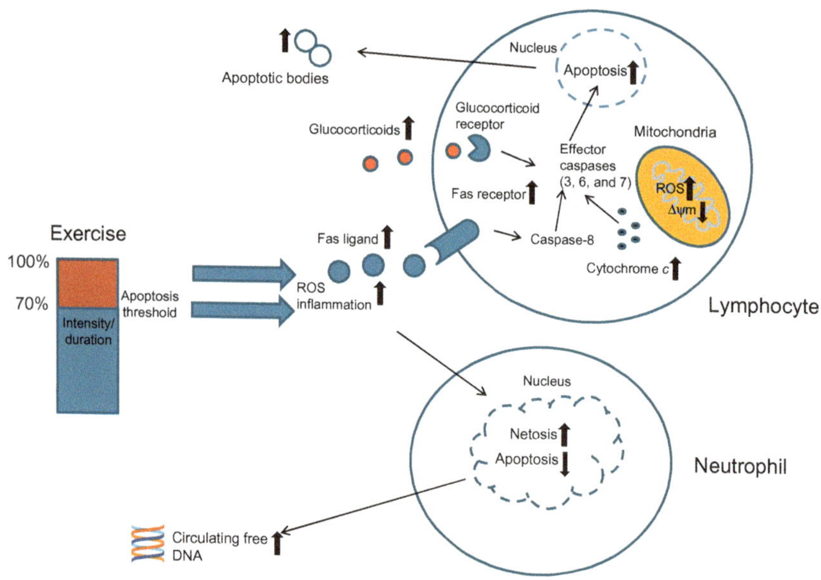

Frank C. Mooren and Karsten Krüger, Figure 1 Suggested mediators and signaling pathways on lymphocyte and neutrophil apoptosis during exercise.

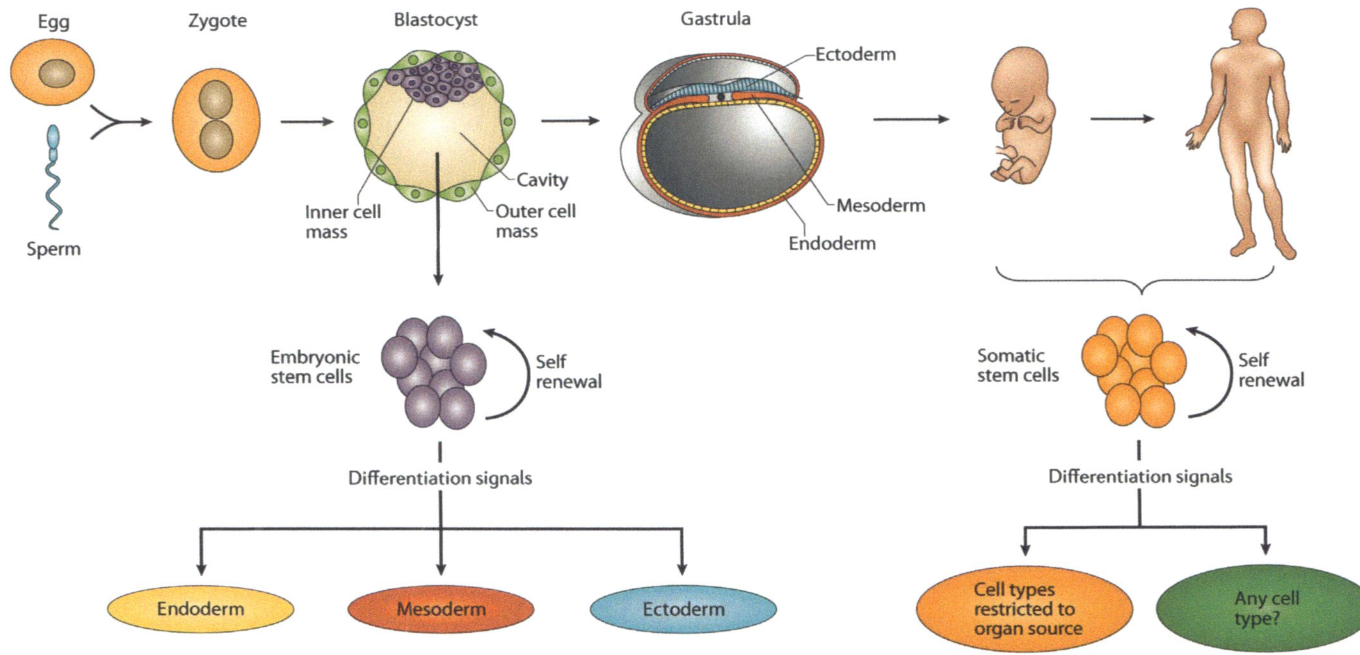

Marni D. Boppart *et al.*, Figure 1 Introduction to stem cells. Embryonic stem cells (ESCs) are important for development of a fetus and may be extracted from the inner cell mass of the blastocyst for the purpose of ESC-based therapies. Adult or somatic stem cells (ASCs) reside in the body after birth for the purpose of routine remodeling or repair and regeneration of tissues following injury. While some ASCs retain multilineage potential, the majority are unipotent and differentiation capacity is limited to the tissue type in which they reside. Positive health outcomes associated with exercise, including enhanced repair and improvements in function, may occur as a result of alterations in ASC quantity and/or function in a variety of tissues. *Reprinted with permission from O'Connor and Crystal.*[1]

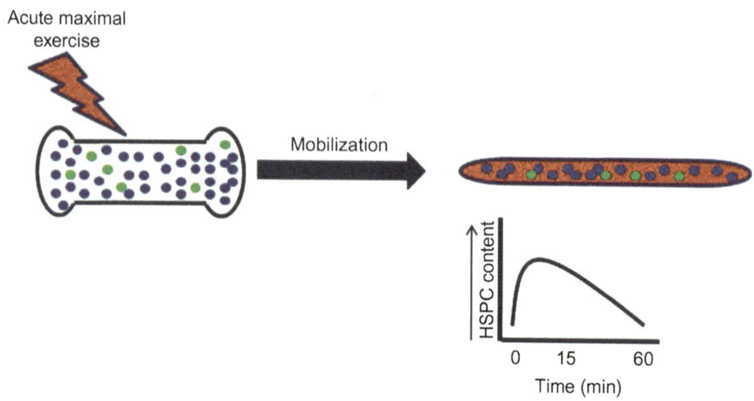

Marni D. Boppart *et al.*, Figure 2 HSPC response to acute exercise. HSPCs (open circles), located in the bone marrow and enriched within the LSK (mouse) and CD34 (human) populations, are mobilized from their bone marrow niche to peripheral blood peaking at around 15 min following an acute, maximal exercise bout. HSPC quantity in peripheral blood returns to preexercise levels rapidly following acute exercise. It is believed that mobilized HSPCs are involved in tissue repair.

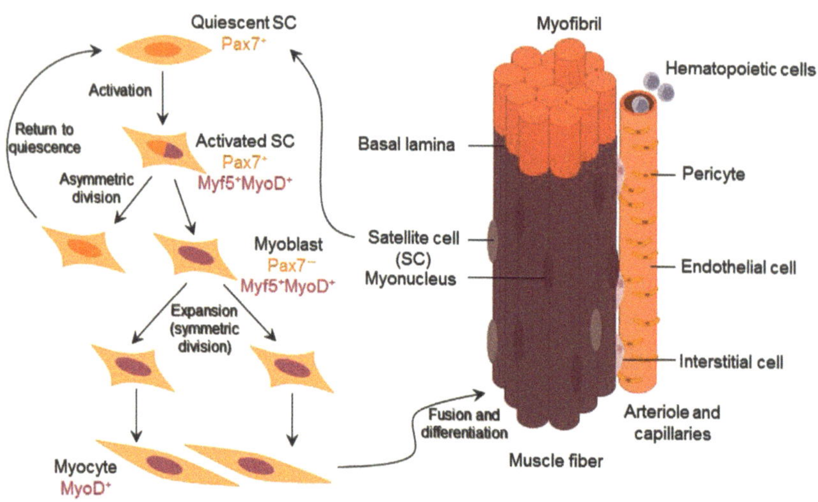

Marni D. Boppart *et al.*, Figure 5 Skeletal muscle resident stem cells. Satellite cells ($Pax7^+$) reside in the space between the sarcolemma and the basal lamina and are required for repair and regeneration of the myofiber following injury. Satellite cell activation in response to injury includes upregulation of the myogenic regulatory factors Myf5 and MyoD and asymmetric cell division. Newly formed myoblasts or myocytes can fuse with damaged fibers or fuse together to form new fibers in the muscle environment. $Pax7^-$ (nonsatellite) stem cells (interstitial cells) and pericytes are located outside the basal lamina in close proximity to vessels. The majority of nonsatellite stem cells in muscle contribute to repair via secretion of factors that allow for satellite cell activation. *Reprinted with permission from Dr. Péter Balogh, University of Pecs, Hungary.*

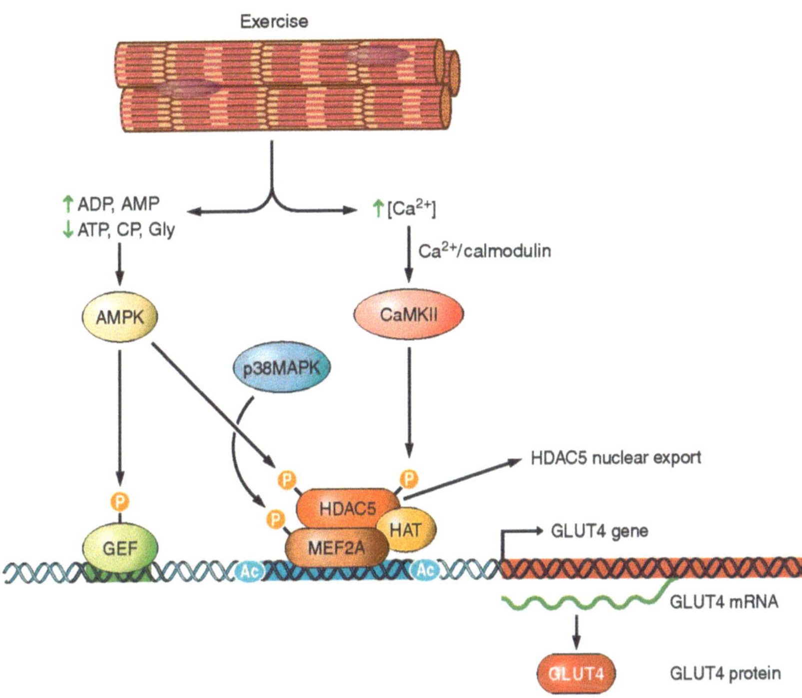

Mark Hargreaves, Figure 2 Overview of the molecular events involved in the exercise-induced increase in GLUT4 expression in human skeletal muscle. *Reproduced from Richter and Hargreaves[33] with permission.*

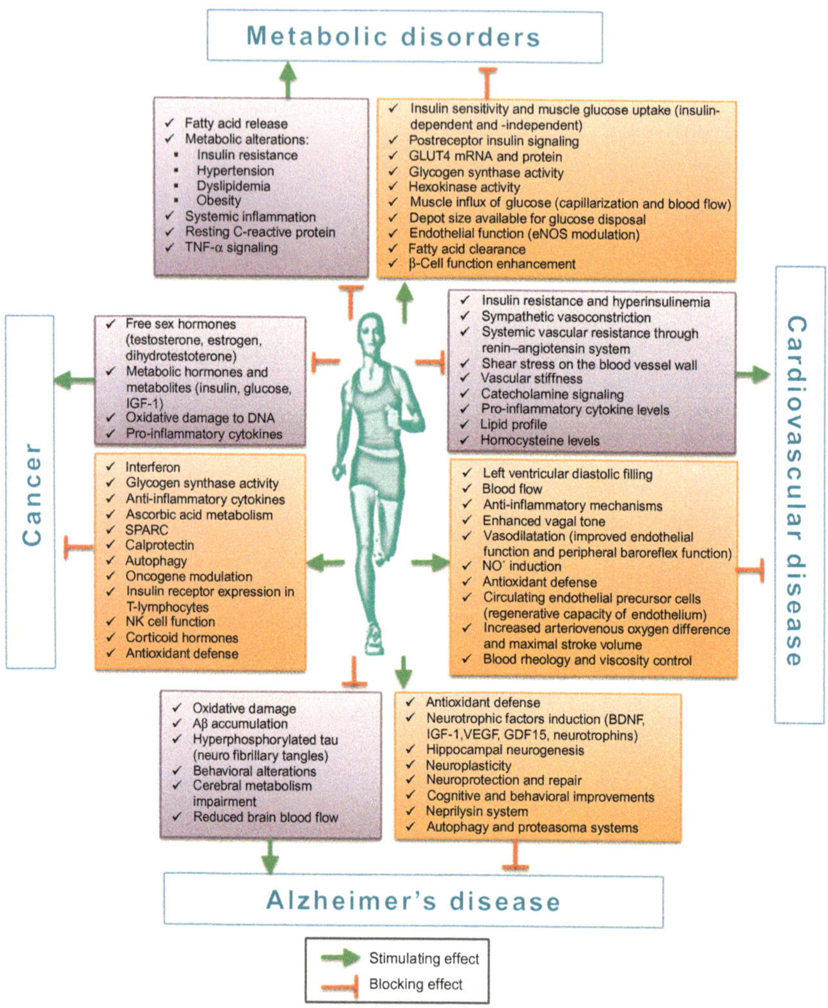

Helios Pareja-Galeano *et al.*, Figure 1 The main biological effects of physical activity on metabolic disorders, cardiovascular disease, Alzheimer's disease, and cancer.

FSC
www.fsc.org
MIX
Papier | Fördert
gute Waldnutzung
FSC® C083411

Kolibri 360 GmbH
Ferdinand-Jühlke-Straße 7
99095 Erfurt, Deutschland
produktsicherheit@kolibri360.de